45—
#35771

Springer Series in Experimental Entomology

Thomas A. Miller, Editor

Sampling Methods in Soybean Entomology

Edited by
M. Kogan and D.C. Herzog

Springer-Verlag
New York Heidelberg Berlin

Series editor:
Thomas A. Miller
Department of Entomology
University of California
Riverside, California 92521
USA

Editors of this volume:
Marcos Kogan
Illinois Natural History Survey and Agricultural Entomology
University of Illinois
172 Natural Resources Building
Urbana, Illinois 61801, USA
and
Donald C. Herzog
Department of Entomology and Nematology
University of Florida
Quincy, Florida 32351, USA

With 252 figures.

Library of Congress Cataloging in Publication Data
Main entry under title:
Sampling methods in soybean entomology
 (Springer series in experimental entomology)
 Includes index.
 1. Soybean—Diseases and pests. 2. Arthropod
populations—Measurement. 3. Arthropoda—
Biological control. I. Kogan, Marcos.
II. Herzog, Donald C. III. Title. IV. Series.
SB608.S7S25   633'.34'970723   80-22342

All rights reserved.
No part of this book may be translated or reproduced in any form without written permission from Springer-Verlag.
The use of general descriptive names, trade names, trademarks, etc., in this publication, even if the former are not especially identified, is not to be taken as a sign that such names, as understood by the Trade Marks and Merchandise Marks Act, may accordingly be used freely by anyone.

© 1980 by Springer-Verlag New York Inc.
Printed in the United States of America.

9 8 7 6 5 4 3 2 1

ISBN 0-387-90446-8 Springer-Verlag New York
ISBN 3-540-90446-8 Springer-Verlag Berlin Heidelberg

Series Preface

Insects as a group occupy a middle ground in the biosphere between bacteria and viruses at one extreme, amphibians and mammals at the other. The size and general nature of insects present special problems to the student of entomology. For example, many commercially available instruments are geared to measure in grams, while the forces commonly encountered in studying insects are in the milligram range. Therefore, techniques developed in the study of insects or in those fields concerned with the control of insect pests are often unique.

Methods for measuring things are common to all sciences. Advances sometimes depend more on how something was done than on what was measured; indeed a given field often progresses from one technique to another as new methods are discovered, developed, and modified. Just as often, some of these techniques find their way into the classroom when the problems involved have been sufficiently ironed out to permit students to master the manipulations in a few laboratory periods.

Many specialized techniques are confined to one specific research laboratory. Although methods may be considered commonplace where they are used, in another context even the simplest procedures may save considerable time. It is the purpose of this series (1) to report new developments in methodology, (2) to reveal sources of groups who have dealt with and solved particular entomological problems, and (3) to describe experiments which might be applicable for use in biology laboratory courses.

Thomas A. Miller
Series Editor

Call to Authors

Springer Series in Experimental Entomology will be published in future volumes as contributed chapters. Subjects will be gathered in specific areas to keep volumes cohesive.

Correspondence concerning contributions to the series should be communicated to:

> Thomas A. Miller, Editor
> Springer Series in Experimental Entomology
> Department of Entomology
> University of California
> Riverside, California 92521
> USA

Foreword

In the field of pest management both researchers and practitioners are constantly faced with the need to answer four questions: (1) What kind of pest is it? (2) How many are there per unit area? (3) How many must there be before economic injury occurs? (4) What will control the pest? The first three of these questions should be answered before the fourth is considered. The twenty-eight chapters in this book, prepared by a group of internationally known scientists, represent a highly successful effort to bring together in one unique publication most of what is known about how to find an answer for the second and third questions as regards arthropods associated with soybean.

Application of the information presented herein will greatly aid pest management specialists in the careful evaluation of pest situations and will most often result in the decision that no control measures are needed. It is imperative that control measures be taken only when absolutely necessary. Soybean, a rapidly expanding crop throughout many countries, is presently attacked by more than 100 species of arthropod pests. Many of these are capable of becoming key pests if populations of their natural control agents are disrupted by unnecessary applications of chemical pesticides. Elevation of a half dozen or so widely occurring insect pests of soybean to key pest status, requiring repetitive applications of broad spectrum chemical pesticides on an annual basis, would be environmentally and economically catastrophic.

Development of control procedures based on application of accurate sampling methods for monitoring populations and establishing economic injury thresholds will help prevent such a catastrophe. The information presented here makes such control procedures possible for most of the world's major arthropod pests of soybean.

Two broad aspects of sampling procedures are considered: highly accurate absolute population estimates and less accurate, although still reliable, relative estimates of population densities. Absolute population estimates are required for establishing economic injury levels and for use in basic population studies. Their establishment is a prerequisite for development of less precise relative population estimates that are adequate for decision-making by practitioners of pest management.

A strong feature of this book is a section devoted to sampling methods for natural control agents. Development and application of integrated control techniques require that populations of these agents also be monitored. In many cases it is as important to have accurate estimates of trends in populations of natural enemy populations as it is for pest populations. The chapter on methods of sampling populations of the little studied pests that attack the plant at ground level and its roots and nodules also helps to fill a long existing need.

Having in one publication the most up-to-date information available on sampling methods for the major arthropod pests of soybean is a significant contribution to the literature of agriculture. It is an especially timely and valuable reference for the rapidly accelerating numbers of scientists concerned with pest management.

L. D. Newsom
Baton Rouge, Louisiana

Preface

Biological phenomena, whether at the cellular or the ecosystem level of organizational hierarchy, are only accessible to investigation through analyses of relatively small samples taken from more or less large populations. Biologists operate on the assumption that observations made on the samples provide representative estimates of the population so that extrapolations can be made with a reasonable degree of certainty. This assumption is only valid if the reliability of the information can be defined within acceptable statistical limits.

Within the rather stable and predictable environment of an organism's body, one may assume that analysis of a small sample reflects the status of the rest of the population, often with better than 99% accuracy. Thus, observations made on several hundred red blood cells can be safely extrapolated to the remaining 25 trillion (25×10^{12}) red blood cells of one person's blood stream. However, a sample of several hundred insects in a crop field is often insufficient to support generalizations concerning populations of the same species in other fields or even in other portions of the same field. Here, the population exists in a nonhomogeneous environment that is much less stable and, consequently, less predictable.

Sampling arthropod populations is the cornerstone of basic research on agricultural ecosystems and the principal tool for building and implementing pest management programs. The dual purpose of sampling, i.e., as a research method for defining the nature and dynamics of communities in agricultural ecosystems and as a means for providing information for pest management decisions, is stressed in almost every chapter in this volume. Morris (R. F. Morris, 1960, Ann. Rev. Entomol. 5:253) describes sampling as a combination of art, science, and drudgery, but he stresses that with the utilization of mathematical procedures for optimal sample allocation, the sampling process is reduced to the latter two components. Questions of sample allocation, for the most part, remain unresolved. However, the development and use of field-tested, sequential sampling plans for management decision-making and for parameter estimation is a step in the right direction.

This book is devoted entirely to discussing the methods used in sampling arthropod populations in a row crop and to summarizing some of the most common analytical manipulations of the sampling data that allow entomologists to achieve in their studies a certain degree of generality with a reasonable degree of reliability. Dedication of this volume to one crop, the soybean, may be interpreted as a case study in the methodology for sampling arthropod populations in agricultural ecosystems. But the reasons to devote an entire volume to sampling methods in soybean entomology go beyond that. The systematic and intensive study of the arthropod fauna associated with soybean is a relatively recent development. Modern soybean entomology has received its greatest impetus since the mid-1960s because of the dramatic expansion of the soybean industry worldwide and because of the concurrent increase in interest and support for research in the various aspects of soybean production, protection, and utilization. To be sure, a valuable literature existed prior to 1960. Much of that work was both isolated and intermittent, however, since most of the earlier researchers in the United States and abroad had professional commitments to crops such as cotton, corn, tobacco, or rice and, thus, could not fully dedicate themselves to the study of soybean arthropods. During the 60s certain land-grant universities and the U.S. Department of Agriculture instituted full time research positions in soybean entomology. As a consequence of this increased research activity in soybean entomology, the literature in the field has proliferated in the past two decades. Much of this literature deals with population studies, and because of discrepancies in sampling procedures it is difficult to compare data from different authors. Standardization of sampling procedures is badly needed. Owing to the ascent of soybean to the largest hectareage harvestable row crop in the United States and its steady expansion in South America and the Orient, soybean entomology is likely to remain a very prolific and dynamic field of entomological research in the foreseeable future, and we feel that publication of this volume is justified at this time.

This book attempts to offer a clear understanding of a methodology that is highly vulnerable to individual bias. All too often entomologists enter into a sampling program blindly, not knowing what or how to sample, what population or portion of it is being sampled, or how many, what size, or how frequently samples are to be taken. Sampling programs are often undertaken in which "favorite" sampling methods are used. This is easily rationalized on the basis of "one method captures more individuals per sampling unit than does another," or "one is used more widely than the rest." However, the adoption of a sampling method must be based on something more concrete, namely sampling variance and cost. We hope that information contained herein will provide a foundation on which a minimum consensus among research entomologists, extension specialists, pest managers, scouts, and farm operators will be achieved; we hope this

Preface

will lead to a uniform expression of data collecting and handling. If we succeed in these objectives, results from various researchers will be more directly comparable. These comparative data can then be used for broader interpretation of trends that transcend the limited boundaries of the regions in which the data were originally generated. This volume represents, to our knowledge, the first attempt to bring together, under one cover, the techniques and results of studies on sampling arthropod populations in any crop ecosystem.

An effort has been made to standardize terminology throughout this book. We may not have been entirely successful, and the reader is asked to forgive possible discrepancies and to call them to our attention. We have departed from a good portion of current population ecological literature by accepting Pielou's (and the statistical ecologists') use of the term "spatial patterns" instead of "population distributions." Pielou (E. C. Pielou, 1977, Mathematical Ecology, John Wiley & Sons, New York, 385 p., see p. 117) clearly addressed this terminological question as follows:

> Much confusion exists in the ecological literature because the word "distribution" is used in both its colloquial and statistical senses, even sometimes in a single sentence, and without any explanation of the meaning intended. Colloquially, "distribution" is synonymous with "arrangement" or "pattern." Statistically, it means the way in which variate values are apportioned, with different frequencies, in a number of possible classes. In this sense, there is no implied reference to spatial arrangement; for instance, we may speak of the distribution of a variate such as tree height without any thought of the location of the trees. Thus, to talk of a population of insects, say, as having "a clumped distribution with a large variance" is nonsense; it is the insects themselves that are clumped or have a clumped pattern, and the large variance pertains to the distribution of the variate, the number of insects per sample unit. To avoid ambiguity in statistical ecology it is most desirable to use the word distribution in its statistical sense only. Then a variate has a distribution, whereas a collection of organisms has a pattern."

We also used the term "dispersion" as a synonym of "distribution" (in the colloquial sense) to be consistent with the terminology defined above. It seems perfectly acceptable to state that a population is randomly dispersed, even if we are only measuring one steady state and not the motions that lead to that specific pattern.

This book is organized into eight major sections. Section I includes (1) descriptions of the crop as the substrate from which arthropods are extracted, (2) the most common sampling procedures in population and in injury estimation, (3) techniques and basic concepts in sampling theory, and (4) sequential sampling programs for parameter measurement and for management decisions. The reader is referred back to this section throughout the entire volume.

Sections II through VIII are devoted to major species or complexes of pests or natural enemies. Within each of the sections there are from one to five chapters on species or species complexes. We have attempted to follow a uniform structure within each chapter by providing the necessary spatial (geographical), temporal (phenological), and biological (life history) background necessary for the development of an intelligent sampling program. Some chapters on species complexes also include practical keys for identification of the most common genera or species within the complex. These keys, however, should be used with caution; and workers, particularly those outside the United States (or the Orient, in the case of the pod borers), should always consult a taxonomist to confirm their identifications. We tried also to provide comprehensive lists of host records whenever appropriate, because much of the sampling information needed to elucidate some aspects of population dynamics must be acquired on those alternate hosts.

The reader will soon realize that chapters differ substantially in the level of detail, breadth, and depth of the coverage. Differences will result in part from the personal style of contributing authors; but more importantly, differences will result owing to the differing amounts of information available on the various species or complexes. By subdividing the subject matter into species or complexes, we have run the risk of a certain amount of overlap, mainly among chapters within a section. We did this by design, however, because we wanted to stress the fact that even among groups that are substantially similar in their overall biological features (e.g., lepidopterous defoliators) there are still peculiarities that need to be taken into account in developing a sampling program.

The coverage of trapping techniques has been dealt with in several chapters (e.g., Chapters 6, 8, 11, 16, 21, and 26). However, we have not made an attempt to summarize trapping techniques in one chapter. The interested reader should refer to basic texts and papers on trapping, particularly those related to cotton insect pests, to supplement this information. One reason for this omission is the fact that far less has been done with trapping methods in soybean entomology than with other relative sampling methods (sweep net, ground cloth, suction net, etc.). Thus, much of the information would have to be borrowed from other crops. Although much of the information contained in the various chapters has been compiled from the literature published over the years, there is a substantial amount of unpublished information. In some cases, special experiments have been conducted to fill gaps in the information for several species or complexes.

The organization and preparation of this volume covered a period of about three years. Many persons, institutions, and agencies should be credited for their support in the various phases of this project. Special credits are given by individual authors following their respective chapters. We must, however, recognize the contributions of those whose efforts, mainly in the last few months of full steam production, made possible the completion of this volume. Cathy Eastman, Charles Helm and the entire crew of the Soybean Entomology Program at Illinois—Barbara Stanger, Susan Post, Suzanne Hart, and Michael Jeffords—helped with the proofreading, while Tzu-Suan Chu carried out much of the lab's work to allow more free time for the others to help with the book.

The Soybean Insect Research Information Center (SIRIC) of the University of

Illinois and Illinois Natural History Survey provided initial lists of references for most chapters. Jenny Kogan, librarian in charge of SIRIC's operations, helped in our attempt to standardize literature citations and in verifying questionable references. John Bouseman, curator of the International Reference Collection of Soybean-Associated Arthropods, and George Godfrey of the Illinois Natural History Survey confirmed validity of scientific names and authorship of species.

Sandy McGary did the original typing of all chapters produced by the Illinois group and the retyping of most other chapters. It was a major effort in speed and concentration, and we are very grateful to Sandy. Finally we want to acknowledge the excellent cooperation we received from Dr. Philip C. Manor, Science Editor, and Ms. Lauren Tresnon Klein, Production Editor, of Springer-Verlag, New York.

Time limitations have allowed for little outside reviewing of chapters. However, we have done extensive cross-reviewing by contributing authors. We are grateful for the support and enthusiasm that all contributing authors brought to this project.

This volume is identified as a contribution of the U. S. Department of Agriculture Regional Project S-74—"Control Tactics and Management Systems for Arthropod Pests of Soybean." Most of the American authors are active members of the Technical Committee of this regional project.

October 29, 1979

Marcos Kogan
Urbana, Illinois

Donald C. Herzog
Quincy, Florida

Contents

SECTION I
Concepts and Techniques 1

Chapter 1
Soybean Growth and Assessment of Damage by Arthropods
M. Kogan and S. G. Turnipseed (With 15 Figures)....................3

Chapter 2
General Sampling Methods for Above-Ground Populations
of Soybean Arthropods
M. Kogan and H. N. Pitre, Jr. (With 11 Figures)....................30

Chapter 3
Introduction to Sampling Theory
W. G. Ruesink (With 1 Figure) 61

Chapter 4
Sequential Sampling Plans for Soybean Arthropods
M. Shepard (With 4 Figures)..................................79

Chapter 5
Sequential Estimation of Soybean Arthropod Population
Densities
W. G. Rudd (With 6 Figures)..................................94

SECTION II
Lepidopterous Defoliators 105

Chapter 6
Sampling Velvetbean Caterpillar on Soybean
D. C. Herzog and J. W. Todd (With 9 Figures) 107

Chapter 7
Sampling Soybean Looper on Soybean
D. C. Herzog (With 9 Figures)................................. 141

Chapter 8
Sampling Green Cloverworm on Soybean
L. P. Pedigo (With 13 Figures)................................. 169

SECTION III
Coleopterous Defoliators 187

Chapter 9
Sampling Mexican Bean Beetle on Soybean
S. G. Turnipseed and M. Shepard (With 6 Figures) 189

Chapter 10
Sampling Bean Leaf Beetles on Soybean
M. Kogan, G. P. Waldbauer, G. Boiteau,
and C. E. Eastman (With 20 Figures) 201

SECTION IV
Other Foliage Feeders 237

Chapter 11
Sampling Aphids in Soybean Fields
M. E. Irwin (With 10 Figures)................................. 239

Chapter 12
Sampling Leafhoppers on Soybean
C. G. Helm, M. Kogan, and B. G. Hill (With 11 Figures).............. 260

Chapter 13
Sampling Phytophagous Thrips on Soybean
M. E. Irwin and K. V. Yeargan (With 13 Figures) 283

Contents

Chapter 14
Sampling Whiteflies on Soybean
S. M. Vaishampayan and M. Kogan (With 2 Figures) 305

Chapter 15
Sampling Mites on Soybean
S. L. Poe (With 4 Figures) 312

SECTION V
Underground Feeders 325

Chapter 16
Sampling Phytophagous Underground Soybean Arthropods
C. E. Eastman (With 15 Figures) 327

SECTION VI
Stem and Axil Feeders 355

Chapter 17
Sampling Coleopterous Stem Borers in Soybean
W. V. Campbell (With 11 Figures) 357

Chapter 18
Sampling *Epinotia aporema* on Soybean
B. S. Correa Ferreira (With 3 Figures). 374

Chapter 19
Sampling Threecornered Alfalfa Hopper on Soybean
A. J. Mueller (With 7 Figures). 382

Chapter 20
Sampling Stem Flies in Soybean
G. A. Gangrade and M. Kogan (With 4 Figures) 394

SECTION VII
Pod Feeders 405

Chapter 21
Sampling *Heliothis* spp. on Soybean
*R. E. Stinner, J. R. Bradley, Jr.,
and J. W. Van Duyn* (With 6 Figures) 407

Chapter 22
Sampling Lepidopterous Pod Borers on Soybean
T. Kobayashi and T. Oku (With 9 Figures)............................422

Chapter 23
Sampling Phytophagous Pentatomidae on Soybean
J. W. Todd and D. C. Herzog (With 20 Figures)....................438

SECTION VIII
Natural Control Agents 479

Chapter 24
Sampling Parasitoids of Soybean Insect Pests
N. L. Marston (With 8 Figures)...................................481

Chapter 25
Sampling Predaceous Hemiptera on Soybean
M. E. Irwin and M. Shepard (With 16 Figures)....................505

Chapter 26
Sampling Ground Predators in Soybean Fields
J. F. Price and M. Shepard (With 3 Figures).....................532

Chapter 27
Sampling Spiders in Soybean Fields
W. H. Whitcomb (With 8 Figures).................................544

Chapter 28
Sampling Pathogens of Soybean Insect Pests
G. R. Carner (With 8 Figures)...................................559

Index 575

Contributors

Boiteau, Gilles, Research Station, Agriculture Canada, Fredericktown, New Brunswick, Canada.

Bradley, J. R., Jr., Department of Entomology, North Carolina State University, Raleigh, North Carolina, 27607, USA.

Campbell, William V., Department of Entomology, North Carolina State University, Raleigh, North Carolina, 27607, USA.

Carner, Gerald R., Department of Entomology and Economic Zoology, Clemson University, Clemson, South Carolina, 29631, USA.

Correa Ferreira, Beatriz Spalding, Departamento de Entomologia, Centro Nacional de Pesquisa da Soja, Caixa Postal 1061, 86.100 Londrina, PR, Brasil.

Eastman, Cathy E., Economic Entomology, Illinois Natural History Survey, 172 Natural Resources Building, Urbana, Illinois, 61801, USA.

Gangrade, Govind A., Operational Research Project on Integrated Control of Rice Pests, Agriculture College, Raipur, Madhya Pradesh, India.

Helm, Charles G., Economic Entomology, Illinois Natural History Survey, 172 Natural Resources Building, Urbana, Illinois, 61801, USA.

Herzog, Donald C., Department of Entomology and Nematology, University of Florida, Quincy, Florida, 32351, USA.

Hill, Bob G., Oklahoma State University, Cooperative Extension Service, Muskogee, Oklahoma, 74401, USA.

Irwin, Michael E., Agricultural Entomology, University of Illinois and Illinois Natural History Survey, 172 Natural Resources Building, Urbana, Illinois, 61801, USA.

Kobayashi, Takashi, Entomological Laboratory, Tohoku National Agricultural Experiment Station, Morioka, Japan.

Kogan, Marcos, Economic Entomology, Illinois Natural History Survey and Agricultural Entomology, University of Illinois, 172 Natural Resources Building, Urbana, Illinois, 61801, USA.

Marston, Norman L., Biological Control of Insects Research, Agricultural Research Service, USDA, Columbia, Missouri, 65201, USA.

Mueller, Arthur J., Department of Entomology, University of Arkansas, Fayetteville, Arkansas, 72701, USA.

Newsom, L. Dale, Department of Entomology, Louisiana State University, Baton Rouge, Louisiana, 70803, USA.

Oku, Toshio, Entomological Laboratory, Tohoku National Agricultural Experiment Station, Morioka, Japan.

Pedigo, Larry P., Department of Entomology, Iowa State University, Ames, Iowa, 50011, USA.

Pitre, Henry N., Jr., Department of Entomology, Mississippi State University, Mississippi State, Mississippi, 39762, USA.

Poe, Sidney L., Department of Entomology, Virginia Polytechnic Institute, Blacksburg, Virginia, 24061, USA.

Price, James F., Agricultural Research and Education Center, Bradenton, Florida, 33508, USA.

Rudd, Walter G., Department of Computer Science and Department of Entomology, Louisiana State University, Baton Rouge, Louisiana, 70803, USA.

Ruesink, William G., Agricultural Entomology, University of Illinois and Illinois Natural History Survey, 172 Natural Resources Building, Urbana, Illinois, 61801, USA.

Contributors

Shepard, Merle, Department of Entomology and Economic Zoology, Clemson University, Clemson, South Carolina, 29631, USA.

Stinner, Ronald E., Department of Entomology, North Carolina State University, Raleigh, North Carolina, 27607, USA.

Todd, James W., Department of Entomology and Fisheries, Georgia Coastal Plain Experiment Station, Tifton, Georgia, 31794, USA.

Turnipseed, Sam G., Edisto Experiment Station, Blackville, South Carolina, 29817, USA.

Vaishampayan, Sharad M., Department of Entomology, Forest Research Institute, Jabalpur 482020, India.

Van Duyn, John W., Department of Entomology, North Carolina State University at Raleigh, Tidewater Research Station, Plymouth, North Carolina, 29762, USA.

Waldbauer, Gilbert P., Department of Entomology, University of Illinois, Urbana, Illinois, 61801, USA.

Whitcomb, Willard H., Department of Entomology, University of Florida, Gainesville, Florida, 32611, USA.

Yeargan, Kenneth V., Department of Entomology, University of Kentucky, Lexington, Kentucky, 40506, USA.

SECTION I

Concepts and Techniques

Chapter 1

Soybean Growth and Assessment of Damage by Arthropods

Marcos Kogan and Sam G. Turnipseed

I. Introduction

It seems fitting to open a book on sampling soybean-associated arthropods with a definition of the habitat in which these arthropods are sampled. There are two main reasons to precisely define the field conditions in a sampling program, whether for research purposes or for scouting to make treatment decisions.

First, the crop represents the substrate from which the arthropod samples are extracted, with the exception of the soil inhabiting arthropods for which the soil is indeed the substrate even of the root and nodule feeding species (see Chapters 10 and 16). In each chapter, in Sections II through VIII, reference is made to the various characteristics of this crop substrate; these characteristics should be clearly understood and properly described.

Second, an understanding of the various ways whereby arthropods injure the crop and the response of the crop to injury is, in the final analysis, the practical justification for a sampling program. This chapter, consequently, sets the stage upon which the "drama" unfolds. A detailed description of all pertinent aspects of the soybean crop at the time samples are taken should be an inseparable part of any sampling program. This description should at least include the variety name, growth stage, height, row width, stand density, erectness of the stand (or degree of lodging), intensity of weed infestation and predominant weeds, and level or degree of injury caused by insects. In an attempt to attain some uniformity and conciseness in terminology, we borrowed from the agronomic literature many of the criteria used in soybean crop description (for a broader treatment of the subject, see Scott and Aldrich 1970, Caldwell 1973, Hill 1976, and Norman 1978).

The second part of this chapter discusses methods for estimating defoliation and pod and seed injury—two of the most common types of injury caused by insects to soybean. The relationship between the amount of injury and the level of damage has been investigated in some detail in the case of defoliators and pod and seed injuring species. Damage due to thinning of the stand that often results from injury by soil inhabiting arthropods (see Chapter 16) or by stem borers (see Chapters 17 and 20) can be indirectly assessed through the rather voluminous agronomic literature on effects on yield of varying soybean plant population density (see references in Pendleton and Hartwig 1973, Minor 1976). An especially acute lack of information exists on damage resulting from injury to roots and nodules. There is also lack of information on the combined effect, on plant growth and yield, of two or more types of injury occurring simultaneously or in succession. Injury caused by stem borers and other more specialized types are discussed in several chapters in Sections IV and V.

A. The Soybean Crop

Familiarity with the growth patterns and morphological characteristics of the soybean plant is necessary for an adequate description of the field conditions at sample time. An excellent review of soybean morphology is given by Carlson (1973). We will limit this section to the description of those crop characteristics that more directly affect the sampling procedures designed to measure interactions of arthropods with the soybean crop. The general pattern of crop dry matter accumulation during the growing season is shown in Fig. 1-1. This pattern reveals the sharp increase in midseason of available food for herbivores and the greater complexity of the substrate as the crop grows. As a consequence, the associated fauna becomes increasingly rich, and larger population densities are more likely to be found later in the season.

1. Soybean Growth Patterns

Most soybean varieties are very sensitive to photoperiod under normal growing conditions (Hartwig 1973). As a consequence of their response to photoperiod, soybean varieties in North America are adapted for full-season growth within latitudinal bands of 180 to 300 km wide. In the northern hemisphere, flowering of a variety adapted to a certain latitudinal band is premature south of this band and delayed north of it. The opposite is true in the southern hemisphere. Altitude has a masking effect on this response probably due to the cooler night temperatures at higher altitudes (Summerfield and Minchin 1976). In the United States varieties are placed into 12 maturity groups—00 to X—with varieties in group 00 being the earliest and in X the latest (Hartwig 1973, Whigham and Minor 1978). Figure 1-2 shows the approximate distribution of maturity bands in North America.

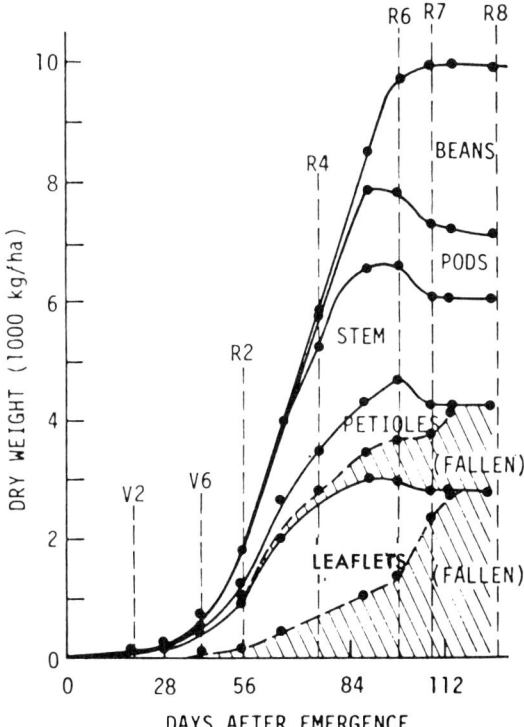

Figure 1-1. Accumulation of dry matter, including fallen leaves, in soybean plants. (Adapted from Hanway 1976).

2. Determinate and Indeterminate Soybean

In general the early maturing varieties (groups 00-IV) typically present an indeterminate type of growth. These varieties first set flowers and pods in the leaf axils near the base of the plants (Fig. 1-3A), and new flowers and pods continue to appear at new nodes as plants grow. New leaves are added after full bloom, as the plants generally have achieved less than half their final height when flowering begins (Fehr and Caviness 1977). Determinate varieties have generally reached their full height when flowering begins, although leaves continue to develop within the canopy for about four weeks after floral initiation. Flowering occurs about the same time along the main stem of the plant; the plant has a terminal leaf on the main stem that is about the same size as lower leaves on the plant, and the terminal node on the main stem bears a raceme that develops a cluster of pods (Fig. 1-3B). Most late-maturing varieties have a determinate type of growth.

3. Stages of Soybean Development

A clear and simple definition of growth stages is essential for the description of crop characteristics at the time samples are taken. We recommend the adoption

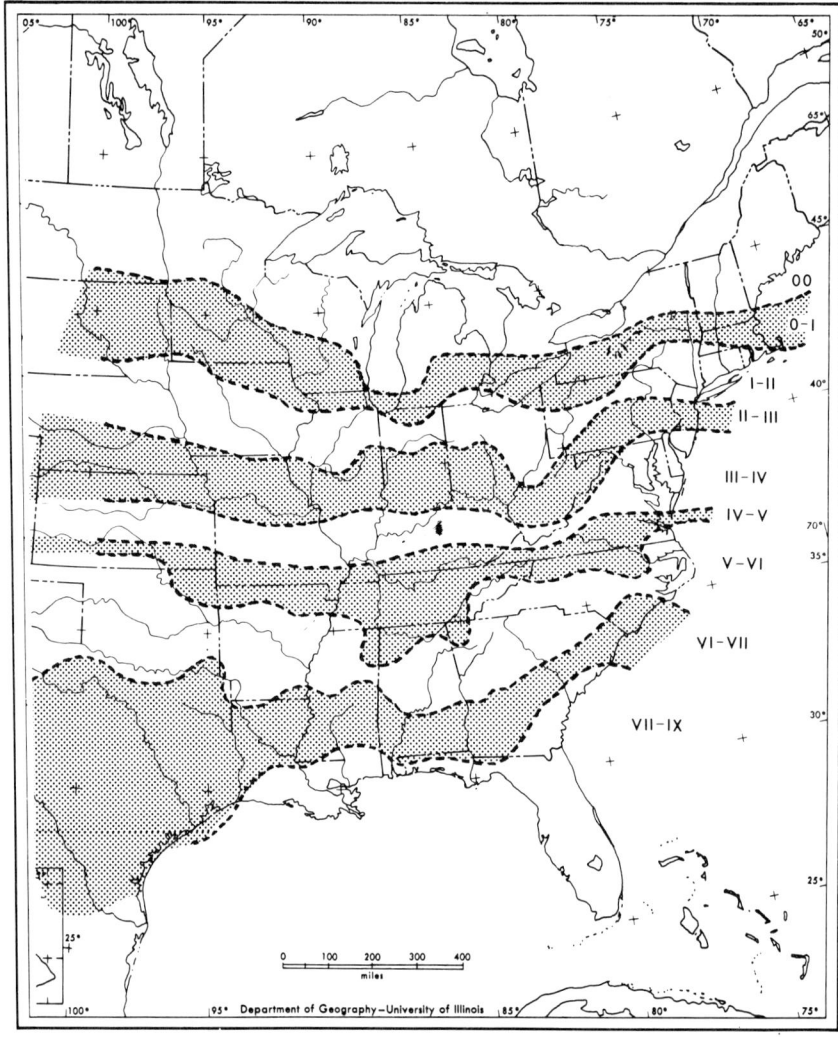

Figure 1-2. Latitudinal bands of maturity groups of soybean varieties adapted to growing conditions in North America. (Based on Barnhart, 1954).

of the method described by Fehr and Caviness (1977) to standardize these descriptions in the entomological literature. The method uses the number of nodes along the main stem to describe the vegetative or V stages. Reproductive stages are designated as R stages. Emerging seedlings are at stage VE and the seedlings with open cotyledons are at stage VC. After the VC stage each node is counted starting with the unifoliolate node. Thus, a plant with one open trifoliolate will be at stage V2. A summary of the method is presented in Table 1-1.

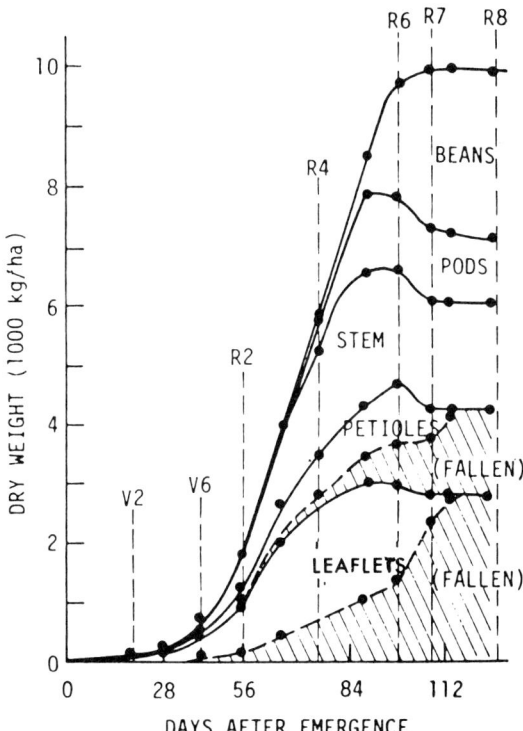

Figure 1-1. Accumulation of dry matter, including fallen leaves, in soybean plants. (Adapted from Hanway 1976).

2. Determinate and Indeterminate Soybean

In general the early maturing varieties (groups 00-IV) typically present an indeterminate type of growth. These varieties first set flowers and pods in the leaf axils near the base of the plants (Fig. 1-3A), and new flowers and pods continue to appear at new nodes as plants grow. New leaves are added after full bloom, as the plants generally have achieved less than half their final height when flowering begins (Fehr and Caviness 1977). Determinate varieties have generally reached their full height when flowering begins, although leaves continue to develop within the canopy for about four weeks after floral initiation. Flowering occurs about the same time along the main stem of the plant; the plant has a terminal leaf on the main stem that is about the same size as lower leaves on the plant, and the terminal node on the main stem bears a raceme that develops a cluster of pods (Fig. 1-3B). Most late-maturing varieties have a determinate type of growth.

3. Stages of Soybean Development

A clear and simple definition of growth stages is essential for the description of crop characteristics at the time samples are taken. We recommend the adoption

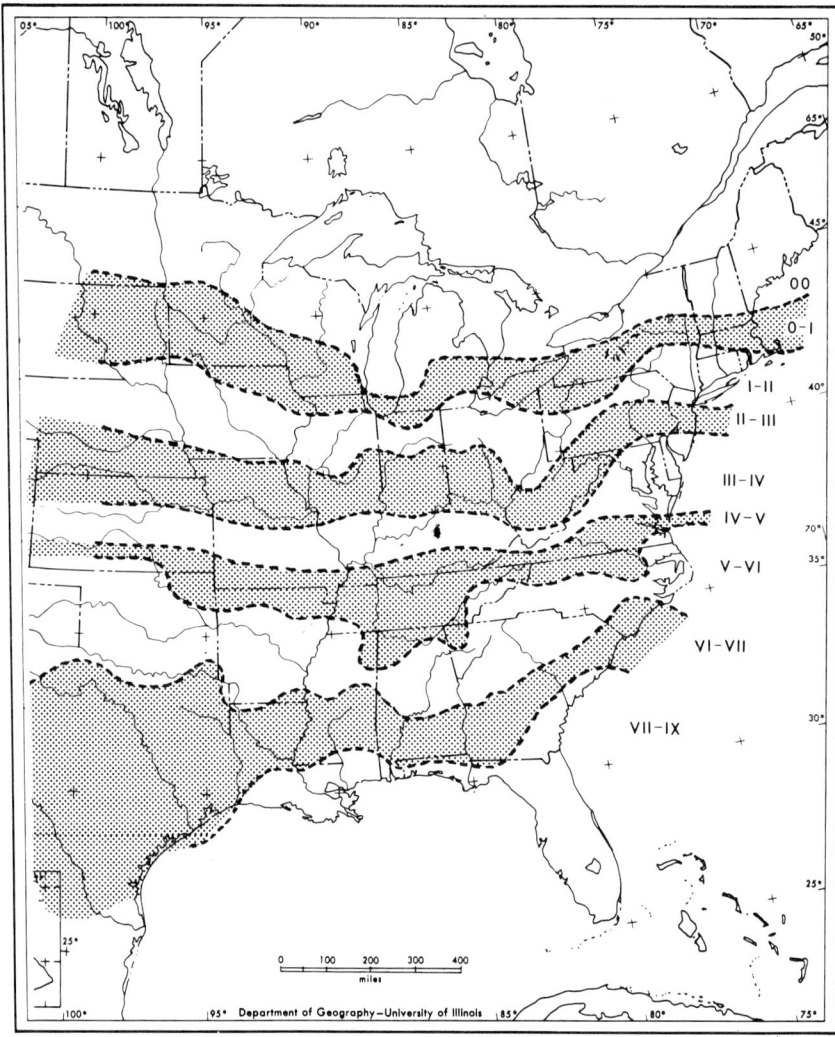

Figure 1-2. Latitudinal bands of maturity groups of soybean varieties adapted to growing conditions in North America. (Based on Barnhart, 1954).

of the method described by Fehr and Caviness (1977) to standardize these descriptions in the entomological literature. The method uses the number of nodes along the main stem to describe the vegetative or V stages. Reproductive stages are designated as R stages. Emerging seedlings are at stage VE and the seedlings with open cotyledons are at stage VC. After the VC stage each node is counted starting with the unifoliolate node. Thus, a plant with one open trifoliolate will be at stage V2. A summary of the method is presented in Table 1-1.

Figure 1-3. (A) Indeterminate growth type of a plant at stage R4; (B) Determinate growth type of a plant at the same growth stage. (Redrawn from Fehr and Caviness, 1977).

4. Assessment of Pod and Seed Development

An adequate definition of pod and seed development is needed in studies of insects feeding on such plant parts. The best visual criterion to indicate the end of the seed filling stage (maximum seed dry weight accumulation or stage R7) has been defined as the loss of green color from the pods (Crookston and Hill 1978). In entomological studies it is often necessary to more fully describe plant development, for instance, the R5 and R6 stages (see Table 1-1). Thus intermediate stages, e.g., R5.5, could be used to indicate that seeds are one-half grown within pods.

For a more quantitative approach to pod development, McWilliams et al. (1977) constructed a simple caliper to measure pod elongation and thickness. Description of the caliper and its use are shown in Fig. 1-4.

5. Estimation of Root Development

Studies of root and nodule feeders may require an adequate expression of the stages of root growth. Unfortunately, there are no easy methods of defining root development. However, much of the injury is caused by insects feeding within a cylinder of soil about 10 cm in diameter and 10-15 cm deep (see Chapters 10 and 16). Nodule counts and dry weight of roots within the cylinder are adequate measures to help estimate the effect of soil inhabiting phytophagous insects (Anderson and Waldbauer 1977). A general description of root

Table 1-1. Description of growth stages of soybean (from Fehr and Caviness 1977)[a]

Stage	Abbreviated stage title	Description
Vegetative Stages		
VE	Emergence	Cotyledons above the soil surface
VC	Cotyledon	Unifoliolate leaves unrolled sufficient so the leaf edges are not touching
V1	First-node	Fully developed leaves at unifoliolate nodes
V2	Second-node	Fully developed trifoliolate leaf at node above the unifoliolate nodes
V3	Third-node	Three nodes on the main stem with fully developed leaves beginning with the unifoliolate nodes
V(n)	nth-node	n number of nodes on the main stem with fully developed leaves beginning with the nodes. n can be any number beginning with 1 for *V1*, first-node stage.
Reproductive Stages		
R1	Beginning bloom	One open flower at any node on the main stem
R2	Full bloom	Open flower at one of the two uppermost nodes on the main stem with a fully developed leaf
R3	Beginning pod	Pod 5 mm long at one of the four uppermost nodes on the main stem with a fully developed leaf
R4	Full pod	Pod 2 cm long at one of the four uppermost nodes on the main stem with a fully developed leaf
R5	Beginning seed	Seed 3 mm long in a pod at one of the four uppermost nodes on the main stem with a fully developed leaf
R6	Full seed	Pod containing a green seed that fills the pod cavity at one of the four uppermost nodes on the main stem with a fully developed leaf
R7	Beginning maturity	One normal pod on the main stem that has reached its mature pod color
R8	Full maturity	Ninety-five percent of the pods that have reached their mature pod color. Five to ten days of drying weather are usually required after R8 before the soybeans have less than 15 percent moisture.

[a] A plant at a reproductive stage of growth should be designated with a combination of the V and R stages. For example, a plant at full bloom with 11 fully developed trifoliolates above the unifoliolate nodes would be at stage V12R2 because the unifoliolate nodes should be counted as one.

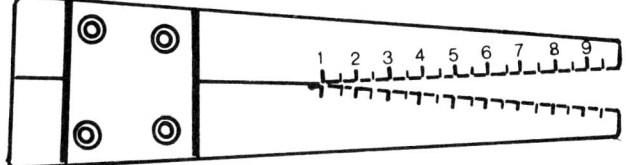

Figure 1-4. A "V" caliper to measure thickness of soybean pods. Graduations are made for 0.5 mm increments in thickness. (Length 15.0 cm; width at base 4.0 cm; width at apex 2.4 cm; angle of jaws 7°10'; graduations are 4 mm apart to measure 0.5 mm increments. Caliper can be built of any rigid material). (Redrawn from McWilliams et al. 1977).

development in soybean is provided in Fig. 1-5 (Fehrenbacher et al. 1969). According to Mitchell and Russell (1971) soybean root growth and development occurs in three phases, each phase corresponding to a specific vegetative or reproductive stage. The phases were characterized by increased root penetration of the soil profile and proliferation of the root system. Early in the season at least 90% of the root dry weight is concentrated in the upper 7.5 cm of the soil profile and later in the season, in the upper 15 cm. A first approximation of root development may be derived from comparisons of top versus root versus root growth (Sivakumar et al. 1977). According to studies conducted with 'Wayne' soybean in Iowa, a top/root ratio of 3.8 was observed with plants at stage V6. The ratio increased to 9.0 by stage V15R2. From these root length and leaf area data it was calculated that root length increased from 630 m/m^2 of leaf area, with plants at the V5 stage, to 1190 m/m^2 at V13R2, and decreased to 345 m/m^2 at V15R2 as the soil dried (Sivakumar et al. 1977). Thus, measurement of the leaf area and determination of growth stage may provide a basis for estimating root development.

B. Physiognomic Description of the Crop

In addition to the growth characteristics that are genetically controlled and rather stable in the various soybean varieties, there are several other features whose expression depends on variable agro-ecological conditions. Some of these conditions, which are mainly a result of cropping procedures, should be described in a sampling program mainly because they may appreciably affect results of the programs.

1. Planting Date, Row Width, Plant Density, and Plant Height

Planting dates vary considerably and may have a profound influence on insect populations. Planting date and row width were shown to affect the relative abundance of predators and incidence of diseases of some serious soybean pests in North Carolina (Sprenkel et al. 1979; see Chapter 23).

Row widths commonly used vary from about 18 cm to 1 m or more. Soybeans

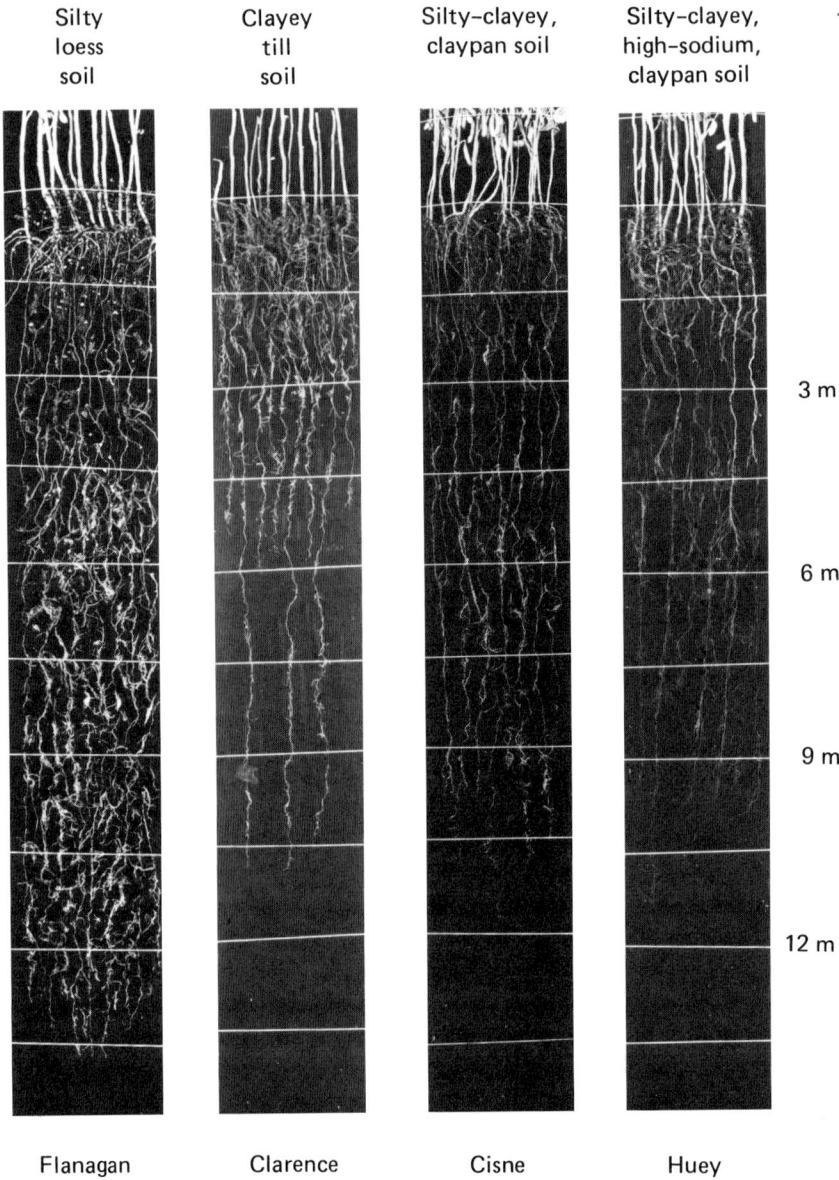

Figure 1-5. Soybean root system in four different types of soil (all fertilized). The compact dense layers in the lower profile of till soils restrict root penetration. Most soybean-root-feeding insects are active in the upper 15 cm. (By courtesy of Dr. J. B. Fehrenbacher, Dept. of Agronomy, University of Illinois, from Fehrenbacher et al. 1969).

are occasionally broadcast with the resulting stand not ordered in rows. Plant height is generally measured from ground level to the growing tip of the main stem. (A measuring tape should be standard equipment of entomological surveyors and scouts.) Plant density is measured by counting the number of plants over a certain length of row (see Chapter 2). Several counts should be made during the growing season because of natural variations in plant populations and because thinning of the stand due to intraspecific competition will decrease plant density as the season progresses.

2. Lodging

Soybean plants grown at high densities under conditions of optimal fertility and high moisture may have a tendency to lodge. Lodging resistance is an important characteristic selected by breeders, but the better yielding varieties do not always have the strongest stems. Although lodging resistance differs with variety, its expression is greatly influenced by environmental conditions. Lodging is usually scored on a scale of 1 (erect) to 5 (prostrate).

3. Weediness

The presence of weeds can greatly influence results of sampling programs. Any description of field conditions is incomplete without identification of prevalent weeds and some indication of weed density. For most practical purposes it may be enough to use descriptive terms (weedy, moderately weedy, weed-free, etc.). Accurate estimates of weed populations, including identification of the species involved, may be necessary in basic ecological studies. Populations of weeds may be measured on the basis of number of plants per unit of row length (Wax 1973). Standard techniques of quadrat analysis developed by plant ecologists may also be adopted to define weed populations in soybean fields.

4. Neighboring Crops

The physiognomic description of the sampling site should include a list of the predominant plant cover (natural or agricultural) of adjacent fields.

II. Assessment of Damage

Injury is the result of those activities of phytophagous insects (feeding, oviposition) that impinge on the integrity of the host plant. Damage, or crop loss, is the consequence of an amount of injury that exceeds the tolerance of the plant (Davidson and Norgaard 1973, Kogan 1976). The assessment of damage caused by insects to soybean is based on measurements of two types of parameters: (a) the nature and amount of injury caused to the various plant parts (roots, stems, foliage, flowers or pods, and seeds) and (b) the relationship between the amount of injury observed and reduction in yield and seed quality. This reduction can be estimated by comparison to uninjured plants grown under similar conditions.

A. Sampling to Estimate Foliage Injury

Defoliation is probably the most common and certainly the most visible type of injury to soybean. Defoliation encompasses all types of injury that result in a functional reduction of the total leaf surface of the plant. Although insects with chewing mouthparts are the ones considered here, those with sucking (e.g., leafhopper) or rasping (e.g., thrips) mouthparts may reduce the photosynthesizing ability of the plant and cause crop damage. Such insects are generally less important than those with chewing mouthparts, but they may contribute to the overall level of defoliation. The methods discussed in this section have been developed primarily for insects with chewing mouthparts, but may also be applied to assess the impact of other defoliating pests on the crop.

As is the case with any other sampling procedure, one must distinguish between the accurate measurement of foliage area needed for research and the gross estimates of defoliation necessary for pest management purposes (Ruesink and Kogan 1975). Sequential sampling plans for insect defoliation on soybean have been proposed by Bellinger and Dively (1978).

1. Measurements of Leaf Area

A review of the principles and methods used in leaf area measurement is found in Sestak et al. (1971). Since that review was published, rapid improvement in instrumentation has occurred. Foremost among these instruments are the electronic area meters that provide a direct digital readout of the area of irregular objects. The techniques used in leaf area measurement can be divided into: (a) photoelectric or electronic, (b) gravimetric, (c) volumetric, (d) planimetric, (e) geometric, and (f) visual estimates. A brief description will be provided of each of them.

(a) *Photoelectric and electronic devices.* Several electronic leaf area-meters are currently available. The LI-COR model LI-3000 portable area meter or the LI-3100 stationary model are examples of such devices (Fig. 1-6). The apparatus uses an electronic method of rectangular approximation with 1 mm^2 resolution. A scanning head provided with an array of light emitting diodes (LED's) perfectly collimated on an opposite array of photodiodes measures the variable width in 1-mm increments equal to the number of LED-photodiode pairs that are blocked. The length of the object is measured by an attached encoding cord (portable systems) or by the speed of the conveyor belt (stationary model). The area is then integrated and displayed for a direct readout in square centimeters.

These electronic devices are rather expensive and may not be readily available to researchers in developing countries. A simpler but reasonably accurate photo-

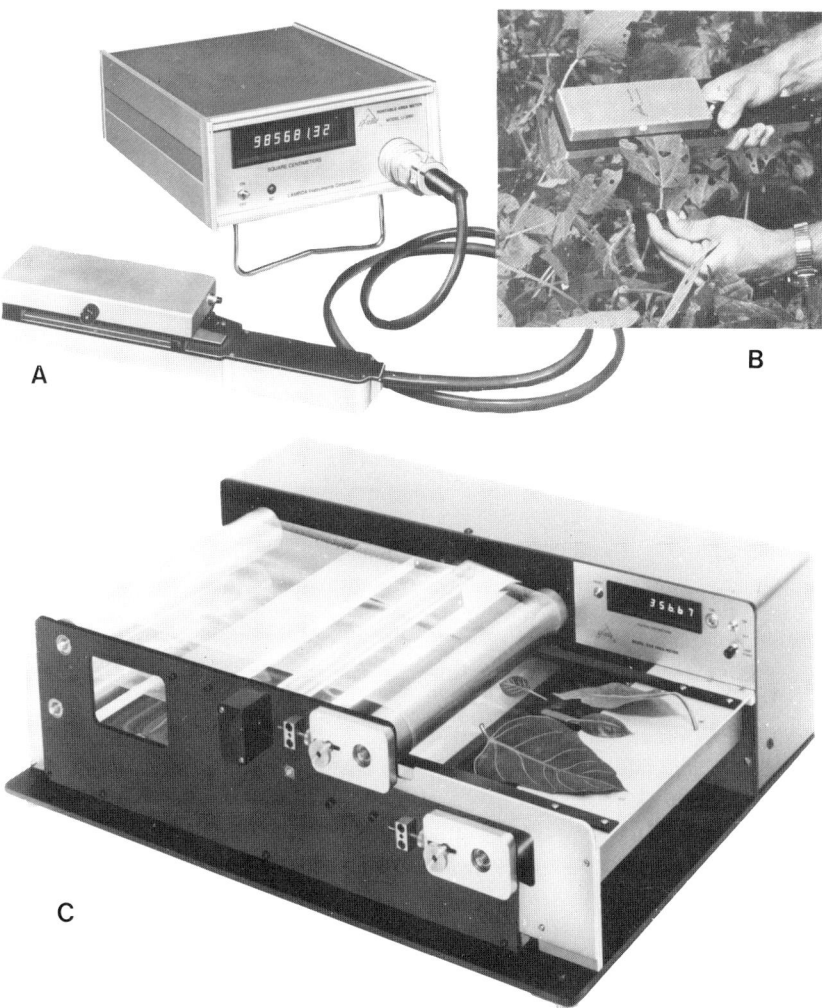

Figure 1-6. Electronic leaf meter. (A) LI-COR model LI-3000 that may be used with a conveyor belt as a stationary unit. (B) Use of model 3000 in the field. (C) Stationary LI 3100 model. (Courtesy of LI-COR, Inc., Lincoln, Nebraska).

planimeter can be built with a set of four photocells that "read" the light passing through a translucent stage illuminated from below by a battery of fluorescent tubes. The current generated in the photocells is "read" by a microamperimeter, and it is proportional to the amount of light passing through the window. The microamperimeter readings are converted to area readings by calibrating the apparatus with a series of dark silhouettes of known area. Such a photoplanimeter was built at a cost about ten times less than that of some of the electronic instruments and was used in studies of leaf area consumption by lepidopterous caterpillars (M. Khalsa and M. Kogan, unpublished). A general view of a Carman

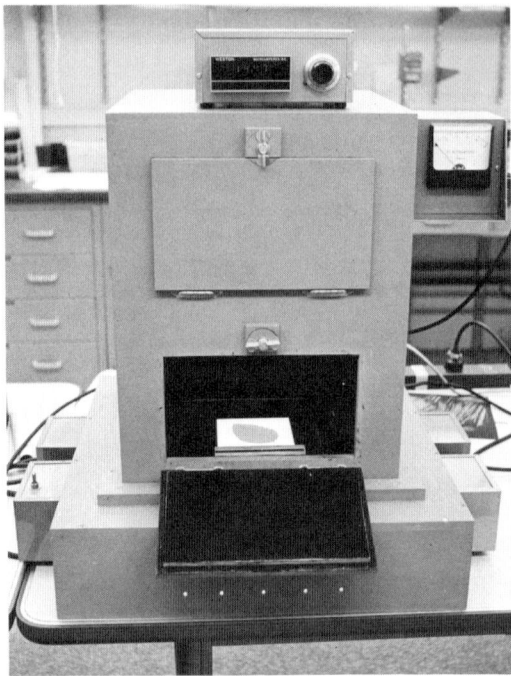

Figure 1-7. Large stage photoplanimeter using a battery of four photocells (Carman 1963 type) (Model built at the University of Illinois; Khalsa and Kogan, unpublished).

type photoplanimeter is shown in Fig. 1-7.

The instruments described above measure actual leaf area present. Estimation of defoliation is then obtained by masking holes or cut-off edges to provide a reading of the "intact" leaf. The amount of defoliation is the difference obtained between the two. The masking of holes may be a time consuming procedure. We have gotten around this difficulty by making duplicator copies of the leaves, cutting out the leaf silhouette, and reading the paper leaf area in the leaf-area meter. By subtracting the area of the actual leaf from the area of the paper leaf, we obtained the amount of leaf area removed.

(b) *Gravimetric method.* A good approximation of the leaf area may be obtained by regressing known leaf areas on the corresponding fresh or dry weight of these areas (Koller 1972, Wiersma and Bailey 1975). A linear equation can then be calculated and leaf weights easily converted into areas. We used a series of discs cut from soybean leaf blades with a cork borer and made successive measurements of 1 to 10 discs. The regression equation was—leaf area in cm^2 = 0.194 + 0.133x (r = .9993), where x is the dry leaf weight in mg of 'Harosoy' soybean grown in the greenhouse (Kogan and Cope 1974). The regression equation—leaf area = .60 + .66x (r^2 = .97)—was obtained with undamaged, field grown leaves (Hatfield et al. 1976). Again, gravimetric measurements cannot provide a direct measure of defoliation, only actual leaf area present. Since the area/dry weight relationship may vary with cultivar and growing conditions, researchers using the gravimetric method should initially derive the conversion equation for their specific set of conditions.

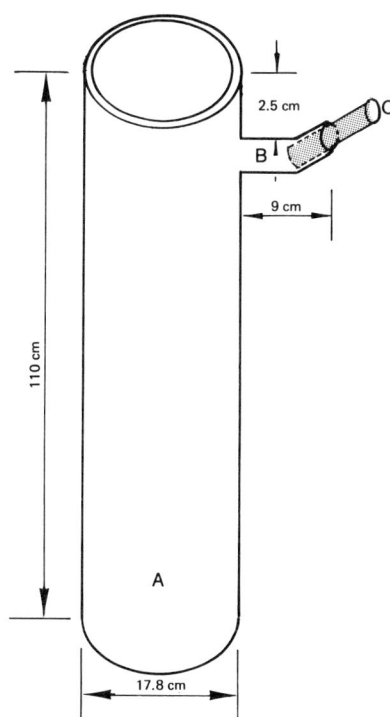

Figure 1-8. Volumetric cylinder with overflow tube to measure biomass of plants. It can be adopted also to measure leaf areas. (A) Plexiglas cylinder (0.6 mm thick walls); (B) Plexiglas overflow tube (16 mm OD, 10 mm ID); (C) Teflon tubing (10 mm OD, 8 mm ID). (Redrawn from Heinzman, Jr. and Eilrich 1977).

(c) *Volumetric method.* A volume displacement cylinder (Fig. 1-8) was used to measure the biomass of potted soybean plants in a nondestructive way (Heinzman, Jr. and Eilrich 1977). The method consists of introducing a plant into a calibrated cylinder filled with water. The excess water that spills over when the plant is introduced is a measure of the plant's volume. The method was originally used to directly correlate vegetative volume of leaves and stems with seed yield. The volumetric technique, however, may be adapted to measure leaf area if a volume to area ratio is computed. In fact, a first approximation to area may be obtained by determining the mean thickness of the leaves.

(d) *Planimetric methods.* Planimeters are mechanical devices to measure area. Accuracy of the measurements is affected by leaf geometry, and application of the method in soybean is rather limited. Various graphic methods (counting dots, squares, etc.), however, can be used if other faster methods are not available. This may be one of the few methods adapted to measuring injury by insects with sucking or rasping mouthparts.

(e) *Geometric methods.* Several conversion factors are available to estimate soybean leaf area from measurements of length and widest point of the central leaflet (Jensen et al. 1977). Conversion of length × width of the central leaflet was obtained by linear regression using leaves of nine soybean varieties. Regression equations were calculated for each individual variety with correlation coefficients ranging from 0.942 to 0.985. The regression equation for the combined results of all varieties was—area of trifoliolate = 0.157 + 0.0182 x, where x = length × width of the central leaflet in mm, and total leaf area is given in cm^2.

The advantage of the geometric method is to provide an estimate of the total leaf area including holes or cut edges. Used in conjunction with a leaf-area meter that excludes holes and cut edges, the geometric method offers a means for calculating percent defoliation on actual leaf area removed.

(f) *Visual estimates.* For pest management purposes it is not important to accurately measure the leaf area but rather to provide a rapid estimate of defoliation. There is no easy or practical way of estimating defoliation of whole plants in the field. The untrained person has a tendency to grossly overestimate defoliation. One method to provide a calibration tool for visual observations was used and extensively tested in a pilot pest management program in Brazil (Kogan et al. 1977). A series of photographs or drawings (Fig. 1-9) displays soybean leaflets with increments of five percent of the total area removed by actual insect feeding. The surveyor picks at random 10-20 leaflets from middle or upper nodes. The leaflets are compared to the series of pictures and are scored within five-percent defoliation classes. The percentages are then averaged, and the mean reflects the estimated level of defoliation in the field.

After practicing with the calibrated pictures, surveyors become quite efficient at estimating defoliation on sight by simply bending a length of row and scanning the plants from top to bottom. A convenient set of slides showing pictures of soybean leaves with various degrees of defoliation has been used to train scouts in the visual estimation of defoliation of standing plants in the field (W. Allen, Virginia Polytechnic Institute, personal communication).

B. Sampling to Estimate Injury to Pods and Seeds

Damage to pods by pod borers and other pod feeding insects is discussed in Section VII. In general a certain number of pods are harvested, and injury is measured in terms of number of injured pods in proportion to the total number of pods harvested.

Injury to seeds is more complex because it must take into account both quantitative and qualitative effects. The effect of injury on the amount (quantity) of seed production is usually measured by comparing yields of insect-free and insect-infested plots (Miner 1966, Todd and Turnipseed 1974, Vincentini and Jimenez 1977). Another method commonly used to express these quantitative effects is the weight of 100 seeds, with samples being taken at random from the bulked seeds of a plot or field. The two measurements do not necessarily reflect the same intensity of attack because a heavy infestation of stink bugs, for instance, may cause the abortion of seeds and the shedding of pods that are never harvested. Therefore, yield measurements are more likely to reflect true quantitative effects, whereas weight of 100 seeds is more useful in reflecting qualitative effects. When expressing measurements on a seed weight basis, it is important to express the weights at a constant moisture content (usually 13%). This can be done by using the formula—seed weight at $x\%$ moisture = initial seed weight $[(100-\text{initial }\% \text{ moisture})/(100-x\% \text{ moisture})]$. The effect of stink bugs and other seed-feeding arthropods on soybean seed quality is usually measured by

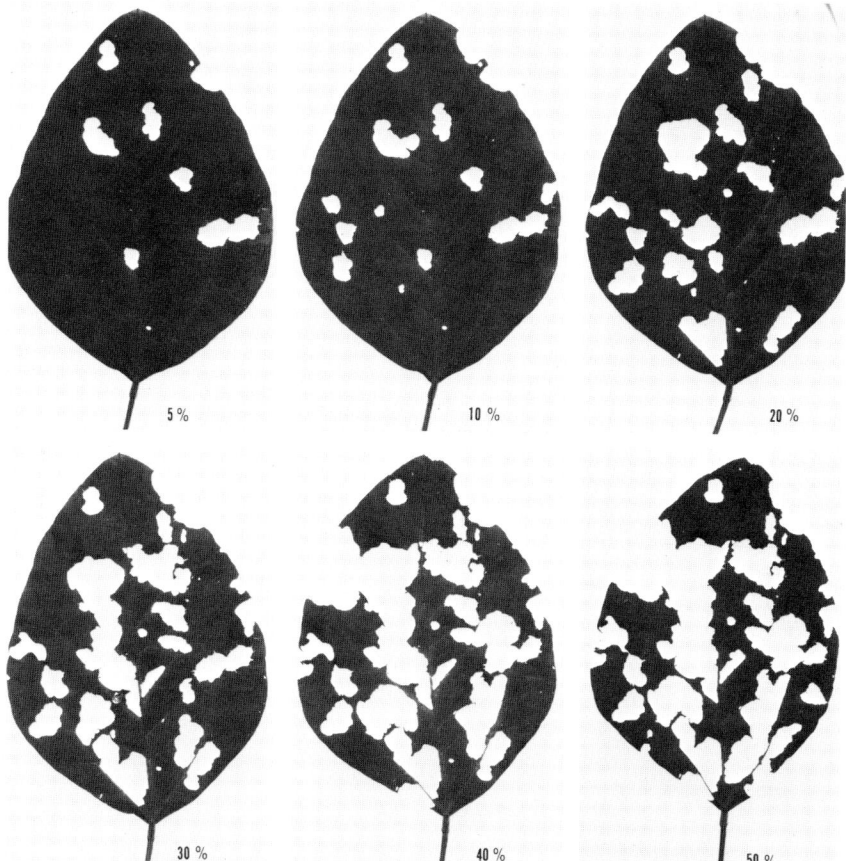

Figure 1-9. Sample of soybean leaflets with various percentages of the blade removed by insect feeding.

one of the following methods: (a) visual ratings of damaged seed, (b) flotation method, (c) differential density method, (d) percent germination, and (e) oil and protein content.

(a) *Visual rating.* This is a quick method of classifying levels of injury to seed. Figure 1-10 shows five classes of seed with various levels of stink bug injury (Todd 1976). One may also quantify the amount of injury by counting the number of seeds in a lot that fall into each class of injury. A weighted index can then be obtained by multiplying the number or percentage of seeds in a given injury class by that class value.

(b) *Flotation method.* A method based on the number of seeds that float in a saturated solution of NaCl was correlated to the amount of stink bug injury (Hart 1970). The method is quick, but seeds damaged by mechanical or other causes will also float, thus overestimating stink bug damage.

(c) *Differential density method.* Hart and Rowan (1970) described a method to

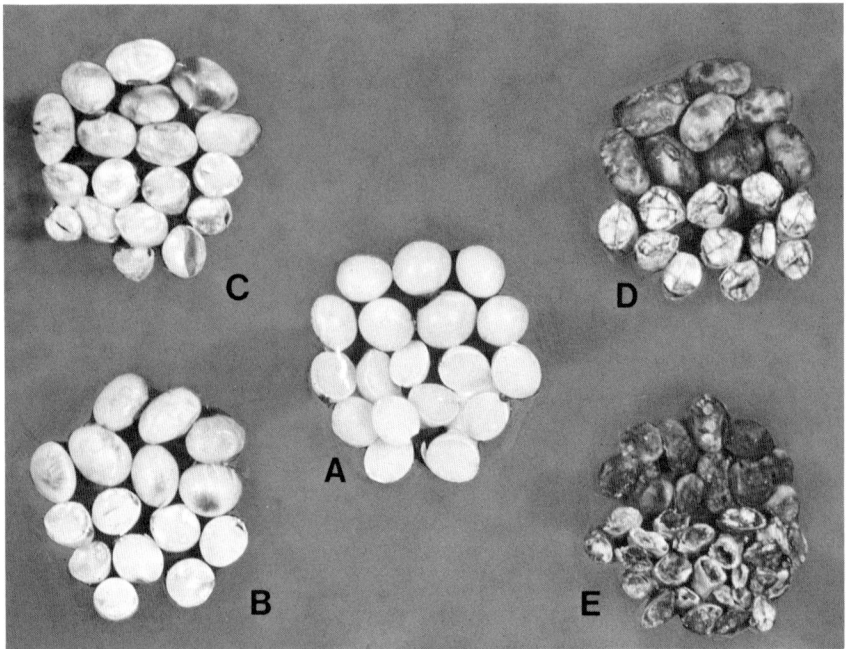

Figure 1-10. Various damage categories of soybeans fed upon by *N. viridula:* (A) no damage, (B) light damage, (C) medium damage, (D) heavy damage, (E) severe damage. Upper half of each group shows beans in cross-section. (By courtesy of J. W. Todd 1976).

measure changes in density of injured seeds. The method consists of measuring the time it takes for individual beans to fall from one fixed point to another through a vertical column of water. An apparatus using a pair of photocells records the passing of the seed between the two fixed points (Fig. 1-11).

(d) *Germination.* Viability of the seed is either measured directly by germination tests or by the tetrazolium test (Delouche 1975).

(e) *Percent oil and protein.* These qualitative parameters have been measured by standard analytical procedures (Association of Official Agricultural Chemists 1960). Rapid estimates of oil and protein concentration are now possible with infrared analysis (Hymowitz et al. 1974, Rinne et al. 1975, Hilliard and Daynard 1976).

III. Estimating the Relationship Between Injury and Damage

Damage or economic loss resulting from injury to the foliage caused by leaf-eating insects is an indirect process that is rather difficult to measure. Foliage-eating pests, therefore, are indirect pests of a crop like soybean in which the

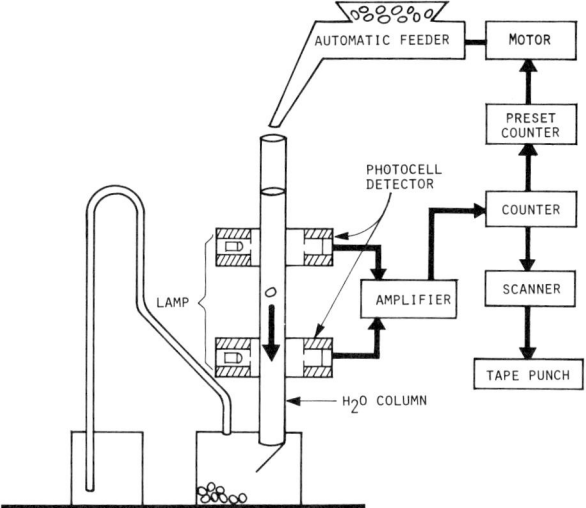

Figure 1-11. Diagram of an instrument to measure soybean seed quality by density; the instrument measures the velocity of descent of soybean in a column of water. (Redrawn from Hart and Rowan 1970).

desired product is the seed or grain. Pod-feeding insects, such as pod borers and certain stink bugs, are direct pests.

The relationship between defoliation and decrease in quantity and quality of grain has been most often approached by simulating insect injury through hand defoliation. Many studies of this sort have been conducted by agronomists— some interested in investigating the economic effect of hail storms, others in studying the leaf area index as related to optimal seed production. There are very few critical studies comparing insect defoliation with hand-simulated defoliation (Poston et al. 1976).

Similarly, effect of seed and pod injury on yield and seed quality has been studied by simulation and by observations of direct insect feeding; much of the work in this area has been conducted on injury caused by stink bugs.

A. Effect of Defoliation on Soybean Yield

Most studies conducted to date have shown that soybean is most tolerant of defoliation occurring up to the beginning of pod setting (stage R3). Prior to this stage plants usually recover from defoliation levels of about 30%. It is generally assumed that recovery depends on post-defoliation growing conditions; however, studies by Turnipseed (1972) indicated that the relationship of defoliation to yield was similar in irrigated and in non-irrigated, and in early and late-planted soybean. There is little information available on the yield effects of lower levels of defoliation followed by other stress factors, such as drought or incidence of diseases.

Despite differences in growth habit there is no evidence that determinate varieties are less tolerant of defoliation than indeterminate varieties except at very high levels of defoliation. Determinate varieties showed a 25, 33, and 10% greater yield reduction than indeterminate varieties when 100% defoliated at stages R2, R4, or R5 respectively (Fehr and Caviness 1977).

Simulated defoliation has been performed by clipping whole leaflets or half leaflets or by punching holes in leaves. The entire level of simulated defoliation is produced at one time, as is the case with hail injury, or a certain level of defoliation may be maintained or deployed in increments over a period of one week to 10 days to more closely reflect insect injury.

It is very difficult to summarize the body of information that has been accumulated on the effect of defoliation on yield. Studies have been conducted with many different varieties, and stages of growth have not always been accurately defined. Variations in plot size and experimental procedures do not allow use of the combined published data to generate an accurate descriptive model of the relationship. There are, however, two sets of data that have allowed the fitting of second degree equations correlating yield reduction with level of defoliation at various growth stages. These two sets were published by Kalton et al. (1949) (as analyzed by Stone and Pedigo 1972), and Thomas et al. (1976) (with regression equations later revised by the authors to force the curves through the origin). Fig. 1-12 shows the relationship of defoliation to decline in yield, as a percentage of the yield of undefoliated controls. Graphs were based on the equations shown in Table 1-2. Although there are differences in the estimated effect of defoliation by these equations, at stage R4 both sets of equations estimate yield effects within what seems to be their confidence intervals (although these have not been computed).

Under most experimental conditions and at all growth stages, yield reductions are not expected to occur below 20% defoliation. At the vegetative stage of growth (prior to bloom) and after seeds are fully developed (stage R6), there is no apparent effect on yield of defoliation levels below 30% (see references on effect of defoliation on yield in Turnipseed and Kogan 1976).

The physiological effects of defoliation are only beginning to be understood. Poston et al. (1976) reported that decreases in net carbon-exchange rates (measured in mg CO_2 dm^{-2} h^{-1}) of excised leaves reached a maximum 12 hr after defoliation occurred but only at 50% defoliation levels. At 25% defoliation there was no effect on net photosynthesis either as a result of insect or of artificial defoliation. Simulated damage caused by using a hole puncher or cutting half-leaflets along the midrib yields results similar to those obtained with actual insect feeding. Cutting leaflets across the midrib increased net photosynthesis and, therefore, was not an adequate method for simulating injury by defoliating insects (Poston et al. 1976). In addition to the effect of defoliation on net photosynthetic rate of soybean plants, it seems that nitrogen fixation is also impaired. Injury to foliage by the soybean looper, *Pseudoplusia includens* (Walker),

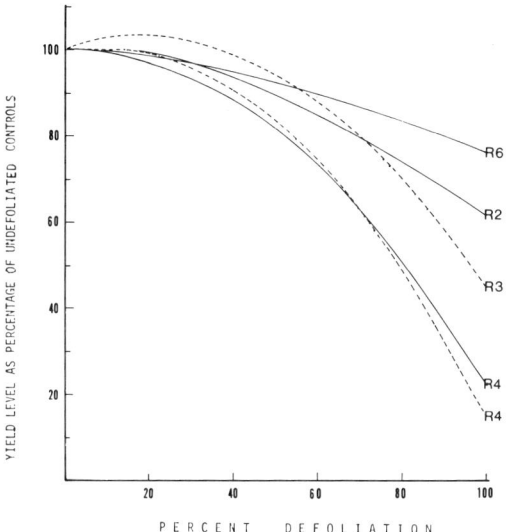

Figure 1.12. Curves correlating percent defoliation with yield decrease expressed as a percentage of yield of undefoliated controls. (Solid line curves based on Stone and Pedigo 1972 from data in Kalton et al. 1949; dashed line curves based on Thomas et al. 1974, with equations revised to pass through the origin.) (Thomas, personal communication).

has been reported to cause substantial reductions in nitrogen fixation (Newsom et al. 1978). It seems, therefore, that it is more important to measure the net photosynthetic rate of the functional leaves remaining on the plant after defoliation than the percentages of the foliage area that is lost. This factor is best expressed as the leaf area index (LAI), or the relationship of the leaf area present per unit of ground area. LAI is a key input in models predicting net assimilation rates (Watson, 1958, Sivakumar 1978). There is, however, some discrepancy in the definition of a critical leaf area index for soybean (see references in Buttery 1970). Also, Turnipseed (1972) demonstrated that yield reduction from mechanical defoliation was not significantly affected by date of planting, use of

Table 1-2. Equations correlating defoliation with yield reduction

Stage of Growth	Equation[1]	Ref.[2]
V2	% Red. = $-0.029X + 0.002X^2$	1
V6-7	% Red. = $-0.142X + 0.003X^2$	1
R2	% Red. = $-0.002X + 0.004X^2$	1
R3	% Red. = $-0.305X + 0.008X^2$	2
R4	% Red. = $0.032X + 0.008X^2$	1
R4	% Red. = $-0.190X + 0.010X^2$	2
R5	% Red. = $0.007X^2$	2
R6	% Red. = $0.050X + 0.002X^2$	1
R7	% Red. = $0.083X$	2

[1] % Red. = percent yield reduction; X = percent defoliation
[2] 1 = Stone and Pedigo 1972, based on data from Kalton et al. 1949.
2 = Thomas et al. 1974, revised to pass through the origin (Thomas, personal communication).

irrigation, or varietal differences. In fact, the largest yield reductions occurred at high production levels in irrigated fields in which the LAI was greatest following defoliation. A better definition of these relationships would greatly advance our understanding of the physiological effect of defoliation, and it would greatly facilitate use of the LAI in estimates of damage.

B. Effect of Defoliation on Seed Quality

Defoliation may interfere with normal processes of seed development with consequent changes in seed quality. Seed size, oil and protein concentration, iodine number, and germination have been used as criteria for seed quality in defoliation studies. In general, protein concentration does not seem to be greatly affected by defoliation, although the tendency is for protein to decrease and for oil to increase with reduction of seed weight resulting from defoliation (Turnipseed 1972). These relationships are shown in Fig. 1-13. It seems that significant qualitative effects are only detected at very high ($>$ 75%) defoliation levels at stages R4 to R6 (Kalton et al. 1949) or as a consequence of sequential defoliation (Thomas et al. 1978).

C. Effect of Pod Injury on Soybean Yield

Simulation of damage to pods has been attempted by removal of pods at various stages of growth (Kincade et al. 1971, Smith and Bass 1972, Turnipseed 1973, Thomas et al. 1974, 1976), and by puncturing pods with a needle (Corso 1977). Up to the R4 stage, plants seem to compensate for considerable depodding; however, removal of pods after the R4 stage greatly affected yields. Thomas et al. (1974) computed a series of equations to estimate the effect of depodding on yield (Table 1-3). These relationships are graphically shown in Fig. 1-14.

D. Effect of Depodding on Seed Quality

Qualitative effects of direct seed injury by stink bugs have been studied in detail (see Chapter 23). Qualitative effects of artificial or simulated depodding or pod injury are also expressed in terms of seed size, chemical composition, and viability of the seed. Few studies were conducted in this area, but in general depodding seems to decrease oil concentration and may slightly increase protein concentration (Turnipseed 1973, Thomas et al. 1976). These effects vary with stage of plant growth at which depodding is inflicted and are more accentuated after the R5 stage. Germination was reduced by high levels (2/3) of depodding at stages R4 and R5 only (Thomas et al. 1976).

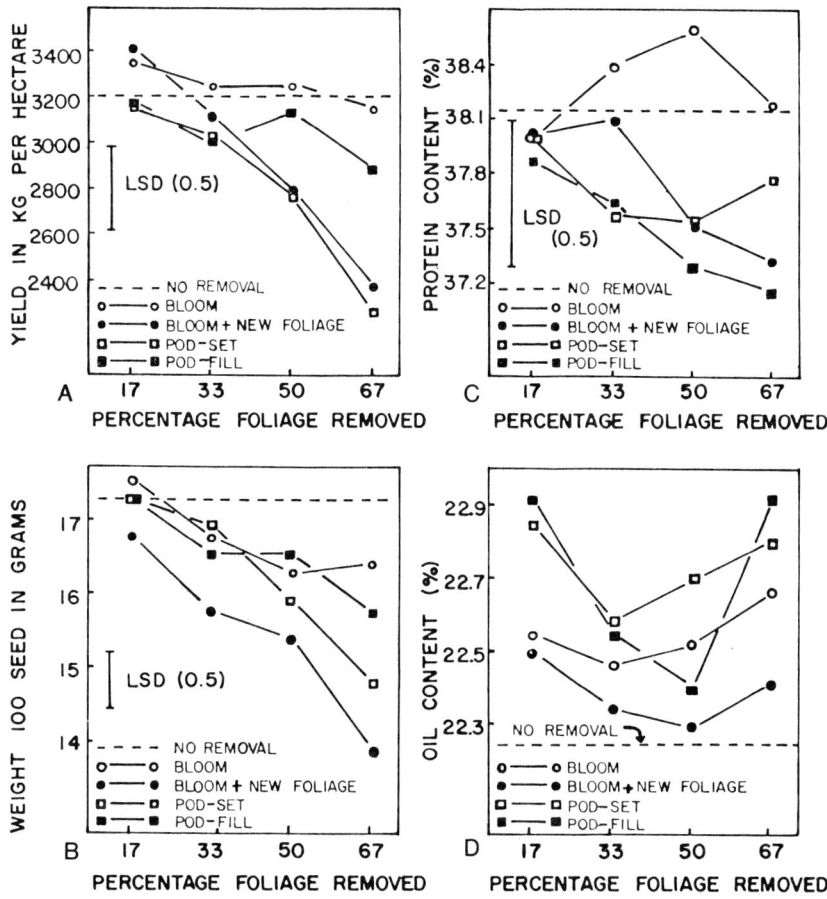

Figure 1-13. (A) Yield; (B) Weight of 100 seeds; (C) Protein concentration; and (D) Oil concentration. 'Jackson' soybean with four levels of defoliation at four stages of plant growth. (From Turnipseed 1972, by permission from the Entomological Society of America)

Table 1-3. Relationship between artificial depodding and yield reduction in the absence of defoliation (from Thomas et al. 1974, equations adjusted to go through zero at the origin by Thomas, personal communication)

Stage of Growth	Equations[1]
R4	% Red. = $0.004090Y^2$
R5	% Red. = $0.330413Y + 0.004275Y^2$
R6	% Red. = $0.455917Y + 0.004899Y^2$
R7	% Red. = $0.817614Y + 0.001782Y^2$

[1] % Red. = percent yield reduction
Y = percent pod removal.

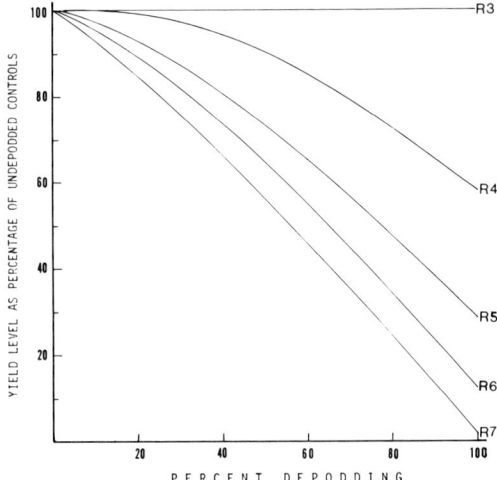

Figure 1.14. Relationship between artificial depodding and yield reduction as a percentage of yield of undepodded controls. 'Clark 63' soybean at five stages of plant growth. (Based on equations in Table 1.3 from Thomas et al. 1974, and personal communication).

E. Interactions of Defoliation and Depodding

Interactions between three levels of defoliation and three levels of depodding at five stages of growth have been reported by Thomas et al. (1974, 1976, 1978). Table 1-4 provides the regression equations calculated by these authors, forcing the curves through the origin (Thomas, personal communication). These equations may be used as a first approximation, and prediction by these equations should be validated under any new set of environmental conditions and with varieties other than those used by Thomas et al. (their work was done with 'Clark 63' soybean in Columbia, Missouri). Figure 1-15 presents a series of curves computed using equations in Table 1-4 and making the levels of defoliation and depodding vary from zero to 80. The curves show the marked effect of the interaction at R3 and R5 with defoliation being more critical at R3 and depodding more critical at R5. At R7 there is no effect of defoliation on yield, and reductions are linearly correlated with levels of depodding. Although the model used in this analysis is rather simplistic, it serves to depict the complexities involved in attempting to combine effects of various types and levels of injury.

IV. Concluding Remarks

Sampling soybean plants to estimate levels of injury is not an easy task. Accurate measurements are very time consuming and for scouting purposes they are pro-

Table 1-4. Equations correlating defoliation and depodding with yield reduction (based on Thomas et al. 1974 and Thomas, personal communication)

Stage of Growth	Equations[1]
R3	% Red. = $-0.304655X + 0.008450X^2$
R4	% Red. = $-0.187923X + 0.010398X^2 + 0.004090Y^2 - 0.003486XY$
R5	% Red. = $0.007309X^2 + 0.330413Y + 0.004275Y^2 - 0.004897XY$
R6	% Red. = $0.290426X + 0.002375X^2 + 0.455917Y + 0.004899Y^2 - 0.004542XY$
R7	% Red. = $0.082978X + 0.817614Y + 0.001782Y^2 - 0.000820XY$

[1] % Red. = percent yield reduction; X = percent defoliation; Y = percent depodding.

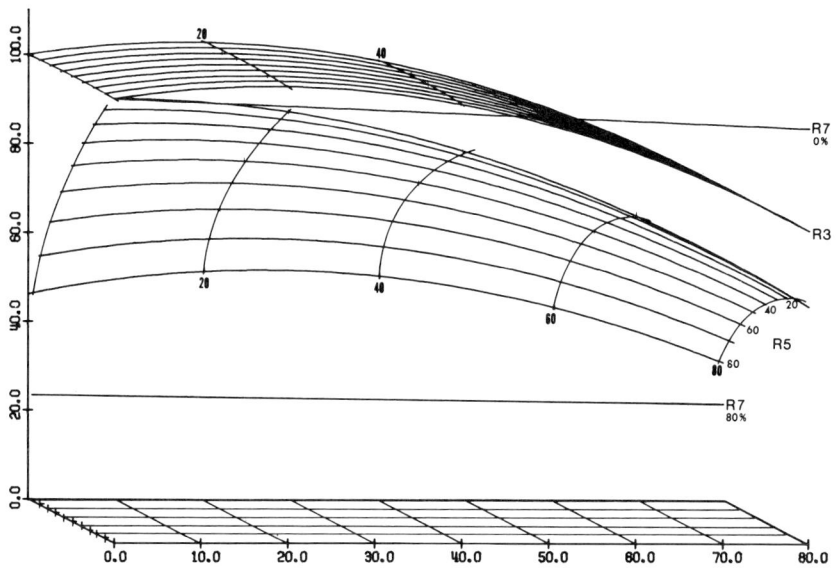

YIELD (R3, R5, R7) AS A FUNCTION OF DEFOLIATION (CURVE FOR EACH DEPODDING LEVEL)

Figure 1-15. Tridimensional plot showing variation of soybean yield (as a percentage of uninjured controls on the vertical axis) resulting from increased levels of defoliation and increased levels of depodding at stages R3, R5 and R7. At stages R3 and R5 the graph shows eight levels of depodding. At stage R7 only 0% and 80% depodding are shown (as straight lines). (Plotted on a CalComp terminal, Univ. of Illinois, using equations in Table 1-4 from Thomas et al. (1974) and Thomas, personal communication).

hibitive and in fact unnecessary. The methods discussed in this chapter are aimed at providing a basis for the researchers in their attempts to establish accurate economic injury levels. These relationships between injury and damage take into account a linear correlation between insect population levels and injury levels (Kogan 1976). Attempts are underway to develop simulation models for the soybean crop growth (Heilman et al. 1977, Rudd 1979). Development of such models will provide a more dynamic tool for investigating interactions of stress factors, physical or biological, on soybean yield and quality. Notwithstanding the level of sophistication and complexity these models may achieve, they will reflect reality only if sufficient field data are available both for development of the models and for their validation. Accurate sampling procedures for measuring injury and the correlation of injury to damage are essential for both phases of model development.

Acknowledgments

We thank Dr. Richard R. Johnson, Department of Agronomy, University of Illinois for critically reviewing the manuscript and Mrs. Ellen Brewer for writing the CalComp program. This paper is based on work supported in part by the National Science Foundation and U.S. Environmental Protection Agency-project "Principles, strategies and tactics of pest population regulation and control in major crop ecosystems," the soybean subproject, through a grant (NSF GB 34718) to the University of California. The opinions expressed herein are those of the authors and not necessarily those of the University of California, NSF, or the EPA.

References

Anderson, T.E. and G P. Waldbauer. 1977. Development and field testing of a quantitative technique for extracting bean leaf beetle larvae and pupae from soil. Environ. Entomol. 6:633-636.
Association of Official Agricultural Chemists. 1960. Methods of analysis. 9th ed., Washington, D.C., 832 p.
Barnhart, F. 1954. Soybeans. F. Barnhart Pub., Caruthersville, Missouri. 290 p.
Bellinger, R.G. and G. P. Dively. 1978. Development of sequential sampling plans for insect defoliation on soybeans. J. N. Y. Entomol. Soc. **86**:278-279.
Buttery, B.R. 1970. Effects of variation of leaf area index on growth of maize and soybeans. Crop Sci. **10**:9-13.
Caldwell, B.E., ed. 1973. Soybeans: Improvement, production, and uses. Amer. Soc. Agron. Pub., Madison, Wisconsin. Agron. Ser. 16. 681 p.
Carlson, J.B. 1973. Morphology. pp. 17-93 in B.E. Caldwell, ed. Soybeans: Improvement, production, and uses. Amer. Soc. Agron. Pub., Madison, Wisconsin. Agron. Ser. 16. 681 p.
Carman, P.D. 1963. A large scale photoelectric planimeter for leaves. Appl. Optics **2**:1317-1322.

Corso, I.C. 1977. Relação entre of efeito associado de percevejos e fungos na produção e qualidade de sementes de soja (*Glycine max* (L.) Merrill), bem como na transmissão de moléstias [Relationship between the associated effect of stink bugs and fungi in yield and quality of soybean seed, as well as in the transmission of diseases]. M.S. thesis, Universidade Federal do Rio Grande do Sul, Porto Alegre, Brazil. 86 p.

Crookston, R.K. and D.S. Hill. 1978. A visual indicator of the physiological maturity of soybean seed. Crop Sci. **18**:867-870.

Davidson, A. and R.B. Norgaard. 1973. Economic aspects of pest control. OEPP/EPPO Bull. **3**:63-75.

Delouche, J.C. 1975. Seed quality and storage of soybeans. pp. 86-107 in D.K. Whigham, ed. Soybean production, protection and utilization. Univ. of Illinois, College of Agriculture, INTSOY Ser. 6. 266 p.

Fehr, W.R. and C.E. Caviness. 1977. Stages of soybean development. Iowa Coop. Ext. Serv. Spec. Rep. **80**:12 p.

Fehrenbacher, J.B., B.W. Ray, and J.D. Alexander 1969. How soils affect root growth. Crops and Soils **21**:14-18.

Hanway, J.J. 1976. Interrelated developmental and biochemical processes in the growth of soybean plants. pp. 5-15 in L.D. Hill, ed. World soybean research. Proc. World Soybean Res. Conf., Interstate Print., Danville, Illinois. 1073 p.

Hart, J.R. 1970. Methods for estimating stink bug damage in soybeans. II. Flotation method. Cereal Chem. **47**:369.

Hart, J.R. and J.D. Rowan. 1970. Methods for determining the extent of stink bug damage in soybeans. I. Density method. Cereal Chem. **47**:49.

Hartwig, E.E. 1973. Varietal development. pp. 187-210 in B.E. Caldwell, ed. Soybeans: Improvement, production, and uses. Amer. Soc. Agron. Pub., Madison, Wisconsin. Agron. Ser. 16. 681 p.

Hatfield, J.L., C.D. Stanley, and R.E. Carlson. 1976. Evaluation of an electronic foliometer to measure leaf area in corn and soybean. Agron. J. **68**:434-436.

Heilman, J.L., E.T. Kanemasu, and G.M. Paulsen. 1977. Estimating dry-matter accumulation in soybean. Can. J. Bot. **55**:2196-2201.

Heinzman, Jr., C.D. and G.L. Eilrich. 1977. Technique for measuring vegetative volume and seed yield in soybeans. Can. J. Plant Sci. **57**:613-614.

Hill, L.D., ed. 1976. World soybean research. Proc. World Soybean Res. Conf., Interstate Print., Danville, Illinois. 1073 p.

Hilliard, J.H. and T.B. Daynard. 1976. Measurement of protein and oil in grains and soybeans with reflected near infrared light. J. Inst. Can. Sci. Technol. Aliment. **9**:11-14.

Hymowitz, T., J.W. Dudley, F.I. Collins, and C.N. Brown. 1974. Estimation of protein and oil concentration in corn, soybean, and oat seed by near-infrared light reflectance. Crop Sci. **14**:713-715.

Jensen, R.L., L.D. Newsom, D.C. Herzog, J.W. Thomas, Jr., B.R. Farthing, and F.A. Martin. 1977. A method of estimating insect defoliation of soybean. J. Econ. Entomol. **70**:240-242.

Kalton, R.C., C.R. Weber, and J.C. Eldredge. 1949. The effect of injury simulating hail damage to soybeans. Iowa Agr. Home Econ. Exp. Sta. Res. Bull. **359**:736-796.

Kincade, R.T., M.L. Laster, and E.E. Hartwig. 1971. Simulated pod injury to soybeans. J. Econ. Entomol. **64**:984-985.

Kogan, M. 1976. Evaluation of economic injury levels for soybean insect pests. pp. 515-533 in L.D. Hill, ed. World soybean research. Proc. World Soybean Res. Conf., Interstate Print., Danville, Illinois. 1073 p.

Kogan, M. and D. Cope. 1974. Feeding and nutrition of insects associated with soybeans. 3. Food intake, utilization, and growth in the soybean looper, *Pseudoplusia includens*. Ann. Entomol. Soc. Amer. **67**:66-72.

Kogan, M., S.G. Turnipseed, M. Shepard, E.B. de Oliveira, and A. Borgo. 1977. A pilot pest management program for soybean in southern Brazil. J. Econ. Entomol. **70**:659-663.

Koller, H.R. 1972. Leaf area-leaf weight relationships in the soybean canopies. Crop Sci. **12**:180-183.

McWilliams, J.M., J.H. Hatchett, and E.A. Stadelbacher. 1977. A hand caliper for measuring thickness of soybean pods. Agron. J. **69**:125-126.

Miner, F.D. 1966. Biology and control of stink bug on soybeans. Ark. Agr. Exp. Sta. Bull. **708**:40 p.

Minor, H.C. 1976. Planting date and plant spacing in soybean production. pp. 56-62 in R.M. Goodman, ed. Expanding the use of soybeans. Proc. of a Conference for Asia and Oceania. Univ. of Illinois, College of Agriculture, INTSOY Ser. 10. 261 p.

Mitchell, R.L. and W.J. Russell. 1971. Root development and rooting patterns of soybean [*Glycine max* (L.) Merrill] evaluated under field conditions. Agron. J. **63**:312-316.

Newsom, L.D., E.P. Dunigan, C.E. Eastman, R.L. Hutchinson, and R.M. McPherson. 1978. Insect injury reduces nitrogen fixation in soybeans. La. Agr. **21**: 15-16.

Norman, A.G., ed. 1978. Soybean physiology, agronomy and utilization. Academic Press, N.Y. 272 p.

Pendleton, J.W. and E.E. Hartwig. 1973. Management. pp. 211-237 in B.E. Caldwell, ed. Soybeans: Improvement, production, and uses. Amer. Soc. Agron. Pub., Madison, Wisconsin. Agron. Ser. 16. 681 p.

Poston, F.L., L.P. Pedigo, R.B. Pearce, and R.B. Hammond. 1976. Effects of artificial and insect defoliation on soybean net photosynthesis. J. Econ. Entomol. **69**:109-112.

Rinne, R.W., S. Gibbons, J. Bradley, R. Seif, and C.A. Brim. 1975. Soybean protein and oil percentages determined by infrared analysis. USDA ARS-NC-**26**: 4 p.

Rudd, W.G. 1979. Simulation of insect damage to soybeans. p. 90 in F.T. Corbin, ed. World Soybean research conference-II. Abstracts. North Carolina State Univ., Raleigh. 117 p.

Ruesink, W.G. and M. Kogan. 1975. The quantitative basis of pest management: Sampling and measuring. pp. 309-351 in R.L. Metcalf and W.H. Luckmann, eds. Introduction to insect pest management. John Wiley and Sons, N.Y. 581 p.

Scott, W.O. and S.R. Aldrich. 1970. Modern soybean production. The Farm Quarterly, Cincinnati, Ohio. 192 p.

Sestak, Z., J. Catsky, and P.G. Jarvis. 1971. Assessment of leaf area and other assimilating plant surfaces. pp. 517-555 in Plant photosynthetic production. Manual of methods. W. Junk, N.V. Pub., The Hague, Holland.

Sivakumar, M.V.K. 1978. Prediction of leaf area index in soya bean. Ann. Bot. 42:251-253.

Sivakumar, M.V.K., K.M. Taylor, and R.H. Shaw. 1977. Top and root relations of field-grown soybeans. Agron. J. 69:470-473.

Smith, R.H. and M.H. Bass. 1972. Relationship of artificial pod removal to soybean yields. J. Econ. Entomol. 65:606-608.

Sprenkel, R.K., W.M. Brooks, J.W. Van Duyn, and L.L. Deitz. 1979. The effects of three cultural variables on the incidence of *Nomuraea rileyi*, phytophagous Lepidoptera, and their predators on soybeans. Environ. Entomol. 8:334-339.

Stone, J.D. and L.P. Pedigo. 1972. Development of economic injury level of the green cloverworm on soybean in Iowa. J. Econ. Entomol. 65:197-201.

Summerfield, R.J. and F.R. Minchin. 1976. An integrated strategy for day length and temperature-sensitive screening of potentially tropic-adapted soyabeans. pp. 186-191 in R. M. Goodman, ed. Expanding the use of soybeans. Proc. of a Conference for Asia and Oceania. Univ. of Illinois, College of Agriculture, INTSOY Ser. 10. 261 p.

Thomas, G.D., C.M. Ignoffo, K.D. Biever, and D.B. Smith. 1974. Influence of defoliation and depodding on yield of soybeans. J. Econ. Entomol. 67:683-685.

Thomas, G.D., C.M. Ignoffo, and D.B. Smith. 1976. Influence of defoliation and depodding on quality of soybeans. J. Econ. Entomol. 69:737-740.

Thomas, G.D., C.M. Ignoffo, D.B. Smith, and C.E. Morgan. 1978. Effects of single and sequential defoliations on yield and quality of soybeans. J. Econ. Entomol. 71:871-874.

Todd, J.W. 1976. Effects of stink bug feeding on soybean seed quality. pp. 611-618 in L.D. Hill, ed. World soybean research. Proc. World Soybean Res. Conf., Interstate Print., Danville, Illinois. 1073 p.

Todd, J.W. and S.G. Turnipseed. 1974. Effects of southern green stink bug damage on yield and quality of soybeans. J. Econ. Entomol. 67:421-426.

Turnipseed, S.G. 1972. Response of soybeans to foliage losses in South Carolina. J. Econ. Entomol. 65:224-229.

Turnipseed, S.G. 1973. Insects. pp. 545-572 in B.E. Caldwell, ed. Soybeans: Improvement, production, and uses. Amer. Soc. Agron. Pub., Madison, Wisconsin. Agron. Ser. 16. 681 p.

Turnipseed, S.G. and M. Kogan. 1976. Soybean entomology. Annu. Rev. Entomol. 21:25-60.

Vincentini, R. and H.A. Jimenez. 1977. El vaneo de los frutos en soja [Seed abortion in soybean pods]. INTA, Estac. Exp. Reg. Agropec. Parana, Entre Rios, Argentina. Ser. Tec. 47:1-30.

Watson, D.J. 1958. The dependence of net assimilation rate on leaf area index. Ann. Bot. 22:37-54.

Wax, L.M. 1973. Weed control. pp. 417-458 in B.E. Caldwell, ed. Soybeans: Improvement, production, and uses. Amer. Soc. Agron. Pub., Madison, Wisconsin. Agron. Ser. 16. 681 p.

Whigham, D.K. and H.C. Minor. 1978. Agronomic characteristics and environmental stress. pp. 77-118 in A.G. Norman, ed. Soybean physiology, agronomy, and utilization. Academic Press, N.Y. 249 p.

Wiersma, J.V. and T.B. Bailey. 1975. Estimation of leaflet, trifoliolate and total leaf area of soybeans. Agron. J. 67:20-30.

Chapter 2

General Sampling Methods for Above-Ground Populations of Soybean Arthropods

Marcos Kogan and Henry N. Pitre, Jr.

I. Introduction

The methods most commonly used to sample above-ground populations of arthropods in row crops have also been extensively employed in soybean entomology research and surveys. Four of the principal methods are (1) direct observations, (2) ground or beat cloth, (3) sweep net, and (4) vacuum or suction net (usually a D-Vac). Two of these methods, the direct observations and the ground cloth, may approach absolute or direct population estimates for some species. The other two are indirect or relative sampling methods. In general, however, these four methods need proper calibration against reliable absolute sampling techniques that permit conversion of population samples to estimates of density. Two other methods, the absolute sampling techniques most commonly used for soybean arthropods, involve the caging and harvesting of whole plants and the fumigation cage. These methods are based on the exhaustive extraction of individuals from a unit of habitat.

Distance to nearest neighbor, recapture of marked individuals, and removal trapping have not been adequately tested with soybean arthropods; therefore, these methods will not be discussed here. The interested reader, however, is referred to Southwood (1978).

For most pest management purposes, it is sufficient to obtain relative population estimates that permit a rapid classification of situations into decision categories (e.g., spray or do not spray) (Ruesink and Kogan 1975). Sequential sampling programs allow the use of relative sampling methods to reach management decisions within defined confidence limits (see Chapters 4 and 5). However, in basic population studies and in establishing the foundation for the practical adoption of population indices in sequential sampling schemes or in scouting for pest management, it is often necessary to obtain absolute population estimates.

The six methods referred to in the first paragraph are described and compared in this chapter. Subsequent chapters on sampling for individual species or species complexes will stress the peculiarities of these six general methods when applied to particular species or complexes. In addition, subsequent chapters will cover in detail the methods, which are not discussed here, that are unique to those species or species complexes.

II. Direct Observations

Visual searches of soybean arthropods were used in two different ways: single plant examination and counts over a measured length of row.

A. Single Plant Examination

Nonadjacent randomly selected plants are initially scanned for large, often fast-moving species. After the initial scan, both sides of each leaf on the plant are searched, as are petioles, axils, and stems. The apical leaves are uncurled and searched for thrips, mites, and other small arthropods. Blossoms and fruits are also searched when present. In the field records are taken of the number of individuals per species, the stage of development, the relative position on the plant, or any other biological data of interest. If the method is used to sample a predetermined set of species, printed forms can be used to facilitate record taking. In some cases it may be advantageous to use computer data sheets or machine-readable forms that save the time and inconvenience of transcribing field data. The method may be used also to assess populations of single species, in which case the search strategy may be simplified by taking advantage of certain behavioral characteristics of the species, e.g., the preference of older green cloverworm, *Plathypena scabra* (F.), larvae for the abaxial side of upper leaves (see Chapter 8).

This method was used to sample cotton arthropods in Texas (Pieters and Sterling 1973) and to study arthropod communities in soybean fields in Illinois (Mayse et al. 1978b).

B. Counts Over a Measured Length of Row

Visual searches for single species over a measured length of row were used to sample adult bean leaf beetles, *Cerotoma trifurcata* (Forster) (Kogan et al 1974) and to sample the potato leafhopper, *Empoasca fabae* (Harris) (see Chapter 12), until plants reached the V4 stage at which time they were large enough to be sampled by other methods. This method is, therefore, recommended for surveys during the early stages of plant development.

Visual searches for certain not very active species feeding on soybean seedlings are made as the surveyor walks along a premeasured length of row and gently turns the leaves ventral side up, after recording those individuals resting or feeding on the upper side of leaves and stems. Species that drop to the ground when disturbed can be easily spotted, as the canopy at this stage is very sparse. Growing tips also have to be carefully examined as several species hide within the furled leaflets.

Slightly different techniques are required for sampling more active insects. For example, adult leafhoppers are counted as the surveyor walks along the premeasured length of row and gently taps the plant to make them jump or fly from the plants. The surveyor must advance against the wind so that the dislodged leafhoppers do not fly ahead to plants not yet sampled; this reduces the chance of counting the same individual more than once. Extreme attention is required to spot all moving leafhoppers. The accuracy of this method is inversely proportional to population size.

C. Characteristics of Direct Observation Sampling Methods

Direct observation probably is the only method among those discussed here that can be used very early in the season when soybean plants are small and tender without seriously damaging the plants. Whole plant counts can be used on larger plants, but these counts over a measured length of row become impractical. Mayse et al. (1978a) found that by using direct observation of single plants they could survey the arthropod fauna on soybean from seedling emergence through physiological maturity without removal of individuals in the populations. Furthermore, direct observation is the only method that allows correct positioning of the sampled organisms in the various strata of the canopy. Trophic relationships can also be observed and recorded (e.g., prey eaten by predators).

This method is extremely dependent on the surveyor's personal ability and visual acuity, factors that certainly limit its usefulness in comparing data gathered by different surveyors. Moreover, the perception of small moving objects is greatly affected by wind. Most sampling methods may be affected by wind, but direct counts are virtually impossible in winds above 12 km/hr. An additional disadvantage of the method is the chance for misidentification of certain species. Since quite often the surveyor has only a glimpse at the sampled specimens, positive identification is often impossible. If the study considers groups of species or species complexes with no need for specific identifications, then the method is less objectionable.

D. Conversion to Absolute Population Estimates

Conversion of visual counts to absolute population estimates is easily obtained if counts are made over a known length of row. The following formulas may be used:

Number of individuals per acre

$$= \frac{\text{(No. individuals in F ft. of row sampled} \times 43{,}560 \times 12)}{\text{(row spacing in inches} \times \text{F ft. of row sampled)}} \qquad (2\text{-}1)$$

Number of individuals per hectare

$$= \frac{\text{(No. individuals in M meters of row sampled} \times 10{,}000)}{\text{(row spacing in meters} \times \text{M meters of row sampled)}} \qquad (2\text{-}2)$$

The conversion of counts of arthropods on nonadjacent single plants to absolute counts is more complex. In general, plant density is more or less uniform, and one may estimate the mean density for a given field of a certain variety. The number of plants in 5-10 random one-m (or one-ft) row samples is counted and the counts are averaged. Knowing the row width one can then compute the plant population per acre or per hectare using formulas 2-3 or 2-4.

Number of plants per acre

$$= \frac{\text{(Number plants per ft of row} \times 43{,}560 \times 12)}{\text{(row spacing in inches)}} \qquad (2\text{-}3)$$

Number of plants per hectare

$$= \frac{\text{(Number of plants per meter of row} \times 10{,}000)}{\text{(row spacing in meters)}} \qquad (2\text{-}4)$$

When making these conversions it should be noted that plant density frequently decreases due to natural thinning during the crop cycle until it becomes relatively stable as the canopy closes. It is therefore necessary to reassess mean plant densities at various stages of plant growth.

The simplest conversion of counts of arthropods on single plants to absolute populations is to multiply the counts per plant by the number of plants per unit of area. In a crop like soybean, however, this may not be that simple because the plants do not occur as discrete units in the crop space. Their branches and leaves criss-cross, and plants in the late vegetative growth stages (V8-V11) (see Chapter 1) usually form a continuous canopy, at least along the rows. For this reason it is difficult to isolate plants for sampling purposes. Counts often represent a sample of the population that inhabits a space above ground that may be 10 or more times greater than the space occupied by the plant at ground level. This space is shared by adjacent plants within the row. Thus, at a normal density of 25 plants per meter of row at ground level, each plant represents 4 cm of row space. The vertical projection of the aerial part, however, occupies about 30-40 cm. The implications of this fact in the conversion procedures have not been studied in soybean. It seems, however, that the simple extrapolation technique may grossly overestimate populations of most soybean arthropods, particularly at lower population levels. Pieters and Sterling (1973), sampling arthropods on cotton, reported that D-Vac samples of single plants consistently resulted in larger population estimates than samples over one meter of row.

III. Ground Cloth

The ground cloth sampling method is also known as the beat cloth, shake cloth, or the plant shake method. This sampling method consists of forcefully displacing the arthropods from the plants onto a sheet spread on the ground between two adjacent rows of plants so that the specimens can be collected and counted. The ground cloth cannot be used to sample insects that display instantaneous escape reactions such as adults of the three cornered alfalfa hopper, *Spissistilus festinus* (Say), and grasshoppers, although it is very adequate for species that drop to the ground when disturbed, such as the bean leaf beetle, *C. trifurcata*. It is also an excellent method for most lepidopterous caterpillars, leaf beetles, stink bugs (particularly nymphs), predatory bugs, and other insects with slow escape reactions. Narrow row spacing may limit the usefulness of this method.

A. Construction of the Ground Cloth and Sampling Procedure (Figure 2-1 A-D)

The technique consists of stretching, between two adjacent rows of plants, a heavy cloth that is held in place by two wooden sticks fastened along the opposite sides of the cloth. The two sticks (e.g., doweling), about 2.0-3.0 cm in diameter, are cut 20 cm longer than the length of the cloth. For instance, if a 1 m long ground cloth is used then the sticks should be cut at 1.2 m. A white or light colored fabric or heavy vinyl is cut to the desired length (usually 1 m).

Figure 2-1. Ground cloth sampling method. (A) Approaching sample site with ground cloth rolled; (B) Unrolled ground cloth spread between two rows of soybean; (C) Bending and shaking plants on the two adjacent rows; (D) Aspirating specimens for identification.

The width of the fabric should be about 20 cm more than the greatest row spacing one may find in the field (usually 1.1 m). The fabric is fastened with tacks or heavy-duty staples to the wooden sticks, or the cloth is folded and sewn tightly around the sticks. Equal length remains free at both ends of each stick to be used as handles in positioning the ground cloth between the rows and rolling the cloth.

This rather long ground cloth can be operated best by two persons. Shorter cloths (0.5-m long) may be used more effectively by one person (Boyer and Dumas 1969).

The sampling procedure consists of approaching a randomly chosen site without disturbing the plants to be sampled. The ground cloth, rolled like a scroll around the sticks (Fig. 2-1A), is carefully laid between two rows (Fig. 2-1B). If the canopy is closed the rolled cloth is gently slid into position at ground level below the foliage. The same procedure is followed if plants are lodged. With one operator at either end, using a 1 m long ground cloth, the cloth is unrolled so that the edges fastened to the sticks touch the bottom of the stems of plants in the two adjacent rows (Fig. 2-1B). It is convenient to kneel down on the corners of the cloth to hold it in position. The surveyors, facing each other, bend the plants of both adjacent rows over the cloth and shake or beat the plants vigorously (Fig. 2-1C). Under certain conditions only one row may be sampled in this manner, and thus the adjacent row is bent away from the cloth. Some prefer to use a certain number of shakes or beats (10 or 15 shakes per sampling site) whereas others shake the vegetation until they feel that all arthropods, except possibly eggs or other sessile organisms, have been dislodged from the plants onto the cloth. The plants are then pushed back from over the cloth to expose the specimens that have fallen. Small arthropods on the cloth can be efficiently collected with an aspirator (Fig. 2-1D). There are no critical studies on the effect of the number of shakes or beats on insect catches. With some practice, however, one arrives at a minimum number of beats above which no additional catches result.

After the insects have been shaken from the plants onto the cloth, two procedures can be followed—(a) the specimens are identified by species, counted, and recorded in the field or (b) the collections are transferred to a properly labeled container for processing in the laboratory at a later time.

The transfer of collections to containers is greatly facilitated if a slick vinyl cloth is used (Shepard and Carner 1976). In this case the cloth is folded in half and one end of the cloth is raised to make the specimens slide into a container held against the opposite end.

B. Characteristics of the Ground Cloth Sampling Method

The ground cloth is ideally suited for sampling slow moving arthropods that are easily dislodged from soybean plants. The method has limited application when plants are small and becomes rather inefficient when plants begin to senesce and shed their leaves. At this stage when plants are shaken many leaves fall on the cloth making it difficult to locate and count the specimens. Small arthropods may remain hidden in the foliage, and others may escape before the fallen leaves have been examined and discarded. The ground cloth is virtually impossible to use in broadcast or in drilled soybean fields. The efficiency of the method is also greatly reduced if plants are badly lodged.

Even when used by different surveyors it is possible to achieve reasonably consistent results because the main factor affecting the procedure is the vigor of the shaking action. A weaker shake, however, can be compensated by increasing the number of shakes. Thus, the vigor of different surveyors seems to have a lesser influence on the effectiveness of this method than in sweep net sampling. Counting the number of shakes or beats may help achieve greater uniformity in the procedure. Differences resulting from different surveyors, however, should not be neglected, and more critical studies are needed in this area.

C. Conversion to Absolute Population Estimates

Since the beat cloth, supposedly removes all arthropods along a measured length of row, it approximates an absolute sampling method for several arthropod species, particularly for certain lepidopterous caterpillars (see Chapters 6-8). If only one row is shaken then the length of the ground cloth represents the length of the row sampled. If two adjacent rows are shaken then the length of row sampled is twice the length of the cloth. Formulas 2-1 and 2-2 can be used to convert number of individuals per foot or meter of row to number of individuals per acre or hectare, respectively. Despite the potential usefulness of this method for obtaining absolute population estimates of certain species, it is advisable to calibrate it against absolute sampling methods such as the fumigation cage or the whole plant harvest method.

IV. Sweep Net

For over a century the sweep net has been the most widely used tool for sampling arthropods on small grain, forage, and many row crops. One reason for this popularity, despite difficulties in standardizing the method for accurate population studies, is that no other method can capture as many insects from vegetation per man hour without increased cost for equipment and damage to the crop (Ruesink and Kogan 1975).

A. Construction of the Sweep Net

The sweep net used for sampling differs from the familiar aerial butterfly net in that it is of rugged construction to withstand the impact against the sampled plant parts. The sweep net consists of three basic components: (a) the net proper—a cone-shaped bag made of medium weight cloth; (b) the ring—a hoop made of heavy gauge wire that keeps the bag open and permits attachment of the bag to the handle; and (c) the handle—a cylindrical extension from the net, made of wood or hard alloy aluminum tubing.

The net. The sweep net most commonly used in soybean entomological research is 38 cm in diameter. The net is made of muslin, cut and sewn as described in Fig. 2-2A-C. The dimensions indicated in the figure result in a net 38 cm in diameter and about 75 cm deep.

The ring and handle. Two of the most common methods of attachment of the ring to the handle are described as follows: (a) a screw is inserted through two loops at the ends of the ring and tightened into a nut embedded in one end of the handle (Fig. 2-2C), or (b) the ends of the ring are bent into an L shape and fitted into an orifice and a groove along one extremity of the handle, and a piece of metal pipe with an internal diameter slightly greater than the diameter of the handle is slipped toward the ring or hoop as the L-bent ends fit snugly into the groove (Fig. 2-2B). The ring must be firmly attached to the handle so that it will not turn or wobble when in use. The hard wood or aluminum tube handle is usually about 2.2 cm in diameter and 60-cm or 90-cm long. Figures 2-3A-D show long (aluminum) handled and short (wood) handled sweep nets with two common methods of attachment of the hoop to the handle. The metal handle usually has a plastic grip at the end. The sweep net with the shorter handle is usually held with one hand while in use, whereas the sweep net with the longer handle can be used more easily with both hands.

One of the difficulties in quantifying and comparing data on sweep net samples stems from the many styles of sweeping. There are, however, some general procedures that, if properly defined, may help achieve a certain degree of standardization.

B. Sampling Procedures

Many different sampling procedures may be adopted with the sweep net. Four of these procedures have been mentioned in the literature and are described here in detail. It is important to identify the particular procedure used in an experiment because each yields different results that are seldom directly comparable.

General Sampling Methods for Above-Ground Populations   37

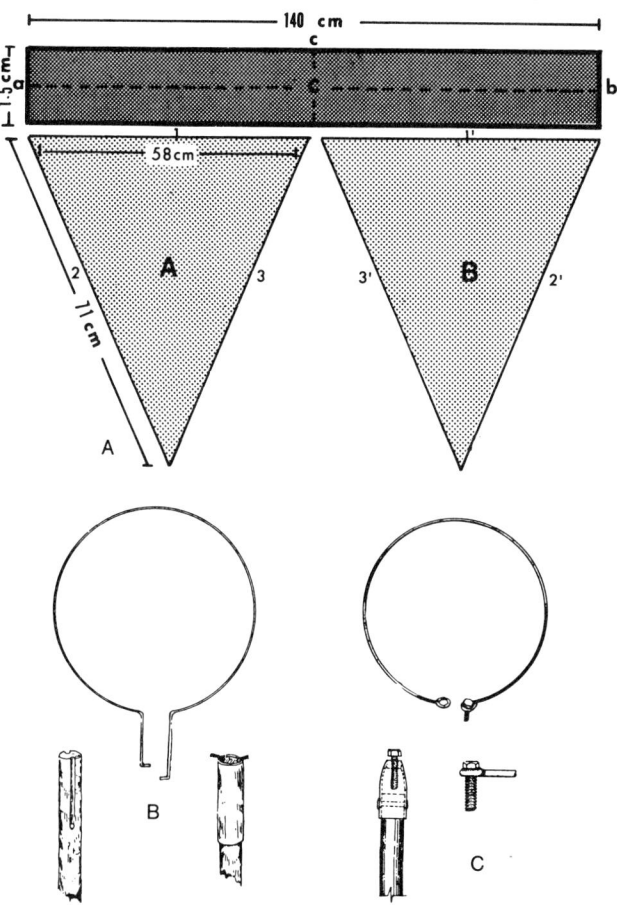

Figure 2-2. (A) Diagram for construction of a 38 cm diameter sweep net. Cut two triangles (A and B) of strong muslin, and a strip C of heavy denim. Fold strip C along line a b and sew the two opposite sides together with sides 1-1' of the triangles (A and B). Now fold the ensemble along line c-d juxtaposing side 2-2' and 3-3' of the two triangles. Sew along sides 2-2' and along sides 3-3' up to the point of attachment of the strip C. The edges of the strip (a-b) should remain open to allow insertion of the metal ring. (B) Sliding sleeve type attachment. (C) Bolt and nut attachment of the hoop to the handle.

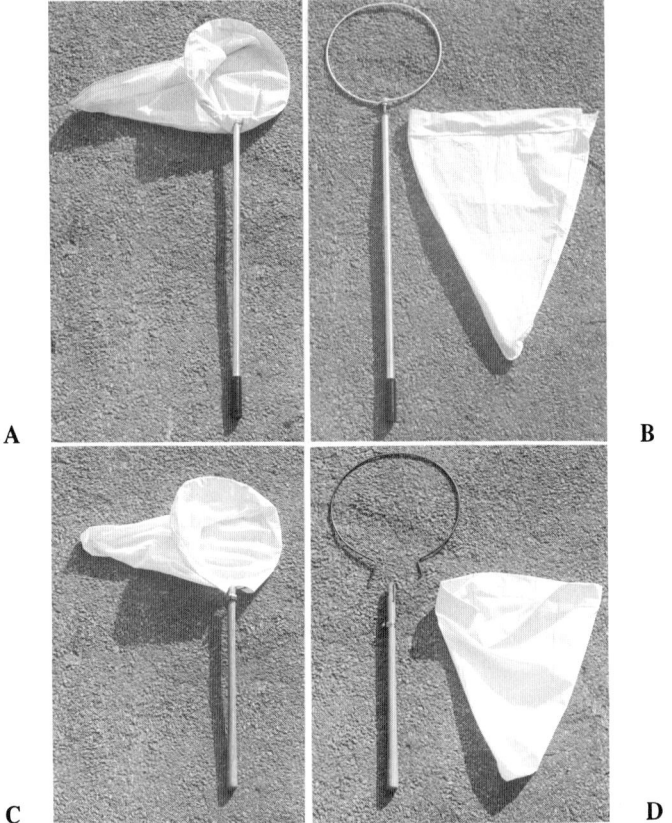

Figure 2-3. (A) Sweep net with long (90 cm) handle; (B) Same as in A disassembled to show the bolt and nut type of attachment; (C) Sweep net with short (60 cm) handle; (D) Same as in C disassembled to show L-bent type of attachment.

1. Sweeps Across One Row (Figures 2-4A-C)

In using the sweep net to sample across one row, the surveyor walks alongside the row, holding the sweep net stretched straight forward with the opening of the net always facing the row of plants to be sampled (Fig. 2-4A). The surveyor moves forward swinging the sweep net so that the opening of the net passes through the foliage (Fig. 2-4B). The net is turned 180° after each sweep as the surveyor advances with each step to swing the sweep net through the foliage from the opposite direction (Fig. 2-4C). The procedure is repeated with each stroke of the net counted as one sweep. One variation of the single row cross sweeping is described by DeLong (1932). In this case, one cross stroke is followed by a quick backstroke over the same vegetation. The rationale for this variation is that the first cross stroke causes the insect to jump and the backstroke is responsible for its capture. DeLong uses this method in his study of leafhopper populations.

Figure 2-4. Sweeps across one row. The sequence from A to C corresponds to three complete sweeps following the movement depicted by the arrow.

2. Sweeps Across Two Rows (Figures 2-5A-F)

The procedure used to sweep across two rows is similar to that used to sweep across one row, but the stroke of the net covers two rows on either side of the surveyor. Again each stroke is counted as one sweep.

3. "Lazy-8" Parallel Sweeps (Figures 2-6A-F)

Using the "lazy-8" parallel sweeps method the surveyor holds the sweep net like an oar, with the hoop below the waist and the net opening perpendicular to the plane of the row to be sampled (Fig. 2-6A). The surveyor moves forward between two rows and forces the net upward through the foliage along one row within the reach of the net (Fig. 2-6B). As the hoop reaches the top of the plant the net is positioned ahead of the surveyor and turned 180° downward with a quick twist of the handle (Fig. 2-6C). With the surveyor's next step the net is brought down and back to the initial position (Fig. 2-6D). With the successive steps, the side of the net describes an inclined figure 8 (Fig. 2-6E). The idea of this procedure is to increase the chances of collecting insects that are distributed in all strata of the canopy irrespective of the plant height. The cross-row sweeping methods are usually limited to sampling the upper 38 cm of the canopy.

One variation of the "lazy-8" method is to make the return stroke of the sweep net over the opposite side of the row previously sampled by the surveyor (Fig. 2-6F). Thus, one half of the figure 8 is made on one side of the row with the other half over the opposite side. This method may be called "lazy-8" alternate sweeping.

4. Pendulum Sweeps

The pendulum sweeping procedure is similar to the "lazy 8" method, except that the hoop is kept about 30-35 cm within the upper canopy of the plants. The net is moved forward along one row, and when it reaches the maximum length of the handle the net is turned 180° and the backward stroke begins with the next step. The net double tracks the path swept with the forward stroke.

There are several possible variations around these basic procedures. One that has been used in the past is sweeping across three rows of plants.

Figure 2-5. (opposite). Sweeps across two rows. Alternate sweeps across rows *A* and *B* correspond to one sweep. The sequence from A to F shows two complete sweeps with the arrow indicating the direction of the third sweep (F).

General Sampling Methods for Above-Ground Populations 43

General Sampling Methods for Above-Ground Populations 45

C. Characteristics of the Sweep Net Sampling Method

Sweep net sampling is perhaps the most convenient sampling procedure for many arthropods, but it is also one of the most difficult to interpret in terms of accuracy and precision.

The sweep net has been used for many decades, but there are few critical studies about its use in quantitative ecology. DeLong (1932) was one of the first to study the efficiency of the sweep net in insect population studies on certain truck and field crops. He recognized the following environmental and plant factors responsible for the variability of results:

a) Temperature—influencing metabolic rates of insects and consequently their speed of escape reactions

b) Temperature and humidity—affecting the microclimate and causing changes in the position on the plants of individuals of the same species

c) Wind velocity—certain species seek shelter under heavy wind; and under moderate wind, collections may vary with the surveyor movement into, with, or across the wind

d) Position of the sun—certain species respond to the shadow cast by the surveyor and escape capture by the net

e) Plant size—very short plants are too fragile to sweep, whereas large plants can only be partly sampled

f) Density of the canopy—dense canopies offer increased resistance to the sweep net movement through the foliage and more protection for the insects

g) Pubescence of leaves and stems—lessens the impact of the stroke or provides additional shelter

If the foliage is wet with rain drops or dew, it becomes very difficult to operate the sweep net, although the same is true for most other sampling procedures. As a general rule regular sampling programs should be suspended under marginal weather conditions.

Much of the variation of sweep net catches, however, stems from human factors. DeLong (1932) considered several of these factors including height of sweeping, rapidity and length of the sweep net stroke, and the sweeping method.

Although the effect of these factors can be intuitively appreciated, there are no critical studies of their quantitative effect on estimates of arthropod populations on soybean. Several comparative studies of sampling methods have been published, and their findings are summarized at the end of this chapter. Although

Figure 2-6. (opposite). "Lazy 8" sweeps. A-C shows a full swing of the net (one sweep forward). D-F shows the return stroke which is counted as a second sweep. The return stroke may be done on the side of the plant opposite to the one where the surveyor stands. ("Lazy 8" alternate method).

none of these studies partitioned the components of the variance as identified by DeLong, they identified some of the most flagrant inadequacies of the sweep net method.

D. Conversion to Absolute Population Estimates

Several methods have been proposed and used to convert sweep net catches to absolute population estimates. Most of these methods use regression techniques to compare population estimates based on the sweep net collection with population densities determined on the basis of some absolute sampling method, such as the fumigation cage and the whole plant harvest sampling techniques that are described later in this chapter.

Another approach is based on the development of empirical models of the sweep net technique. Various intervening climatic, plant, insect, and human factors are taken into consideration and mathematically defined. The resulting equations are used to convert the sweep net catches to absolute estimates when conditions at the time of collection are carefully recorded. Since these conversion methods are also applicable to the beat cloth, the D-Vac, and other relative sampling procedures, they will be discussed after the description of the absolute sampling methods.

V. Suction Nets

Devices that generate suction have been used either as stationary traps for sampling from the air or as portable or mobile units for sampling from vegetation. One of the best known of these devices is the D-Vac suction net (Dietrick et al. 1959, Dietrick 1961).

A. D-Vac Suction Net

There are two models of the D-Vac sampler—a portable, hand carried model and a backpack model. Both machines consist of:

1. portable blower
2. small single cylinder gasoline engine
3. flexible air conduit
4. collecting head and bag
5. reducing cone
6. frame

General Sampling Methods for Above-Ground Populations

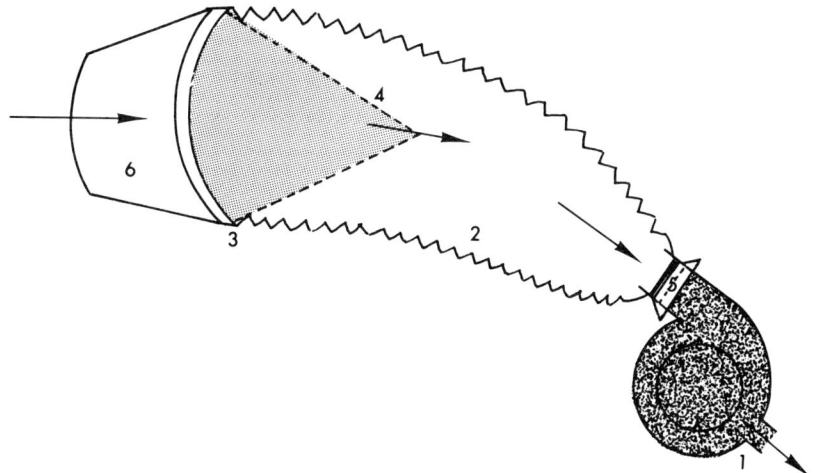

Figure 2-7. Simplified diagram of a portable suction net: 1. Blower; 2. Flexible air conduit; 3. Head cone holder; 4. Organdy bag for collecting the arthropods; 5. Wide mesh screen to protect the blower; 6. Reducing cone used to adjust the opening to the desired diameter. Structural frame and gasoline engine are not illustrated. (Based on Dietrick et al. 1959).

Detailed descriptions of these devices are found in the papers by Dietrick et al. (1959) for the hand-carried model and Dietrick (1961) for the backpack model.

The operational principle of the suction sampler is diagrammed in Fig. 2-7. Both models operate similarly, but there are no critical studies comparing their efficiencies. In most cases it is convenience and availability that have dictated which model was used in a particular research.

B. Sampling Procedures

Just as the sweep net is used in different ways by different researchers, so can the suction net be operated in various different modes. Each sampling procedure proposed by surveyors has been suited for specific sampling conditions. No one procedure is best for sampling all arthropods. When a sampling program is initiated and is based on suction net sampling, preliminary tests should be conducted to determine which procedure yields the most desirable results. The procedures described here can be used with either the hand-carried or the backpack model D-Vac.

1. Spot Sampling

The surveyor approaches the sampling site against the wind direction and with the air exhaust of the blower directed away from the sampling site. The sampler head cone is oriented downward, placed over a plant or clump of plants and pushed down as far as possible toward the ground (Fig. 2-8). The area of the opening of the sampler head, ca. 900 cm^2, represents the unit area sampled. This technique was used by Marston et al. (1976) in faunal studies on soybean.

2. Sampling a Measured Length of Row: Uniform Height

In sampling a specific area the surveyor walks a premeasured distance between rows or counts a certain number of steps. The sampler head cone is held horizontally with the plane of the cone at a 45° angle toward the row and at a constant height either just above or at the height of the plants. The plants are scraped by the front edge of the head cone as the surveyor advances, thus exposing and dislodging the arthropods. Most small arthropods that try to escape, particularly by flying, are sucked into the net. This procedure has been used to sample spiders in soybean fields in Illinois at various strata of the plant canopy (LeSar and Unzicker 1978).

3. Sampling a Measured Length of Row: Variable Height

The procedure for sampling at variable heights on the plant is similar to the procedure for sampling at uniform height, but the sampler head cone is moved up

Figure 2-8. Backpack model of the D-Vac suction net. Procedure for spot sampling.

General Sampling Methods for Above-Ground Populations   49

and down the entire height of the plants as the surveyor advances. A sample of the full depth of the plant canopy is obtained with this procedure.

C. Characteristics of the Suction Net Sampling Method

The suction net is ideally suited for collecting certain arthropods that are small enough to be sucked into the net and not excitable enough to jump or fly by virtue of the turbulence and noise created by the fan motor. It gives good results for adult leafhoppers (see Chapter 12), threecornered alfalfa hoppers, certain flies, many small lepidopterous larvae, nymphs, and adults of certain small Hemiptera, spiders, etc. (Pedigo et al. 1972, Hillhouse and Pitre 1974, Shepard et al. 1974, Turnipseed 1974, Marston et al. 1976). Strong flying insects, e.g., grasshoppers, usually fly away from the advancing sampler head cone and generally escape capture.

The vacuum or suction net is primarily a research tool. Because it is cumbersome and has a rather high initial purchase cost, it cannot be widely recommended for scouting programs. Perhaps the greatest limitation of the suction net, even in research, is maintenance. To limit the weight of the hand-carried model, a very small, delicate gasoline engine is used. With operation under rough conditions in the field, engine troubles are inevitable. For this reason more reliable service has been obtained in our research programs with the backpack model. This model has a larger engine which is normally kept at a more level position; therefore, it is less prone to malfunction. If at all possible one should attempt to have a backup unit and a supply of spare parts to assure that the research program will not be foiled by mechanical difficulties with the equipment. Inadequately tuned D-Vac engines may result in a loss of suction leading to variations within the sample collections. If the air flow through the suction net varies, one may expect differences in total arthropod catches, although no critical studies have been conducted in this area.

D. Conversion to Absolute Population Estimates

The suction net used in spot sampling approaches an absolute sampling method for a few species. With this method the area of the sampler head cone opening corresponds to the area of the field sampled. The residual population, after the sample is taken, can be determined by direct observations; but a better calibration is the comparison with results of absolute methods. The conversion of suction net collections over a measured length of row to population densities is affected by the same variables that affect conversions of sweep net collections. If, however, it is assumed that the suction air flow into the sampler cone is uniform, then the variability due to some of the environmental and human factors is somewhat reduced. There are no critical studies of the effect of variable factors on the efficiency of the D-Vac in sampling arthropods on row crops.

VI. Absolute Methods

Accurate estimations of population densities are needed both for basic ecological research and sound pest management programs (Gonzalez 1970, Bottrell and Adkisson 1977). The four basic sampling methods discussed above are intensely subject to environmental, human, and other biological (plant or animal) factors. The validity of studies made with these techniques can only be judged on the basis of an analysis of their efficiency vis-a-vis a comparatively more reliable, if also costlier, sampling method. Review of absolute sampling methods and additional references can be found in Southwood (1978) and Ruesink and Kogan (1975). Only two methods that have had rather general application will be described—the fumigation cage and whole plant harvest. Both methods are based on isolating a population in a known surface area—the area of a cage. The two methods differ only in the technique used to dislodge the insects from the plants.

A. Fumigation Cage

Cages of various sizes and shapes can be constructed with wooden or metal frames and plastic or metal sides. A large plastic garbage can has been used instead of a framed cage (Fig. 2-9). The height of the cage should be at least equal to the maximum height of the plants. The area can vary with the desired size of the sampling unit. Prior to sampling with the fumigation cage, a plywood or sheet metal base plate is placed on either side at the base of the plants to be sampled. Suitcase latches are used to lock the base plate halves together (see Pedigo et al. 1972 for details of construction). The area of the base plate is slightly greater than that of the bottom of the fumigation cage. Base plates should be set in position 24 hours prior to the time of the actual collection to allow the population to equilibrate after being disturbed by positioning the plates. The sites of the plates should be marked with flags so they are visible above the canopy. At collection time the surveyor approaches the site and carefully holds the cage above the plant canopy level. At the sampling site the cage is dropped on the base plate, and all sides are carefully sealed with soil to keep arthropods from escaping.

The cage should be provided with a sleeve or a small opening to allow fumigation of the interior. An aerosol canister containing 20 percent pyrethrin makes

General Sampling Methods for Above-Ground Populations   51

Figure 2-9. Fumigation cage using a large heavy-duty garbage can. Usually a base plate is placed at the sampling site about 24 hr prior to making the collection. (Base plate not shown in these pictures.)

an excellent fumigant. Usually a 5-8 second jet is enough to knock down all arthropods inside the cage. Without removing the cage the surveyor inserts one arm through the sleeve and vigorously shakes the plant. The cage is then lifted and the arthropods are collected on the base plate. The plants should be thoroughly searched for individuals still attached.

It has been a common practice to take the sweep net, beat cloth, D-Vac or other relative samples to be calibrated in the immediate vicinity of the cage (Pedigo et al. 1972). This precaution seeks to reduce the influence of irregular spatial patterns within the field. If, however, the population under study follows a Poisson distribution or is only slightly clumped, this precaution may be superfluous because even with relatively small sampling units adjacent sites in the field may have markedly different population densities.

B. Whole Plant Harvest

In the whole plant harvest sampling procedure there is no need to position a base plate. The most convenient cage is a $1.8 \times 1.8 \times 1.8$ m saran cage mounted on a cubic frame of hard alloy aluminum tubing (Fig. 2-10). One side of the cage has a vertical opening with zipper to permit access into the cage. The cage is positioned on the sampling site (two persons are needed) and sealed around the bottom with soil. One surveyor enters the cage and carries an aspirator, clippers, and plastic bags. All arthropods on the plants and on the inside top or sides of the cage are collected with the aspirator. Subsequently all plants are clipped at ground

Figure 2-10. Whole plant harvest sampling method. (A) Cage being carried to sampling site; (B) Cage in position; (C) Surveyor inside cage collecting arthropods on the ground after plants have been clipped and bagged; (D) Base area after cage is removed.

level, placed in large plastic bags and tightly sealed. All the debris and loose leaves on the ground are also collected and placed in a separate bag. The cage is left in place for 1-2 hr to allow individuals that fell into crevices or that were covered by dust to be detected on the ground or as they move up the sides of the cage. Arthropods on the bagged plants and in the debris are separated and identified in the laboratory. Berlese funnels or washing techniques can be used in this separation (Mayse et al. 1978b).

This method was used by Ruesink and Haynes (1973) for cereal leaf beetle adults on grain and by Kogan et al. (in preparation) for bean leaf beetle adults on soybean. Hillhouse and Pitre (1974) used a one row cage fitted on a base plate, but they clipped and removed the plants for separation of the sample. The "clam trap" (Leigh et al. 1970, Marston et al. 1976, Mayse et al. 1978b) is a variation of this method.

VII. Use of Absolute Estimates in Calibration of Relative Sampling Methods

Although a complete census of all arthropods in a field is practically impossible, the absolute sampling methods described here may approach a level of accuracy that is acceptable for most population studies. Even these general methods have many limitations, and certain arthropod species require development of special sampling procedures. These are dealt with in subsequent chapters.

Conversion of relative to absolute population estimates have used simple ratios, regressions, or conversion models. Models have been developed for the sweep net but are not available for the other techniques. Conversion ratios and regression techniques are equally applicable to direct observations, ground cloth, sweep net, or suction net.

A. Regression and Ratio Methods

Romney (1945) used correlation coefficients to estimate the accuracy of his sweep net samples of insects on stands of Russian thistle in New Mexico. Simple ratios were used by Kretszchmar (1948) to compare the efficiency of the sweep net with that of a cylindrical cage as an absolute population measuring device in sampling insects on soybean. Fenton and Howell (1957) used percentage taken by the sweep net from a population calculated on the basis of absolute counts per unit of area in alfalfa. The regression method uses the model:

$$y = \beta x + \alpha \quad (2\text{-}5)$$

where y = number caught by the relative sampling method (ground cloth, sweep net, etc.), x = number caught by the absolute sampling method (fumigation cage or whole plant harvest), β = regression coefficient, α = intercept. In most cases the use of paired sampling units within a field does not yield a significant regression analysis because of the patchy spatial patterns of many insect populations. As a consequence of these patterns it is unlikely for samples in a pair to be taken from areas with similar densities. Regression analyses have been successfully used, however, in comparing means of collections by different sampling methods in different fields (Rudd and Jensen 1977).

If we force the regression line through the origin in the logical assumption that at a zero population level both sampling methods will yield a zero value, then the model is reduced to the form:

$$y = \beta x \quad (2\text{-}6)$$

and β can be computed by formula 2-7:

$$\beta = \frac{\Sigma y}{\Sigma x} \qquad (2\text{-}7)$$

or the ratio of the catches by the two methods (Bliss 1967). These ratios have been commonly used in most studies that require conversion. From the scatter about the lines with slope equal the ratio $\Sigma y/\Sigma x$, the variance is computed as:

$$S^2 = \frac{\Sigma(y^2/x) - (\Sigma y)^2/\Sigma x}{n-1} \qquad (2\text{-}8)$$

(Bliss 1967). Another method to compute the variance of a ratio $\bar{y}/\bar{x}$ is provided by Yates (1965):

$$S^2(\bar{y}/\bar{x}) = (\bar{y}/\bar{x})^2 (S_y^2/\bar{y}^2 + S_x^2/\bar{x}^2) \qquad (2\text{-}9)$$

where $\bar{y}$ and $\bar{x}$ = mean catches by the two methods, S_y^2 and S_x^2 = variances of the two methods.

B. Conversion Models

A conversion of sweep net collections to density was proposed by Menhinick (1963) using the equation:

$$M = (S/N) \times (T/A), \qquad (2\text{-}10)$$

where M = number of sweeps necessary to catch all insects in 1 m^2, S = number of net strokes, N = number of insects caught in the S strokes, T = number of insects in the area A, and A = area of the field in m^2. An estimate of T is based on some absolute population sample. This model was used to compute M for several insect species on lespedeza clover. The value of M varied from 2.3 to 10.8 depending on the species sampled and weather conditions.

Ruesink and Haynes (1973) developed a model to predict cereal leaf beetle, *Oulema melanopus* (L.), density from sweep net catches. A generalized model stated that:

$$D = C/dLbp, \qquad (2\text{-}11)$$

where D = density (number of individuals per unit area), C = number of individuals per sweep, d = diameter of the sweep net opening, L = length of each stroke of the net, b = proportion of the population that is in the path of the sweep net, and p = probability of getting into the sweep net if the individual is in the path of the net. Units have to be consistent. If all linear measurements are in meters, D is given in number of individuals per m^2.

There are serious difficulties in estimating b and p. Ruesink and Haynes (1973) achieved this in part for the cereal leaf beetle larvae and adults, and their work should be consulted for possible application of the mathematical and experimental approach to insect species on soybean and other row crops.

Most researchers have concluded that the sweep net, and most other relative sampling methods, are valuable in certain comparative surveys and in providing a relative estimate of population density. The conversion of relative to absolute population densities, however, has to be approached with great reservation. Conversion models or factors must be computed on a species-by-species basis, for each developmental stage of a species, for specific plant growth stages and crop characteristics, and for very precisely defined environmental conditions.

VIII. Comparison of Sampling Techniques

The choice of a sampling technique is usually based on certain criteria related to the levels of precision required and cost limitations of the sampling program. Precision and cost depend on the sampling unit size, the number of samples, the pattern of sampling, the time of day and season when samples are taken, and the method used to count or collect the arthropods (Pieters and Sterling 1973). Chapter 3 discusses methods for computing the number of samples necessary to achieve a certain level of reliability. Sequential sampling plans are discussed in Chapters 4 and 5. The following is a summary of some techniques that have been used to compare various sampling methods. Results of some comparisons reported by various authors are tabulated (Table 2-1).

A. Graphic Comparisons

When only relative population estimates are needed, an expedient way of comparing various sampling methods is by plotting the mean or the total number of individuals against time or unit of effort. If general trends detected by some absolute method are reproduced by some less time consuming relative method, then the latter may be selected. This method may be used when determining the relative abundance of a species or studying seasonal life histories in which only the detection of general trends is sufficient. For example, Shepard et al. (1974) showed graphically that the ground cloth, sweep net, and D-Vac adequately

Table 2-1. Comparison of sampling methods based on published data. Means expressed as number of individuals per meter of row

Species	Sampling Method[a]					Ref.[b]
	Absolute $\bar{x}$ / RV	SW-1 $\bar{x}$/RV	SW-2 $\bar{x}$ / RV	GC $\bar{x}$ /RV	D-Vac $\bar{x}$ / RV	
A. gemmatalis (larvae)	--	--	1.4/21.8	6.3/21.5	--	(1)
P. includens (larvae)	--	--	2.0/20.1	10.6/19.9	.8/31.1	(1)
P. scabra (larvae)	--	--	2.9/17.3	14.5/ 6.8	2.3/15.5	(1)
	35.5/8.3	--	--	37.6/8.5	--	(2)
	11.2/8.8	5.6/11.8	--	--	--	(2)
H. zea (larvae)	--	--	3.1/11.4	23.2/12.0	1.8/30.5	(1)
E. varivestis (adults)	--	--	.6/27.5	4.9/20.7	.1/57.7	(1)
E. varivestis (larvae)	--	--	1.6/28.0	20.0/20.7	.4/36.4	(1)
C. trifurcata (adults)	38.7/17.5	6.9/9.4	5.7/17.0	8.9/8.7	--	(3)[c]
S. festinus	21.6/7.8	4.3/7.1	7.2/30.3	2.4/4.1	--	(3)[c]
Thysanoptera	11.9/24.3	4.3/52.6	--	11.7/21.8	16.6/44.2	(4)
Araneida	6.12/18.2	1.9/27.2	--	3.4/15.6	5.2/14.3	(4)
Anthocoridae	22.4/11.8	1.9/40.8	--	3.6/23.9	5.4/23.7	(4)
Nabidae	3.02/22.0	.2/72.5	--	1.0/31.8	3.1/13.6	(4)
	--	--	--	5.5/20.4	--	(1)
Geocoridae	--	--	--	3.5/26.1	.4/14.7	(1)

[a] SW-1 sweep net across one row; SW-2 sweep net across two rows; GC ground cloth

[b] (1) Turnipseed (1974), (2) Pedigo et al. (1972), (3) Hillhouse and Pitre (1974), (4) Marston et al. (1976).

[c] Computed on data from late pod development in Hillhouse and Pitre (1974).

revealed seasonal trends of small green cloverworm larval populations but that the D-Vac was inadequate for large larvae (Fig. 2-11).

B. Statistical Methods

Relative variation "RV" (2-12) is a widely used statistic to compare the efficiency of various sampling methods (Pedigo et al. 1972, Hillhouse and Pitre 1974, Southwood 1978, see also Chapter 3):

$$RV = 100 \, S_{\bar{x}}/\bar{x} \qquad (2\text{-}12)$$

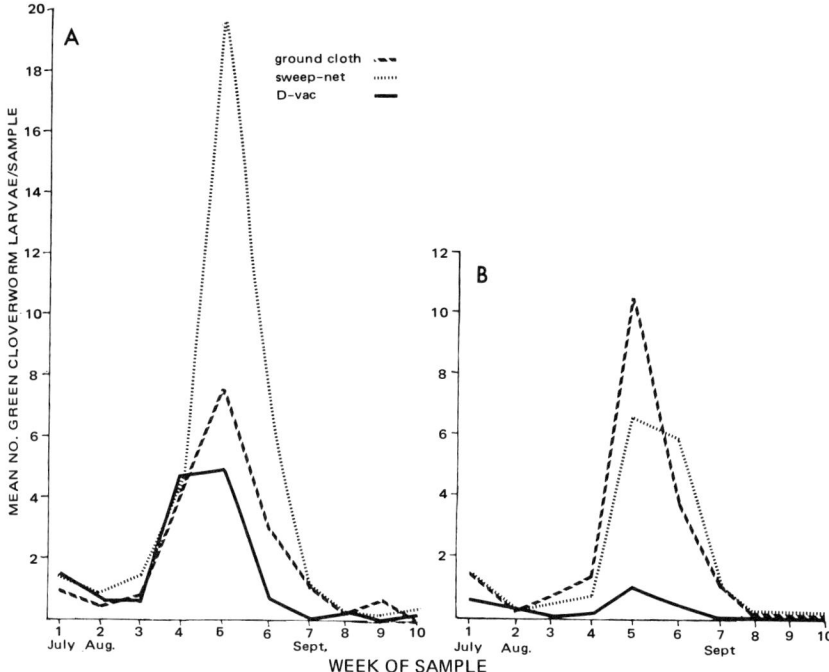

Figure 2-11. Mean number of green cloverworm larvae collected from soybean using ground cloth, sweep net, and D-Vac. (A) Small larvae (< 1.25 cm); (B) Large larvae (> 1.25 cm). (From Shepard et al. 1974, by courtesy of the authors and the Entomological Society of America.)

where $S_{\bar{x}}$ = standard error of the mean ($S_{\bar{x}} = s/\sqrt{n}$, s = standard deviation, and n = number of observations). According to Southwood (1978) an RV < 25 is usually adequate for most extensive sampling programs, but certain intensive sampling programs used for parameter evaluation may require an RV < 10. Turnipseed (1974) used instead the coefficient of variability (2-13) and arithmetic means to compare efficiency of various sampling methods.

$$CV = 100 \, s/\bar{x} \tag{2-13}$$

The relative net precision (RNP)(2-14) was used by several authors when cost of sampling was taken into consideration (Pedigo et al. 1972, Hillhouse and Pitre 1974, Southwood 1978):

$$RNP = 1/\overline{RV} \, C_s, \tag{2-14}$$

where $\overline{RV} = \sum_{i=1}^{n} RV_i/n$, mean relative variation, C_s = time consumed per sampling unit in man hours.

Precision of a sampling method is greatly affected by the spatial patterns of the sampled species. A highly clumped population is likely to produce a much higher RV than a randomly distributed population for the same number of sampling units. For this reason statistics that take into account these spatial patterns have been proposed and could be used to compare sampling methods. One such statistic (2-15) was indicated by Young et al. (1976).

$$D = \sqrt{(1/k_i)(1/\bar{x}_i)/n} \ , \tag{2-15}$$

where D = RV as in (2-12) but expressed in decimal form, k = parameter of the negative binomial distribution (see Chapter 3, and Southwood 1978); n = number of samples.

Analysis of variance and tests for differences of means are used in connection with RV, CV, or RNP. In most of these analyses the data (x_i) are transformed into $\sqrt{x_i + 0.5}$, to stabilize the variance. The precise transformations recommended in various kinds of distributions are discussed by Southwood (1978) (see also Chapter 3). In common practice, however, he suggests that it is adequate to transform data from a regularly dispersed population by using x^2, data from a slightly contagious one by using $\sqrt{x}$, and data from distinctly aggregated or contagious populations by using log x. Correlation analysis has also been used in comparisons of sampling methods (Kogan et al. 1974, Shepard et al. 1974, Rudd and Jensen 1977).

Comparisons of sampling procedures were made by several researchers using various species associated with soybean. Table 2-1 is a compilation of these results expressed as a mean number of individuals per meter of row followed by the relative variation (RV). Data were extracted from Pedigo et al. (1972), Hillhouse and Pitre (1974), Turnipseed (1974), and Marston et al. (1976).

From Table 2-1 we computed the overall mean RV and standard deviation for each sampling method. Thus, the absolute methods yield $\overline{RV}$ = 14.8 ± 6.5, sweep net across one row = 36.9 ± 22.5, sweep net "lazy 8" = 21.7 ± 6.5, ground cloth = 18.7 ± 6 5, and D-Vac horizontal = 23.7 ± 10.7. Thus, the overall $\overline{RV}$ for the ground cloth was 3.9 percentage points above the absolute $\overline{RV}$, whereas the sweep net "lazy 8", the D-Vac, and the sweep net across one row were respectively 6.9, 8.9, and 22.1 percentage points above the absolute $\overline{RV}$. This may indicate that the ground cloth is a good method for research and survey of those arthropod species associated with soybean that are included in the table. However, the merits of each method have to be assessed by its performance towards achievement of one's specific goals in population studies.

Acknowledgments

Thanks are due to Dr. Kenneth Yeargan, University of Kentucky, for very helpful comments. Photographs and mounts for illustration of the various sampling methods were provided by Mr. Larry Farlow and Mr. Lloyd LeMere, Illinois Natural History Survey. Charlie Helm graciously agreed to pose for those photographs.

References

Bliss, C. I. 1967. Statistics in biology. vol. 1. McGraw Hill, N.Y. 558 p.
Bottrell, D. G., and P. L. Adkisson. 1977. Cotton insect pest management. Annu. Rev. Entomol. 22:451-481.
Boyer, W. P., and B. A. Dumas. 1969. Plant-shaking methods for soybean insect survey in Arkansas. pp. 92-94 in Survey methods for some economic insects. USDA ARS 81-31:140 p.
DeLong, D. M. 1932. Some problems encountered in the estimation of insect populations by the sweeping method. Ann. Entomol. Soc. Amer. 25:13-17.
Dietrick, E. J. 1961. An improved backpack motor fan for suction sampling of insect populations. J. Econ. Entomol. 54:394-395.
Dietrick, E. J., E. I. Schlinger, and R. van den Bosch. 1959. A new method for sampling arthropods using a suction collecting machine and modified Berlese funnel separator. J. Econ. Entomol. 52:1085-1091.
Fenton, F. A., and D. E. Howell. 1957. A comparison of five methods of sampling alfalfa fields for arthropod populations. Ann. Entomol. Soc. Amer. 50:606-611.
Gonzalez, D. 1970. Sampling as a basis for pest management strategies. Proc. Tall Timbers Conf. Ecol. Anim. Contr. Habitat Manage. 2:83-101.
Hillhouse, T. L., and H. N. Pitre. 1974. Comparison of sampling techniques to obtain measurements of insect populations on soybeans. J. Econ. Entomol. 67:411-414.
Kogan, M., and C. G. Helm. 1979. Soybean leaf pubescence and the relative abundance and injury levels of the potato leafhopper, *Empoasca fabae*. (In preparation).
Kogan, M., W. G. Ruesink, and K. McDowell. 1974. Spatial and temporal distribution patterns of the bean leaf beetle, *Cerotoma trifurcata* (Forster), on soybeans in Illinois. Environ. Entomol. 3:607-617.
Kogan, M., G. P. Waldbauer, C. E. Eastman, and C. G. Helm. 1979. Absolute estimates of bean leaf beetle adult populations on soybean. (In preparation).
Kretzschmar, G. P. 1948. Soybean insects in Minnesota with special reference to sampling techniques. J. Econ. Entomol. 41:568-591.
Leigh, T. F., D. Gonzalez, and R. van den Bosch. 1970. A sampling device for estimating absolute insect populations in cotton. J. Econ. Entomol. 63:1704-1706.

LeSar, C. D., and J. D. Unzicker. 1978. Soybean spiders: Species composition, population densities, and vertical distribution. Ill. Natur. Hist. Surv. Biol. Notes 107:1-14.

Marston, N. L., C. E. Morgan, G. D. Thomas, and C. M. Ignoffo. 1976. Evaluation of four techniques for sampling soybean insects. J. Kansas Entomol. Soc. 49:389-400.

Mayse, M. A., M. Kogan, and P. W. Price. 1978a. Sampling abundance of soybean arthropods: Comparison of methods. J. Econ. Entomol. 71:135-141.

Mayse, M. A., P. W. Price, and M. Kogan. 1978b. Sampling methods for arthropod colonization studies in soybean. Can. Entomol. 110:265-274.

Menhinick, E. F. 1963. Estimation of insect population density in herbaceous vegetation with emphasis on removal sweeping. Ecology 44:617-621.

Pedigo, L. P., G. L. Lentz, J. G. Stone, and D. F. Cox. 1972. Green cloverworm populations in Iowa soybeans with special reference to sampling procedure. J. Econ. Entomol. 65:414-421.

Pieters, E. P., and W. L. Sterling. 1973. Comparison of sampling techniques for cotton arthropods in Texas. Texas Agr. Exp. Sta. Misc. Publ. 1120:8 p.

Ruesink, W. G., and D. L. Haynes. 1973. Sweep net sampling for the cereal leaf beetle, *Oulema melanopus*. Environ. Entomol. 2:161-172.

Ruesink, W. G., and M. Kogan. 1975. The quantitative basis of pest management: Sampling and measuring. pp. 309-351 in R. L. Metcalf and W. H. Luckmann, eds. Introduction to insect pest management. John Wiley and Sons, N.Y. 587 p.

Romney, V. E. 1945. The effect of physical factors upon catch of the beet leafhopper (*Eutettix tenellus* (Baker)) by a cylinder and two sweep net methods. Ecology 26:135-147.

Rudd, W. G., and R. L. Jensen. 1977. Sweep net and ground cloth sampling for insects in soybeans. J. Econ. Entomol. 70:301-304.

Shepard, M., and G. R. Carner. 1976. Distribution of insects in soybean fields. Can. Entomol. 108:767-771.

Shepard, M., G. R. Carner, and S. G. Turnipseed. 1974. A comparison of three sampling methods for arthropods in soybeans. Environ. Entomol. 3:227-232.

Southwood, T. R. E. 1978. Ecological methods. Halsted Press, N.Y. 524 p.

Turnipseed, S. G. 1974. Sampling soybean insects by various D-Vac, sweep, and ground cloth methods. Fla. Entomol. 57:217-223.

Yates, F. 1965. Sampling methods for censuses and surveys. 3rd ed., Imp. Charles Griffin and Co., London, 440 p.

Young, J. H., E. K. Johnson, and R. G. Price. 1976. Sampling cotton and other field crops for insects. Okla. Agr. Exp. Sta. Bull. B-723:10 p.

Chapter 3

Introduction to Sampling Theory

William G. Ruesink

I. Introduction

One of the first things that a field entomologist learns is how to sample an insect population, for it is a tenet of the discipline that until one knows what species are present and how many there are, nothing is known. Some entomologists spend a major portion of their time working on sampling techniques or on interpreting sampling data. It is easy for entomologists to become convinced that the object of their professional existence is to work on sampling. As one drifts in this direction, it is beneficial to consider a statement by Morris (1960): "Sampling has no intrinsic merit, but is only a tool which the entomologist should use to obtain certain information, provided there is no easier way to get the information." Sampling is a means to an end. Only by having a clear understanding of the end can the sampling program be optimized.

There are two fundamentally different reasons for sampling insect populations: one is to determine if the economic threshold has been exceeded and, hence, decide that control measures should be begun; the other is to estimate population density with some predetermined degree of reliability, typically for research purposes. In the first case, considerable time and expense can be saved by developing sequential sampling programs. These programs minimize the number of samples taken to reach the decision. Chapters 4 and 5 explain sequential sampling in detail, but before a sequential plan is developed, one should understand the material presented in the remainder of this chapter.

Measurements taken to estimate population density in field crops fall into three groups: absolute methods, relative methods, and population indices. Absolute methods provide estimates of density per unit of land area. Relative methods give density per some unit other than land area. The sweep net is an example. Absolute density can be estimated from relative density only when enough is known about the effects of the habitat and the weather on the catch. Population indices do not count insects at all, but rather are measures of insect products (e.g., frass) or effects (e.g., plant damage).

Of the many sampling methods presented in this book, most will be relative methods. For the majority of applications, relative estimates should be converted to absolute estimates, otherwise it is impossible to know whether differences in catch represent differences in density or a change in the efficiency of the sampling method. On the other hand, there are cases in which relative estimates are quite adequate, such as when samples are taken weekly during two entire growing seasons and the data are used to say which year's population was greater.

A. Definitions

Statisticians and biologists have developed a technical vocabulary to facilitate communication about sampling programs. Unfortunately, many of the words in this vocabulary also have a nontechnical meaning, and some authors have been ambiguous in their use of these. Some of the most commonly used technical terms will be presented here so that they need not be repeatedly redefined in subsequent chapters.

The following definitions were adapted from Bliss (1967), Hansen et al. (1953), Karandinos (1976), Pedigo et al. (1972), Snedecor and Cochran (1967) and Southwood (1978).

A "sample" consists of a small collection drawn from a larger "population" about which information is desired. It is the sample that is observed, but it is the population that is studied. The number of observations (n) in the sample is referred to as "sample size," while the physical composition and magnitude of a single sample is called the "sample unit." For example, the sample unit might be 10 sweeps with a sweep net; a sample size of 20 would then imply 20 sets of 10 sweeps each.

The arithmetic mean ($\bar{x}$) is the most commonly used measure of central tendency and is computed by

$$\bar{x} = \frac{1}{n} \Sigma x_i \tag{3-1}$$

where x_i represents the ith of the n observations. The common measure of spread among the observations is the "standard deviation" (s), which is computed by

Introduction to Sampling Theory

$$s = \sqrt{\frac{\Sigma x_i^2 - \frac{1}{n}(\Sigma x_i)^2}{n-1}} \qquad (3\text{-}2)$$

The "variance" (s^2) is simply the square of the standard deviation, while the "standard error" ($S_{\bar{x}}$) is

$$S_{\bar{x}} = \frac{s}{\sqrt{n}} . \qquad (3\text{-}3)$$

While the standard deviation is an expression of the spread among observations, the standard error is a measure of the distance of $\bar{x}$ to the true mean of the population being sampled.

Several ratios are in use that compare the spread among observations to the observed mean. The "coefficient of variation" (CV) is defined as

$$CV = s/\bar{x}. \qquad (3\text{-}4)$$

A more useful ratio is the $S_{\bar{x}}$ to the $\bar{x}$, which entomologists usually refer to as "relative variation" (RV):

$$RV = (S_{\bar{x}}/\bar{x})(100). \qquad (3\text{-}5)$$

This book will use CV and RV as defined here, but the reader should be aware that these terms often are used in the literature to describe slightly different concepts. These ratios are popular because for many sampling programs s increases as $\bar{x}$ increases, and they are unitless and hence comparable regardless of the sample unit used.

"Accuracy" refers to the difference between a sample result and the true population value. Although accuracy is desired, generally no independent knowledge of the true population value is available, so instead the "precision" or "reliability" of a sample result is evaluated. These two terms are used as synonyms and refer to the difference between a sample result and the result that would be obtained from a complete census taken under the same conditions. The distinction between accuracy and reliability is subtle, but generally entomologists deal only with reliability.

Reliability can be measured using relative variation or formal probabilistic statements and confidence intervals. A "confidence interval" brackets the estimate, usually symmetrically, and states that the true population value lies within the interval with specified probability. When n is large, say 30 or more, or for some other reason normality can be assumed, a confidence interval for the mean can be written as

$$\bar{x} - t_\alpha s/\sqrt{n} \leq \mu \leq \bar{x} + t_\alpha s/\sqrt{n} \tag{3-6}$$

where t_α is the value of the t-distribution with probability level α and degrees of freedom (df) as in s. The degrees of freedom are n - 1 if s is estimated from the n samples; if added information is used to better estimate s, the degrees of freedom will exceed n. When sampling insect populations, the errors are seldom distributed normally for small n, and a symmetrical confidence interval is hence a poor choice. Procedures for computing asymmetrical intervals to account for the underlying statistical distribution are available, but because of the computational complexity they have found very little application. An introduction to the literature of asymmetrical intervals appears in Bliss (1967, pp. 199-203).

A relatively uncomplicated method is available for one special case of an asymmetrical interval. This relates to a sampling result in which every one of the n samples produces zero individuals. If there is justification for assuming that the underlying distribution be binomial, Poisson, or negative binomial, it is possible to compute a one-sided confidence interval and state that the population mean is less than some upper limit with probability level α. The general formulas in Table 3-1 were derived from concepts of basic probability theory while the examples indicate that the upper limit is greater for the more clumped distributions.

Whenever the choice between two candidate sampling programs depends also on costs (either dollars or effort), our knowledge about reliability must be combined somehow with the cost of obtaining the information. One program is said to be more "efficient" than another if under specified conditions it yields more reliable results per unit cost. "Relative net precision" (RNP) is one measure of efficiency and is defined by

$$RNP = \frac{100}{(RV)(C_s)} \tag{3-7}$$

where C_s is the total cost of the n samples used to compute RV. Another means of comparing programs is presented at the end of this chapter.

Table 3-1. One-sided confidence interval for three common statistical distributions given that no individuals were found in n samples; k is the parameter of the negative binomial and α is the level of confidence desired

Distribution	Upper limit for the mean (individuals/sample)	
	General form	Example (n=10, α=.05, k=2)
Binomial	$1 - \alpha^{1/n}$	0.26
Poisson	$-(\ln\alpha)/n$	0.30
Negative binomial	$k(\alpha^{(-1/nk)} - 1)$	0.32

Introduction to Sampling Theory

B. Characteristics of Sampling Statistics

Several other features of a sampling statistic (e.g., mean or standard deviation) are important enough to deserve mention here. For one, it should be unbiased. If its average value over all possible random samples is equal to the true value in the population, the estimate is called "unbiased." It is called "consistent" if the proportion of sample estimates within a small fixed amount of the true population value approaches 100% as n increases. Of several similar statistics, the most "efficient" one is least variable for a given sample size, and a "sufficient" statistic captures all the information about the population feature in question that is contained in the sample observations, regardless of sample size.

II. Spatial Patterns, Statistical Distributions, and Measures of Aggregation

The spatial pattern of individuals in the habitat has a tremendous influence on the sampling plan. To understand this relationship, it is necessary to understand the concept of randomness as used by statisticians.

In two dimensions a spatial pattern is random if every point on the surface has an equal probability of being occupied by an individual. An alternate way of phrasing this definition is to say that knowing the location of one individual on the surface provides no information as to the location of any other individual.

Spatial patterns can deviate from randomness in either of two directions (Fig. 3-1). If the presence of an individual at one point increases the probability of another individual being nearby, then the spatial pattern will be clumped, whereas if it decreases the probability of another being nearby, the pattern will be more uniform than random.

If a population exhibits a random spatial pattern, and if the act of taking one sample does not bias the numbers taken in any subsequent samples, the number of individuals per sample unit will follow the Poisson statistical distribution. The probability that a unit will contain n individuals is

$$P_n = \frac{\lambda^n e^{-\lambda}}{n!} \tag{3-8}$$

where λ is the single parameter of the Poisson distribution. An important feature of this distribution is that the mean and the variance are equal and are defined as the parameter λ.

If a population exhibits an aggregated spatial pattern, and if samples are taken

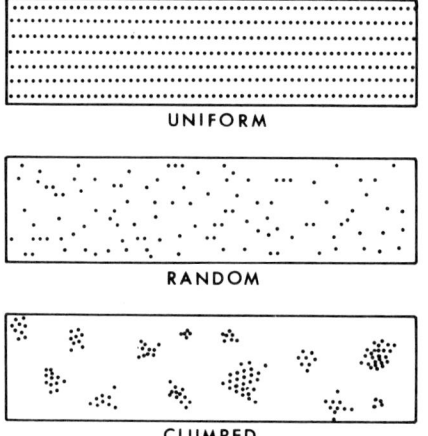

Figure 3-1. Types of spatial pattern.

from randomly distributed points in space, the number of individuals per sample unit will follow one of the so-called contagious statistical distributions. One important feature of all contagious distributions is that the variance exceeds the mean. In entomology the most frequently used contagious distribution is the negative binomial, which is described by two parameters—the mean $\bar{x}$ and the exponent k. The expected frequency of zeros is

$$p_0 = \left(\frac{k}{\bar{x}+k}\right)^k \tag{3-9}$$

and the probability that a sample will contain exactly n individuals (n>o) is

$$p_n = p_0 \frac{k(k+1)\cdots(k+n-1)}{n!} \left(\frac{\bar{x}}{\bar{x}+k}\right)^n. \tag{3-10}$$

The parameter k can be estimated from the mean and variance of the sample data using

$$k = \frac{\bar{x}^2}{s^2 - \bar{x}}. \tag{3-11}$$

In the case that $\bar{x}$ is greater than 0.5, the above equation works well only for populations that are very highly aggregated. If $\bar{x}$ is large and the aggregation is only moderate, other more complex methods must be used to estimate k. Anscombe (1949) and Southwood (1978) discuss several alternate approaches.

Other discrete contagious distributions include the Neyman type A, Poisson

Introduction to Sampling Theory 67

binomial, Poisson with added zeros, and double Poisson. There are several theoretical reasons why the negative binomial usually fits entomological data as well as or better than any other distribution (Williamson 1963), but in some cases others are preferred (e.g., McGuire et al. 1957). Katti (1966) discusses 22 discrete distributions while Gates and Ethridge (1972) present mathematical equations and methods to fit eight of those.

Even though a set of data may agree with the negative binomial better than the Poisson, it is possible that the difference is not biologically significant. As a general rule whenever k exceeds 8.0 the Poisson will serve as an adequate substitute for the mathematically more complex negative binomial. In addition, on occasion the normal distribution adequately approximates the best-fitting distribution. For example, if one is interested only in the right-hand half of the distribution, the normal is close to the Poisson whenever λ exceeds 0.25, while for the negative binomial, the approximation is good for x̄ greater than 5.0 if k equals 2.0, but never holds for k less than about 1.0. This approximation allows a simple estimation of the probability that a sample will contain fewer than a specified upper limit.

An alternative method is available for determining whether the Poisson or negative binomial is preferred for a given situation. For each data set, the mean (x̄) and the variance (s^2) are computed. The variance can be related to the mean by the power law:

$$s^2 = a\bar{x}^b . \tag{3-12}$$

As few as three or four sets of data are sufficient to estimate the coefficient a and the exponent b, if the sets cover a wide range of mean densities. According to Taylor (1961) the coefficient a varies with sampling technique and habitat, while the exponent b is constant for the species. The most convenient method for estimating a and b is via linear regression of log s^2 on log x̄. If both a and b are equal to 1.0 (i.e., not statistically different from 1.0), the Poisson distribution is an adequate description of the data. Often b is greater than 1.0, with $1.4 < b < 2.0$ being typical for field crop pests.

Several methods (see Southwood 1978) are available for estimating the common k of the negative binomial from multiple data sets. The methods are adequate when the range of means is small, say one order of magnitude. But it has been shown many times that k varies with mean density, and, therefore, whenever large ranges in the mean are to be considered, some method is required that will estimate a density related k. Equations (3-11) and (3-12) are easily used for this purpose; by substituting $s^2 = a\bar{x}^b$ into (3-11) and dividing both numerator and denominator by x̄, we get:

$$k = \frac{\bar{x}}{a\bar{x}^{b-1} - 1} . \tag{3-13}$$

Equation (3-13) is recommended instead of a common k whenever mean densities vary more than an order of magnitude.

III. The Normal Distribution and Transformation of Data

With most statistical analyses the assumption is made that the sampled population is normally distributed. Some analyses are sufficiently robust to give good results even when the distributions depart considerably from normality. For those that are not, it is often possible to find a mathematical formula (transformation) which, when applied to the individual original observations, will produce a nearly normal distribution (Wadley 1967).

Since the normal distribution is treated thoroughly in most introductory statistics texts, only a brief discussion is presented here. This familiar, bell-shaped symmetrical curve is described by the function

$$Z = \frac{1}{\sigma \sqrt{2\pi}} e^{-\frac{1}{2}[(y-\mu)/\sigma]^2} \qquad (3\text{-}14)$$

where μ and σ are the parametric mean and standard deviation, respectively, of the population being sampled. As with all probability distribution functions, the area under the curve integrates to 1.0, while the area to the left of an arbitrary point y can be interpreted as the probability that a randomly selected sample from the population will be less than y. For example, if y is chosen as equal to the mean (μ), the probability of drawing a sample containing less than μ is exactly 0.5.

Two situations commonly arise for which data transformations may be desirable. One relates to the assumption of variance homogeneity in analysis of variance. It is well known that ANOVA is quite insensitive to violations of the equal variances assumption (e.g., Bliss 1967, p. 237), but at times additional power can be gained by transforming data to improve homogeneity. The $\sqrt{x}$ and log x transformations are most commonly used with population data while the $\sin^{-1} x$ is usually best for percentage data. A second situation that may call for data transformation involves computation of asymmetrical confidence intervals. The statement that 96% of the observations fall within two standard deviations of the mean is based on the assumption of normality. Any skewness in the underlying distribution invalidates this interpretation.

A guide to selecting the best transformation for population data was published by Taylor (1961), where he suggested plotting log s^2 vs log $\bar{x}$ for several populations of varying mean density. The slope (b) of the linear regression thus obtained indicates the best transformation:

Introduction to Sampling Theory

$$y = x^{(1-\frac{1}{2}b)} \quad \text{if } b \neq 2 \qquad (3\text{-}15)$$

$$\text{or} \quad y = \log x \quad \text{if } b = 2. \qquad (3\text{-}16)$$

Thus if b = 1, the best transformation is $y = \sqrt{x}$, and if b = 1.5, the best transformation is $y = x^{\frac{1}{4}}$.

IV. Converting Relative to Absolute Estimates

While the results of a sampling program utilizing relative methods will provide some information, it is generally true that an absolute estimate of population density is more valuable. As costs associated with most absolute methods are high, entomologists have developed conversion factors for certain relative methods. Multiplying the catch per sample by the conversion factor produces an estimate of absolute density. Much research still needs to be done in developing conversion factors for the more popular relative methods of sampling (see Chapter 2).

The research program for developing a conversion factor involves very intensive sampling using both the relative method and an absolute method. The sample size for each method should be large enough to obtain RV ≤ 10 for a given population and condition (usually a given field during a short period of time). This insures that the conversion factor (i.e., the ratio of absolute density to relative catch) will be estimated rather precisely for the conditions under which the sample was taken. Developing conversion factors with sufficient generality usually requires obtaining 20 or more independent observations over a broad range of conditions and utilizing multivariate statistical analysis to produce an equation that indicates how conversion factors vary with environmental conditions such as temperature, light intensity, wind speed, and crop height. Pedigo et al. (1972), sampling the green cloverworm with a sweep net, were successful in establishing a conversion factor that did not vary with environment; however, they did not indicate the range of conditions covered by their data. Hillhouse and Pitre (1974) and Marston et al. (1976) looked at several relative methods for several insect species and found that the conversion factors varied significantly among sampling dates when sweep net, ground cloth, and suction net were used to sample various insect species. Although they did not report the environmental conditions associated with each set of samples, it is likely that much of the variation could have been accounted for if crop and climatic conditions had been considered.

Two studies are recorded in the literature in which a formula was developed for a conversion factor as a function of environmental conditions. Ruesink and Haynes (1973) studied the sweep net for sampling ceral leaf beetle adults in small grains and obtained an equation that included temperature, solar radiation, wind speed, and crop height (see Chapter 2). Their equation accounted for 87% of the variation observed in the conversion factor. Cherry et al. (1977) proposed a conversion factor for sweep net sampling of the potato leafhopper in alfalfa. Their equation included only temperature and wind speed, but they did not report how much of the variation was accounted for by these factors.

V. Sample Size Determination

The material presented in this section is taken largely from Karandinos (1976). Similar material can be found in many other sources, but as Karandinos points out, some of the published formulas contain seemingly minor errors that can greatly affect results. For example, it is fairly common for a published formula to accidentally exclude a square root sign or an exponent; such an omission can often change the result by a factor of 10 or more.

As the sample size increases, so does the reliability of the estimated density. The problem is to decide how much time and effort to put into sampling, that is, to find the proper balance between the reliability of the estimate and the cost of obtaining it. The solution to the problem depends in part on one's definition of reliability and very much on the statistical distribution of the sample data.

There are two common ways of expressing reliability, and both will be discussed here. The first states that the standard error of the estimated mean should be within a certain value of the mean. The "certain value" can be expressed as a specific numerical quantity or as a percent of the mean. For example, one might want the standard error to be one larva per foot of row, or perhaps 20% of the mean density. Depending on which is more appropriate for a given situation, the corresponding formula must be selected. The second way to express reliability uses a formal probabilistic statement and states that the estimated mean should be within a certain value of the true mean density. Again, the certain value may be expressed as either a specific numerical quantity or as a percent of the mean.

A. Sample Size Based on Reliability Defined in Terms of Standard Error

In research work it is most commonly required that the standard error be within a percentage of the mean, while in decision making it is more frequent that the criterion is that the estimated mean be within a specific value of the true mean. Since these generalities do not hold in all cases, it is important to consider each case as it arises and to choose the expression of reliability that is most appropriate.

In the formulas that follow, these variables and coefficients are used:

Introduction to Sampling Theory

μ = population or "true mean"
$\bar{x}$ = sample mean (the estimate of μ)
σ^2 = population variance
s^2 = sample variance (the estimate σ^2)
s = standard deviation
$S_{\bar{x}}$ = standard error
k = dispersion parameter of the negative binomial distribution
h = half length of probabilistic confidence interval
g = half length of standard error interval
α = probability of not satisfying the criterion
n = sample size
c = coefficient for $S_{\bar{x}}$ as a fraction of $\bar{x}$
d = coefficient for probabilistic interval as a fraction of $\bar{x}$
$Z_{\alpha/2}$ = upper $\alpha/2$ point of the cumulative standard normal distribution
a = coefficient in Taylor's power law
b = exponent in Taylor's power law

Looking first at the case in which reliability is expressed in terms of the standard error, solve the equation $S_{\bar{x}} = \sqrt{s^2/n}$ for n to obtain

$$n = s^2/S_{\bar{x}}^2 . \qquad (3\text{-}17)$$

In the case that reliability is defined by requiring the standard error to equal some fixed constant (g), $S_{\bar{x}} = g$ which when substituted into (3-17) gives

$$n = s^2/g^2 . \qquad (3\text{-}18)$$

For the Poisson distribution $s^2 = \bar{x}$, so

$$n = \bar{x}/g^2 \qquad (3\text{-}19)$$

and for the negative binomial $s^2 = \bar{x} + \bar{x}^2/k$ which gives (when simplified)

$$n = (\bar{x}k + \bar{x}^2)/kg^2 . \qquad (3\text{-}20)$$

If the relationship from Taylor's power law ($s^2 = a\bar{x}^b$) is used, without determining the underlying distribution represented,

$$n = a\bar{x}^b/g^2 . \qquad (3\text{-}21)$$

In the case that reliability is defined by requiring the standard error to be equal to a fraction of the mean, $S_{\bar{x}} = c\bar{x}$ which when solved for n gives

$$n = s^2/(c^2 \bar{x}^2). \qquad (3\text{-}22)$$

Table 3-2. Formulas for sample size (n) when reliability is defined in terms of the standard error

Assumed underlying distribution	To obtain a standard error equal to a	
	constant (g)	Fraction (c) of $\bar{x}$
General	$n = \dfrac{s^2}{g^2}$	$n = \dfrac{s^2}{c^2 \bar{x}^2}$
Poisson	$n = \dfrac{\bar{x}}{g^2}$	$n = \dfrac{1}{c^2 \bar{x}}$
Negative binomial	$n = \dfrac{\bar{x}k + \bar{x}^2}{kg^2}$	$n = \dfrac{k + \bar{x}}{c^2 k \bar{x}}$
Taylor's power law	$n = \dfrac{a\bar{x}^b}{g^2}$	$n = \dfrac{a\bar{x}^{b-2}}{c^2}$

By substituting the same expressions for s^2, it is simple to generate equations analogous to (3-19), (3-20) and (3-21) for this case. Table 3-2 summarizes the eight formulas for sample size when reliability is expressed in terms of the standard error.

B. Sample Size Based on Reliability Defined in Terms of Probabilistic Statements

When it is preferable to express reliability in terms of formal probabilistic statements, the computations are slightly more complex, but still rather reasonable as long as it can be assumed that one is working with the normal distribution. Regardless of the underlying probability distribution, as the sample size becomes large normality is approached, and, therefore, the confidence interval can be approximated by

$$\bar{x} \pm Z_{\alpha/2} \sqrt{s^2/n} \ . \qquad (3\text{-}23)$$

In most entomological sampling situations this approximation is satisfactory for n equal to or greater than 30. For smaller sample sizes the exact confidence interval may be asymmetrical, with a shorter tail on the zero side of $\bar{x}$ and a longer one on the ∞ side. A verbal interpretation of the above equation is that the true mean is likely to be between $\bar{x} - Z_{\alpha/2}\sqrt{s^2/n}$ and $\bar{x} + Z_{\alpha/2}\sqrt{s^2/n}$ with probability approximately $(1 - \alpha)$. The commonly used 95% confidence interval requires using $\alpha = 0.05$ for which $Z_{.025} = 1.96$.

Introduction to Sampling Theory

In the case that reliability is defined by requiring that $\bar{x}$ be within a fixed value (h) of μ with probability $(1 - \alpha)$, $h = Z_{\alpha/2} \sqrt{s^2/n}$ which when solved for n gives

$$n = \left(\frac{Z_{\alpha/2}}{h}\right)^2 s^2 . \tag{3-24}$$

For the Poisson distribution ($s^2 = \bar{x}$) this becomes

$$n = \left(\frac{Z_{\alpha/2}}{h}\right)^2 \bar{x} , \tag{3-25}$$

and for the negative binomial ($s^2 = \bar{x} + \bar{x}^2/k$),

$$n = \left(\frac{Z_{\alpha/2}}{h}\right)^2 \left(\frac{\bar{x}k + \bar{x}^2}{k}\right) . \tag{3-26}$$

If the underlying distribution is undetermined, but a and b of Taylor's power law are known, use

$$n = a \left(\frac{Z_{\alpha/2}}{h}\right)^2 \bar{x}^b . \tag{3-27}$$

In the case that reliability is defined by requiring that $\bar{x}$ be in a fraction (d) of the mean with probability of $(1 - \alpha)$, $d\bar{x} = Z_{\alpha/2} \sqrt{s^2/n}$ which when solved for n becomes

$$n = \left(\frac{Z_{\alpha/2}}{d}\right)^2 \frac{s^2}{\bar{x}^2} . \tag{3-28}$$

Again, expressions for s^2 can be substituted into equation (3-28) to obtain formulas for n in terms of $\bar{x}$ for the Poisson and negative binomial distributions and for Taylor's power law. These are summarized in Table 3-3.

VI. Selecting a Best Sampling Method

Ultimately this is a subjective decision, but something can be done to make the choice more objective. As mentioned in an earlier section, there are basically two conflicting objectives: reliability and economy. For any given sampling method,

Table 3-3. Formulas for sample size (n) when reliability is defined by formal probabilistic statements

Assumed underlying distribution	To obtain a confidence interval equal to twice[a] the value of a	
	Constant (h)	Fraction (d) of $\bar{x}$
General	$n = \left(\dfrac{Z_{\alpha/2}}{h}\right)^2 s^2$	$n = \left(\dfrac{Z_{\alpha/2}}{d}\right)^2 \dfrac{s^2}{\bar{x}^2}$
Poisson	$n = \left(\dfrac{Z_{\alpha/2}}{h}\right)^2 \bar{x}$	$n = \left(\dfrac{Z_{\alpha/2}}{d}\right)^2 \dfrac{1}{\bar{x}}$
Negative binomial	$n = \left(\dfrac{Z_{\alpha/2}}{h}\right)^2 \left(\dfrac{\bar{x}k + \bar{x}^2}{k}\right)$	$n = \left(\dfrac{Z_{\alpha/2}}{d}\right)^2 \left(\dfrac{k + \bar{x}}{k\bar{x}}\right)$
Taylor's power law	$n = a\left(\dfrac{Z_{\alpha/2}}{h}\right)^2 \bar{x}^b$	$n = a\left(\dfrac{Z_{\alpha/2}}{d}\right)^2 \bar{x}^{b-2}$

[a] i.e., h or d $\bar{x}$ on each side of $\bar{x}$.

reliability can be increased by taking more samples, but the law of diminishing returns soon makes that an expensive operation.

As the simplest case, assume that a method for estimating the absolute density of bean leaf beetle adults is needed and that data on two candidate techniques are available. Method A is unbiased, samples 1 m of row, has the mean and variance related according to $s^2 = 3.5\,\bar{x}^{1.50}$, and costs \$2.50 per sample to operate. Method B on the average misses 10% of the population, sampling 2 m of row, has the mean and variance related by $s^2 = 6.2\,\bar{x}^{1.54}$, and costs \$1.20 per sample to operate. The setup costs of acquiring the sampling equipment and training people to use it will be ignored.

Table 3-4 summarizes the reasoning for densities of 0.3 and 30 beetles per m of row using as the reliability criterion that the standard error should be within 20% of the mean. For Method A, a density of 0.3 will produce a mean catch per sample of 0.3; using the formula from Table 3-2 for obtaining n (c = 0.2):

$$n = \frac{a\bar{x}^{b-2}}{c^2}$$

$$n = \frac{3.5(.3)^{-.50}}{(.2)^2}$$

Table 3-4. Comparison of the cost of obtaining similar information by two hypothetical sampling methods at two population densities

Quantity	Method A	Method B
When absolute density is 0.3 per m of row:		
Mean	0.30	0.54
Sample size	160	206
Cost	$400.00	$247.00
When absolute density is 30 per m of row:		
Mean	30.0	54.0
Sample size	16	25
Cost	$40.00	$30.00

$$n = \frac{3.5(1.8257)}{.04}$$

$$n = 160.$$

This number of samples, at $2.50/sample, will cost $400.00. When the density is 30/m of row, 16 samples are needed to satisfy the reliability criterion and the cost is $40.00.

For Method B, a density of 0.3 will produce a mean catch per sample of 0.54. Using the same equation from Table 3-4, it is seen that n = 206 at a cost of $247.00. When the density is 30/m of row, 25 samples are needed at a cost of $30.00. Comparing the two methods, Method A is 30-60% more expensive than Method B. However, Method B has the drawback of a 10% bias, and nothing has been said about setup costs. The selection of a sampling method is still subjective, but the costs have been identified.

In those cases in which absolute density estimates are not required, the comparison becomes even more subjective. The basis for comparison should still consider the cost of acquiring the desired information; the difficulty arises in specifying the criteria for reliability.

In no case is it sufficient to determine which method catches the most insects, nor is it sufficient to look at the coefficient of variation. The only purpose of sampling is to obtain information, and the choice of sampling method should depend on the cost of obtaining that information.

VII. Regional Population Estimation

The preceding relates to estimating density within a single habitat, such as within a field, but there are several situations in which regional population estimates offer the only means for addressing a problem. An example would be a field study of adult bean leaf beetle mortality during the overwintering generation. Estimates of total population size would be needed for several successive points in time—perhaps at emergence from the soil as new adults, upon first entering overwintering sites, again in overwintering sites in late winter, and two or more times in the spring after returning to the fields.

Since this insect moves among habitats during its lifetime, it is not sufficient to follow the population in a single field. The study area would have to be large enough to include a sufficient quantity of all essential habitats so that it could be assumed that migrations into and out of the study area were equal. A minimum size would probably be 1,000 ha.

One approach to estimating regional population size involves dividing the study area into subareas according to habitat. For example, the 1,000 ha might consist of 300 ha of soybeans, 500 ha of other row crops, 100 ha of alfalfa or clover, 75 ha of woods, 6 ha of woods edge, 9 ha of field borders or roadsides, and 10 ha uninhabitatable (roadways, water, building sites, etc.). If the density per ha in each habitat is estimated, the total population size can be found by simply adding together the parts. If the standard error is known for each density estimate, a standard error can be computed for the total using the rules (Yates 1953) that (1) the standard error of a multiple of an estimate is the same multiple of the standard error of the estimate (i.e., $S_{\bar{x}}(\ell y) = \ell S_{\bar{x}}(y)$), and (2) the standard error of the sum of a number of independent estimates is the square root of the sum of the squares of the standard errors of the estimates:

Table 3-5. Computation of the number of bean leaf beetles in a hypothetical 1,000 ha study area on October 14

Habitat	ha	Estimated density per m²			Computed density per ha[a]			Computed population in habitat[b]		
		$\bar{x}$	±	$S_{\bar{x}}$	$\bar{x}$	±	$S_{\bar{x}}$	$\bar{x}$	±	$S_{\bar{x}}$
Soybeans and soybean stubble	300	0.50	±	0.10	5.0	±	1.0	1.50	±	0.30
Other row crops	500	0.02	±	0.02	0.2	±	0.2	0.10	±	0.10
Alfalfa and clover	100	0.80	±	0.25	8.0	±	2.5	0.80	±	0.25
Woods	75	0.20	±	0.08	2.0	±	0.8	0.15	±	0.06
Woods edge	6	35.00	±	20.00	350.0	±	200.0	2.10	±	1.20
Field borders and roadsides	9	15.00	±	10.00	150.0	±	100.0	1.35	±	0.90
Uninhabitatable	10	0.0	±	0.00	0.0	±	0.0	0.0	±	0.0
	1,000							6.00	±	1.55

[a] In thousands of beetles

[b] In millions of beetles.

$$S_{\bar{x}}(y_1 + y_2 + y_3 + \cdots) = \sqrt{S_{\bar{x}}^2(y_1) + S_{\bar{x}}^2(y_2) + S_{\bar{x}}^2(y_3) + \cdots} \quad (3\text{-}29)$$

Table 3-5 illustrates this approach using hypothetical bean leaf beetle data. Sawyer and Haynes (1978) discuss the question of how to optimally allocate limiting sampling resources among habitats in such a stratified sampling program. Even without using the formal approach they present, it is clear from inspecting the $S_{\bar{x}}$ terms in the right hand column of Table 3-5 that the greatest benefit would be obtained by increasing the sample size in the woods edge and that the best place to decrease sample size, if needed, would be the woods.

References

Anscombe, F. J. 1949. The statistical analysis of insect counts based on the negative binomial distribution. Biometrics **5**:165-173.

Bliss, C. I. 1967. Statistics in biology. vol. 1. McGraw Hill, N.Y. 558 p.

Cherry, R. H., K. A. Wood, and W. G. Ruesink. 1977. Emergence trap and sweepnet sampling for adults of the potato leafhopper from alfalfa. J. Econ. Entomol. **70**:279-282.

Gates, C. E., and F. G. Ethridge. 1972. A generalized set of discrete frequency distributions with FORTRAN program. Int. Assoc. Math. Geol. **4**:1-24.

Hansen, M. H., W. N. Hurwitz, and W. G. Madow. 1953. Sample survey methods and theory. vol. 1. John Wiley & Sons, N.Y. 638 p.

Hillhouse, T. L., and H. N. Pitre. 1974. Comparison of sampling techniques to obtain measurements of insect populations on soybeans. J. Econ. Entomol. **67**:411-414.

Karandinos, M. G. 1976. Optimum sample size and comments on some published formulae. Bull. Entomol. Soc. Amer. **22**:417-421.

Katti, S. K. 1966. Interrelations among generalized distributions and their components. Biometrics **22**:44-52.

Marston, N. L., C. E. Morgan, G. D. Thomas, and C. M. Ignoffo. 1976. Evaluation of four techniques for sampling soybean insects. J. Kans. Entomol. Soc. **49**:389-400.

McGuire, J. U., T. A. Brindley, and T. A. Bancroft. 1957. The distribution of European corn borer larvae *Pyraustra nubilalis* (HBN), in field corn. Biometrics **13**:65-78.

Morris, R. F. 1960. Sampling insect populations. Annu. Rev. Entomol. **5**:243-264.

Pedigo, L. P., G. L. Lentz, J. G. Stone, and D. F. Cox. 1972. Green cloverworm populations in Iowa soybeans with special reference to sampling procedure. J. Econ. Entomol. **65**:414-421.

Ruesink, W. G., and D. L. Haynes. 1973. Sweepnet sampling for the cereal leaf beetle, *Oulema melanopus.* Environ. Entomol. **2**:161-172.

Sawyer, A. J., and D. L. Haynes. 1978. Allocating limited sampling resources for estimating regional populations of overwintering cereal leaf beetles. Environ. Entomol. **7**:62-66.

Snedecor, G. W., and W. G. Cochran. 1967. Statistical methods. Iowa State Univ. Press, Ames, Iowa. 593 p.

Southwood, T. R. E. 1978. Ecological methods with particular reference to the study of insect populations. Halsted Press, N.Y. 524 p.

Taylor, L. R. 1961. Aggregation, variance and the mean. Nature **189**:732-735.

Yates, F. 1953. Sampling methods for censuses and surveys. Charles Griffin and Co., London. 440 p.

Wadley, F. M. 1967. Experimental statistics in entomology. Graduate School Press, USDA, Washington, D.C. 133 p.

Williamson, E. 1963. Tables of the negative binomial probability distribution. John Wiley, London. 275 p.

Chapter 4

Sequential Sampling Plans for Soybean Arthropods[1]

Merle Shepard

I. Introduction

The need for rapid yet reliable sampling methods for insect pests becomes more apparent as emphasis on implementation of pest management programs increases. Design of a sampling program in which allocation of sampling resources and cost are considered is one of the first and most important steps in developing effective control strategies. The soybean producer must decide, or have someone decide for him, whether soybean pests are present in sufficient numbers to justify the expense of specific management tactics (e.g., insecticide application). Also, assessment of the densities of the naturally occurring biological control agents is becoming more important in the decision-making processes.

Until workable and reliable computer simulation models are developed to predict accurately soybean insect pest population densities, routine season-long sampling programs will be the bases on which management decisions are made. Even with the most sophisticated simulation models, insect and damage censuses will be needed to provide the necessary inputs to initialize the simulation.

As was pointed out in Chapter 2, the objective of a sampling program may be to study certain aspects of insect population dynamics; but for management decisions, information about a population should be collected in the shortest amount of time, at the lowest possible cost, yet with a high level of reliability. Sequential sampling has been found to possess these and other merits that make it especially useful for making pest management decisions.

---

[1] Technical contribution no. 1523, published by permission of the Director, South Carolina Agricultural Experiment Station.

Although sequential sampling was developed early in the 1940s (Wald 1943), its utility as a tool in pest surveys was not realized until almost 10 years later (Stark 1952, Morris 1954). Initially, sequential analysis was developed and used to classify war research problems and restrictions were placed on the use of this technique until 1945 (Wald 1945). Waters (1955) provided a thorough discussion of sequential sampling as it relates to entomological use. Since his publication relating to forest insect surveys, several sequential sampling models have been developed for many agriculturally important insects (Harcourt 1966, Gonzalez 1971, Allen et al. 1972, Ingram and Green 1972, Sevacherian and Stern 1972, Shepard 1973, Pieters and Sterling 1974,1975, and others) including some that occur on soybean (Waddill et al. 1974, Hammond and Pedigo 1976, Strayer et al. 1977). The rationale of sequential sampling with emphasis on its use in pest management was discussed by Onsager (1976) and a literature review of sequential sampling plans for insects has been published (Pieters 1978).

A major reason for the increased interest in sequential sampling is that the sampler can rapidly classify population densities into broad categories such as low, medium, or high or into levels requiring treatment or no treatment. Acceptance of these categories with a high level of reliability precludes the necessity for intensive surveys for estimating population parameters. Additional advantages to the use of sequential sampling are that the sample size is not fixed but is dependent on the population density, with fewer samples required for sparse and dense populations. Also, sampling by sequential methods allows one to assess the impact of a particular control tactic on the pest population.

Although few time-budget studies have been conducted comparing sequential sampling with other methods, Sterling (1975) reported a savings of 76% in time using the plan for cotton arthropods, yet the correct decision was reached 95% of the time as compared to conventional sampling techniques. Wald (1945) and Waters (1955) had reported earlier more than a 50% savings in time when sequential methods were used.

Briefly, sequential sampling is a procedure in which samples are taken in sequence with decisions made after each sample, based on cumulative information obtained. When the population density of the species involved is at or near the damage threshold or at some other level set a priori, the decision may be to continue sampling. Thus, with the sequential method, the flexibility of the sample size usually results in large savings in time because intensive sampling may be needed only at intermediate densities. With most sequential decision plans, a minimum number of samples is usually arbitrarily set. This is usually based on the sampler's experience or intuition to insure that decisions are not made from an unrealistically low number of samples.

Time and location of sample sites are other general considerations that must be addressed. Although no thorough studies have been made on the minimum number of sequential samples that should be taken in different size fields, experience and knowledge of possible clumping or "edge effects" often are helpful in this regard. With respect to the sampling pattern, sampling the cotton fleahopper, *Pseudotomoscelis seriatus* (Reuter), using an X pattern through the field required less time and fewer samples to reach a decision than did sampling a circular or diamond-shaped pattern (Pieters and Sterling 1974).

Besides sampling pest insects, it is also possible to sequentially sample the damage incurred by the plant. For example, a sequential plan for cotton buds (squares) damaged by boll weevils and *Heliothis* spp. has been developed (Pieters and Sterling 1975). Conceivably, stink-bug damaged soybean seeds could be sampled in a similar manner provided the sampler was trained to recognize this type of damage.

Sequential sampling plans for predators may also be useful. Waddill et al. (1974) published plans for *Nabis* spp. and *Geocoris* spp., which are two of the most important beneficial genera of predators in soybeans. Although formulation of sequential sampling schemes ordinarily requires some knowledge of a damage threshold for pests, classification of population levels of the above predators into low, medium, and high was accomplished by examining population data collected for several years. Knowledge of this kind of information could be helpful when treatment for a pest is only marginally justified. A decision may be made to postpone the treatment or return in a few days and sample again.

II. Developing the Sequential Sampling Plan

Certain essential information is needed before a sequential sampling plan can be developed. Later in this chapter a sequential plan will be discussed whereby more precise population estimations can be made. Sequential programs of this kind require more time and effort than do the more conventional plans used in making pest management decisions.

There are three fundamental pieces of information necessary for sequential model development: (1) the mathematical distribution of the insects or their damage in the field, (2) the damage threshold or some other means of setting class limits, and (3) values of α and β, in equations given later, that simply incorporate the level of risk (of making a wrong decision) that the sampler is willing to accept.

Knowledge of the mathematical distributions (see Chapter 3) of a soybean insect is not only essential to construction of sequential sampling plans but determination of these spatial patterns are also necessary to apply the best statistical procedure for analysis of field data. Several mathematical distributions have been proposed that fit dispersal patterns of insects in the field. Distributions of some of the most abundant insect species in soybean fields are presented in Table 4-1.

Table 4-1. Distribution patterns of some insect species in soybeans

Soybean insect	Distribution	Reference
Green Cloverworm, *Plathypena scabra*	Poisson	Hammond and Pedigo, 1976 Shepard and Carner, 1976
Velvetbean Caterpillar, *Anticarsia gemmatalis*	Poisson	Shepard and Carner, 1976 Strayer et al., 1977
Heliothis spp.	Poisson	Shepard and Carner, 1976
Mexican bean beetle, *Epilachna varivestis*	Neyman's Type A and Negative Binomial	Shepard and Carner, 1976
Soybean looper, *Pseudoplusia includens*	Poisson	Shepard and Carner, 1976
Bean leaf beetle, *Cerotoma trifurcata*	Negative Binomial	Kogan et al., 1974
Damsel bugs, *Nabis* spp.	Poisson	Waddill et al., 1974
Big eyed bugs, *Geocoris* spp.	Poisson	Waddill et al., 1974

Counts of several species of soybean insects have been tested for goodness of fit, using chi square, the Poisson, Poisson with zeros, Poisson binomial, Neyman's type A, negative binomial, and Thomas double Poisson distributions (Shepard and Carner 1976). For most soybean insect species only the Poisson, the negative binomial, and binomial need to be considered for sequential sampling plans. Waters (1955) provided equations for calculating sequential models from these three distributions.

The Poisson distribution arises when counts of insects (or their damage) are randomly arranged within a field. The spatial pattern of most species in soybean fields are more closely approximated by this distribution (Kogan et al. 1974, Waddill et al. 1974, Hammond and Pedigo 1976, Shepard and Carner 1976). A discussion of the Poisson distribution is found in Chapter 3. Interestingly, when the population density is low, the Poisson often typifies the distributional pattern even if certain species are known to clump.

Equations (4-1) and (4-2) are used to develop decision lines (or tables) for counts that are distributed as a Poisson (Waters 1955).

$$d_1 = b_n - h_1 \text{ (lower line)} \qquad (4\text{-}1)$$

$$d_2 = b_n + h_2 \text{ (upper line)} \qquad (4\text{-}2)$$

By calculating $d_{1,2}$ (the cumulative number of insects) for a certain number of samples (n), parallel lines can be drawn on a graph (Fig. 4-1). For the above

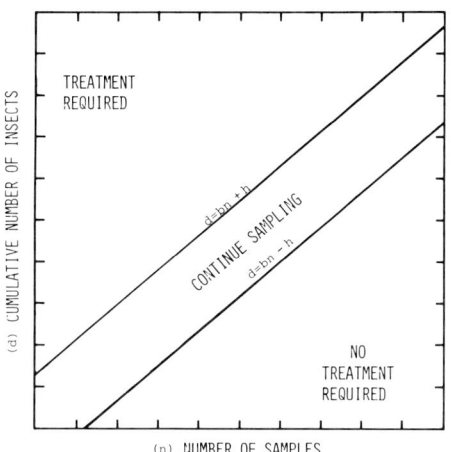

Figure 4-1. Typical sequential sampling plans for insect pests of soybean.

equation, b is the slope and h_1 and h_2 are the intercepts. The risk of suggesting no treatment when treatment is required is α. Likewise, β is the risk of recommending treatment when no treatment is needed. The damage threshold values on designated class limits are $\bar{x}_1$ and $\bar{x}_2$. In other words $\bar{x}_1$ might be the level designated as a light (no treatment required) population and x_2 as a dense (treatment required) population. Formulas used to calculate b, h_1, and h_2 are:

$$b = \frac{\bar{x}_2 - \bar{x}_1}{\ln \bar{x}_2 - \ln \bar{x}_1} \quad (4\text{-}3)$$

$$h_1 = \frac{\ln \frac{1-\alpha}{\beta}}{\ln \bar{x}_2 - \ln \bar{x}_1} \quad (4\text{-}4)$$

$$h_2 = \frac{\ln \frac{1-\beta}{\alpha}}{\ln \bar{x}_2 - \ln \bar{x}_1} \quad (4\text{-}5)$$

Hammond and Pedigo (1976) developed sampling plans for the green cloverworm, *Plathypena scabra* (F.), on soybean in Iowa (see Chapter 8). The above formulas were used to develop the plan that could accommodate three different general plant growth stages. The proposed economic injury levels for these stages were 36, 54, and 45 larvae/m of row for growth stages V7-R1, R2-R3, and R4-R5, respectively. The risk of misclassifying the population was set at $\alpha=\beta=0.10$.

Strayer et al. (1977) constructed sampling plans for the velvetbean caterpillar,

Anticarsia gemmatalis (Hübner), in soybean. The differential response of soybean at different growth stages to defoliation was incorporated into the model by providing separate sampling schemes for populations occurring at prebloom and postbloom (Fig. 4-2). Tables are more easily utilized for making sequential sampling decisions in the field than are figures or graphs. An example is presented in Table 4-2 (Strayer et al. 1977) for sampling the velvetbean caterpillar on soybean in Florida (see Chapter 6).

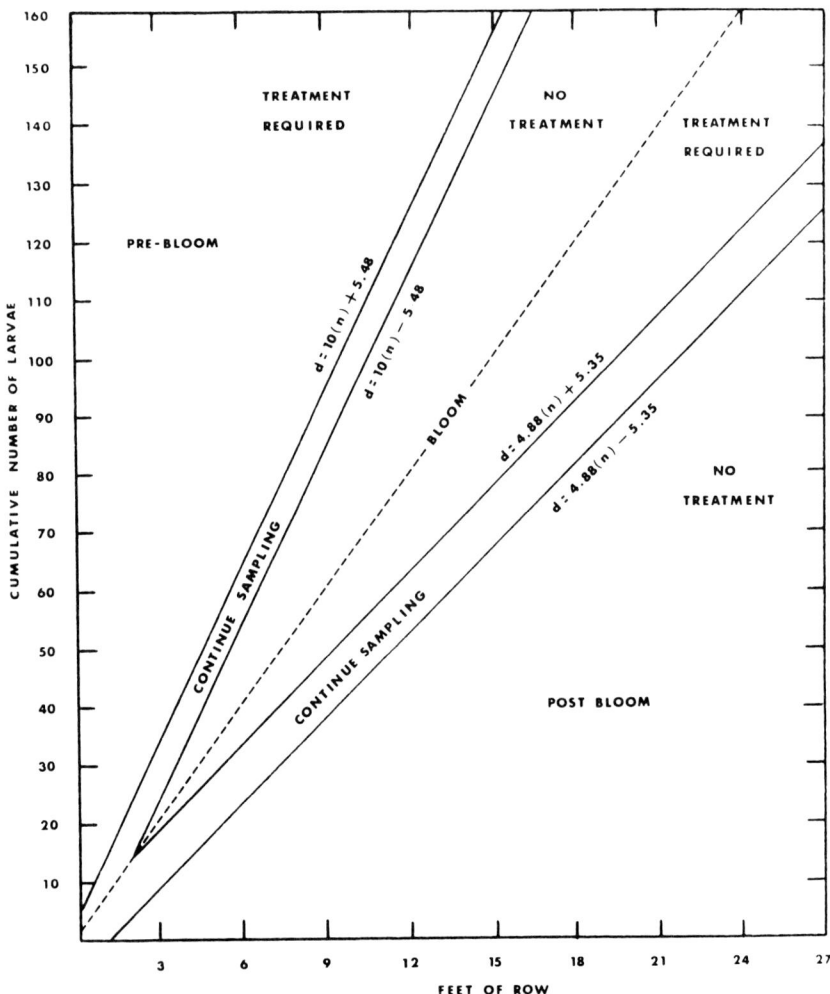

Figure 4-2. Sequential sampling model for use in making prebloom and postbloom treatment decisions for velvetbean caterpillar larvae > 1.27 cm (½ inch). (Reproduced with permission from Strayer et al. 1977).

In order to understand the workings of the sequential sampling plan, operating characteristic curves, which provide information about the probability of a correct decision, may be designed (Fig. 4-3). The relationship between the number of samples required at different population densities is represented by the average sample number (ASN) curve (Fig. 4-4). These curves show how risk of a correct decision and the number of samples required relate to population density. The formula for calculating the operating characteristic curve is,

$$\bar{x} = \frac{\frac{b}{h} \ln \frac{L(\bar{x})}{1 - L(\bar{x})}}{\left(\frac{L(\bar{x})}{1 - L(\bar{x})}\right)^{\frac{1}{h}} - 1} \quad (4\text{-}6)$$

Table 4-2. Sequential sampling table for treatment decisions for velvetbean caterpillars ($\geq$ 1.27 cm in length). (Modified after Strayer et al. 1977)

Sample number	Cumulative ft. (30.5 cm) of row sampled	Cumulative number of larvae	
		Prebloom[1]	
6	18	175	185
7	21	205	215
8	24	235	245
9	27	265	275
10	30	295	305
11	33	325	335
12	36	355	365
13	39	385	395
14	42	415	425
15	45	445	455
		Postbloom[2]	
6	18	83	93
7	21	97	108
8	24	112	123
9	27	126	137
10	30	141	152
11	33	156	166
12	36	170	181
13	39	185	196
14	42	200	210
15	45	214	225

[1] 8-12 larvae per 30.5 cm of row
[2] 4-6 larvae per 30.5 cm of row.

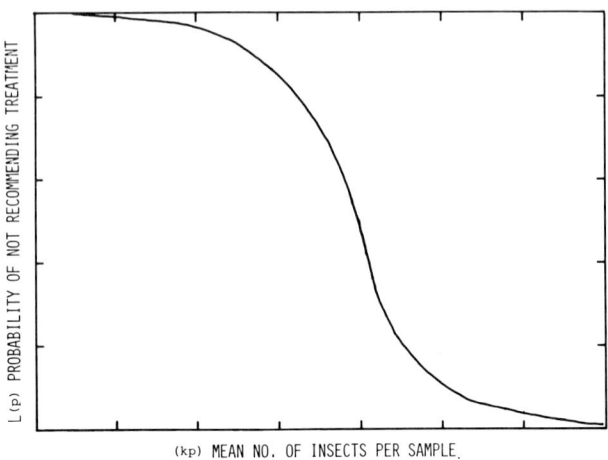

Figure 4-3. Typical operating characteristic curve for soybean insect pests.

where $\bar{x}$ = mean number of insects and $L(\bar{x})$ = the probability of a correct decision.

For the average sample number curve,

$$E_{(n)} = \frac{L(\bar{x})(h_1 + h_2) - h_2}{b - \bar{x}}, \qquad (4\text{-}7)$$

where $E_{(n)}$ = the average number of samples required for various population levels.

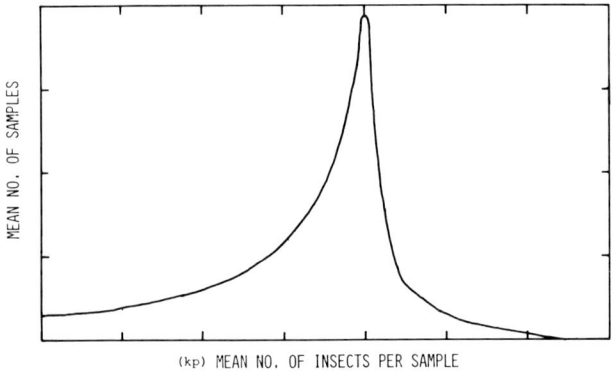

Figure 4-4. Average sample number curve for soybean insect pests.

A word of caution here. The ASN curves are often unrealistic with respect to the minimum number of samples required particularly at very low and very high population levels. In general, for most soybean fields of 15 hectares or less, a minimum of four samples should be taken for species whose spatial pattern follows the Poisson distribution before consulting the sequential table (M. Shepard, M. Sullivan, and S. Turnipseed, unpublished).

The negative binomial distribution may describe a species that exhibits clumping in the field, such as eggs and larvae of the Mexican bean beetle, *Epilachna varivestis* (Mulsant), (Shepard and Carner 1976, see Chapter 9) and the bean leaf beetle, *Cerotoma trifurcata* (Forster), (Kogan et al. 1974, see Chapter 10). This distribution has two parameters, the mean ($\bar{x}$) and the exponent k. The latter parameter measures the degree of aggregation in a population (Waters 1959, see Chapter 3). Several methods are available for calculating k (Southwood 1978, see Chapter 3) but rather than using the iterative method of Bliss and Owen (1958), which is greatly expedited by computer, a good approximation may be obtained by $k = \bar{x}^2/(S^2 - \bar{x})$.

The sequential sampling decision equations described earlier may be used by substituting the formulas (4-8) to (4-12) to develop the plan for insects dispersed following a negative binomial distribution.

$$b = k \frac{\ln \frac{q_2}{q_1}}{\ln \frac{p_2 p_1}{p_1 q_2}} \qquad (4\text{-}8)$$

$$h_1 = \frac{\ln \frac{\beta}{1-\alpha}}{\ln \frac{p_2 q_1}{p_1 q_2}} \qquad (4\text{-}9)$$

$$h_2 = \frac{\ln \frac{1-\beta}{\alpha}}{\ln \frac{p_2 q_1}{p_1 q_2}} \qquad (4\text{-}10)$$

$$p_1 = \frac{\bar{x}_1}{k} \text{ and } q_1 = 1 + p_1 \qquad (4\text{-}11)$$

$$p_2 = \frac{\bar{x}_2}{k} \text{ and } q_2 = 1 + p_2 . \qquad (4\text{-}12)$$

The binomial distribution has not been used as frequently in developing sequential sampling plans for crop insect pests as have the Poisson and negative binomial. However, based on the binomial distribution, Sterling and Pieters (1974, 1975) developed a sequential sampling package for key arthropods of cotton in Texas, and Sterling (1976) formulated sequential decision plans for management of cotton arthropods in Southeast Queensland, Australia. Models constructed from the binomial distribution allowed certain improvements in the sequential scheme because all that was required was to determine whether the sample unit was infested or uninfested. In other words, counts of arthoropods within a sample unit were not necessary, because a polynomial regression equation was found to describe the relationship between mean number of arthropods per plant and the proportion of plants infested (Ingram and Green 1972, Sterling 1976). it is possible that a similar relationship exists between the density of arthropods in soybean and the proportion of ground cloth (or sweep net) samples containing the species in question. However, it is likely that the sampling unit that is selected for use in soybean fields may not be the ground cloth sample. The unit must be sufficiently small so that no more than 5-10 individuals are present per unit (Sterling and Pieters 1978). This is not usually true when pest populations reach damaging levels except perhaps for insects with a very low economic threshold such as stink bugs (see Chapter 23) and *Heliothis* spp. (see Chapter 21). Thus when the ground cloth method is used even at low population densities, one may find over 8-10 individuals while at high densities over 100 individuals of the same species may be present. A few sweep net samples in several locations may suffice in this case.

Equations for the slope and intercept for the sampling plan based on the binomial distribution are:

$$b = \frac{\ln\left[\frac{1-\bar{x}_1}{1-\bar{x}_2}\right]}{\ln\left[\frac{\bar{x}_2}{\bar{x}_1}\right]\left[\frac{1-\bar{x}_1}{1-\bar{x}_2}\right]} \tag{4-13}$$

$$h_1 = \frac{\ln\left[\frac{1-\alpha}{\beta}\right]}{\ln\left[\frac{\bar{x}_2}{\bar{x}_1}\right]\left[\frac{1-\bar{x}_1}{1-\bar{x}_2}\right]} \tag{4-14}$$

$$h_2 = \frac{\ln\left[\frac{1-\beta}{\alpha}\right]}{\ln\left[\frac{\bar{x}_2}{\bar{x}_1}\right]\left[\frac{1-\bar{x}_1}{1-\bar{x}_2}\right]} \tag{4-15}$$

As mentioned before for other distributions, $\bar{x}_1$ is the population level at which treatment is required, and $\bar{x}_2$ is the level at which no treatment is needed.

Determination of damage thresholds is one of the most perplexing yet challenging objectives of research on soybean insect pests (see Chapter 1). Without this information, sequential sampling plans cannot be successfully developed for insect pests because no precise "class limits" can be set. Sometimes treatment thresholds, in terms of numbers of insects per meter-of-row, are available from the literature. These may be used to construct sequential sampling plans that can be tested in the field. When no data are available relative to damage thresholds, experimentation, experience, or empirical observation may be necessary. Because thresholds may vary according to geographical locations, phenology of the crop, etc., no attempt will be made to report them here for various insect pests of soybean. A logical first step in testing suggested damage thresholds that are to be incorporated into sequential plans would be to do so in several experimental test sites over time and compare yield results (see also Chapters in Sections II-VIII).

Because soybean plants vary in their susceptibility to defoliation according to their stage of development (Turnipseed 1972, Hammond and Pedigo 1976, Strayer et al. 1977, see Chapter 1), sequential models must be adjusted accordingly. At the more susceptible stages of development, sequential sampling plans for the velvetbean caterpillar (see Chapter 6) and loopers (see Chapter 7) incorporated a recommendation to return to the field at closer intervals.

As with any sampling procedure, there is risk involved. Deciding on acceptable (predetermined) levels of precision is the third essential bit of information needed for sequential model development. Most published reports on sequential sampling of pest insects whose populations patterns follow the Poisson or negative binomial distributions, set the risk levels (α and β) at 0.10. This means that only 1 out of every 10 samples may be misclassified. This has been found to be an acceptable level considering the more intensive sampling necessary to obtain higher levels of precision. With population patterns following the binomial distribution, treatment level is determined by the percent of the uninfested sampling units. Thus, the values for α and β may be somewhat larger (Sterling 1976).

III. Sequential Sampling with a Fixed Coefficient of Variation of the Mean

Although conventional sequential sampling plans classify populations in broad categories (e.g., low or high; treat or don't treat), they do not allow precise estimation of the population mean. Anscombe (1949) and Kuno (1969) reported a sequential plan that estimates the mean density of a population relative to a fixed coefficient of variation of the mean. This method obviously requires that more sampling be carried out before reaching a decision about the population density; but several advantages make it appealing, particularly if one can accept the fact that precision can be expressed in terms of the coefficient of variation of the mean that can be set a priori. Samples can then be taken sequentially until the specified level of coefficient of variation is reached.

Allen et al. (1972) referred to this approach as a sequential counting plan and used the technique in sampling the bollworm, *Heliothis zea* (Boddie), on cotton. Anscombe (1949) and Kuno (1969) proposed that the sample variance of population was a function of the mean according to the relationship: $s^2 = f(\bar{x})$; and Kuno (1969) defined a linear relationship between "mean crowding" and the mean. Mean crowding was defined as "mean number per individual of other individuals in the same quadrat," and expressed as $\overset{*}{m} = \bar{x} + (s^2/\bar{x} - 1)$ (Lloyd 1967). From this relationship Kuno (1969) suggested that $s^2 = (\alpha + 1)\bar{x} + (\beta + 1)\bar{x}^2$, which as Allen et al. (1972) pointed out, is a least squares fit to the quadratic equation $s^2 = a_1 \bar{x} + a_2 \bar{x}^2$, where a_1 and a_2 are constants.

After determination of these constants for the species in question, the following formula is used to draw "stop lines" for desired levels of coefficients of variation:

$$C = \sqrt{\frac{n \frac{a_1 T_n}{n} + \frac{a_2 T_n^2}{n^2}}{T_n^2}} \tag{4-16}$$

Then T_n (cumulative number of insects) is solved for by:

$$T_n = \frac{a_1}{C^2 - \frac{a_2}{n}} . \tag{4-17}$$

Examples of the sequential counting plan are given by Kuno (1969).

As with the more conventional sequential sampling plan, it is not necessary to take preliminary samples to obtain information for sample size determination. Obviously smaller coefficients of variability will require more samples.

This method has not been thoroughly tested on a large-scale basis and while it may play an important role in developing pest management strategies, more research is needed. Because of its precision, it may be particularly useful as a method for providing data to initialize computer simulation models.

IV. Concluding Remarks

Sequential models designed for specific pests should be developed from a knowledge of pest population dynamics and a phenology of the crop. As indicated by Strayer et al. (1977), when sampling occurred at seven-day intervals, sequential sampling was not able to provide treatment information in time to truncate the peaks of velvetbean caterpillar infestation. In such cases, adjustments in the sampling scheme are needed, such as sampling for smaller larvae or shortening sampling intervals.

Most sequential sampling plans previously reported usually do not consider field size, minimum number of samples, or costs of taking samples. However, Sterling (1976) recommends a minimum of 10 samples and Allen et al. (1972) suggests that not over 20 acres be sampled for one sequential decision for the bollworm in cotton. These and other considerations will have to be approached on a "trial and error" basis using experimental test fields. A general discussion of sample size is provided by Ruesink (see Chapter 3). Although sequential sampling was not employed, reliable treatment decisions were made for soybean insect pests in Brazil using six samples per 10-20 hectares (Kogan et al. 1977).

Other modifications that may refine sequential sampling methods include developing models that incorporate early as well as late instars of pests and combine groups of pests into "ecological guilds." This would greatly simplify sampling schemes that must consider a pest complex that occurs concurrently, with each species contributing to a damage threshold. Finally, further development of sequential plans that include biological control agents is in the realm of sequential sampling possibilities.

In summary, development of sequential sampling plans requires knowledge of (1) the mathematical distribution of the insects or their damage, (2) damage thresholds or treatment levels, and (3) acceptable levels of risk. Once these factors are determined, sequential sampling plans can be developed with ease. Computer programs that fit frequency counts of insects to several discrete distributions greatly facilitate determination of the insects' spatial patterns (Gates and Ethridge 1972). Sequential tables are easily calculated from a computer program (Talerico and Chapman 1970) and when the insect patterns follow a negative binomial distribution, calculations of the parameter k may be carried out using the iterative method (Bliss and Owen 1958) for which computer programs are available (M. Shepard, unpublished). Risk levels (α and β) that usually range from 0.1 to 0.2 may vary according to willingness to accept the probability of an incorrect decision. As expected, decreasing the risk involves taking more samples.

As entomologists gain familiarity with the sequential sampling plan, its use in pest management programs will undoubtedly increase. Sequential plans for management decisions allow classification of populations with a predetermined level of precision. But the main advantage to using this approach is the substantial savings in time required to make reliable decisions about whether or not to apply a control tactic. These reliable sampling plans have resulted in a reduction of 50% of the pesticide load applied to some crops without losses in yields (Casey et al. 1975).

References

Allen, J., D. Gonzalez, and D. V. Gokhale. 1972. Sequential sampling plans for the bollworm, *Heliothis zea.* Environ. Entomol. **1**:771-780.

Anscombe, F. J. 1949. Large-sample theory of sequential estimation. Biometrika **36**:455-458.

Bliss, C. I., and A. R. G. Owen. 1958. Negative binomial distributions with a common k. Biometrika **45**:37-58.

Casey, J. E., R. D. Lacewell, and W. Sterling. 1975. An example of economically feasible opportunities for reducing pesticide use in commercial agriculture. J. Environ. Qual. **4**:60-61.

Gates, C. E., and F. G. Ethridge. 1972. A generalized set of discrete frequency distributions with Fortran program. J. Int. Assoc. Math. Geol. **4**:1-7.

Gonzalez, D. 1971. Sampling as a basis for pest management strategies. Proc. Tall Timbers Conf. Ecol. Anim. Contr. Habitat Manage. **2**:83-101.

Hammond, R. B., and L. P. Pedigo. 1976. Sequential sampling plans for green cloverworm in Iowa soybeans. J. Econ. Entomol. **69**:181-185.

Harcourt, D. G. 1966. Sequential sampling for the imported cabbageworm, *Pieris rapae* (L.). Can. Entomol. **98**:741-746.

Ingram, W. R., and S. M. Green. 1972. Sequential sampling for bollworms on raingrown cotton in Botswana. Cotton Grow. Rev. **49**:265-275.

Kogan, M., W. G. Ruesink, and K. McDowell. 1974. Spatial and temporal distribution patterns of the bean leaf beetle, *Cerotoma trifurcata* (Forster), on soybeans in Illinois. Environ. Entomol. **3**:607-617.

Kogan, M., S. G. Turnipseed, M. Shepard, E. B. de Oliveira, and A. Borgo. 1977. Pilot insect pest management program for soybean in southern Brazil. J. Econ. Entomol. **70**:661-663.

Kuno, E. 1969. A new method of sequential sampling to obtain the population estimates with a fixed level of precision. Res. Popul. Ecol. **11**:127-136.

Lloyd, M. 1967. 'Mean crowding.' J. Anim. Ecol. **36**:1-30.

Morris, R. F. 1954. A sequential sampling technique for spruce budworm egg surveys. Can. J. Zool. **32**:303-313.

Onsager, J. A. 1976. The rationale of sequential sampling, with emphasis on its use in pest management. USDA Tech. Bull. **1526**:19 p.

Pieters, E. P. 1978. Bibliography of sequential sampling plans for insects. Bull. Entomol. Soc. Amer. **24**:372-374.

Pieters, E. P., and W. L. Sterling. 1974. A sequential sampling plan for the cotton fleahopper, *Pseudatomoscelis seriatus.* Environ. Entomol. **3**:102-106.

Pieters, E. P., and W. L. Sterling. 1975. Sequential sampling cotton squares damaged by boll weevils or *Heliothis* spp. in the coastal bend of Texas. J. Econ. Entomol. **68**:543-545.

Sevacherian, V., and V. M. Stern. 1972. Sequential sampling plans for *Lygus* bugs in California cotton fields. Environ. Entomol. 1:704-710.

Shepard, M. 1973. A sequential sampling plan for treatment decisions on the cabbage looper on cabbage. Environ. Entomol. 2:901-903.

Shepard, M., and G. R. Carner. 1976. Distribution of insects in soybean fields. Can. Entomol. 108:761-771.

Southwood, T. R. E. 1978. Ecological methods with particular reference to the study of insect populations. Halsted Press, N.Y. 524 p.

Stark, R. W. 1952. Sequential sampling of the lodgepole needle miner. For. Chron. 28:57-60.

Sterling, W. L. 1975. Sequential sampling of cotton insect populations. Beltwide Cotton Prod. Res. Conf. Proc. pp. 133-136.

Sterling, W. L. 1976. Sequential decision plans for the management of cotton arthropods in south-east Queensland. Aust. J. Ecol. 1:265-274.

Sterling, W. L., and E. P. Pieters. 1974. A sequential sampling package for key cotton arthropods in Texas. Tex. Agr. Exp. Sta. Tech. Rep. 74-32:28 p.

Sterling, W. L., and E. P. Pieters. 1975. Sequential sampling for key arthropods of cotton. Tex. Agr. Exp. Sta. Tech. Rep. 75-24:121 p.

Sterling, W. L., and E. P. Pieters. 1979. Sequential decision sampling. S. Coop. Ser. Bull. 231:85-101.

Strayer, J., M. Shepard, and S. G. Turnipseed. 1977. Sequential sampling for management decisions on the velvetbean caterpillar, *Anticarsia gemmatalis* (Hubner), on soybeans. J. Ga. Entomol. Soc. 12:220-227.

Talerico, R. L., and R. C. Chapman. 1970. SEQUAN. A computer program for sequential analysis. U.S. For. Serv. Res. Note NE 116:6 p.

Turnipseed, S. G. 1972. Response of soybeans to foliage losses in South Carolina. J. Econ. Entomol. 65:224-229.

Waddill, V. H., B. M. Shepard, S. G. Turnipseed, and G. R. Carner. 1974. Sequential sampling plans for *Nabis* spp. and *Georcoris* spp. on soybeans. Environ. Entomol. 3:415-419.

Wald, A. 1943. Sequential analysis of statistical data: theory. Columbia Univ. Stat. Res. Grp. Rep. 75 and Off. Sci. Res. Develop. Rep. 1998.

Wald, A. 1945. Sequential tests of statistical hypothesis. Ann. Math. Stat. 16:117 186.

Waters, W. E. 1955. Sequential sampling in forest insect surveys. For. Sci. 1:68-79.

Waters, W. E. 1959. A quantitative measure of aggregation in insects. J. Econ. Entomol. 52:1180-1184.

Chapter 5

Sequential Estimation of Soybean Arthropod Population Densities[1]

Walter G. Rudd

I. Introduction

A common problem for field research entomologists is the accurate determination of insect population densities. Such determinations are necessary, for example, to test treatment programs; to investigate reproductive, mortality, and dispersal patterns; and to validate model results. This survey estimation problem is to be contrasted with the management sampling problem, in which the objective is to determine whether or not the population of a pest species is above a prescribed treatment threshold.

Field measurements of any kind are expensive and time-consuming. It is, therefore, important to make the most efficient use of available sampling resources. To do so the objective should be to take the smallest sample required, in terms of the number of observations, to satisfy prescribed requirements on the statistical accuracy of the measurement.

Sequential sampling plans for the management sampling problem are well-known (Wald 1945,1947, see Chapter 4). Here the investigator specifies ahead of time a level of confidence to be associated with the decision made based on the sample taken. For example, the researcher might insist on a 90% confidence level on the decision that the population is above a given threshold. Sequential sampl-

---

[1] Research supported in part by the Graduate Research Council of Louisiana State University and by the National Science Foundation and the Environmental Protection Agency, through a grant (NSF GB34718) to the University of California. The findings, opinions, and recommendations expressed herein are those of the author and not necessarily those of the University of California, the National Science Foundation, or the Environmental Protection Agency.

ing plans provide "stopping rules" that tell the researcher when enough observations have been taken. These stopping rules usually are based on the cumulative numbers of individuals captured in the sample. Typical applications are described in Oakland (1950), Waters (1955) and Allen et al. (1972) (see also Chapters 4, 6, 7, 8, 10, and 23).

The survey estimation problem, with the exception of work by Kuno (1969) that provides results valid in the large sample limit, has received almost no attention. The problem is to determine the sample size necessary to provide an estimation of a population mean with prechosen limits on the lengths of error bars relative to the mean at prescribed confidence limits. The sample size depends on the sampling distribution of the population being measured. Only two distributions, the Poisson and the negative binomial, will be considered because they are mathematically tractable and because it appears that one or the other will provide an adequate description of the sampling distribution of most arthropod species associated with soybean (see Chapter 3).

In the following mathematical derivation, equations are developed and confidence intervals are computed at various levels $1 - \alpha$ for the two distributions. The resulting equations are then used to produce the desired sequential estimation plans for the distributions.

II. Sequential Estimation

The probability of finding r individuals in a single observation is described by the discrete probability distribution $p(r; m,s)$ in which m is a parameter for which the sample mean is an efficient estimator and s is a set of parameters with known values. Let $P_N(\hat{K}; m,s)$ be the sampling distribution corresponding to $p(r; m,s)$. $P_N(\hat{K}; m,s)$ is then the probability of finding $\hat{K}$ individuals in a sample consisting of N observations from the same population.

Following Cramer (1946), given $\hat{K}$ individuals found in a sample of size N, compute the $1 - \alpha$ level confidence interval $(m_1(K,N), m_2(K,N,s))$ by solving the equations

$$\sum_{j=0}^{\hat{K}} P_N(j; m_2, s) = \alpha/2 \qquad (5\text{-}1)$$

and

$$\sum_{j=0}^{\hat{K}} P_N(j; m_1, s) = 1 - \alpha/2 \ . \qquad (5\text{-}2)$$

We choose as a measure of precision of the population estimate, the ratio of the length of the confidence interval to the sample mean

$$\rho_\alpha(\hat{K},N,s) = N[m_2(\hat{K},N,s) - m_1(\hat{K},N,s)]/\hat{K} \qquad (5\text{-}3)$$

or, equivalently,

$$\rho_\alpha(\hat{K},N,s) = \frac{K_2(\hat{K},N,s) - K_1(\hat{K},N,s)}{\hat{K}} \qquad (5\text{-}4)$$

where $K_i = Nm_i$.

To design a sequential estimation plan for the distribution p, select an α and choose an acceptable value ρ_0 for ρ. For a given N, solve equation (5-3) or (5-4) for $\hat{K}(N)$. The next integer larger than $\hat{K}$ is the number that must be found to insure that the measured $\rho \leq \rho_0$ for the sample size N.

In use, whenever the number measured becomes greater than $\hat{K}(N)$, sampling may be discontinued with assurance that the $1 - \alpha$ level confidence interval will be shorter than $\rho_0 \hat{K}$. Continued sampling can only produce smaller values of ρ.

III. The Poisson Distribution

The Poisson distribution has the form

$$p(r;m) = m^r e^{-m}/r! \qquad (5\text{-}5)$$

with sampling distribution

$$P_N(\hat{K};m) = (Nm)^{\hat{K}} e^{-Nm}/\hat{K}! = K^{\hat{K}} e^{-K}/\hat{K}! \qquad (5\text{-}6)$$

where $K = Nm$.

Confidence limits for the Poisson distribution for various values of α are shown in Fig. 5-1. These were obtained via an interval-halving solution technique for equations (5-1) and (5-2). To determine confidence limits, draw a horizontal line through the measured value of $\hat{K}$. The confidence limits are the $\bar{K}$ values corresponding to the intersection of this line with the curves for the desired α value.

For the Poisson distribution, $\rho_\alpha(\hat{K},N)$ is independent of N. Table 5-1 shows $\rho_\alpha(\hat{K})$ for various values of α. For $\hat{K}$ larger than 50, the normal approximation to the Poisson distribution (Ricker 1937) provides confidence intervals and ρ values accurate to three or more decimal places. The normal approximation for ρ is

$$\rho_\alpha(\hat{K}) = Z_{\alpha/2} \sqrt{4\hat{K} + Z^2_{\alpha/2}} / \hat{K} \qquad (5\text{-}7)$$

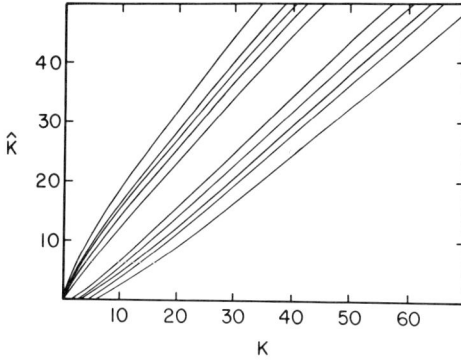

Figure 5-1. Confidence limits for the Poisson distribution. From inside out: $\alpha = 0.4, 0.2, 0.1, 0.05, 0.01$.

where

$$\Phi(Z_{\alpha/2}) = \alpha/2 \tag{5-8}$$

and $\Phi(x)$ is the cumulative normal distribution.

For smaller ρ values than Table 5-1 offers, equation (5-7) can be solved for the stopping values of $\hat{K}$:

$$\hat{K} = [\beta(2 + \sqrt{4 + \rho^2}\,)] \tag{5-9}$$

where $\beta = Z^2_{\alpha/2}/\rho^2$ and the square brackets indicate the next integer larger than the quantity within the brackets.

IV. The Negative Binomial Distribution

The negative binomial distribution function

$$p(r;m,k) = (1 + \frac{m}{k})^{-k} \frac{\Gamma(k+r)}{r!\,\Gamma(k)} \left(\frac{m}{m+k}\right)^r \tag{5-10}$$

depends on the parameters m, for which the sample mean is an efficient estimator (Bliss and Fisher 1953), and k, a measure of the tendency of the individuals to "cluster." It is assumed that k is a known index of the behavior of the species sampled and, hence, does not vary from sample to sample. Bliss and Owen (1958) show that this is often a reasonable assumption.

Anscombe (1950) shows that the sampling distribution is

$$P_N(\hat{K};m,k) = p(\hat{K};Nm,Nk) = p(\hat{K};K,Nk) \tag{5-11}$$

Table 5-1. ρ Values for the Poisson distribution

K	$\alpha = .4$	$\alpha = .2$	$\alpha = .1$	$\alpha = .05$	$\alpha = .01$
1	2.169 9	3.358 6	4.389 8	5.332 4	7.346 1
2	1.372 3	2.110 5	2.739 5	3.303 1	4.468 0
3	1.072 9	1.645 3	2.129 7	2.559 4	3.441 7
4	0.908 0	1.390 2	1.796 5	2.154 7	2.879 7
5	0.800 5	1.224 5	1.580 3	1.893 1	2.522 5
6	0.723 6	1.106 2	1.426 3	1.707 8	2.272 9
7	0.665 2	1.016 5	1.309 8	1.567 0	2.082 1
8	0.618 9	0.945 5	1.217 6	1.456 6	1.932 8
9	0.581 1	0.887 2	1.142 2	1.365 6	1.808 3
10	0.549 4	0.838 6	1.079 2	1.290 0	1.707 5
11	0.522 4	0.797 3	1.025 9	1.225 9	1.622 7
12	0.499 0	0.761 3	0.979 4	1.169 8	1.546 9
13	0.478 4	0.729 9	0.938 7	1.121 6	1.481 7
14	0.460 3	0.702 0	0.902 9	1.078 3	1.425 0
15	0.444 0	0.677 2	0.870 8	1.039 6	1.375 0
16	0.429 3	0.654 8	0.842 0	1.005 1	1.327 3
17	0.416 0	0.634 4	0.815 6	0.973 9	1.285 3
18	0.403 9	0.615 9	0.791 7	0.944 8	1.247 2
19	0.392 7	0.598 8	0.769 7	0.918 7	1.212 5
20	0.382 4	0.583 1	0.749 4	0.894 7	1.181 2
21	0.373 0	0.568 6	0.731 0	0.872 3	1.150 0
22	0.364 1	0.555 1	0.713 5	0.851 4	1.123 9
23	0.355 9	0.542 7	0.697 3	0.831 8	1.096 7
24	0.348 2	0.530 5	0.682 2	0.813 8	1.072 9
25	0.341 0	0.519 8	0.668 0	0.797 0	1.051 0
26	0.334 3	0.509 4	0.654 7	0.781 0	1.029 8
27	0.327 8	0.499 8	0.642 0	0.765 7	1.009 3
28	0.321 8	0.490 4	0.630 1	0.751 8	0.990 2
29	0.316 1	0.481 7	0.618 9	0.738 1	0.973 3
30	0.310 7	0.473 5	0.608 2	0.725 6	0.956 7
31	0.305 5	0.465 5	0.598 2	0.713 5	0.939 5
32	0.300 5	0.458 0	0.588 5	0.701 8	0.925 0
33	0.295 9	0.450 9	0.579 4	0.690 9	0.909 8
34	0.291 5	0.444 1	0.570 5	0.680 3	0.897 1
35	0.287 2	0.437 6	0.562 1	0.670 5	0.882 9
36	0.283 1	0.431 4	0.554 1	0.660 9	0.870 1
37	0.279 1	0.425 3	0.546 5	0.651 5	0.859 5
38	0.275 4	0.419 6	0.539 0	0.642 8	0.847 4
39	0.271 8	0.414 1	0.531 9	0.634 5	0.835 9
40	0.268 3	0.408 8	0.525 2	0.626 3	0.825 0
41	0.265 0	0.403 7	0.518 5	0.618 3	0.814 6
42	0.261 7	0.398 8	0.512 2	0.610 7	0.804 8
43	0.258 6	0.394 0	0.506 1	0.603 5	0.795 3
44	0.255 6	0.389 5	0.500 2	0.596 6	0.786 4
45	0.252 7	0.385 1	0.494 6	0.589 9	0.777 2

46	0.249 9	0.380 8	0.489 0	0.583 2	0.767 4
47	0.247 2	0.376 7	0.483 7	0.576 9	0.759 6
48	0.244 6	0.372 7	0.478 5	0.570 7	0.752 1
49	0.242 1	0.368 8	0.473 6	0.564 8	0.742 9
50	0.239 6	0.365 1	0.468 7	0.559 0	0.736 0

The k dependence of the sampling distribution means that the confidence intervals (K_1, K_2) depend on N, in contrast to the case with Poisson distribution.

Confidence intervals for the negative binomial distribution, k = 1, α = 0.2, are shown in Fig. 5-2. In the calculations leading to these and the subsequent results, the recursion relations were used:

$$P_N(0; m,k) = \left(1 + \frac{m}{k}\right)^{-Nk} \tag{5-12}$$

$$P_N(r + 1; m,k) = \left(\frac{m}{m+k}\right)\left(\frac{Nk + r}{r + 1}\right) P_N(r; m,k) \tag{5-13}$$

For large Nk, the negative binomial sampling distribution approaches the Poisson sampling distribution. Results shown in Fig. 5-2 and others in this study indicate that Nk products on the order of one hundred or more are necessary before the Poisson distribution can be substituted for the negative binomial distribution with reasonable accuracy.

To avoid the necessity of recomputing confidence intervals for different values of k, the fact has been used that if k' = βk then

$$P_N(\hat{K}; K, k') = P_{N/\beta}(\hat{K}; K, k) \tag{5-14}$$

Therefore, all results reported herein are based on calculations for k = 1. Translation to other k values is done simply by adjusting N by a scale factor.

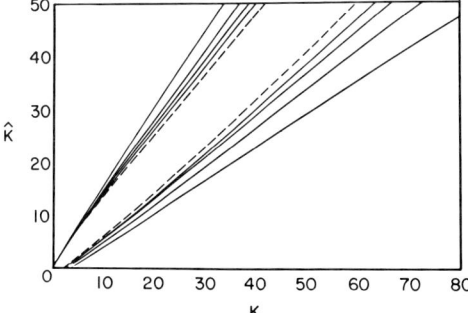

Figure 5-2. Confidence limits for the negative binomial distribution, α = 0.2 (solid line). Confidence limits for the Poisson distribution, α = 0.2 (dashed line).

V. Design and Use of the Sequential Estimation Plans

Use of any sequential plan first requires selection of the appropriate form of the statistical population function. To do this, a large number of measurements of populations of the species under investigation must be obtained, recording counts from individual measurements and keeping the results from each sample separate. For each sample, use a maximum likelihood technique to obtain sample estimates of the parameters for each population distribution function that is being tried. For the Poisson distribution, simply compute the mean ($\bar{x}$); for the negative binomial distribution compute the sample mean ($\bar{x}$) and a value for k. Then compute a predicted frequency distribution for each of the theoretical distributions and use a goodness-of-fit test (comparing r^2 values works better than the χ^2 test) to choose that distribution that shows the best fit with the measured frequency distribution for the sample. Repeat this procedure for all the available samples and select that distribution that appears to perform the best over all the samples. This latter decision is usually clear-cut; it is often found that, for example, one distribution function provides the highest r^2 for 85% of the samples.

If necessary, as is the case of the negative binomial distribution, compute a global or common value of the population parameters that are to remain fixed from sample to sample. For the negative binomial distribution, in which it is assumed k is a parameter that depends only on reproductive characteristics of the species and not on the population, simply carry out a regression analysis to attempt to find a correlation between k and the sample mean. If the population is truly negative binomial, there should be no correlation between k and the sample mean and the result of the regression analysis is a common value for k. The results, using this technique, for those species found to have the negative binomial distribution do not differ significantly from those obtained by using the technique described in Anscombe (1950) and Bliss and Owen (1958).

If the data are not available to compare distribution functions as described, one might temporarily choose between the Poisson and negative binomial distributions by considering the reproductive characteristics of the species. If eggs are laid singly at random by a given species, a first guess should be that the distribution is Poisson. If the eggs are grouped together, either because they are laid in clusters or because egg-laying individuals congregate, chances are the dispersion of progeny will follow the negative binomial distribution. In the latter case, some preliminary estimates of k can be obtained from sample means and variances as described in Anscombe (1950) and Bliss and Fisher (1953) and a common k can be determined as discussed above.

To design a sequential estimation plan, select an appropriate α (.1 corresponds to a 90% confidence level, for example) and a value for ρ_0, the desired ratio of the length of the confidence interval to the sample mean. As is always the case, the choice of values for these parameters is highly subjective and depends on qualitative criteria determined by research objectives, resources available, and the species measured. These values are determined by the answer to the question "How accurately do we need to know the population mean?"

For the Poisson distribution, once values have been determined for ρ_0 and α, refer to Table 5-1 to determine the stopping rule. Starting from the top of the table, search down the column headed by the chosen α value until the first ρ smaller than ρ_0 is found. The $\hat{K}$ value for that ρ is the smallest cumulative number of individuals captured that will yield a ρ smaller than ρ_0. In other words, the stopping rule is to continue sampling until at least $\hat{K}$ individuals have been caught.

For example, if $\rho_0 = 1$ and $\alpha = .1$ are chosen, continue sampling until at least 44 individuals have been counted.

Use equation (5-9) to determine $\hat{K}$ for values of ρ_0 and for α not available in Table 5-1.

For the negative binomial distribution, select from Figs. 5-3 to 5-5 the plot corresponding to the chosen value of α. Draw a horizontal line through the desired value of ρ and determine $\hat{K}$ for each N by dropping perpendiculars from the intersections of the horizontal line with the curves for the various values of N. The result is a table of cumulative numbers counted that are needed in order to be able to stop at a given sample size N.

Figures 5-3 to 5-5 were computed for k = 1. For other values of k, substitute N/k for each value of N.

By plotting $\hat{K}$ versus N a graph is obtained like one of the curves in Fig. 5-6, in which plans are shown for $\alpha = 0.1$, $\rho = 1$ for k = 0.5, 1 and 2.

Simply keep a count of the cumulative number of individuals $\hat{K}$ caught. When $\hat{K}$ first becomes larger than the required number for a given N, stop. Graphically, this involves plotting $\hat{K}$ versus N continuously until a point is plotted that lies above the curve for the plan.

VI. Application to Soybean Insect Studies

A computer study based on over 200 samples taken in the years 1973-75 (W. Rudd, R. Jensen, D. Newsom and D. Herzog, unpublished), using the techniques described in Section V, showed the following:

1. soybean looper, *Pseudoplusia includens* (Walker), is dispersed according to the Poisson distribution;
2. velvetbean caterpillar, *Anticarsia gemmatalis* (Hübner), is dispersed according to the Poisson distribution;
3. Southern green stinkingbug, *Nezara viridula* (L.), is dispersed according to the negative binomial distribution. In particular, "combined" populations— third instar through adult stage—have a k value of 1.023.

For several years, researchers at Louisiana State University have been using management sampling plans based on these findings. Also in use is a sequential

Figure 5-3

Figure 5-4

Figure 5-5

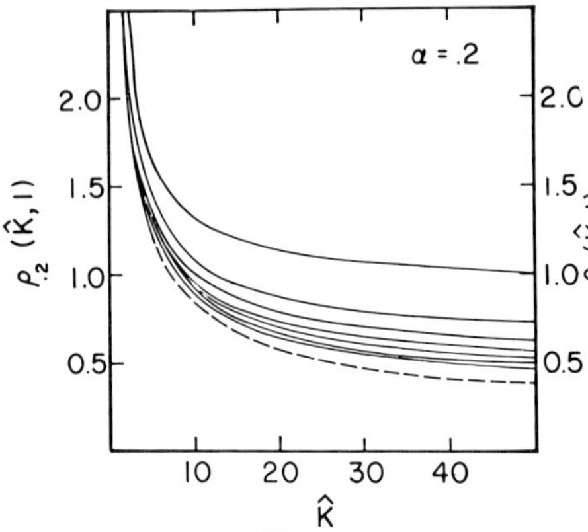

Figure 5-3. Negative binomial index of precision ρ for $\alpha = 0.2$. From top to bottom: N = 10, 20, 30, 40, 50, 70, 90 (solid line). Poisson index of precision, $\alpha = 0.2$ (dashed line).

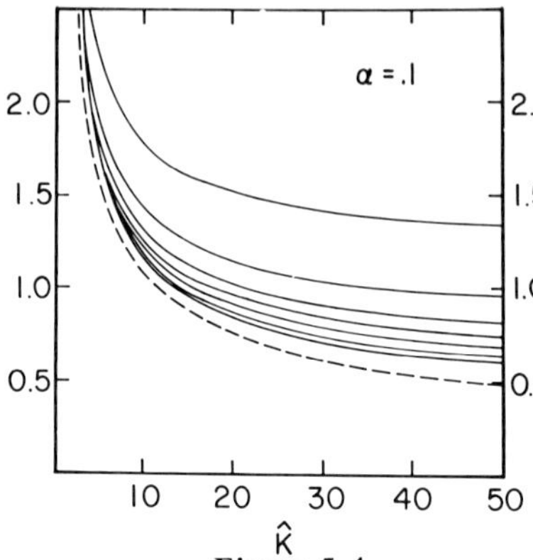

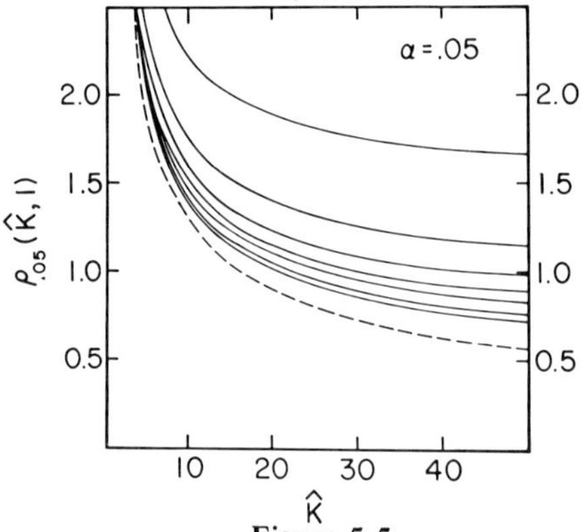

Figure 5-4. Negative binomial index of precision ρ for $\alpha = 0.1$. From top to bottom: N = 10, 20, 30, 40, 50, 70, 90 (solid line). Poisson index of precision, $\alpha = 0.1$ (dashed line).

Figure 5-5. Negative binomial index of precision ρ for $\alpha = 0.05$. From top to bottom: N = 10, 20, 30, 40, 50, 70, 90.

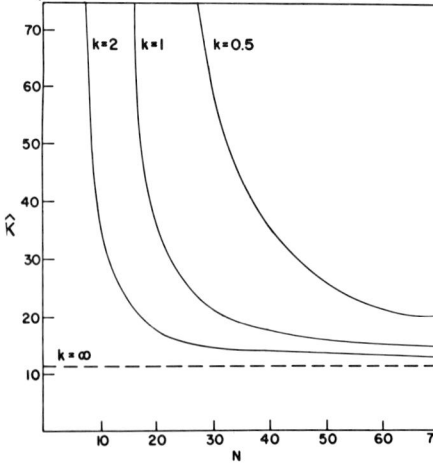

Figure 5-6. Sequential estimation plan for $\rho_{.1}(K,1) = 1$ for the negative binomial distribution. $k = \infty$ is the Poisson limit.

estimation plan for *N. viridula* combined populations. Researchers in Florida (D. Herzog, personal communication) have been using sequential estimation plans based on these findings for *N. viridula* and *A. gemmatalis*.

The plan has been tested using *N. viridula* field data, and it has been determined that the plan works. Using $\alpha = 0.1$ and $\rho_0 = 1$, measured ρ values were less than 1 in approximately 93% of the samples. This percentage value exceeds 90% because in practice it is difficult to stop at exactly the right point when surveyors in a sampling team are widely dispersed over a soybean field. It pays to designate one person to keep a count of cumulative individuals caught and to signal the others to stop when the plan says to stop.

VII. Concluding Remarks

When this sequential estimation plan was first used it was thought that some sampling time would be saved because previous sampling programs were based on a fixed sample size. However, the plan dictated that much larger samples be taken early in the season and much smaller samples late in the season than had been taken previously. The result was, while the total sampling effort remained the same over the season, fewer fields were sampled early and more fields late. Through the use of the plan, it seems possible to solve with confidence the problem of the number of sample units in a sampling program for research or pest management purposes.

Acknowledgments

I wish to thank Drs. Richard P. O'Neill and William E. Waters for their helpful suggestions concerning this work.

References

Allen, J., D. Gonzalez, and D. V. Gokhale. 1972. Sequential sampling plans for the bollworm, *Heliothis zea*. Environ. Entomol. 1:771-780.

Anscombe, F. J. 1950. Sampling theory of the negative binomial logarithmic series distributions. Biometrika 37:358-381.

Bliss, C. I., and R. A. Fisher. 1953. Fitting the negative binomial distribution to biological data. Biometrics 9:176-200.

Bliss, C. I., and A. R. G. Owen. 1958. Negative binomial distribution with a common k. Biometrika 45:37-58.

Cramer, H. 1946. Mathematical methods of statistics. Princeton, N.J. p. 509-513.

Kuno, E. 1969. A new method of sequential sampling to obtain the population estimates with a fixed level of precision. Res. Popul. Ecol. 11:127-136.

Oakland, G. B. 1950. An application of sequential analysis to whitefish sampling. Biometrics **6**:59-67.

Ricker, W. E. 1937. The concept of confidence or fiducial limits applied to the Poisson frequency distribution. J. Amer. Stat. Assoc. **32**:349-356.

Wald, A. 1945. Sequential tests of statistical hypotheses. Annu. Math. Stat. **16**:117-186.

Wald, A. 1947. Sequential analysis. John Wiley & Sons, Inc., N.Y. 212 p.

Waters, W. E. 1955. Sequential sampling in forest insect surveys. For. Sci. **1**:68-79.

SECTION II
Lepidopterous Defoliators

Chapter 6

Sampling Velvetbean Caterpillar on Soybean

Donald C. Herzog and James W. Todd

I. Introduction

The velvetbean caterpillar, *Anticarsia gemmatalis* (Hübner) (Lepidoptera:Noctuidae), is a key defoliator of soybean in the Gulf Coast area of the United States (Nickels 1926, Douglas 1930, Hinds and Osterberger 1931, Greene 1976, Turnipseed and Kogan 1976) through Mexico (Pacheco 1970), Surinam (Van Dinther 1956), Colombia (Posada et al. 1970), Venezuela (Guagliumi 1966), Brazil (Ferreira 1970), and southward into Argentina (Rizzo 1972) (Fig. 6-1). This species is known by the common name "gusano terciopelo" in Mexico (Pena and Sifuentes 1972), and in Brazil it is referred to as "lagarta da soja" (Panizzi et al. 1977).

Although it is widely distributed in the Western Hemisphere, economic infestations rarely occur in the United States north of South Carolina, Georgia, and the states bordering the Gulf of Mexico. Adults of the species, however, have been collected as far north as Wisconsin (Smith 1893) and Ontario, Canada (Watson 1916a).

A. Host Plants and Nature of Injury

The velvetbean caterpillar is a defoliating species that habitually feeds as a larva in the upper one-half to one-third of the soybean leaf canopy. The young caterpillar feeds first on the shell of the egg from which it has just emerged, then begins to strip the epidemis and mesophyll from the lower surface of the tender upper leaves and shoot growth. This feeding process continues until near the end

Figure 6-1. Distribution of map of *Anticarsia gemmatalis*. Boundaries marked with a dashed line indicate uncertainty of records.

of the second instar when the larva begins to skeletonize the leaf by eating all the soft material but leaving the veins intact. After the second instar, it consumes the whole leaf except the larger veins and midrib (Watson 1915). Douglas (1930) described the type of injury by velvetbean caterpillar to soybean as follows:

> They begin feeding on the tender leaves near the top of the plant. After the top leaves have been destroyed, the older leaves near the bottom of the plant are devoured. Still further, when all the leaves have been eaten, the tender parts of the stems are eaten away. When a large infestation occurs, the larvae hollow out the ends of the stems, eat the buds from the stalks, chew off the branches and small bean pods completely, and sometimes gnaw the outside wall from the main stalk. The plants are, of course, ultimately killed.

Watson (1916a) found that the caterpillars feed continuously on velvetbean, stopping only to molt, and that fourth, fifth, and sixth instar larvae consume vegetation equal to their own weight in 15 to 16 hours. In more recent studies, Boldt et al. (1975) reported that the velvetbean caterpillar consumed 84 ± 4.1

cm² of soybean leaf area at 30°C during its larval development. Based on foliage consumption reported by Reid (1975) and larval development reported by Watson (1915), the total foliage consumption is calculated to be ca. 121.7 cm². These data are summarized in Table 6-1. Numerous fields in Georgia, Florida, Alabama, and Louisiana were observed that had been completely stripped of foliage within 5 to 7 days after populations of late instars reached levels of 80 to 100/row-m. The velvetbean caterpillar is considered the most important soybean defoliator in the western hemisphere (Turnipseed and Kogan 1976).

The velvetbean caterpillar feeds primarily on the Leguminosae (Ford et al. 1975). At least 34 species of the Leguminosae have been reported to serve as hosts for the larvae. Besides the legumes, only three other families—Begoniaceae, Gramineae, and Malvaceae—with a total of five species, have been recorded as hosts (Table 6-2). Nickels (1926) first reported *A. gemmatalis* damage to soybean in the southern counties of South Carolina in the late summer and autumn of 1925. He also noted that in the same year, this insect has been reported as a soybean pest in Florida. Although first reported as an economic pest of velvetbean (Chittenden 1905), both Hinds (1930) and Ellisor et al. (1938) found that the insect showed a preference for soybean when soybean and velvetbean were grown in adjacent rows. Ellisor (1942) reported that following maturation of soybean in late season, the larvae preferred alfalfa to soybean, and that when shortages of both of these species occurred, larvae would defoliate cowpea and peanut. The aforementioned species appear to be among the most highly preferred hosts of the velvetbean caterpillar, and sampling effort expended in these crops will probably detect the earliest infestation of larvae in an area.

B. Life Cycle and Phenology

The life history of the velvetbean caterpillar was initially described from observations of individuals infesting velvetbean at Gainesville, Florida (Watson 1916a,b), and subsequently on soybean in Louisiana (Douglas 1930, Ellisor 1942). During the summer months in Florida the preoviposition period averaged

Table 6-1. Foliage consumption by velvetbean caterpillar larvae (after Watson and Reid 1975)

Stage	Foliage Consumption (cm²/day)	Duration (days)	Total Consumption (cm²)
2nd instar	0.3	3.6	1.1
3rd instar	1.5	3.5	5.3
4th instar	3.9	3.7	14.4
5th instar	8.1	3.5	28.4
6th instar	14.4	5.0	72.0
Total for all Development Stages			121.2

Table 6-2. Cultivated and wild plants recorded as hosting larvae of the velvetbean caterpillar

Family	Scientific Name	Common Name	Reference
Leguminosae	Aeschynomenes sp.	joint vetch	DPI[a]
	Agati grandiflora	Australian corkwood tree	Wolcott (1936)
	Arachis hypogaea	peanut	Anonymous (1928)
	Cajanus cajan	pigeon pea	McCord (1974)
	C. indicus	pigeon pea	DPI[a]
	Canavalia gladiata	sword bean	Wolcott (1936)
	C. maritima		Buschman et al. (1977)
	C. rosea		Tietz (1972)
	C. sp.	"horse bean"	Watson (1916a)
	Cassia fasciculata	partridge pea	Herzog (unpublished)
	C. obtusifolia	coffeeweed	Buschman et al. (1977)
	Desmodium floridanum	Florida beggarweed	Tietz (1972)
	Dolichos lablab	hyacinth bean	Buschman et al. (1977)
	Galactia speciformis	milk pea	Buschman et al. (1977)
	Glycine max	soybean	Nickels (1926)
	Indigofera hirsuta	hairy indigo	Tietz (1972)
	Lespedeza sp.		USDA (1954a)
	Medicago sativa	alfalfa	Ellisor and Graham (1937)
	Pachyrhizus erosus	yam bean	Buschman et al. (1977)
	Phaseolus lathryoides	wild bean	Wolcott (1936)
	P. limensis	lima bean	Tietz (1952)
	P. semierectus		Tietz (1972)
	P. speciosus	sweet pea vine	Buschman et al. (1977)
	P. vulgaris	garden bean	Wolcott (1936)
	Pisum sativum	English (garden) pea	DPI[a]
	P. sp.	cabbage (field) pea	DPI[a]
	Pueraria lobata	kudzu	Watson (1915)

	Scientific name	Common name	Reference
	P. phaseoloides	tropical kudzu	Telford and Childers (1947)
	Rhyncosia minima	snout bean	Buschman et al. (1977)
	Robinia pseudoacacia	black locust	Hinds and Osterberger (1931)
	Sesbania emerus	long pod	DPI[a]
	S. exaltata	hemp sesbania	Tietz (1972)
	S. macrocarpa	coffee weed	Hinds and Osterberger (1931)
	Stizolobium deeringianum	Florida velvetbean	Chittenden (1905)
	Tephrosia (sp.)		USDA (1954b)
	Vigna luteola	deer pea	Buschman et al. (1977)
	V. repens	cowpea	DPI[a]
	V. sinensis	cowpea	Cotton (1918)
	[b]	wooly pyrol	Ballou (1912)
Begoniaceae	*Begonia* sp.	begonia	DPI[a]
Gramineae	*Oryza sativa*	rice	Rosetto et al. (1971)
	Triticum sp.	wheat	Wille (1940)
Malvaceae	*Gossypium hirsutum*	cotton	Douglas (1930)
	Hibiscus esculentus	okra	Todd (unpublished)

[a] Host records on file Florida Department of Agriculture and Consumer Services, Division of Plant Industry (DPI), Gainesville, Fla.
[b] No scientific name available.

3 days (Watson 1915,1916a). Eggs are laid singly on the undersides of leaves and leaf petioles. They are usually placed close to the leaf surface between the plant hairs and adhered tightly to the plant (Greene et al. 1973). Eggs are hemispherically shaped, 1-1.5 mm in diameter, and sculptured with a series of ridges converging at a very prominent micropyle (Ellisor 1942); they have been variously described as white to pale green when first laid, changing to delicate pink, cryptic green, orange, or reddish-brown before hatching (Watson 1916b, Douglas 1930, Hinds 1930, Ellisor 1942, Greene et al. 1973).

The average duration of each of the first five larval stages is listed in Table 6-1. The duration of the sixth larval stage ranges from 5 to 20 days and the average duration of the pupal stage was ca. 7 days (Watson 1915,1916a). Thus, the development time from egg to adult ranged from 31 to 46 days. The optimum temperature for larval development in the laboratory was $26.7°C$ (Leppla et al. 1977). The incubation period for *A. gemmatalis* eggs in Louisiana during the months of July, August, and September was usually 3 days, and the duration of the larval stages inclusive was ca. 3 weeks (Ellisor 1942). Similar results have been reported from Florida (Reid 1975).

The feeding, mating, and ovipositional behavior of the adult were observed in the field by Watson (1916a) and Greene et al. (1973). Daytime flight was nondirectional, rapid, and for only short distances when the moths were disturbed. After sunset, flight movement became slower but more controlled and directional. Moths have been observed feeding or drinking from dew droplets on bahiagrass seed heads. Oviposition appeared to be related to temperature decrease and humidity increase, although the egg-laying activity diminishes with dew accumulation on plants (Greene et al. 1973). Leppla (1976) further described the circadian periodicity of locomotor activity, feeding, mating, and oviposition for *A. gemmatalis* observed in the laboratory confirming earlier observations in the field. Females mated one to six times but no males mated more than twice. A mean of 402 eggs was deposited during the lifetime of a mated female. Oviposition peaked on the fifth day and the reproductive life of the species was 12 to 14 days (Leppla 1976).

The dorsal wing coloration of the moth (Fig. 6-2A) is highly variable, but the ventral wing surface is less variable, usually a cinnamon brown with a submarginal row of light spots with a medium dark line. A diagonal black line across the wings is usually present on the dorsal surface. Wing coloration of laboratory-reared moths ranged from light brown in females to near black in males (Leppla et al. 1977). Melanism was found to be intensified and wing scale patterns more pronounced as developmental temperature decreased. Color of field-collected moths is highly variable, and pigmentation is not a particularly dependable morphological character for either species or for sexual differentiation. However, Greene (1974) described a sexual dimorphism of leg scales that facilitates rapid and accurate sexual identification. Tufts of long scales are present on femora of prothoracic and tibiae of methathoracic legs of males but absent on female legs.

The larvae (Fig. 6-2B), like the adults, are highly variable in color and mark-

Figure 6-2. *Anticarsia gemmatalis:* (A) Adult moth; (B) Larva. (Courtesy of Merle Shepard).

ings, particularly after the second instar. Larvae generally exhibit dark-colored longitudinal lines bordered by lighter colored narrower lines on a background of light to dark green. On some the longitudinal lines are faint or even absent. These larvae are usually pale or light yellowish-green, though some may be light brown. The pale forms may show the markings of the dark forms in later instars. All gradations of color and markings occur and nothing has been recorded to explain this phenomenon (Watson 1915).

Apparently, the velvetbean caterpillar overwinters in the tropical areas of southern Florida, through the Caribbean Islands, Central America, and most of northern and central South America.

It has been suggested that the inability of *A. gemmatalis* to overwinter in northern Florida is due to the absence of host plants during the winter (Anonymous 1927). However, it has also been reported that *A. gemmatalis* larvae appear to be sensitive to direct exposure to cold temperatures since all larvae were apparently killed by a $-3°C$ freeze in November 1974 in north central Florida (Buschman et al. 1977). It has been proposed that *A. gemmatalis* could survive the winter as pupae with only a slight increase in the pupal period since pupae survived when placed on the ground overnight during a $-6°C$ freeze in north central Florida (Watson 1916a). It is unlikely that the insect survives the cold winters of South Carolina or Louisiana because they usually pupate less than 2.5 cm below the soil surface and succumb easily to severe weather (Nickels 1926, Douglas 1930, Hinds 1930, Ellisor 1942). So, it appears that temperature limits its ability to overwinter successfully north of the proposed $28°$ latitude northern overwintering boundary, and the southern latitudinal limit is not known. Availability of suitable hosts in more northern latitudes during the winter does not seem to be the limiting factor. *Vigna luteola* (Jacq.), *Pueraria lobata* (Willd.), and alfalfa remain green and survive in good condition in areas north of central Florida, yet the velvetbean caterpillar does not overwinter there even though these are favorable hosts (Buschman et al. 1977, Newsom et al. 1980).

It is from these tropical overwintering reservoirs that yearly seasonal migrations occur into the southern soybean producing areas of the United States and the northern areas of South America during summer (Anonymous 1927, Watson 1916b, 1932, Buschman et al. 1977). Currently, it is not known whether the velvetbean caterpillar migrates across the Gulf of Mexico or through Florida and along the Texas and Louisiana Gulf Coasts. However, it has been postulated that immigrating moths from Central and South America fly directly across the Gulf of Mexico and invade soybean producing areas along the Louisiana and Texas Gulf Coasts (Newsom et al. 1980). This suggests that the overwintering population in Florida contributes little, if any, to populations of the species in other Gulf Coast states.

A generalized scheme for a typical seasonal migration of velvetbean caterpillar is given in Fig. 6-3. Northward migration proceeds slowly during March, April, and May then accelerates in June, July, and August as soybean, peanut, and

Figure 6-3. Generalized schema showing first occurrences of velvetbean caterpillar larvae on soybean and/or wild hosts and in parentheses the average date of occurrence of damaging populations (adapted from Watson 1916a and Greene 1976).

other suitable hosts become available. The first occurrence of velvetbean caterpillar in the spring is dependent on the availability of suitable cultivated and wild hosts. Moths are first captured in central Florida between the last of June and the middle of August (Watson 1916a, Anonymous 1927), and in south Louisiana moths were present in light trap collections from July to December (Chapin and Callahan 1967).

Most authors report three to four generations per year in the southeastern United States with the third generation considered the most destructive (Douglas 1930, Hinds and Osterberger 1931, Ellisor 1942, Dugas and Gray 1944, Greene 1972, Wuensche 1976). Watson (1916a), however, reported that generations were indistinguishable in Florida as velvetbean caterpillars in all stages of development could be found at any time after August. A computer simulation model of velvetbean caterpillar population dynamics on Florida soybean (Menke 1973) has been validated under field conditions (Menke and Greene 1976). Laboratory research has been facilitated by the development of an artificial medium and mass rearing procedure (Greene et al. 1976).

Synchronization of host plant and pest phenologies is an important consideration in the characterization of pest populations. This is particularly important for an oligophagous species such as the velvetbean caterpillar since in addition to having a relatively narrow host range, the species is restricted somewhat in its temporal distribution due to host plant phenology and condition. Ovipositing moths must not only find the proper plant species, but also those in the proper stage of phenological development (Moscardi 1979). Figure 6-4 schematically shows the phenological relationship of velvetbean caterpillar and some of its hosts. The wild host plants thought to contribute most significantly to the overwintering survival of velvetbean caterpillar populations in south Florida include *V. luteola, Phaseolus lathyroides* (L.), *Dolichos lablab* (L.) (hyacinth bean), and *P. lobata* (kudzu) (Buschman et al. 1977).

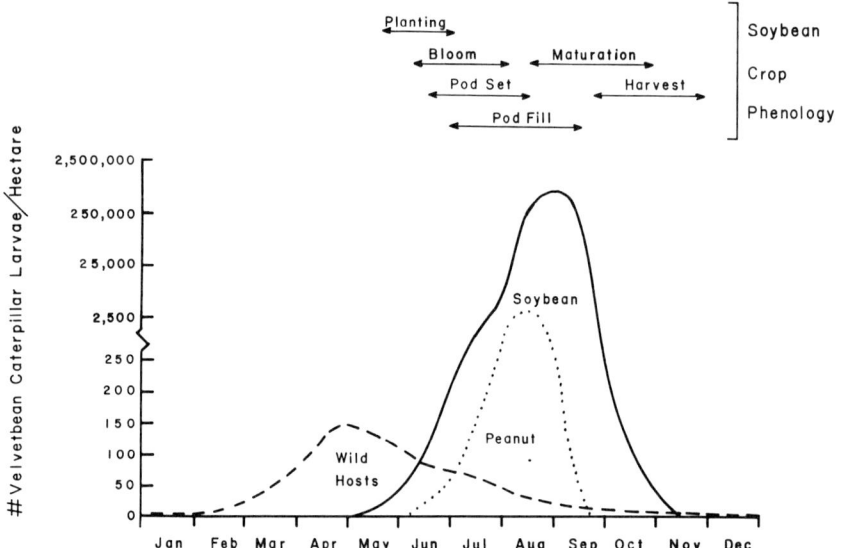

Figure 6-4. Seasonal host sequence and relative abundance of velvetbean caterpillar in the southeastern United States (adapted from Greene 1976).

II. Sampling Adult Populations

No absolute method has yet been developed for the measurement of velvetbean caterpillar adult populations. However, a method under investigation for use with green cloverworm (see Chapter 8) that may prove useful is the "flushing" of moths from the crop foliage by one or more persons while walking through a delimited area.

Relative sampling methods for velvetbean caterpillar adults are usually based on trapping. Although some have observed that velvetbean caterpillar moths are not sampled effectively with blacklight traps, (E. R. Mitchell, personal communication), blacklight has been found to be a useful tool for surveying moth populations. Table 6-3 gives the mean nightly catch of velvetbean caterpillar adults in a walk-in blacklight trap in Gadsen Co., Florida in 1973-75 and 1977-78 (D. C. Herzog and G. L. Greene, unpublished). Chapin and Callahan (1967) also reported catches of velvetbean caterpillar in blacklight traps from July through December in 1957-60.

Table 6-3. Mean nightly catch of velvetbean caterpillar adults in a walk-in blacklight trap (Gadsen Co., FL)

Week Ending	Mean # adults/night				
	1973	1974	1975	1977	1978
July 2	–	–	–	0.0	0.0
9	–	–	–	0.0	0.0
16	–	–	–	0.3	0.0
23	–	0.3	0.3	0.0	0.0
30	–	0.0	17.1	0.0	0.0
Aug. 6	–	0.1	21.7	3.4	0.0
13	10.7	3.7	34.4	3.1	0.3
20	24.3	13.1	28.1	3.1	2.3
27	46.1	67.7	60.1	14.0	15.7
Sep. 3	333.9	8.7	73.0	66.4	33.7
10	170.4	145.3	37.9	33.9	110.9
17	14.9	116.6	10.0	95.8	193.3
24	25.4	32.0	8.1	68.9	158.3
Oct. 1	145.7	8.6	10.0	40.3	142.1
8	27.0	0.0	13.3	58.0	100.4
15	8.4	0.0	10.3	44.0	17.6
22	5.9	0.0	–	0.1	7.4
29	1.1	0.1	–	5.6	8.0
Nov. 5	0.9	0.1	–	6.4	–
12	1.1	–	–	0.0	–
19	0.6	–	–	0.0	–
26	2.4	–	–	0.0	–
Dec. 3	0.4	–	–	0.0	–

The Entomological Society of America has recommended standards for blacklight traps used in general insect surveys (Harding et al. 1966). These standards should be met when adopting blacklight trapping as a survey tool to insure that catches from the same area or even different areas are comparable. In addition, when operating several light traps simultaneously, care should be taken that each is located outside the visible range of the others (Myers and Pedigo 1977). Further, traps should be oriented so that all have equal exposure to the crop, preferably 180° or more.

When utilizing blacklight as a survey tool for velvetbean caterpillar, captured moths must be properly identified. Adults of the genus *Mocis* and of *Pseudaletia unipunctata* (Haworth) are superficially similar to those of the velvetbean caterpillar. The most convenient field character for distinguishing these species is the appearance of the wings (Fig. 6-5).

The presence of a sex pheromone has been demonstrated in female velvetbean

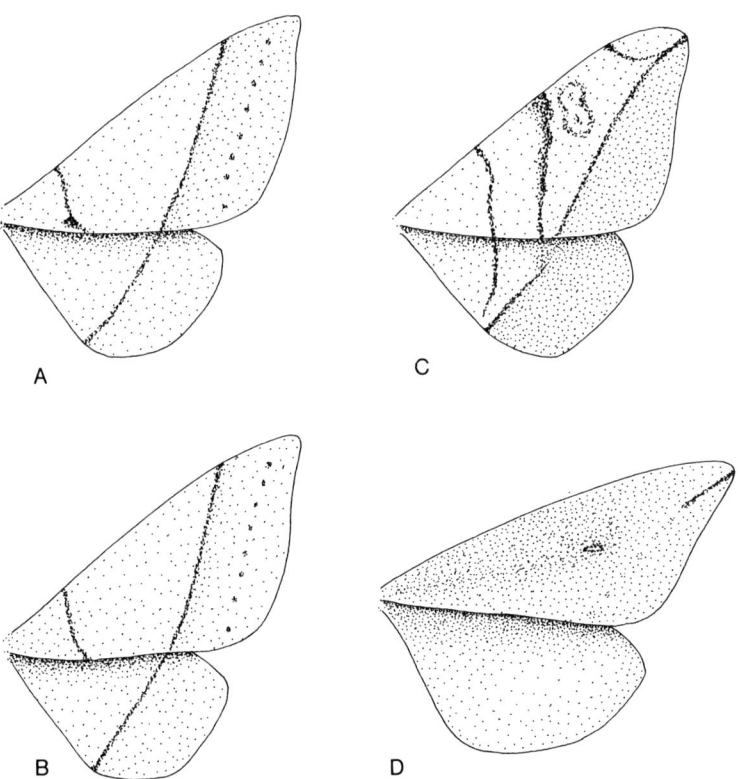

Figure 6-5. Generalized representation of field characters allowing separation of: (A) - (B) *Mocis* spp., (C) *A. gemmatalis*, (D) *Pseudaletia unipunctata*.

caterpillar moths (Johnson 1977). While the attractive component(s) has not yet been identified and so is not available in synthetic form, caged virgin females may be utilized effectively as a pheromone source. The electric grid trap of Mitchell et al. (1972) was the best of the traps tested though it was found to be less than 100% efficient in capturing males attracted to virgin females confined within the traps. Pheromone augmentation of blacklight traps was not tested, but since blacklight alone provides significant catch it is probable that the incorporation of pheromone would provide an additive effect similar to that achieved with the soybean looper (see Chapter 7) and that this combination would furnish a most efficient tool for velvetbean caterpillar adult survey.

Several factors elucidated by Johnson (1977) allow the most efficient use of female-baited electric grid traps. Virgin females capture more males than mated females, but there is only slight difference favoring wild over laboratory-reared virgin females. There is no apparent advantage of three over one virgin female per cage. Virgin females capture significant numbers of males through seven days of age with peak catch occurring at two and three days. There is no great difference in numbers of males captured at hourly intervals through the night. Significantly larger catch (2X) is achieved with traps located 15 m within the field perimeter than with traps located 15 m outside the field. It is probable that traps located in the field margin would yield catch comparable to those located within the field.

Potential uses of pheromone trapping programs for velvetbean caterpillar have been enumerated by Johnson (1977) as a means of: defining migration routes, detecting the time of arrival of immigrant moths, monitoring populations for timing of control measures, and as a means of control, either before or after populations have moved to the soybean producing area. An additional use is as a signal for the initiation of an intensive pest management sampling program.

When undertaking a pheromone trapping program it would be wise to consult the recommendations of Roelofs and Cardé (1977) concerning experimental design of field trapping research as relates to lepidopteran sex pheromone response.

An additional survey sampling method not found reported in the literature as having been used for velvetbean caterpillar adults, but which has been used by Herzog (unpublished) on several occasions is the "sugarline." This method, which was developed for monitoring populations of *Heliothis* spp. on cotton (Lincoln et al. 1967), involves spraying a concentrated sugar solution on a portion of the crop to serve as an attractant and flight arrestant for moths feeding in the field being sampled. Sugar sprays (187 g/1 of water is recommended for *Heliothis*) are applied in the late afternoon and after dark. Moths observed with the aid of a flashlight while walking along the sugarline are collected by hand or with a net. For *Heliothis* spp. this survey is made between deep dusk and midnight. Feeding of the velvetbean caterpillar continues through all hours of the night, but peak activity is from sundown to midnight (Greene et al. 1973). For velvetbean caterpillar, this time period would be the most advantageous for "sugarline" surveys.

The sugarline may draw moths from some distance. The method was more efficient for sampling populations of migrating *Heliothis* moths than it was for sampling those actively feeding and ovipositing in the field. It may be more advantageous under some circumstances to survey a natural nectar line such as a nectar-laden weed bank adjacent to a soybean field. *Cassia obtusifolia* (L.) (sicklepod) would be an excellent nectar source (Herzog, unpublished). Methods and time period for survey by this alternative method are identical to those used with the sugarline.

A modification of the above method involves crushing grapes on top of a screen and observing moth feeding (Greene et al. 1973). The moths attracted are primarily female. This technique could also be adapted as a survey tool.

The "sugarline" and its variations, while useful in establishing the presence of moths in a field and in the establishment of periods of nocturnal activity for the species may be used as a survey tool to estimate relative moth abundance.

III. Sampling for Eggs

The visual survey of soybean plants for velvetbean caterpillar eggs has been considered impractical due to difficulty in detecting the small, green structures even when resulting larval populations are high (Greene et al. 1973). If egg surveys are undertaken, sampling must be accomplished by direct visual examination of plants. Absolute egg population measurements can be made by examining all above-ground plant parts along a stretch of row. While the usual oviposition site is the undersides of leaves (Greene et al. 1973), both leaf surfaces as well as other plant parts (petioles, blossoms, pods, etc.) must be examined until it is established that the leaf undersurface is the sole oviposition site or that a predictable portion of the eggs are deposited there.

Relative sampling methods for velvetbean caterpillar eggs can be based on some modification of the whole plant examination. It has not yet been established whether oviposition within some level of the canopy is favored, and, therefore, it is essential that leaves be sampled from all foliar strata within the crop. When sampling leaves within the canopy, it is difficult to choose leaves in an absolutely random manner. Furthermore, a strictly random method of sampling is not necessarily the most efficient for minimizing sample variable since, by chance, the majority of samples may come from a restricted portion of the sampling universe (Southwood 1978), such as from either the upper, middle, or lower portions of the crop canopy. The use of some form of restricted sampling method may be preferred. The two most obvious and probably most convenient are stratified random and stratified systematic samples of leaves from the crop canopy. The canopy may be arbitrarily stratified into upper, middle, and lower portions, each comprised of approximately one-third of the canopy foliage. With random stratification, equal numbers of leaves (or leaflets) may be chosen at random from the three foliage strata and examined for the presence of eggs. With systematic stratification, single leaves (or leaflets) are chosen at random from the three strata of single plants either at random or systematically within the crop. The requirements and disadvantages of such a sampling procedure are given by Poole (1974).

In early season 1000 leaves or more may be required per observation (with multiple observations made) to insure an adequate estimate of the egg population, while at peak population levels an observational unit decreased by one order of magnitude with smaller numbers of observations may be adequate to supply the information needed.

IV. Sampling Larval Populations

When sampling lepidopterous larvae on soybean, regardless of the sampling objective, it is necessary to correctly identify the species captured. Larvae of the velvetbean caterpillar can be readily separated from larvae of the soybean looper and green cloverworm by the presence of four pairs of fleshly abdominal prolegs, as opposed to two and three, respectively, in the other species. However, the first pair of abdominal prolegs on early instar velvetbean caterpillar larvae are minute and very difficult to distinguish. Therefore, care should be taken that small larvae are not misidentified as green cloverworm. Other larvae that also possess four abdominal prolegs are the corn earworm, tobacco budworm, fall armyworm, and beet armyworm; these are often encountered on soybean. Perhaps the easiest and most effective field-expedient method of separateing velvetbean caterpillar larvae from those of the other species named is to observe the behavior of the larvae when disturbed. Velvetbean caterpillar larvae, like those of the green cloverworm, wiggle violently when disturbed, often appearing to "spring" from the plants. By contrast, *Heliothis* spp. larvae curl their bodies tightly into a coil and remain relatively motionless. The corn earworm and tobacco budworm larvae vary in color from light green or pink to nearly black and have small dark spines and alternating light and dark stripes along the length of their bodies. Velvetbean caterpillar larvae are usually green in color, but dark melanic forms may be found in late season when populations are high. These are often mistaken for and identified as "armyworms" by growers. Their identity as velvetbean caterpillars can be determined by their activity when disturbed, as described. Further, the armyworms (*Spodoptera* spp.) may be identified by the presence of a white inverted "Y" on the front of the head.

Absolute measurements of velvetbean caterpillar larval populations can be made using whole plant examinations or fumigation cage methods described in Chapter 2. Whole plant samples are taken by examining all above-ground plant parts along a prescribed row length. This method is particularly tedious and, because velvetbean caterpillar larvae are easily disturbed and fall from the plant, the sampler runs the risk of disturbed individuals dropping from the plant and escaping detection. Therefore, the whole plant examination method of absolute population measurement is not recommended.

Three general relative sampling methods are commonly used to estimate velvetbean caterpillar larval abundance. These are the ground cloth, the sweep net, and the vacuum net.

A. Comparison of Methods

The ground cloth, sweep net, and vacuum net methods were used by various investigators in comparative studies of sampling efficiency for studies of velvetbean caterpillar ecology and management. Several variations of the basic sampling procedures outlined in Chapter 2 were used. Sweep samples consisted of the following: (a) sweeps across one or two rows (Hillhouse and Pitre 1974, Turnipseed 1974, Rudd and Jensen 1977); (b) "lazy-8" sweeps, across and down (Hillhouse and Pitre 1974, Turnipseed 1974); and (c) sweeps upward the foliage of one row (Hillhouse and Pitre 1974). Ground cloth samples consisted of sections of either one or two adjacent rows, 1 to 2.4 m long (Hillhouse and Pitre 1974, Turnipseed 1974, Rudd and Jensen 1977). The vacuum net was used along a certain stretch of row or vertically over a group of plants.

Hillhouse and Pitre (1974) compared a whole plant capture fumigation cage and the ground cloth methods with several variations of the sweep net method. They found that the fumigation cage method collected significantly more larvae than did the ground cloth or sweep net methods (Table 6-4), collecting an average of three times as many larvae as did the other methods. For the velvetbean caterpillar, samples taken by the ground cloth method did not approach absolute population measurements as was proposed by Rudd and Jensen (1977). Hillhouse and Pitre (1974) concluded that the relative methods used were inefficient compared to the absolute method. Despite the fact that there were no statistically significant differences among the ground cloth and sweep net variations used, they further concluded that "the best collections were obtained by sweeping upward against the foliage of one row and by using the ground cloth-shake

Table 6-4. Comparison of relative estimates of the mean number of velvetbean caterpillar larvae determined by various sample methods in Mississippi (adapted from Hillhouse and Pitre 1974)

Sample method	# larvae/sample
16 sweeps across 1 row	39.1 b
8 sweeps across 2 rows	46.9 b
16 sweeps upward against foliage of 1 row	68.8 b
16 sweeps across and down 1 row in "figure 8"	54.1 b
6.1 m ground cloth shake method	62.7 b
6.1 m fumigation cage method	183.1 a

method." They stated, however, that the ground cloth-shake method was less efficient in collecting velvetbean caterpillar larvae than *Heliothis* larvae because the former tended to "hold tenaciously to the plant when disturbed." This observation is at variance with those of others and was apparently made to help explain the unexpected inefficiency of the ground cloth method. A consensus among many researchers is that velvetbean caterpillar larvae are sampled easily and quite efficiently by most methods because they inhabit the upper, more accessible portion of the canopy and because of their characteristic activity when disturbed. For further comparison of the methods, Hillhouse and Pitre (1974) grouped all lepidopterous larvae and, hence, no further direct inference can be made concerning the velvetbean caterpillar.

Turnipseed (1974) compared the D-Vac, ground cloth, and sweep net sampling methods for larvae of the velvetbean caterpillar. The ground cloth method collected significantly more larvae than did the D-Vac method (Table 6-5). However there was no significant difference between the D-Vac and sweep net methods or between the sweep net and ground cloth methods. In a second experiment two variations of the sweep net and ground cloth methods were introduced. The ground cloth beat method and sweeps across two rows collected significantly more larvae than did sweeps along the row (Table 6-5). However, there was no significant difference in numbers of larvae collected between the ground cloth shake and any of the other variations studied.

Coefficients of variation were virtually identical for ground cloth beat and sweeps across two rows (27.0 and 29.6, respectively) and for ground cloth shake and sweeps along one row (43.0 and 43.6, respectively).

Turnipseed (1974) concluded that the D-Vac sampling method was not an efficient tool for sampling larvae of the velvetbean caterpillar. Sweeping across two rows was an effective method (CV = 29.6 compared to CV = 27.0 for the ground cloth), but processing of samples in the laboratory was time consuming. He speculated that if, for example, velvetbean caterpillar and a few other target species "were counted directly from the net, efficiency of this method would be greatly improved." This would be the case when the sampling objective is to make management decisions. The ground cloth method captured the largest number of larvae per sample. Time required in and cost of sampling were not

Table 6-5. Mean numbers of velvetbean caterpillar larvae collected by D-Vac, sweep, and ground cloth methods in South Carolina (adapted from Turnipseed 1974)

Sample method	# larvae/sample	
	Experiment 1	Experiment 2
D-Vac horizontal — 6.1 m	0.8 a	
15 sweeps across 2 rows	4.0 ab	17.8 b
15 sweeps along 1 row		8.3 a
Ground cloth beat — 1.2 m	6.3 b	
Ground cloth beat — 2.4 m		21.3 b
Ground cloth shake — 2.4 m		15.3 ab

discussed, but two samplers were used with the ground cloth method—one at either end of the net—while only one person was required to collect and process samples taken by the other two methods. Assuming that the times required for all methods were comparable, cost of sampling by the ground cloth method would be twice that of the other two methods if two samplers were used.

Rudd and Jensen (1977) compared the two most commonly used relative sampling methods: the sweep net and the ground cloth. A comparison of CV ratios obtained for velvetbean caterpillar larvae shows that 21% more information is gained per observation with the sweep net than with the ground cloth or 21% more observations are necessary to obtain the same information. However, the time required for ground cloth sampling was 2.3 times that required with the sweep net. This represents a sampling cost differential of 2.75. Even if the sampling times reported by Hillhouse and Pitre (1974) of 6 and 10 minutes for the two methods, respectively, are used, the cost of obtaining comparable information with the ground cloth would be twice that of the sweep net.

B. Calibration of Sampling Procedures

For sampling larvae of lepidopterous insects, such as the velvetbean caterpillar, numbers of individuals captured by the ground cloth method has historically been considered as approaching absolute numbers (Rudd and Jensen 1977), but research has shown that this is not necessarily true (Hillhouse and Pitre 1974). Relative sampling methods should be calibrated to some reliable absolute population assessment method. Failing this, calibration of other relative methods to the ground cloth, on which most pest management control recommendations are based, should be accomplished. Only Rudd and Jensen (1977) thus far have attempted to do this for the velvetbean caterpillar. Results of regression analysis on an observation/observation basis yielded a rather low correlation ($r = 0.432$). However, when calculated on a per field basis regression analysis showed that numbers of velvetbean caterpillar larvae captured by the two methods were highly correlated ($r = 0.950$, $P < 0.01$). When two sampling methods are so highly correlated it is possible to compute factors for the calibration of one method with another. Louisiana recommends a treatment threshold for velvetbean caterpillar of 26 larvae/row-meter (Newsom et al. 1975). By Rudd and Jensen's (1977) calculations, 77 larvae/25 sweeps = 26 larvae/row-meter; this reduces to 3.08 larvae/sweep. Extrapolation to numbers/50 sweeps and /100 sweeps yields 154 and 308 larvae, respectively (may be rounded to 150 and 300, respectively, introducing only 3% error).

To make a direct comparison of the results obtained by Rudd and Jensen (1977) and Hillhouse and Pitre (1974) it is necessary to determine what portion of the larvae present in a 1-m row section is collected by one sweep of a net. In Rudd and Jensen's (1977) study 77 larvae/25 sweeps = 26/row-meter. Therefore, 3.08 larvae/sweep = 26/row-meter. Each sweep captured 12.8% of the larvae captured by ground cloth from 1 row-meter. In Hillhouse and Pitre's (1974) study with sweeps across one row 39.1 larvae/16 sweeps = 62.7/6.1 row-meters. Therefore, 2.44 larvae/sweep = 10.3/row-meter, and each sweep captured 26.0% of the larvae captured by ground cloth from 1 row-meter. So, there apparently was some variation in technique between the two studies.

There is an apparent inconsistency in the results obtained by Turnipseed (1974). In one experiment, 15 sweeps across captured 63% as many larvae as did the 1.2-m ground cloth beat. Intuitively, one would expect that an increase of the ground cloth sample unit from 1.2 to 2.4 row-meters would result in an advantage over the sweep net method in which the sample unit was maintained constantly. However, examination of the data reveals that 15 sweeps across captured 84% as many larvae as did the 2.4 m ground cloth beat. Considering the fact that the two experiments were conducted by the same individual using the same relative methods it is disturbing to note that the sweep net method captured a significantly greater portion of the population in one experiment than in another. Clearly, the two experiments may not be directly compared.

C. Spatial Patterns

A knowledge of the population sampling distribution function (the dispersion characteristic) permits the development of sequential sampling plans that allow us to make a management decision or accurately describe population parameters according to prespecified levels of risk (see Chapters 4 and 5). Since the spatial pattern or dispersion of individuals in the field dictates which of several mathematical distributions are used to construct the sampling plans, it is essential that this relationship be understood (see Chapter 3).

All analyses conducted to date reveal that the spatial patterns of the velvetbean caterpillar in soybean are best described by the Poisson distribution (Shepard and Carner 1976, Strayer et al. 1977, W. G. Rudd personal communication). However, many data sets are apparently adequately described by several frequency distributions (Table 6-6). Based on existing studies, the velvetbean caterpillar larval pattern is considered to follow the Poisson series for the purpose of calculation of sequential sampling plans. It should be noted, however, that results of intensive sampling conducted by Greene (1973) suggested that the negative binomial distribution provided the best model.

D. Sequential Sampling Plans for Pest Management Decisions

Sampling programs are conducted to answer questions about populations. Frequently it would be helpful to know how many samples (observations) will be required to answer questions concerning the population. The number of samples required depends on the following: (1) the reason for sampling, (2) the sampling

Table 6-6. Percentage fit of counts of velvetbean caterpillars from soybean fields to discrete frequency distributions, 1972-1974 (adapted from Shepard and Carner 1976)

Frequency distribution	Percentage fit of:	
	Large larvae	Small larvae
Poisson	73	81
Poisson with zeros	61	70
Neyman's type A	51	56
Negative binomial	47	51
Thomas double Poisson	43	47
Poisson binomial	39	51
# of data sets	51	53

distribution (dispersion), (3) the true population level, and (4) the level or precision desired. The two major objectives of sampling are the detection of general population trends or changes in population size and the in-depth investigation of population dynamics. For management decisions concerning the imminent occurrence of economic injury, one needs only to determine if the population or the damage produced is above or below some critical range.

Construction of a sequential sampling plan requires knowledge of the population distribution function for the species and the establishment of an economic threshold of infestation (or damage). Velvetbean caterpillar larval population levels presently recommended for control in most states are 26 larvae/row-m (Kogan 1976) during the critical podset through pod filling period, i.e., growth stages R3 through R5 of Fehr and Caviness 1977 (see Chapter 1). However, in Florida the recommended treatment level is 13 larvae/row-meter during the pod filling period, but 33 larvae/row-meter in vegetative through full bloom (R2) stages (Strayer and Greene 1974). As illustrated above, 26 larvae/row-meter = 300/100 sweeps; 13 larvae/row-meter, therefore, would be 150/100 sweeps; and 33 larvae/row-meter would be 375/100 sweeps.

Another factor to be considered before a sequential sampling plan is developed is standardization of the sampling method to be used. This is not as necessary with the ground cloth method because regardless of the row length sampled, conversion can easily be made to numbers per meter of row. With sweep net sampling, however, standardization is essential. Jensen (1976) found seven variations of the sweep net sampling method reported in the literature as being used in soybean. He advocated the development of a standardized procedure for sampling a crop with a sweep net and described the method in use on soybean in Louisiana (see Chapter 2). The adoption of this standardized sweep net method by those using a sweep net to sample soybean is advocated here.

Using methods and formulas supplied by Waters (1955) and Onsager (1976) and based on unpublished data (Herzog and Todd, unpublished), sequential sampling plans for management have been constructed that can be used either with shake cloth or sweep net sampling methods. The mechanical and statistical development of the sequential sampling plan are discussed in Chapter 4.

The first sampling plan is constructed for use with a ground cloth that simultaneously samples 1-m sections of two adjacent rows (2 row-m). The economic threshold recommended for defoliating caterpillars by most states was used— 26 larvae/row-m (larvae 1 cm in length or larger). The population level below which economic damage is not expected to occur before the next sampling period and at which a control decision can be deferred is set at three-fourths of the economic threshold or 19.5 larvae/row-m. Since the velvetbean caterpillar is capable of rapidly building to extremely damaging population levels, the level of risk involved in making a control decision was set at $\alpha = \beta = 0.01$. The fact that velvetbean caterpillar larvae are distributed at random permits the use of such a low risk level without requiring an excessive number of samples to reach a control decision.

As velvetbean caterpillar populations often increase very rapidly in the field, it may be advantageous to return to the field before the routine weekly time interval has elapsed. A second set of parallel decision lines can be constructed and incorporated into the sampling plan which will instruct samplers to return to the field in three days (½ the regular sampling interval) to monitor the progress of the infestation. If the larval population exceeds 13 larvae/row-m, the field should be resampled in three days; but if it falls below 10 larvae/row-m the weekly interval is sufficient. This decision is not as critical as the control decision and the risk level was set at $\alpha = \beta = 0.05$. The resulting sampling plan is shown in Fig. 6-6. A tabular form of the plan is often more convenient for use in the field. This is presented in Table 6-7.

Also, since velvetbean caterpillar larval populations are capable of building rapidly to economic levels, it is essential that numbers of small larvae be monitored closely. In the event that large numbers of early instar larvae are found in the field, it will be necessary to shorten the sampling interval to three days to check on the progress of the infestation.

Velvetbean caterpillar larval populations are often decimated in the field by epizootics of the fungus *Nomuraea rileyi* (Farlow) Samson (Allen et al. 1971, Kish and Allen 1978). The rapid advance of an epizootic is dependent on the availability of fungal inoculum, the larval population density, and the occurrence of optimal environmental conditions. Therefore, the sampler should be aware of the occurrence of white- or green-encrusted larvae indicating an impending outbreak of the disease within the population (see Chapter 28). Having observed the onset of an epizootic, the producer or management specialist may elect to postpone the application of chemical control measures to populations at or above threshold levels. If this election is made, the sampling interval must be shortened so that the progress in reducing the population can be monitored and appropriate control decision made.

A similar sequential sampling plan was constructed for use with the sweep net sampling method, with 50 sweeps as the observational unit. All parameters used

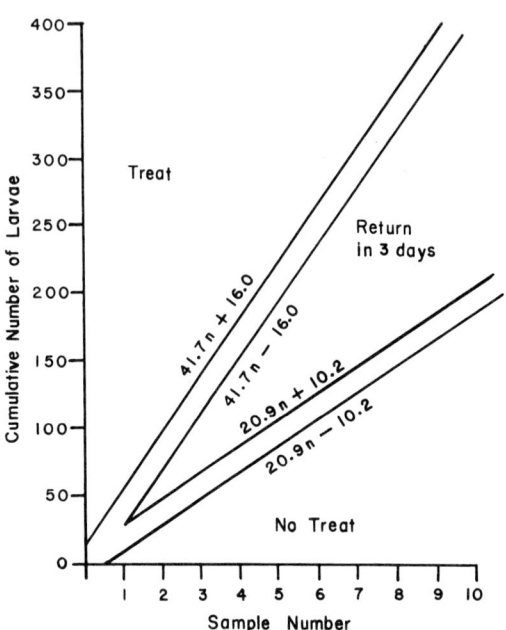

Figure 6-6. Sequential sampling plan for velvetbean caterpillar larvae using the shake cloth sampling method.

Table 6-7. Sequential table for sampling velvetbean caterpillar larvae on soybean by the ground cloth method

No. 2-m samples	Cumulative number of velvetbean caterpillars								
	$\geqslant$		$\leqslant$			$\geqslant$		$\leqslant$	
1	59		—			—		11	
2	99		67			52		32	
3	141		109			73		53	
4	183	INITIATE CONTROLS	151	CONTINUE SAMPLING	CONTROL NOT WARRANTED RETURN IN 3 DAYS	94	CONTINUE SAMPLING	73	CONTROL NOT WARRANTED RETURN IN 1 WEEK
5	225		193			115		94	
6	266		234			136		115	
7	308		276			157		136	
8	350		318			177		157	
9	391		359			198		178	
10	433		401			219		199	

in calculation of the plan were identical to those used in the preceding plan with the exception that the decision lines were based on thresholds converted to numbers per 50 sweeps. This plan is shown in graphic form in Fig. 6-7 and in tabular form in Table 6-8.

The choice of the 50 sweep observational unit will facilitate further comparison between sampling methods as relates to the number of samples required to reach a decision using sweep net and ground cloth sampling methods. This is possible because times required to sample by the two methods are comparable according to Rudd and Jensen (1977) (time required to take two 25-sweep samples is approximately equal to the time required to take one 2-m shake cloth sample). With a sequential sampling plan there is no fixed sample size, but, rather, as sampling progresses, a level is reached at which the infestation has been categorized with the desired level of accuracy. The average sample number curves shown in Fig. 6-8 give the average expected number of samples that would be required to reach a management decision at any level of population. The upper curve gives the average sample number for ground cloth sampling and the lower curve for the sweep net. The ascending portion of the first peak pertains to a decision that controls are not warranted. The descending portion of the first peak and ascending portion of the second peak relate to a decision that no control is warranted, but that the situation is tenuous and the sampling interval should be shortened to three days. On the descending portion of the second peak, controls are necessary.

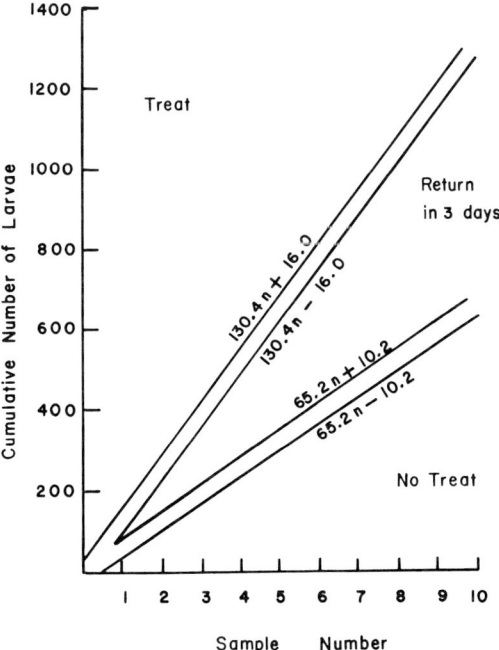

Figure 6-7. Sequential sampling plan for velvetbean caterpillar larvae using the sweep net sampling method.

Table 6-8. Sequential table for sampling velvetbean caterpillar larvae on soybean by the sweep net method

No. 50-sweep samples	Cumulative number of velvetbean caterpillars						
	≥	≤			≥	≤	
1	147	114			76	55	
2	277	244			141	120	
3	488	375			206	185	
4	538	505	INITIATE CONTROLS	CONTINUE SAMPLING	271	250	CONTROL NOT WARRANTED RETURN IN 3 DAYS
5	668	636			337	315	
6	799	766			402	381	
7	929	896			467	446	
8	1060	1027			532	511	
9	1190	1157			597	576	
10	1320	1288			663	641	

(Columns after ≤ indicate: INITIATE CONTROLS / CONTINUE SAMPLING / CONTROL NOT WARRANTED RETURN IN 3 DAYS; and CONTINUE SAMPLING / CONTROL NOT WARRANTED RETURN IN 1 WEEK)

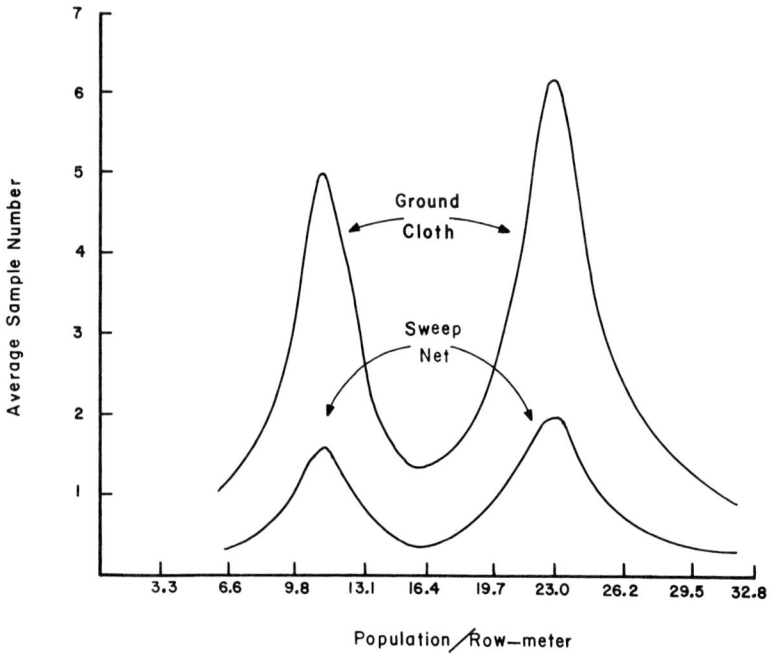

Figure 6-8. Average sample number curves for velvetbean caterpillar collected by shake cloth and sweep net sampling methods.

Examination of the average sample number curves in Fig. 6-8 reveals that at all population levels a larger number of observations are required to reach a management decision with ground cloth sampling than with the sweep net. Since both sequential sampling plans are based on the same parameters in the same proportions, the large difference in the expected average sample number required results from the difference in the number of individuals captured per sample to detect the same population level, rather than from a bias in favor of one of the sampling methods. It is a characteristic of sequential sampling plans that the smaller the population mean measured per sample (52/sample with the ground cloth vs. 150/sample with the sweep net at the threshold level), the larger the number of samples required to reach a management decision. The use of the shake cloth sampling method results in the need for taking almost three times as many observations to reach a management decision than would be the case with the sweep net. Since the times required in sample acquisition are comparable, it is easily seen that the cost of making control decisions based on sweep net sampling under a sequential sampling plan will be about one-third the cost of the ground cloth method.

Strayer (1973) also constructed several sequential sampling plans for management of velvetbean caterpillar populations that were designed for use with the shake cloth sampling method. The plans constucted were based on upper-lower population decision levels of 4.9-9.8, 9.8-16.3, 26-33, and 33-49 larvae/row-meter. The level of risk was set at $\alpha = \beta = 0.1$. Strayer et al. (1977) tested two of the sequential sampling models under Florida field conditions. Their results showed that populations were sufficient to recommend a control decision with both the 4.9-9.8 and 9.8-16.3 larvae/row-meter models. Although controls were applied as directed by the models, "the information was too late to truncate the peak." There were no differences in defoliation among sequentially sampled (treated) plots and the untreated controls. They indicated that a weekly sampling interval was "not adequate to prevent economic loss" and advocated the use of a three- to five-day sampling interval because in their study such intervals "would have provided the necessary information for adequate treatment decision . . . ." They recommended that the plan based on 4.9-9.8 larvae/row-m should not be used because it would invoke a treatment decision too early. They further concluded that "Based on defoliation and economic injury data, the prebloom sequential sampling model for Florida growing conditions should be based on 8 to 12 large larvae/30.5 cm of row and the postbloom model should include 4 to 6 large larvae/30.5 cm of row" (see Chapter 4, Fig. 4-2).

There are three major inconsistencies in the conclusions derived from this study. First, they presented no yield data to justify the statement that the weekly sampling interval was inadequate in preventing economic loss. Second, an examination of population counts from sequentially sampled plots in Fig. 6-9 and comparison of this data with the sequential sampling plan in Fig. 4-2 reveal that the plan based on 13-20 larvae/row-meter would result in a control decision made at the same time as that derived from the 4.9-9.8 and 9.8-16.2

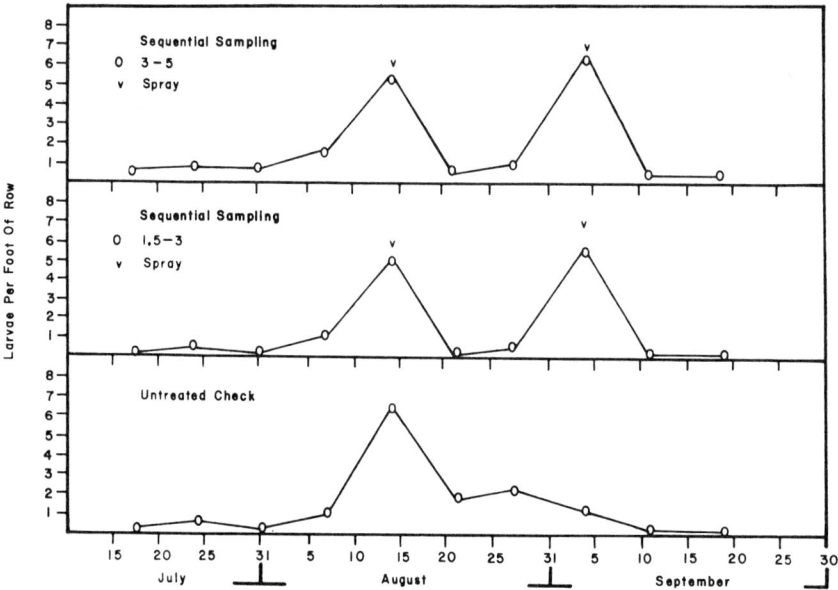

Figure 6-9. Populations of velvetbean caterpillar resulting from evaluation of Strayer's sequential sampling plans (adapted from Strayer 1973).

larvae/row-m plans. It is probable that the model based on 20-26 larvae/row-m would be more realistic. Finally, examination of population counts from untreated plots in Fig. 6-9 and comparison with populations in sequentially sampled plots reveal that at the time the first insecticide application was made, the population was declining and that the control measures were unnecessary. Further, the first insecticide application resulted in a resurgence of the population to a level above initial pretreatment levels and almost six times that present in untreated plots on the same date. So, in this case, the use of the sequential sampling plans recommending control of larvae at 5-16/row-m resulted in two unnecessary insecticide applications. Again, the plan recommending control of larvae at 26/row-m may have been more realistic.

E. Sequential Sampling Plans for Parameter Definition

The quantification of population parameters presents an entirely different sampling problem from the sequential sampling for pest population management. While the objective of the latter is simply the categorization of a population as either exceeding or falling below some critical level, the former is directed at the acquisition of point estimates of the population at some specified level of precision. This type of information is needed for in-depth studies of population dynamics such as for the construction, refinement, or validation of insect population dynamics simulation models. Rudd, in Chapter 5 of this book, provides methods and formulas for developing sequential estimation plans that would enable a researcher to measure a population mean at a predetermined level of precision by collecting a specified number of individuals from the population. When collecting a species distributed according to a Poisson series, this number is a constant regardless of the number of sampling units, the sampling method, or the size of the observational unit. The method provides that at a given level of probability, a population mean can be estimated by collecting $\hat{K}$ individuals from the population. Table 6-9 is a partial listing of results of calculations of the sampling plan with the confidence interval proportional to the mean and at various levels of probability.

While it is a fact that every researcher would like to characterize a population with the greatest precision obtainable, often a compromise must be made between the ideal and that which is feasible. For example, to achieve a 90% probability ($\alpha = 0.1$) that the confidence interval length is 10% of the mean ($\rho = 0.1$), one must collect 1084 individual velvetbean caterpillar larvae from the population. When collecting velvetbean caterpillar larvae from a population present at the threshold level of 26/row-m, the required four observations using the shake cloth method sampling 2 row-m are completely practicable. However, the collection of 1084 larvae in the early season at a population level of 250/ hectare would be physically impossible. It is probable that the researcher will choose a sliding scale with precision of population estimates increasing as available manpower permits, but also subject to population size and sampling method utilized. In the latter example, the researcher may be forced, due to economic constraints, to accept a 90% probability that the confidence interval is equal to the mean with 100 sweep samples (47 observations required) or a 50% probability that the confidence interval is equal to the mean with 2-m shake cloth samples (49 observations required).

Table 6-9. Number of velvetbean caterpillar larvae that must be collected by any sampling method to derive a confidence interval of length ρ at varying levels of risk α

Confidence interval length (ρ)	Risk level (α)				
	0.5	0.2	0.1	0.05	0.01
1.0	2	7	12	17	29
0.5	8	27	44	63	108
0.25	17	106	174	247	427
0.1	182	658	1084	1538	2656
0.05	730	2631	4331	6148	10619

V. Sampling Pupal Populations

Sampling of velvetbean caterpillar pupae has not been discussed in the literature. However, since this species pupates in the soil just below the surface, any absolute or relative measurements of velvetbean caterpillar pupal populations must be made by soil sampling (see Chapter 16). Soil samples excavated to a depth of no more than 8 cm should be sufficient to obtain reliable, if not absolute, estimates of the population. The size of the sample (surface area) required to efficiently measure pupal abundance will be population-dependent. A 900 cm^2 sample size would be an inefficient means of measuring the pupal population produced by 1 larvae/3 row-m. Likewise, a 10 m^2 sample size would be excessive for measuring the pupal population produced by 150 larvae/per row-m.

Soil samples should be excavated to a depth of 8 cm. If taken from light (sand or loam) soils, they may be sifted through a 6- or 8-mesh screen that will retain only the larger debris, large insects, and any pupae that may be present. If taken from heavy (clay) soils, the samples may have to be "washed" through a screen or chemically dispersed as described by Eastman (1976) to break up the soil aggregates.

Other noctuid pupae of similar appearance are likely to be encountered in soil samples. If pupae are encountered, pupal descriptions are available to aid in their identification: velvetbean caterpillar (Watson 1916a), fall armyworm (Luginbill 1928), beet armyworm (Wilson 1934), corn earworm and tobacco budworm (Quaintance and Brues 1905, Neunzig 1960).

VI. Indirect Sampling Procedures

Indirect sampling procedures pertaining to velvetbean caterpillar population assessment involve the estimation or measurement of defoliation (see Chapter 1). Ruesink and Kogan (1975) stated that experienced field entomologists are able to visually estimate insect defoliation with great accuracy. However, the subjective estimates of different specialists in the field are usually significantly different.

The visual estimation of defoliation is truly a subjective decision-making process, based on an arbitrary personal decision concerning the severity of damage. For most practical purposes defoliation classes are usually set at 10% intervals; however, often 5% increments would be more useful, but more difficult to separate.

When assessing defoliation by means of visual estimation, it is essential that the foliage of the canopy as a whole be examined. Since the velvetbean caterpillar initially infests the upper strata of the canopy, a superficial examination of the upper canopy across the field will lead to an inflated estimate of the amount of foliage lost to insect feeding. Two methods are in general use for insuring the estimation of defoliation of the canopy as a whole. The first involves parting the foliage of two rows and examining the plants from top to bottom against either the natural background or white cloth background. The second involves random removal of plants. The plants are then examined using the sky as a background. Whichever method of visual estimation is used, it is essential that several observations be made in a field before a decision is reached. For a discussion of the more empirical methods of estimating defoliation see Chapter 1.

VII. Concluding Remarks

The method used in sampling velvetbean caterpillar populations is not of major importance so long as entomologists are able to relate counts to absolute population levels and ultimately to potential crop loss. However, when cost becomes an important factor, the efficiency, precision, and number of observations required to attain a sampling goal should be known for each available sampling method so that each sampler can make an objective decision concerning the use of the most cost-efficient method.

Several additional studies are needed, not all specifically concerned with sampling per se, so that population assessment of the velvetbean caterpillar can be conducted most efficiently. At present, entomologists are not able to relate blacklight or pheromone trap catch to field populations of moths. To accomplish this, some absolute sampling method must be developed for velvetbean caterpillar moths. A related problem is the determination of the relationship between light or pheromone trap catch and potential larval populations in the field, as was attempted by Myers and Pedigo (1977) for the green cloverworm. A knowledge of this relationship could provide an early warning system for impending economic infestations and could be used as a signal for the initiation of larval population monitoring. Finally, the eggs of all lepidopterous species found in soybean should be described and compared. This will allow direct species differentiation. It is essential when studying egg parasitism or when the incidence of egg parasitism is high because hatching to rear larvae for identification is impossible.

References

Allen, G. E., G. L. Greene, and W. H. Whitcomb. 1971. An epizootic of *Spicaria rileyi* on the velvetbean caterpillar, *Anticarsia gemmatalis,* in Florida. Fla. Entomol. **54**:189-191.

Anonymous. 1927. Factors determining northern limits of *Anticarsia gemmatalis.* Fla. Entomol. **11**:10-12.

Anonymous. 1928. The velvetbean caterpillar, a peanut pest in the Everglades. Fla. Entomol. **12**:39-40.

Ballou, H. A. 1912. Insect pests of the lesser Antilles. West Indies Imp. Dep. Agr. Pamph. Ser. **71**:210 p.

Boldt, P. E., K. D. Biever, and C. M. Ignoffo. 1975. Lepitopteran pests of soybeans: consumption of soybean foliage and pods and development time. J. Econ. Entomol. **68**:480-482.

Buschman, L. L., W. H. Whitcomb, T. M. Neal, and D. L. Mays. 1977. Winter survival and hosts of the velvetbean caterpillar in Florida. Fla. Entomol. **60**:267-273.

Chapin, J. B., and P. S. Callahan. 1967. A list of the Noctuidae (Lepidoptera, Insects) collected in the vicinity of Baton Rouge, Louisiana. Proc. La. Acad. Sci. **30**:39-48.

Chittenden, F. H. 1905. The caterpillar of *Anticarsia gemmatalis* injuring velvetbean. USDA Bur. Entomol. Bull. **54**:77-79.

Cotton, R. T. 1918. Insects attacking vegetables in Porto Rico. J. Dep. Agr. P. R. **2**:265-313.

Douglas, W. A. 1930. The velvetbean caterpillar as a pest of soybeans in southern Louisiana and Texas. J. Econ. Entomol. **23**:684-690.

Dugas, A. L., and J. Gray. 1944. Velvetbean caterpillar control on soybeans and peanuts. La. Agr. Exp. Sta. Annu. Rep. 1943/1944:83-86.

Eastman, C. E. 1976. Infestation of root nodules of soybean by larvae of the bean leaf beetle, *Cerotoma trifurcata* (Forster), and the platystomatid fly, *Rivellia quadrifasciata* (Macquart). Ph.D. diss., Louisiana State University, Baton Rouge. 113 p.

Ellisor, L. O. 1942. Notes on the biology and control of the velvetbean caterpillar, *Anticarsia gemmatalis* (Hbn.). La. Agr. Exp. Sta. Bull. **350**:17-23.

Ellisor, L. O., and L. T. Graham. 1937. A recent pest of alfalfa. J. Econ. Entomol. **30**:278-280.

Ellisor, L. O., H. J. Gayden, and E. H. Floyd. 1938. Experiments on the control of the velvetbean caterpillar, *Anticarsia gemmatalis* (Hbn.). J. Econ. Entomol. **31**:739-742.

Fehr, W. R., and C. E. Caviness. 1977. Stages of soybean development. Iowa Coop. Ext. Serv. Spec. Rep. **80**:12 p.

Ferreira, E. 1970. Pragas da soja no Rio Grande do Sul. Symp. Brasileiro de Soja **1**:1-17.

Ford, B. J., J. R. Strayer, J. Reid, and G. L. Godfrey. 1975. The literature of the arthropods associated with soybeans. IV. A bibliography of the velvetbean caterpillar, *Anticarsia gemmatalis* (Hübner) (Lepidoptera:Noctuidae). Ill. Natur. Hist. Surv. Biol. Notes **92**:15 p.

Greene, G. L. 1972. Biology and control of arthropods on soybeans. Fla. Agr. Exp. Sta. Annu. Rep. 1972:209.

Greene, G. L. 1973. Biology and control of arthropods on soybeans. Fla. Agr. Exp. Sta. Annu. Rep. 1973:244.

Greene, G. L. 1974. Sexual dimorphism of *Anticarsia gemmatalis* leg scales. Fla. Entomol. **57**:180.

Greene, G. L. 1976. Pest management of the velvetbean caterpillar in a soybean ecosystem. pp. 602-610 in L. D. Hill, ed. World soybean research. Proc. World Soybean Res. Conf., Interstate Print., Danville, Illinois. 1073 p.

Greene, G. L., J. C. Reid, V. N. Bount, and T. C. Riddle. 1973. Mating and oviposition behavior of the velvetbean caterpillar in soybeans. Environ. Entomol. 2:1113-1115.
Greene, G. L., N. C. Leppla, and W. A. Dickerson. 1976. Velvetbean caterpillar: a rearing procedure and artificial medium. J. Econ. Entomol. 69:487-488.
Guagliumi, P. 1966. Insetti e aracnidi delle piante comuni del Venezuela segnalati nel periodo 1938-1963. Inst. Agron. per l'Oltremare, Firenze. Rel. Monogr. Agr. Subtrop. Trop. (n.s.) 86:391 p.
Harding, W. C., J. G. Hartsooke, and G. G. Rohwer. 1966. Blacklight trap standards for general insect surveys. Bull. Entomol. Soc. Amer. 12:31-32.
Hillhouse, T. L., and H. N. Pitre. 1974. Comparison of sampling techniques to obtain measurements of insect populations on soybeans. J. Econ. Entomol. 67:411-414.
Hinds, W. E. 1930. Occurrence of *Anticarsia gemmatilis* as a soybean pest in Louisiana in 1929. J. Econ. Entomol. 23:711-714.
Hinds, W. E., and B. A. Osterberger. 1931. The soybean caterpillar in Louisiana. J. Econ. Entomol. 24:1168-1173.
Jensen, R. L. 1976. Are you doing a complete job of sweeping the field? Agri-Fieldman 32:31-32,34.
Johnson, D. W. 1977. Behavioral studies of the velvetbean caterpillar in response to its sex pheromone. M. S. thesis, University of Florida, Gainesville. 40 p.
Kish, L. P., and G. E. Allen. 1978. The biology and ecology of *Nomuraea rileyi* and pilot program for predicting its incidence on *Anticarsia gemmatalis* in soybean. Fla. Agr. Exp. Sta. Tech. Bull. 795:48 p.
Kogan, M. 1976. Soybean disease and insect pest management. pp. 114-121. in R. M. Goodman, ed. Expanding the use of soybeans. Proc. of a Conference for Asia and Oceania. Univ. of Illinois, College of Agriculture, INTSOY Ser. 10. 261 p.
Leppla, N. C. 1976. Circadian rhythms of locomotion and reproductive behavior in adult velvetbean caterpillars. Ann. Entomol. Soc. Amer. 69:45-48.
Leppla, N. C., T. R. Ashley, R. H. Guy, and G. D. Butler. 1977. Laboratory life history of the velvetbean caterpillar. Ann. Entomol. Soc. Amer. 70:217-220.
Lincoln, C., J. R. Phillips, W. H. Whitcomb, G. C. Dowell, W. P. Boyer, K. O. Bell, Jr., G. L. Dean, E. J. Matthews, J. B. Graves, L. D. Newsom, D. F. Clower, J. R. Bradley, Jr., and J. L. Bagent. 1967. The bollworm-tobacco budworm problem in Arkansas and Louisiana. Ark. Agr. Exp. Sta. Bull. 720:66 p.
Luginbill, P. 1928. The fall armyworm. USDA Tech. Bull. 34:91 p.
McCord, E. 1974. Survey and control of some lepidopterous larvae destructive to the pigeon pea, *Cajanus cajan* (L.) Millspaugh. M. S. thesis, University of Florida, Gainesville. 66 p.
Menke, W. W. 1973. A computer simulation model: the velvetbean caterpillar in the soybean agroecosystem. Fla. Entomol. 56:92-105.
Menke, W. W., and G. L. Greene. 1976. Experimental validation of a pest management model. Fla. Entomol. 59:135-142.
Mitchell, E. R., J. C. Webb, A. H. Baumhover, R. W. Hines, J. W. Stanley, R. G.

Endris, D. A. Lindquist, and S. Masuda. 1972. Evaluation of cylindrical electric grids as pheromone traps for loopers and tobacco hornworms. Environ. Entomol. 1:365-386.
Moscardi, F. 1979. Effect of soybean crop phenology on development, food consumption, and oviposition of *Anticarsia gemmatalis* (Hubner). Ph.D. diss., University of Florida, Gainesville. 138 p.
Myers, T. V., and L. P. Pedigo. 1977. Seasonal fluctuation in abundance, reproductive status, sex ratio, and mating of the adult green cloverworm. Environ. Entomol. 6:225-228.
Neunzig, H. H. 1960. The pupae of *Heliothis zea* and *Heliothis virescens* (Lepidoptera:Noctuidae). Ann. Entomol. Soc. Amer. 53:551-552.
Newsom, L. D., R. L. Jensen, D. C. Herzog, and J. W. Thomas. 1975. A pest management system for soybeans. La. Agr. 184:10-11.
Newsom, L. D., M. Kogan, F. D. Miner, R. L. Rabb, S. G. Turnipseed, and W. H. Whitcomb. 1980. General accomplishments toward better pest control in soybean. In: C. B. Huffaker, ed. New technology of pest control. John Wiley & Sons, N.Y.
Nickels, C. B. 1926. An important outbreak of insects infesting soybeans in lower South Carolina. J. Econ. Entomol. 14:614-618.
Onsager, J. A. 1976. The rationale of sequential sampling, with emphasis on its use in pest management. USDA Tech. Bull. 1526:10 p.
Pacheco, F. 1970. Plagas del Valle del Yaqui. Mex. Inst. Nac. Invest. Agr. SAG Circ. CIANO 53:124 p.
Panizzi, A. R., B. S. Correa, D. L. Gazzoni, E. B. de Oliveira, G. G. Newman, and S. G. Turnipseed. 1977. Insetos da soja no Brasil. EMBRAPA Cent. Nac. Pesq. Soja Bol. Tec. 1:20 p.
Pena, M. R., and J. A. Sifuentes. 1972. Lista de nombres cientificos y comunes de plagas agricolas en Mexico, 1972. Agr. Tec. Mex. 3:132-144.
Poole, R. W. 1974. An introduction to quantitative ecology. McGraw-Hill Book Co., N.Y. 532 p.
Posada, O. L., I. Z. dePolaniz, E. S. deArevalo, A. Saldarriaga, F. A. Garcia Roa, and Y. R. Cardenas. 1970. Lista de insectos daninos y otras plagas en Colombia. Programa Nacional de Entomologia Publ. Misc. 17:202 p.
Quaintance, A. L., and C. T. Brues. 1905. The cotton bollworm. USDA Div. Entomol. Bull. 50:155 p.
Reid, J. C. 1975. Larval development and consumption of soybean foliage by the velvetbean caterpillar, *Anticarsia gemmatalis* (Hübner) (Lepidoptera: Noctuidae), in the laboratory, Ph.D. diss., University of Florida, Gainesville. 125 p.
Rizzo, H. F. 1972. Enemigos animales del cultivo de la soja. Rev. Inst. Bolsa. Cereales 2851:1-6.
Roelofs, W. L., and R. T. Carde. 1977. Responses of Lepidoptera to synthetic sex pheromone chemicals and their analogues. Annu. Rev. Entomol. 22:377-405.
Rosetto, C. J., S. Silveira Neto, D. Link, J. G. Vieira, E. Amante, D. M. Souza, N. V. Banzatto, and A. M. Oliveira. 1971. Pragas de arroz no Brasil: con-

tribuições técnicas da delegação brasileira. Reun. Com. Arroz. Amer. Com. Intern. Arroz 2:149-238.
Rudd, W. G., and R. L. Jensen. 1977. Sweep net and ground cloth sampling for insects in soybeans. J. Econ. Entomol. 70:301-304.
Ruesink, W. G., and M. Kogan. 1975. The quantitative basis of pest management: sampling and measuring. pp. 309-351 in R. L. Metcalf and W. H. Luckmann, eds. Introduction to insect pest management. John Wiley & Sons, N.Y. 587 p.
Shepard, M., and G. R. Carner. 1975. Distribution of insects in soybean fields. Can. Entomol. 108:767-771.
Smith, J. B. 1893. A catalogue, bibliographical and synonymical, of the species of moths of the lepidopterous superfamily Noctuidae, found in Boreal America with critical notes. Bull. U. S. Nat. Mus. 44:424 p.
Southwood, T. R. E. 1978. Ecological methods with particular reference to the study of insect populations. Halsted Press, N.Y. 524 p.
Strayer, J. R. 1973. Economic threshold studies and sequential sampling for management of the velvetbean caterpillar, *Anticarsia gemmatalis* (Hübner), on soybeans. Ph.D. diss., Clemson University. 97 p.
Strayer, J., and G. L. Greene. 1974. Soybean insect management. Fla. Coop. Ext. Serv. Cir. 395:15 p.
Strayer, J., M. Shepard, and S. G. Turnipseed. 1977. Sequential sampling for management decisions on the velvetbean caterpillar on soybeans. J. Ga. Entomol. Soc. 13:220-227.
Telford, E. A., and N. F. Childers. 1947. Tropical kudzu in Puerto Rico. P. R. Fed. Exp. Sta. (Mayaguez) Circ. 27:1-30.
Tietz, H. M. 1952. The Lepidoptera of Pennsylvania. The Pennsylvania State College School of Agriculture, State College, Penn. 194 p.
Tietz, H. M. 1972. An index to the described life histories, early stages and hosts of the Macrolepidoptera of the continental United States and Canada. The Allyn Museum of Entomology, Sarasota, Fla. 1:1-536.
Turnipseed, S. G. 1974. Sampling soybean insects by various D-vac, sweep, and ground cloth methods. Fla. Entomol. 57:217-223.
Turnipseed, S. G., and M. Kogan. 1976. Soybean entomology. Annu. Rev. Entomol. 21:247-282.
USDA. 1954a. Cereal and forage insects. USDA Coop. Econ. Insect. Rep. 4:565-573.
USDA. 1954b. Cereal and forage insects. USDA Coop. Econ. Insect. Rep. 4:881-885.
Van Dinther, J. B. M. 1956. Soybean insects. Entomol. Ber. (Amsterdam) 16(6):104-109.
Waters, W. E. 1955. Sequential sampling in forest insect surveys. For. Sci. 1:68-79.
Watson, J. R. 1915. The velvetbean caterpillar (*Anticarsia gemmatalis*). Fla. Agr. Exp. Sta. Annu. Rep. 1914/1915:49-64.
Watson, J. R. 1916a. Life-history of the velvetbean caterpillar [*Anticarsia gemmatalis* (Hübner)]. J. Econ. Entomol. 9:521-528.

Watson, J. R. 1916b. Control of the velvetbean caterpillar. Fla. Agr. Exp. Sta. Bull. **130**:45-58.

Watson, J. R. 1932. Further notes on the velvetbean caterpillar. Fla. Entomol. **16**:24.

Wille, J. E. 1940. Department de Entomologia. Memoria del jefe del Departamento. La Molina Estac. Exp. Agr. Mem. **12**:177-210.

Wilson, J. W. 1934. The asparagus caterpillar: its life history and control. Fla. Agr. Exp. Sta. Bull. **271**:1-26.

Wolcott, G. N. 1936. "Insectae Borinquenses." A revision of " 'Insectae Portoricensis', a preliminary annotated check-list of the insects of Porto Rico, with descriptions of some new species," and "First supplement to Insectae Portoricensis." J. Agr. Univ. P. R. **20**:1-600.

Wuensche, A. L. 1976. Relative abundance of seven pest species and three predacious genera in three soybean ecosystems in Louisiana. M. S. thesis, Louisiana State University, Baton Rouge. 384 p.

Chapter 7

Sampling Soybean Looper on Soybean

Donald C. Herzog

I. Introduction

The soybean looper, *Pseudoplusia includens* (Walker) (Lepidoptera:Noctuidae), is the most abundant of the plusiine caterpillars attacking soybean over much of North and South America and all of Central America. It is limited in its distribution to the western hemisphere. While it ranges throughout much of the United States from New York to California (Fig. 7-1), economic infestations of this pest on soybean seldom occur north of Texas, Arkansas, Mississippi, Alabama, Georgia, and South Carolina.

Although this species was reported as an economic pest of soybean in Virginia as early as 1957 (Morris 1958), it was not generally recognized as a soybean pest until the 1960s. Prior to that time populations of the soybean looper had been misidentified as the cabbage looper, *Trichoplusia ni* (Hubner). This situation was clarified following the reports of Hensley et al. (1964) and Canerday and Arant (1966) which confirmed that the looper complex in Louisiana and Alabama contained both *P. includens* and *T. ni*. In both cases the soybean looper was the predominant species.

A. Host Plants and Nature of Injury

The soybean looper is a defoliating species that habitually feeds in the lower one-half to two-thirds of the soybean crop canopy. Young larvae have been found to select highly digestible leaves of low fiber content (Kogan and Cope 1974), while older larvae are less discriminating and, in fact, prefer the older,

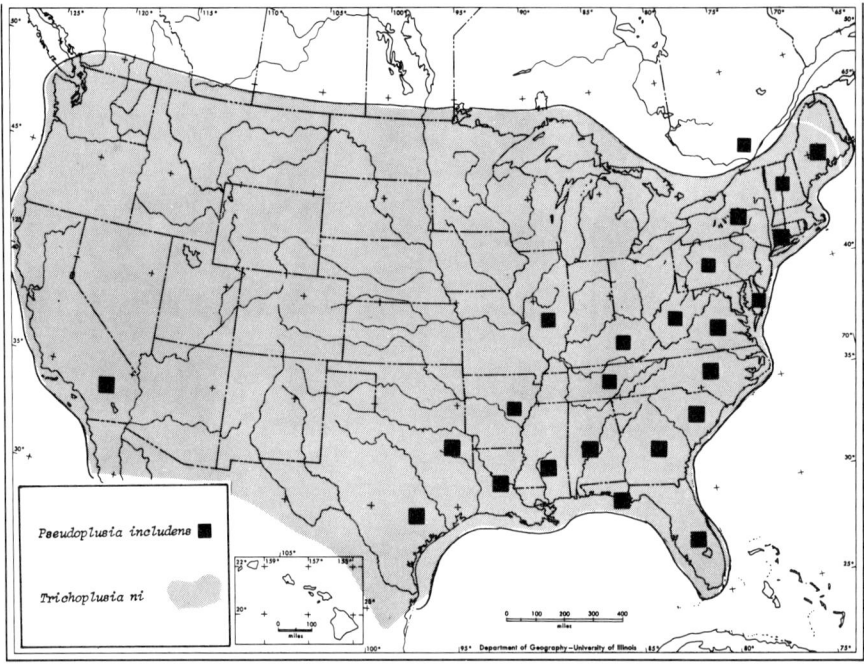

Figure 7-1. Distribution map of *Pseudoplusia includens* and *Trichoplusia ni* in North America. (Based mainly on Eichlin and Cunningham 1978).

more mature leaves (Strayer and Greene 1974). As increased populations defoliate the lower canopy, larvae migrate to and consume the upper leaves. Soybean looper larvae feed from the undersurface of the leaves. First and second instars do not cut completely through the leaf but leave the upper cutinous layer of the epidermis intact, giving the damaged area of the leaf a distinct silvered appearance when viewed from above. Third through sixth instars consume large, irregular areas of leaves, characteristically leaving the larger lateral leaf veins intact. This gives severely defoliated leaves a lace-like appearance. Surface area of leaves consumed by soybean looper larvae through their developmental period has been reported from a minimum of 82 to a maximum of 207 cm^2 (Reid and Greene 1973, Kogan and Cope 1974).

The soybean looper has been recorded in the literature as attacking a wide range of agricultural, vegetable, and floricultural crops, and an additional group of noncultivated plants have been reported as hosts. The host list contains members of 28 plant families (Table 7-1). Although predominantly a defoliator, this species has occasionally been found feeding on soybean pods and those of table legumes.

B. Life Cycle and Phenology

The life cycle of the soybean looper consists of the egg, six larval instars, the pupa, and the adult (Figs. 7-2, 7-3). Duration of the larval stages reported in the literature ranges from 13 to 20 days and the minimum generation time has been reported as ranging from 27 to 34 days (Mitchell 1967, Caneday and Arant 1967). There are three to four annual generations on soybean in Louisiana (Burleigh 1972).

The soybean looper is known to overwinter in the United States only in southern Florida and southern Texas. Invasion of other areas results from annual emigrations of moths arising from these reservoirs or from Central or South America or the islands of the Caribbean. Soybean looper moths have been captured in pheromone-baited blacklight traps in the winter at various locations in Florida with almost continuous moth activity recorded in several of these locations (Mitchell et al. 1975). The soybean looper reproduces throughout the year in the Rio Grande Valley of Texas (Harding 1976), but it reproduces during the winter only at a reduced rate in south Florida (Endris 1973).

In the southeast the soybean looper survives the winter in Florida and increases there during the spring months. As temperatures become favorable for development, moths migrate northward and, based on light trap catches, arrive in Georgia in June and July and in South Carolina in August and September (Mitchell et al. 1975). Adult moths were collected in Alabama from May to October (Caneday and Arant 1966) and in Louisiana from March to December (Chapin and Callahan 1967), indicating that this species is not active through the winter in either state. In Louisiana, larvae were collected from soybean in late May or early June (Herzog unpublished, Burleigh 1972). Unidentified plusiine larvae were collected

Figure 7-2. Scanning electron photomicrograph of *P. includens* egg. (Courtesy of Gerald Carner).

Table 7-1. Cultivated and wild plants recorded as hosting larvae of the soybean looper

Family	Scientific Name	Common Name	Reference
Amaranthaceae	*Amaranthus* sp.	pigweed	Harding (1976)
Araceae	*Philodendron* sp.	philodendron	DPI[a]
Araliaceae	*Schefflera actinophylla*	schefflera	DPI[a]
Begoniaceae	*Begonia* sp.	lettuce-leaf begonia	J. W. Todd (1978, personal communication)
Caryophyllaceae	*Dianthus caryophyllus*	carnation	Morishita et al. (1967)
Chenopodaceae	*Chenopodium album*	lambsquarter	Harding (1976)
Commelinaceae	*Commelina pendula*	wandering Jew	Crumb (1956)
	Zebrina pendula		Tietz (1972)
Compositae	*Aster* sp.	aster	Morishita et al. (1967)
	Calendula officinalis	calendula	Morishita et al. (1967)
	Chrysanthemum spp.	chrysanthemum	Morishita et al. (1967)
	Eupatorium sp.		Tietz (1972)
	Erigeron canadensis	horseweed	Harding (1976)
	Gerbera jamesonii	Transvaal daisy	Morishita et al. (1967)
	Helianthus annuus	sunflower	Teetes et al. (1970)
	H. spp.	wild sunflower	Harding (1976)
	Lactuca sativa	lettuce	Crumb (1956)
	L. sp.	wild lettuce	Harding (1976)
	Parthenium sp.	parthenium	Harding (1976)
	Senecio cineraria	dusty miller	Morishita et al. (1967)
	Solidago spp.	goldenrod	Eichlin and Cunningham (1969)
	Sonchus spp.	sow thistle	Harding (1976)
	Xanthium pennsylvanicum	cocklebur	Martin et al. (1976)
Convolvulaceae	*Ipomoea batatas*	sweet potato	Hensley et al. (1964)
	I. purpurea	tall morning-glory	Herzog (unpublished)
Cruciferae	*Brassica oleracea*	broccoli	Harding (1976)

Family	Scientific name	Common name	Reference
Cruciferae	B. oleracea	cabbage	Harding (1976)
	B. oleracea	collards	Crumb (1956)
	Lepidium virginicum	pepperweed	Harding (1976)
	Matthiola incana	stock	Morishita et al. (1967)
	Nasturtium officinale	watercress	DPI[a]
Cucurbitaceae	Citrullus vulgaris	watermelon	DPI[a]
Euphorbiaceae	Croton capitatus	woolly croton	Crumb (1956)
	Poinsettia pulcherrima	poinsettia	Morishita et al. (1967)
Geraniaceae	Geranium spp.	geranium	Crumb (1956)
	Pelargonium sp.	geranium	Tietz (1972)
Gesneriaceae	Saintpaulia ionantha	African violet	Morishita et al. (1967)
Gramineae	Zea mays	corn	Endris (1973)
	Z. mays	sweet corn	Janes and Greene (1970)
Labiateae	Coleus hybridus	coleus	Morishita et al. (1967)
	Mentha sp.	mint	USDA (1956)
Lauraceae	Persea americana	avocado	Eichlin and Cunningham (1978)
Leguminosae	Arachis hypogaea	peanut	Canerday and Arant (1966)
	Cyamopsis tetragonoloba	guar	Teetes et al. (1970)
	Glycine max	soybean	Morris (1958)
	Medicago sativa	alfalfa	Crumb (1956)
	Phaseolus limensis	lima bean	Genung (1958)
	P. vulgaris	garden bean	Genung (1958)
	Pisum sativum	garden pea	Canerday and Arant (1966)
	Vigna sinensis	cowpea	Genung (1958)
Liliaceae	Allium sativum	garlic	DPI[a]
	Asparagus retrofractus	asparagus	Morishita et al. (1967)
Malvaceae	Gossypium hirsutum	cotton	Folsom (1936)
	Hibiscus esculentus	okra	Bottimer (1926)
Passifloraceae	Passiflora incarnata	passion flower vine	Harding (1976)

Table 7-1. (cont.)

Family	Scientific Name	Common Name	Reference
Polygonaceae	*Rumex* spp.	dock	Harding (1976)
Portulacaceae	*Portulaca oleracea*	purslane	Harding (1976)
Rubiaceae	*Ixora coccinea*	flame of the woods	DPI[a]
Saxifragiaceae	*Hydrangea* sp.	hydrangea	Morishita et al. (1967)
Solanaceae	*Caspicum grossum*	bell pepper	Harding (1976)
	Cyphomandra betacera	tree tomato	DPI[a]
	Lycopersicum esculentum	tomato	Canerday and Arant (1966)
	Nicotiana rustica	wild tobacco	Harding (1976)
	N. tabacum	tobacco	Crumb (1956)
	Peperomia obtusifolia	oval-leaf peperomia	DPI[a]
	Physalis sp.	groundcherry	Harding (1976)
	Solanum gracile		DPI[a]
	S. tuberosum	potato	USDA (1963)
	S. sp.	husk tomato	Harding (1976)
	S. sp.	purple nightshade	Harding (1976)
Umbelliferae	*Apium graveolens*	celery	Janes and Genung (1977)
Verbenaceae	*Lantana montevidensis*	lantana	DPI[a]
	Verbena spp.	vervain	Harding (1976)

[a]Host records on file Florida Department of Agriculture and Consumer Services, Division of Plant Industry (DPI), Gainesville, FL.

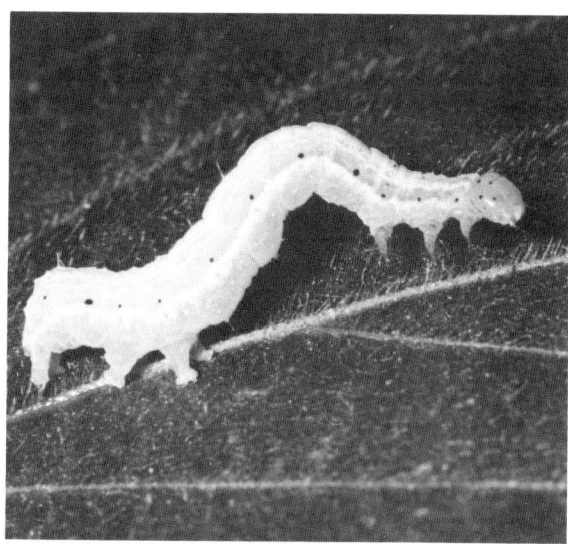

Figure 7-3. *P. includens* larvae. (Courtesy of Merle Shepard).

from soybean in July in both South Carolina (Carner et al. 1974) and North Carolina (Deitz et al. 1976).

Under most circumstances *P. includens* populations on soybean remain below economically damaging levels. Outbreak populations most frequently develop in ecosystems where cotton and soybean are grown in close proximity (Burleigh 1972) probably because cotton nectar provides an adequate carbohydrate food source for adults (Jensen et al. 1974). The availability of cotton nectar enables a female to develop a normal complement of eggs and results in an approximate 60-fold increase in egg production, giving rise to the development of extremely large populations. In addition, mistimed or unwarranted applications of highly toxic pesticides may severely disrupt the predator-parasite-pest equilibrium in the field and result in the induction of the soybean looper to economic pest status in areas where this species seldom poses an economic problem.

Since the soybean looper exhibits two distinct population growth curves dependent on the adequacy of adult food, it is possible to construct two generalized representations of population phenology. Figure 7-4 depicts soybean looper population dynamics in an agricultural ecosystem that has both cotton and soybean as component crops, while Fig. 7-5 represents a population in an ecosystem in which cotton is not grown, but noncultivated plants such as *Cassia fasciculata* (Michx.) provide a subadequate nectar source for adults.

II. Sampling Adult Populations

No absolute method is available for sampling soybean looper adult populations. However, the "flushing" method developed for green cloverworm moths (see

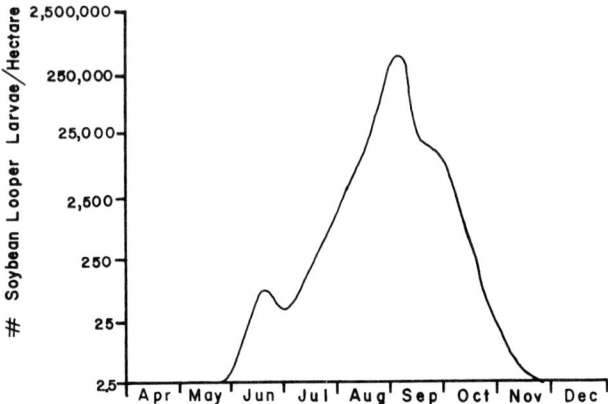

Figure 7-4. Phenology of soybean looper populations in the presence of an adequate adult food source. Adapted from Burleigh (1972), Brousseau (1974), Jensen (1975 personal communication), Herzog (1976), Wuensche (1976), Husin (1978), Rudd et al. (1980), and Herzog (unpublished).

Chapter 8) may prove applicable to the soybean looper.

Relative methods of population measurement are generally based on trapping. Three general types of traps have been used to measure adult populations of *P. includens:* blacklight, cylindrical electric grid and sticky traps. Standards recommended by the Entomological Society of America should be met when utilizing blacklight traps to insure that catches from the same area or even different areas are comparable (Harding et al. 1966). This is particularly important in

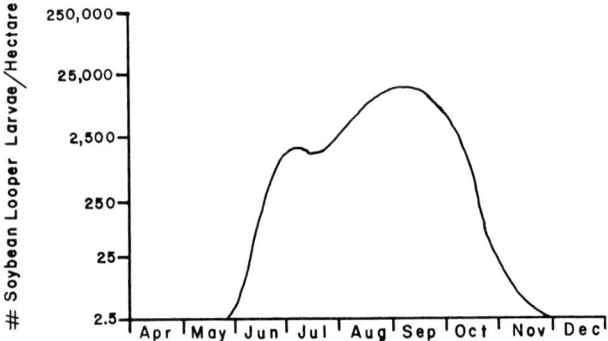

Figure 7-5. Phenology of soybean looper populations with a subadequate adult food source. Adapted from Burleigh (1972), Herzog (1976), Wuensche (1976), Husin (1978), and Herzog (unpublished).

studies that cover a broad geographic area (e.g., Mitchell et al. 1975). When operating several light traps simultaneously, care should be taken that each is located outside the visible range of the others (Myers and Pedigo 1977). Further, each trap should be oriented so that all have equal exposure to the crop, preferably 180° or more.

A characteristic of blacklight trapping that complicates its use in population assessment is a relatively large number of closely related species are attracted and captured. Although *P. includens* and *T. ni* comprise more than 90% of the plusiine moths captured in light trapping, significant numbers of *Rachiplusia ou* (Guenee), *Trichoplusia oxygramma* (Geyer), *Syngrapha falcifera* (Kirby), *Autographa biloba* (Stephens), *Argyrogramma basigera* (Walker), *Argyrogramma verruca* (F.), and *Allagrapha aerea* (Hubner) are also often captured (Chapin and Callahan 1967; Canerday and Arant 1967; Eichlin and Cunningham 1969, 1978).

Most soybean looper trapping experiments have been conducted in conjunction with sex pheromone research or have used pheromone as a complementary attractant to the blacklight. Earliest work used virgin female soybean loopers confined within either blacklight or cylindrical electric grid traps (Mitchell 1972). The cabbage looper sex pheromone, cis-7-dodecen-1-ol acetate, synthesized by Berger (1966), is also the attractive component of the soybean looper sex pheromone (Tumlinson et al. 1972). This compound, under the common name "looplure," has been used in moth sex pheromone and trapping research.

There is a definite additive effect achieved with the incorporation of blacklight and sex pheromone trapping of soybean looper adults. However, this combination retains one of the hinderances of blacklight alone. Virgin female soybean loopers attract large numbers of *P. includens* and *T. oxygramma* in addition to *T. ni* (Mitchell et al. 1972), and there are reports that the cabbage looper sex pheromone also stimulates males of *R. ou, Autographa biloba,* and *A. californica* (Speyer) (Shorey et al. 1965, Berger and Canerday 1968). Although there are specific periods of nocturnal activity of *P. includens, T. ni,* and *T. oxygramma,* the calling periods overlap and, therefore, they are not sufficiently distinct to insure that even hourly catches would be composed of a large majority of one of the species (Mitchell 1973).

The electric grid trap has been used in combination with sex pheromone. Moths are captured by electrocution after they have been attracted to the trap. Both blacklight and electric grid traps are limited to their applicability by the availability of a source of current. However, battery powered blacklight traps are available. Sex pheromones are also used in combination with sticky traps, but this method has the disadvantage of rendering many specimens unidentifiable except by dissection and examination of the genitalia.

Experimental designs of field trapping research as related to lepidopteran sex pheromone response require replication of treatments (levels, etc.) to help compensate for effects of trap placement due to a nonhomogenous distribution of infestation. Replications allow assignation of blocks to areas of more homogenous population groupings. Frequent rerandomization of treatments within blocks effectively compensate for nonuniform populations and even interactions among contiguous treatments (Roelofs and Carde 1977).

The "sugarline" (see Chapter 6) is an additional survey sampling method used only on one occasion with the soybean looper (Herzog, unpublished). The peak period of feeding activity of soybean looper moths occurs from 2 000-2 300 h with a smaller peak of activity from 0 300-0 500 h (Lingren et al. 1977). For soybean looper, these times would be the most advantageous for "sugarline" surveys. Under certain circumstances it may be more advantageous to survey a natural nectar line such as the edge of a blooming cotton field or nectar-laden weed bank adjacent to a soybean field.

Whatever method of survey is used in measuring soybean looper moth populations, it is necessary to have at hand characters for separating plusiine species that are commonly found in soybean. A key, adapted from Habeck (1968), is supplied in Appendix 7-1 to assist in the identification of plusiine moths likely to be encountered in soybean looper adult sampling and survey.

III. Sampling for Eggs

Any sampling of eggs of this species must be accomplished via direct visual examination of plants. Absolute population measurements are made by whole plant examination along a stretch of row. Harding (1976) examined a minimum of 1.5 m of row (5 ft)/sample, and took a maximum of four samples/field. Martin et al. (1976) collected eggs from 5 to 27 m of row of soybean.

Relative sampling methods for soybean looper eggs can be based on some modification of the whole plant examination. Soybean looper moths lay the majority of their eggs on the undersurface of soybean leaves, but it has not yet been established whether oviposition within some level of the canopy is favored. For this reason it is essential that leaves are sampled from all strata within the crop (see Chapter 6 for a discussion of stratification of egg sampling).

The greatest problem likely to be encountered in sampling for eggs is the identification of the eggs collected. Apparently no research has yet provided characters that would allow field identification of the eggs of the plusiine species (Canerday and Arant 1967), and even the eggs of other nonplusiine species may cause confusion. It will, therefore, probably be necessary to take the eggs to the laboratory, allow them to hatch, and rear the larvae on leaves or artificial media to a size that will allow identification.

IV. Sampling Larval Populations

When sampling lepidopterous larvae on soybean, it is necessary to correctly identify the species captured. Larvae of the soybean looper can easily be separated from other nonplusiine larvae encountered in soybean by the presence of two pair of fleshly abcominal prolegs plus an anal pair (Fig. 7-2). However, larvae of six other looper species—*Trichoplusia ni, Rachiplusia ou, Argyrogramma verruca, Autographa precationis* (Guenee), *Allagrapha aerea* (Hubner), and *Autoplusia egena* (Guenee)—are often encountered in complex with *P. includens* on soybean with varying degrees of frequency (Hensley et al. 1964, Canerday and Arant 1966, Kogan 1975, Martin et al. 1976, Eichlin and Cunningham 1978, Herzog unpublished). Under most circumstances in sampling, for making management decisions, such mixed populations will contain small proportions of other looper species and the population can be treated as if it contained exclusively soybean looper. However, in population dynamics studies species identify is essential. A key, adapted from Eichlin and Cunningham (1969 and 1978), is supplied in Appendix 7-2 to assist in the identification of plusiine larvae likely to be encountered in sampling for soybean looper larvae on soybean.

Absolute measurements of soybean looper larval populations can be made using whole plant examinations (Martin et al. 1976, Harding 1976) or fumigation cage methods (see Chapter 2). Whole plant samples are taken by examining all above-ground plant parts along a prescribed row length. This method is particularly tedious and the sampler runs the risk of disturbed individuals dropping from the plant and escaping detection.

Relative measurements of soybean looper larval populations are sufficient for almost all types of sampling plans regardless of the objective, provided that factors are available for calibration of sampling methods or conversion to absolute populations.

Three methods of sampling to obtain relative population estimates are commonly used by researchers and management specialists: the beatcloth or plant shake, sweep net, and vacuum net. The reliability of each of these sampling methods should be judged on the basis of sample variability, cost, similarity to absolute population estimates, degree of correlation with absolute or other relative population estimates, and ease of calibration with absolute or other relative population estimates. In the end, however, the choice of a relative sampling method is usually based on the personal bias of the surveyor.

A. Comparison of Methods

The sweep net, ground cloth, and vacuum net methods were used by various investigators in comparative studies of sampling efficiency for studies of soybean looper ecology and management. Several variations of the basic sampling procedures described in Chapter 2 were used. Sweep samples consisted of the following: (a) sweeps across one or two rows (Hillhouse and Pitre 1974, Shepard et al. 1974, Turnipseed 1974, Marston et al. 1976, Rudd and Jensen 1977); (b) "lazy 8" sweeps across and down (Turnipseed 1974, Hillhouse and Pitre 1974, see also Chapter 2), and (c) sweeps upward against the foliage of one row

(Hillhouse and Pitre 1974). Ground cloth samples consisted of either sections of one or two adjacent rows, 1 to 2.4 m long (Shepard et al. 1974, Turnipseed 1974, Rudd and Jensen 1977). The vacuum net was used along a certain stretch of row or vertically over a group of plants.

A general conclusion reached by Turnipseed (1974) and Shepard et al. (1974) was that the ground cloth method captured more larvae than did either the sweep net or vacuum net methods (Table 7-2) and that the ground cloth method permitted detection of peak looper populations three to four weeks earlier than did the sweep net method. In these studies, almost without exception, the largest coefficients of variation were produced by the vacuum net sampling variations. Ground cloth and sweeps across were least variable for sampling small looper larvae. Sweeping appears to be equal to and in some cases superior to the ground cloth method from the standpoint of sample variability when sampling soybean looper larvae, and both would be similarly efficient if samples were processed in the field (Turnipseed 1974). This would be the case in which samples are taken with the objective of making management decisions. The vacuum net sampling methods were not efficient tools for sampling larvae of the soybean looper.

The ground cloth method was tested with beating and with shaking motions (Turnipseed 1974). The beat variation collected significantly more soybean looper larvae than did the shake variation (Table 7-3). Two possible explanations were advanced for this difference: (1) greater numbers of loopers, which hold rather tenaciously onto foliage, were dislodged by the vigorous beating of plants downward compared to shaking in a less vertical motion; (2) shaking resulted in more lateral dislodging than beating downward, causing some insects to fall on the soil away from the ground cloth.

The soybean looper-cabbage looper complex showed much variation over plant growth stages and Hillhouse and Pitre (1974) regarded their data concerning these species as inconclusive. However, all relative sampling methods studied adequately and efficiently estimated absolute populations of the lepidopterous larval

Table 7-2. Mean numbers of *P. includens* larvae sampled from soybean using D-Vac, sweep, and ground cloth methods (adapted from Turnipseed 1974)

	$\bar{X}$ soybean loopers/sample[a]		
	Experiment 1	Experiment 2	
D-Vac	0.5 a	0.3 a	1.0 a
Sweep	2.8 b	2.8 b	2.8 ab
Ground cloth[b]	7.5 c	7.0 c	5.3 b

[a]Means followed by the same letter do not differ significantly at the 5% level (Duncan's multiple range test)

[b]Ground cloth observation consisted of 2.4/row-m in Experiment 1 and 1.2/row-m in Experiment 2.

Table 7-3. Mean numbers of *P. includens* larvae sampled from soybean by variations of D-Vac, sweep, and ground cloth methods (adapted from Turnipseed 1974)[a]

	Field 1		Field 2					
	Large + Small		Large		Small		Large + Small	
	$\bar{X}$	C.V.	$\bar{X}$	C.V.	$\bar{X}$	C.V.	$\bar{X}$	C.V.
D-Vac vertical (7.6 m)	0.3 a	200	3.0 a	61	4.5 a	46	7.5 a	46
D-Vac horizontal (7.6 m)	0.3 a	200	2.3 a	43	4.0 a	71	6.5 a	62
15 Sweeps across	2.0 bc	41	13.0 b	0	10.3 bc	29	23.3 ab	13
15 Sweeps along	0.8 ab	128	5.0 a	33	7.3 ab	73	12.3 a	40
Ground cloth beat (2.4 m)	8.0 d	27	18.3 b	59	25.5 d	29	43.8 c	41
Ground cloth shake (2.4 m)	3.0 c	61	11.0 b	43	14.8 c	40	25.8 b	40

[a]Means followed by the same letter do not differ significantly at the 5% level (Duncan's Multiple Range Test).

complex. Relative net precision was greatest for all sweep methods tested. They speculated that the inability of the ground cloth to more completely estimate populations of the more tenacious larvae, such as loopers (which they said were able to reestablish a foothold on the plants instead of falling to the ground), accounted for the inefficiency of this method in estimating absolute populations of the grouped lepidopterous larvae.

Rudd and Jensen (1977) compared the two relative sampling methods—sweep net and ground cloth. Results of regression analysis showed that numbers of soybean loopers captured by the two methods were highly correlated (r = .987, $P < 0.01$). With two sampling methods that are so highly correlated, it is possible to compute factors for calibration of one method with another. Louisiana recommends a treatment threshold for soybean looper of 26 larvae/row-m (8/row-ft) (Newsom et al. 1975). By Rudd and Jensen's (1977) calculations, 38.4 larvae/25 sweeps = 26 larvae/row-m; this reduces to 1.54 larvae/sweep. Extrapolation to numbers—50 sweeps and 100 sweeps yields 77 and 154 larvae, respectively (may be rounded to 75 and 150, respectively, introducing only 3% error). A ratio of the standard deviation to the mean $s/\bar{x}$ was calculated as a "dimensionless indicator of the variability of a given set of data . . . ." From their data, this ratio was 1.14 for the ground cloth and 1.12 for the sweep net. They indicate that the least variable method will provide more information per sample if sample sizes are equal, or require smaller samples to obtain the same information. Since there is only a 2% difference, samples of soybean looper taken by the two methods provide essentially the same amount of information. However, the time required for ground cloth sampling was twice that of the sweep net; and if two samplers were required with the ground cloth, as was the case with Turnipseed (1974), this would result in a four-fold difference in the cost of obtaining the same information.

Sampling soybean looper larvae by means of a sweep net may be disadvantageous for some types of experiments. Newman and Carter (1975) have shown that samples of soybean looper larvae collected with a sweep net have a higher level of infection by the pathogen *Entomophthora gammae* Weiser than comparable samples of larvae collected with a shake cloth. They speculated that collections made with the sweep net overestimated the infection level of the field population because infected larvae moved to the upper portions of the plants where they were more easily captured with the sweep net. However, soybean looper larvae collected by the two methods showed no difference in infection by *Nomuraea rileyi* (Farlow) Samson.

B. Calibration of Sampling Procedures

For sampling larvae of lepidopterous insects, such as the soybean looper, numbers of individuals captured by the ground cloth method has historically been considered approaching absolute numbers (Rudd and Jensen 1977), however, research has not yet shown this to be true. For most pest management purposes it will be sufficient to consider this as being the case. What remains, then, is to calibrate other commonly used sampling methods to the ground cloth method. Only Rudd and Jensen (1977) thus far have attempted to do this for the soybean looper. They found, as stated above, that 1.5 larvae/sweep is equal to 26 larvae/m of row.

To attempt a direct comparison of the results obtained by Rudd and Jensen (1977) and Turnipseed (1974), one must assume that sweeps across two rows capture twice as many larvae as sweeps across one row. In the Rudd and Jensen (1977) study 38.4 larvae/25 sweeps = 26 larvae/row-m; this reduces to 1.54/sweep. In the Turnipseed (1974) study numbers of larvae captured per sweep across two rows equalling 26 larvae/m ranged from 0.96 to 2.16.

Since there are so many variations of the sweep net method and since even with a standardized sweeping method individual variation exists as to force of the sweep, distance between sweeps, etc., it would be advisable for samplers adopting the sweep net method of sampling to calibrate their individual variation with the ground cloth method. The same would hold true for anyone adopting the D-Vac sampling method.

To accomplish this calibration it is necessary to take a number of paired comparison samples. It is difficult, even impossible, to achieve a high degree of correlation when comparing methods on an observation/observation basis but this difficulty is overcome by making comparisons on a per field basis (Rudd and Jensen 1977). Regression analysis is conducted using means calculated for each method in each field on each sampling date (see Chapter 2).

C. Spatial Patterns

Knowledge of the population sampling distribution function (dispersion characteristics) warrants the design of sequential sampling plans that allow management decision making subject to prespecified risk levels (see Chapters 4 and 5). Since the spatial pattern of individuals in the sampling universe has a very significant influence on the sampling plan, it is essential that this relationship be understood.

The sampling distribution of the soybean looper and the cabbage looper on soybean seems to follow a Poisson series (Shepard and Carner 1975, Shepard and Greene 1980). Indeed, further analysis of the data presented by Rudd and Jensen (1977) indicated that this species was distributed at random in all situations studied (W. G. Rudd personal communication).

The dispersion characteristics of soybean looper larval populations sampled in Louisiana in 1976 are shown in Table 7-4 (Herzog, unpublished). In only one instance did Poisson series analysis indicate that the population was not dispersed at random. These samples were taken in a field that had received two applications of the foliar systemic fungicide benomyl. Since this fungicide ad-

Table 7-4. Dispersion characteristics of soybean looper larval populations in Louisiana, 1976

Date	N	Mean # Larvae/ 50 sweeps	s^2	$s^2/\bar{x}$	Poisson d.f.	x^2
Rapides Parish						
7/8/76	120	0.001	0.001	1.00	1	0.002
7/22/76	180	0.050	0.059	1.18	1	3.20
7/29/76	180	0.167	0.173	1.04	1	0.26
8/5/76	90	0.256	0.215	0.84	1	1.45
8/12/76	76	0.579	0.673	1.16	1	1.93
8/19/76	72	1.819	1.840	1.01	3	3.64
St. Landry Parish						
7/26/76	96	0.031	0.031	1.00	1	0.48
8/2/76	96	0.208	0.251	1.21	1	3.75
8/9/76	48	1.458	1.790	1.23	3	1.55
	48	0.875	1.261	1.44	1	2.74
8/24/76	24	13.417	30.265	2.26	6	3.18
	24	17.817	25.174	1.41	6	1.52
8/30/76	24	7.250	17.114	2.36	4	1.27
	24	9.458	30.242	3.20	6	17.31

versely affects pathogenic fungi that infect soybean looper, it is possible that improper application may have resulted in a nonrandom dispersion of larvae. In all other cases analyses revealed that populations were dispersed in random fashion. Therefore, for the calculation of sequential sampling plans, the soybean looper larval pattern is considered to follow the Poisson series.

D. Sequential Sampling Plans for Pest Management Decisions

In addition to the distribution pattern a second piece of information needed for the construction of a sequential sampling plan is the economic threshold of infestation—that level of infestation above which economic loss would exceed the cost of control. Soybean looper larval population levels presently recommended for control in most states are 26 larvae/m of row (Kogan 1976) during the critical pod-set through pod-filling period (soybean plant growth stages R3 through R5 of Fehr and Caviness 1977, see Chapter 1). However, in Florida the recommended treatment level is 13 larvae/m or row during the pod-filling period, but 33 larvae/m of row in vegetative through full bloom (R2) stages (Strayer and Greene 1974). As illustrated above, 26 larvae/m of row = 150/100 sweeps; 13 larvae/m of row, therefore, would be 75/100 sweeps; 33 larvae/m of row would be 188/100 sweeps.

Using methods and formulas supplied by Waters (1955) and Onsager (1976) and based on this author's unpublished data, sequential sampling plans for management have been constructed that can be used either with ground cloth or sweep net sampling techniques. Mechanical and statistical development of such plans are discussed in Chapter 4.

The first sampling plan, constructed for use with a ground cloth that simultaneously samples 1 m sections of two adjacent rows (2 row-m) is identical to that constructed for the velvetbean caterpillar (see Chapter 6). A similar plan was constructed for use with sweep net sampling, with 50 sweeps as the observational unit. All parameters used in calculation of the plan were identical to those used in the preceding plan with the exception that the decision lines are based on thresholds converted to numbers per 50 sweeps. This plan is shown in Fig. 7-6 and in tabular form in Table 7-5.

Since soybean looper larval populations are capable of building rapidly to economic levels, it is essential that numbers of small larvae are monitored closely. In the event that large numbers of early instar larvae are found in the field it will be necessary to shorten the sampling interval to three days to check on the progress of the infestation.

Another very important point is relevant to sampling for management. Soybean looper larval populations are often controlled in the field by epizootics of the fungus *Entomophthora gammae* (Brousseau 1974). With this in mind, the

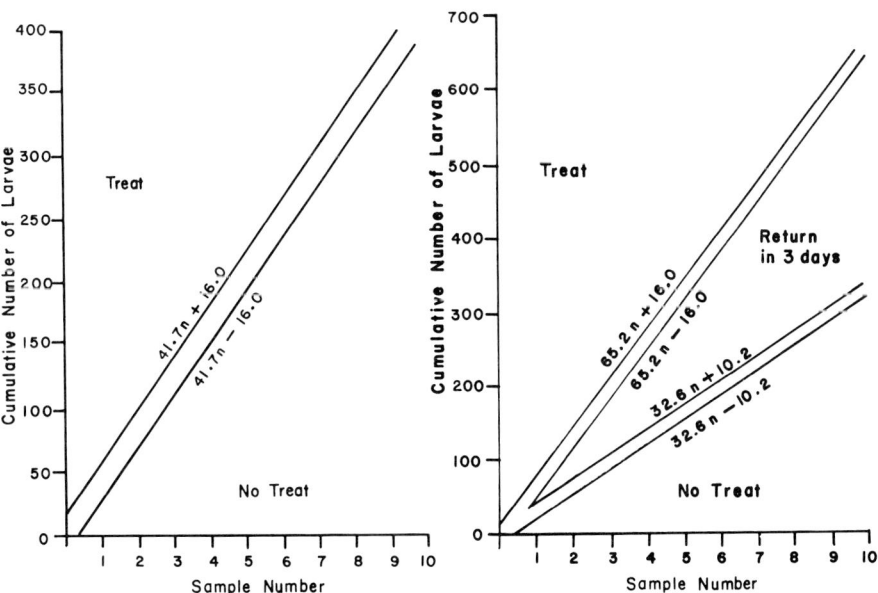

Figure 7-6. Sequential sampling plan for soybean looper larvae using the sweep net sampling method.

Table 7-5. Sequential table for sampling soybean looper on soybean by the ground cloth method

No. 1.8-m samples	Cumulative number of soybean looper			
	≤	≥	≥	≤
1	58	–	–	11
2	99	67	52	32
3	141	109	73	53
4	183	151	94	73
5	225	193	115	94
6	266	234	136	115
7	308	276	157	136
8	350	318	177	157
9	391	359	198	178
10	433	401	219	199
	INITIATE CONTROL	CONTINUE SAMPLING	CONTROL NOT WARRANTED BUT RETURN IN 3 DAYS	CONTINUE SAMPLING	CONTROL NOT WARRANTED

Table 7-6. Sequential table for sampling soybean looper on soybean by the sweep net method

No. 50-sweep samples	Cumulative number soybean loopers			
	≤	≥	≤	≥
1	81	–	–	22
2	146	114	75	55
3	212	180	108	88
4	277	245	141	120
5	342	310	173	153
6	407	375	206	185
7	472	440	238	218
8	538	506	271	251
9	603	571	304	283
10	668	636	336	316
	INITIATE CONTROLS	CONTINUE SAMPLING	CONTROL NOT WARRANTED BUT RETURN IN 3 DAYS	CONTROL NOT WARRANTED

sampler should be alert for evidence of incipient outbreaks of disease within the pest population. Having observed the onset of disease, the management specialist can often postpone insecticlal applications to populations at or above threshold levels. Of course, here again, the sampling interval should be shortened so that the progress of the epizootic in reducing the population can be monitored.

A further comparison of sampling methods may be made by observing the average sample number (ASN) curves shown in Fig. 7-7. Since times required for sample acquisition are comparable (Rudd and Jensen 1977) the ASN curves provide an indirect measure of time required to reach a management decision for the sweep net and ground cloth methods.

Examination of these curves reveals that at all population levels a larger number of observations (hence more time) is required to reach a decision with ground cloth sampling than with the sweep net. For the sake of comparison, if times required per observation by the two methods are similar, and if observations taken by the two methods provide the same amount of information, then the sweep net should be the method of choide because the ground cloth sampling plan requires 56% more samples and, therefore, 56% more time to reach a decision.

The relationship described above can be illustrated by the calculation of a sampling plan with reduced sample size. The plan illustrated in Fig. 6-6 can be

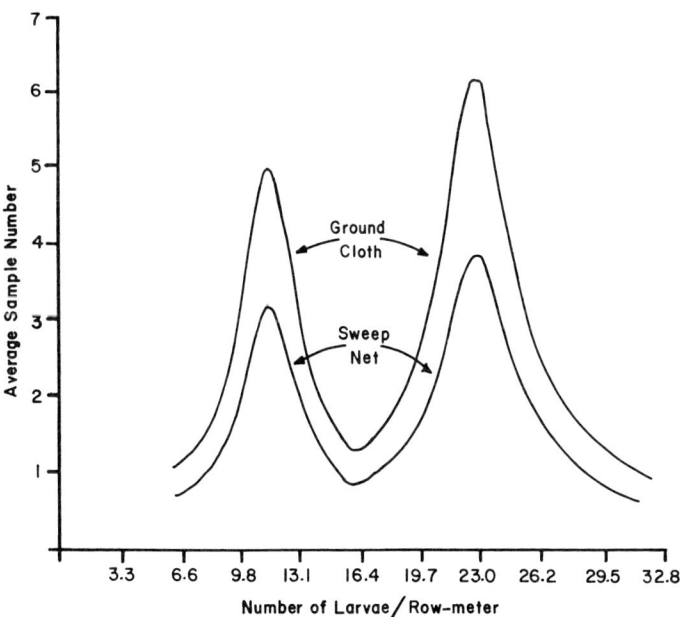

Figure 7-7. Average sample number curves for soybean looper collected by ground cloth and sweep net sampling methods.

converted to use with 1 m ground cloth samples by dividing the cumulative number on the vertical axis by a factor of two. At the same time the average sample number on the vertical axis of Fig. 7-7 must be multiplied by two. The reverse would be true if the sample size were increased.

A similar relationship results if the threshold is increased by a factor of two to a level of 13 larvae/row-m. In this instance the cumulative number of larvae on the vertical axis in Fig. 7-6 should be reduced by a factor of two. The average sample number on the vertical axis is increased by a factor of two, but in this situation the mean larval population on the horizontal axis must be reduced by a factor of two.

E. Sequential Sampling Plans for Parameter Definition

Sequential sampling plans, such as those described, are useless for studies in which point estimates of the population at a given level of precision are required. This is the case with sampling populations with the objective of construction or validation of insect population dynamics models. Formulas for developing a sequential estimation scheme were developed in Chapter 5, and their use in parameter definition for randomly distributed lepidopterous larvae is discussed in Chapter 6.

V. Sampling Pupal Populations

Sampling of soybean looper pupae has not been discussed in the literature. However, since this species pupates in silk-like cocoons attached to the leaves of their host plants, absolute or relative measurements of soybean looper pupal populations may be made by whole plant examinations along a given section of row or, with broadcast plantings, within a given area.

Here again, the pupae of several species of Plusiinae are likely to be encountered. But differences in pupal color are sufficient to allow separation of four species (Canerday and Arant 1967). Pupae of *P. includens* are lightest in color ranging from white to creamy-white to milky-green with irregular black spots. The pupae of *Trichoplusia ni* are uniformly brown, those of *Autographa biloba* are darker brown, and those of *Rachiplusia ou* are dark brown to black.

VI. Indirect Sampling Procedures

Indirect sampling procedures pertaining to soybean looper population assessment involve the estimation and/or measurement of defoliation. When assessing defoliation by means of visual estimation, it is essential that the foliage of the canopy as a whole be examined. Since the soybean looper initially infests the lower strata of the canopy, a superficial examination of the upper canopy across the field will give no indication of impending economic damage. A further discussion of estimation/measurement of defoliation is presented in Chapter 1.

VII. Concluding Remarks

Because soybean looper is difficult to control with many of the currently available pesticides and since resistance to some of the previously efficacious materials is beginning to develop (e.g., methomyl, J. B. Graves personal communication) the development and utilization of sampling plans for their species is assuming added importance. Sampling for decision making should probably give greater attention to numbers of small larvae present than has been in the past. The producer, or the person giving the producer guidance, must be able to identify this species when it is encountered in sampling. If mistaken for more easily controlled species or if the reverse is true, serious consequences often result.

Conclusions presented for velvetbean caterpillar (see Chapter 6) also pertain to soybean looper.

Acknowledgments

Original research reported here was conducted by the author while employed as Assistant Professor, Department of Entomology, Louisiana State University, Baton Rouge, LA. Research supported in part by the National Science Foundation and the Environmental Protection Agency, through a grant (NSF GB-34718) to the University of California. The findings, opinions, and recommendations expressed herein are those of the author and not necessarily those of the University of California, the National Science Foundation or the Environmental Protection Agency.

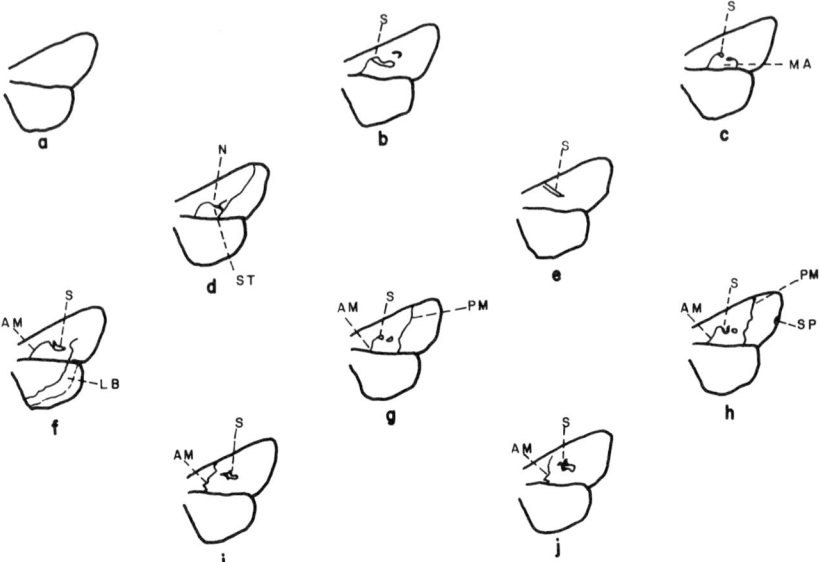

Figure 7-8. Wings of Plusiinae (redrawn after Habeck, 1968). AM, antemedial line; LB, light band; MA, medial area; N, notch; PM, postmedial line; S, stigma; SP, spot; ST, streak.

APPENDIX 7-1: Key for the Identification of Plusiinae moths most commonly found in soybean fields in the United States
(Adapted from Habeck 1968)

1. Forewing without a silver stigma or streak (Fig. 7-8a). . . . .
   . . . . *Allagrapha aerea* (Hübner)
   Forewing with a silver stigma or streak 2
2. Stigma large, extending more than one fourth length of wing, deeply incised above (Fig. 7-8b). . . . . *Autographa biloba* (Stephens)
   Stigma smaller and shaped otherwise. 3
3. Lower medial area and terminal area deep brilliant gold (Fig. 7-8c). . . . . .
   . . . . *Argyrogramma verruca* (F.)
   Lower medial and terminal areas not noticeably more brilliant than rest of forewing 4
4. Forewing with an inconspicuous silver streak located at posterior edge of notch on anterior margin of lower medial area (Fig. 7-8d). . . . .
   . . . . bean leaf skeletonizer, *Autoplusia egena* (Guenee)
   Forewing with a conspicuous silver streak or stigma. 5
5. Stigma an oblique trapezoid, continued back to costa as a pale stripe (Fig. 7-8e) . . . . *Trichoplusia oxygramma* (Geyer)

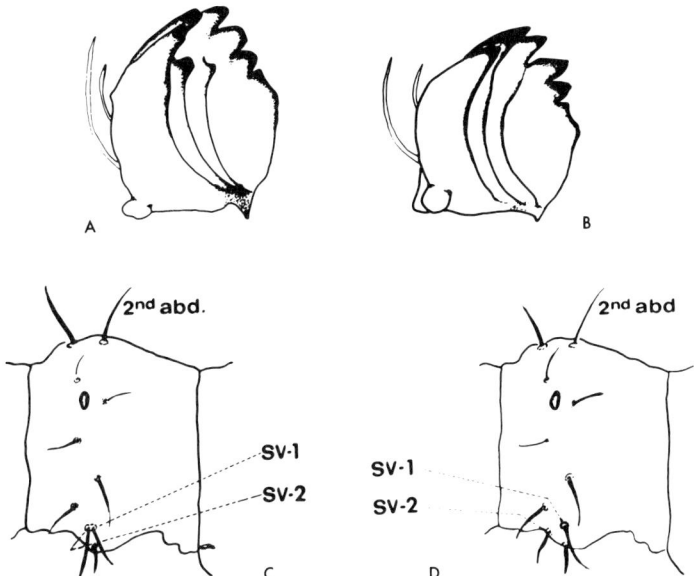

Figure 7-9. Right mandibles (mesal view): (A) *Pseudoplusia includens*, (B) *Trichoplusia ni*. Larval chaetotaxy: (C) pinacula of SV-1 and SV-2 fused on abdominal segment 2 (lateral view), (D) pinacula of SV-1 and SV-2 separated. (Reproduced by permission from Eichlin and Cunningham 1978).

    Stigma not trapezoidal 6
6. Antimedial line distinct, even oblique, nearly straight or evenly curved (Fig. 7-8f,g,h) 7
    Antimedial line indistinct, waved or scalloped (Fig. 7-8i,j). 9
7. Hind wing with a distinct light band (Fig. 7-8f) ....
    .... celery looper, *Syngrapha falcifera* (Kirby)
    Hind wing without a light band, gradually darker toward border. 8
8. Postmedial line dark and well defined, evenly curved, ground color pinkish-brown, no dark spot on margin of forewing (Fig. 7-8g) ....
    .... *Argyrogramma basigera* (Walker)
    Postmedial line not well defined and not evenly curved, ground color dull brown or gray, dark spot on margin of forewing (Fig. 7-8h). ....
    .... soybean looper, *Pseudoplusia includens* (Walker)
9. Hind tibia prominently spined; forewing with dark and light shaded areas, anterior opening of stigma V-shaped, usually narrowest at base (Fig. 7-8i). .
    .... *Rachiplusia ou* (Guenee)
    Hind tibia without visible spines; forewing nearly concolorous through-

out, anterior opening of stigma not V-shaped, wider basally than at neck (Fig. 7-8j) . . . . cabbage looper, *Trichoplusia ni* (Hübner)

If external characteristics for identification are obliterated, as is often the case where sticky traps are used, moths will have to be dissected and the morphology of the genitalia examined. McDunnough (1944) has illustrated those of males and females of *P. includens, T. ni, R. ou, S. falcifera, A. egena,* and *A. verruca* and the males only of *A. biloba*. Callahan (1961) illustrated these species in more detail and included *T. oxygramma* and *A. basigera*. Eichlin and Cunningham (1978) have presented extensive keys to the known North American plusiine species based on genitalic morphology of both males and females. These papers should be consulted when it is necessary to identify specimens by internal morphology.

APPENDIX 7-2: Key for Identification of Plusiinae larvae most frequently encountered in soybean fields in the North America
(Adapted from Eichlin and Cunningham 1978)

1. Vestigial prolegs present on third and fourth abdominal segments. 2
   Vestigial prolegs absent on third and fourth abdominal segments 4
2. Second and third ribs of mandible terminating in processes before reaching cutting margin (Fig. 7-9a). 3
   Second and third ribs of mandible not terminating in processes but continuing to cutting margin (Fig. 7-9b). . . . .
   . . . . cabbage looper, *Trichoplusia ni* (Hübner)
3. Sclerotized ridge of hypopharynx with 10 ridges. . . . .
   . . . . soybean looper, *Pseudoplusia includens* (Walker)
   Sclerotized ridge of hypopharynx with 13 ridges. . . . .
   . . . . *Argyrogramma verruca* (F.)
4. At least one rib of mandible terminating in process before reaching cutting margin. 5
   No ribs on mandible terminating in processes but continuing to cutting margin. . . . . *Rachiplusia ou* (Guenee)
5. Pinacula of setae SV-1 and SV-2 at least partially fused on second abdominal segment (Fig. 7-9c) second and third ribs of mandible terminating in processes before reaching margin; head with a black, lateral stripe including ocelli 1-6. . . . . *Autographa precationis* (Guenee)
   Pinacula of setae SV-1 and SV-2 separated on second abdominal segment (Fig. 7-9d). 6
6. Second and third ribs of mandible terminating in processes before reaching cutting margin . . . . *Allagrapha aerea* (Hübner)
   Second rib of mandible only terminating in process before reaching cutting margin. . . . . bean leaf skeletonizer, *Autoplusia egena* (Guenee)

References

Berger, R. S. 1966. Isolation, identification and synthesis of the sex attractant of the cabbage looper, *Trichoplusia ni*. Ann. Entomol. Soc. Amer. **59**:767-771.
Berger, R. S., and T. D. Canerday. 1968. Specificity of the cabbage looper sex attractant. J. Econ. Entomol. **61**:452-454.
Bottimer, L. J. 1926. Notes on some Lepidoptera from eastern Texas. J. Agr. Res. **33**:797-819.
Brousseau, D. E. (1974). A field study of the seasonal history of *Entomophthora gammae* (Weiser) and its relationship to the soybean looper, *Pseudoplusia includens* (Walker). M. S. thesis, Louisiana State University, Baton Rouge. 62 p.
Burleigh, J. G. 1972. Population dynamics and biotic controls of the soybean looper in Louisiana. Environ. Entomol. **1**:290-294.
Callahan, P. S. 1961. A morphological study of spermatophore placement and mating in the subfamily Plusiinae (Noctuidae, Lepidoptera). 11th Int. Congr. Entomol. **1**:339-345.
Canerday, T. D., and F. S. Arant. 1966. The looper complex in Alabama. J. Econ. Entomol. **59**:742-743.
Canerday, T. D., and F. S. Arant. 1967. Biology of *Pseudoplusia includens* and notes on biology of *Trichoplusia ni, Rachiplusia ou,* and *Autographa biloba*. J. Econ. Entomol. **60**:870-871.
Carner, G. R., M. Shepard, and S. G. Turnipseed. 1974. Seasonal abundance of insect pests of soybeans. J. Econ. Entomol. **67**:487-493.
Chapin, J. B., and P. S. Callahan. 1967. A list of the Noctuidae (Lepidoptera, Insect) collected in the vicinity of Baton Rouge, Louisiana. Proc. La. Acad. Sci. **30**:39-48.
Crumb, S. E. 1956. The larvae of the Phalaenidae. USDA Tech. Bull. **1135**:1-356.
Deitz, L. L., J. W. Van Duyn, J. R. Bradley, Jr., R. L. Rabb, W. M. Brooks, and R. E. Stinner. 1976. A guide to the identification and biology of soybean arthropods in North Carolina. North Carolina Agr. Exp. Sta. Tech. Bull. **238**:264 p.
Eichlin, T. D., and H. B. Cunningham. 1969. Characters for identification of some common plusiine caterpillars of the Southeastern United States. Ann. Entomol. Soc. Amer. **62**:507-510.
Eichlin, T. D., and H. B. Cunningham. 1978. The Plusiinae (Lepidoptera:Noctuidae) of American north of Mexico, emphasizing genitalic and larval morphology. USDA Tech. Bull. **1567**:1-122.
Endris, R. G. 1973. The biology and overwintering of the cabbage looper and soybean looper in Florida. Unpublished M. S. thesis, University of Florida, Gainesville. 59 p.
Fehr, W. R., and C. E. Caviness. 1977. Stages of soybean development. Iowa Coop. Ext. Serv. Spec. Rep. **80**:12 p.
Folsom, J. W. 1936. Notes on little-known cotton insects. J. Econ. Entomol. **29**:282-285.
Genung, W. G. 1958. Investigations for control of insects attacking the pods of table legumes. Fla. State Hort. Soc. Proc. **71**:25-29.

Habeck, D. H. 1968. Annotated key to the Plusiinae moths of Florida (Lepidoptera:Noctuidae). Fla. Dep. Agr. Div. Plant Ind. Entomol. Circ. 72:1-2.

Harding, J. A. 1976. Seasonal occurrence, hosts, parasitism and parasites of cabbage and soybean loopers in the lower Rio Grande Valley. Environ. Entomol. 5:672-674.

Harding, W. C., J. G. Hartsook, and G. G. Rohwer. 1966. Blacklight trap standards for general insect surveys. Bull. Entomol. Soc. Amer. 12:31-32.

Hensley, S. D., L. D. Newsom, and J. Chapin. 1964. Observations on the looper complex of the noctuid subfamily Plusiinae. J. Econ. Entomol. 57:1006-1007.

Herzog, D. C. 1976. Impact of foliar fungicide applications on the incidence of entomopathogenic fungi in defoliating caterpillar populations on soybean. Program of the 3rd Annual Meeting of the Southern Soybean Disease Workers' Council, Baton Rouge, La., Mar. 18, 1976. 28 p.

Hillhouse, T. L., ana H. N. Pitre. 1974. Comparison of sampling techniques to obtain measurements of insect populations on soybeans. J. Econ. Entomol. 67:411-414.

Husin, A. R. B. 1978. Effects of foliar fungicides on the entomopathogenic fungi, *Nomuraea rileyi* (Farlow) Samson and *Entomophthora gammae* (Weiser), and the abundance of defoliating caterpillar populations on soybean. M. S. thesis, Louisiana State University, Baton Rouge. 139 p.

Janes, M. J., and W. G. Genung. 1977. Evaluation of insecticide sprays for control of vegetable leaf miner (Agromyzidae), black cutworm, granulate cutworm, cabbage looper, and soybean looper (Noctuidae) on celery. Belle Glade (FL) Agr. Res. Ed. Cent. Res. Rep. (Mimeo).

Janes, M. J., and G. L. Greene. 1970. An unusual occurrence of loopers feeding on sweet corn ears in Florida. J. Econ. Entomol. 63:1334-1335.

Jensen, R. L., L. D. Newsom, and J. Gibbens. 1974. The soybean looper: effects of adult nutrition on oviposition, mating frequency, and longevity. J. Econ. Entomol. 67:467-470.

Kogan, M. 1975. Plant resistance in pest management. pp. 103-146 in R. L. Metcalf and W. H. Luckmann, eds. Introduction to insect pest management. John Wiley & Sons, N.Y. 587 p.

Kogan, M. 1976. Soybean disease and insect pest management. pp. 114-121 in R. M. Goodman, ed. Expanding the use of soybeans. Proc. of a Conference for Asia and Oceania. Univ. of Illinois, College of Agriculture, INTSOY Ser. 10. 261 p.

Kogan, M., and D. Cope. 1974. Feeding and nutrition of insects associated with soybeans. 3. Food intake, utilization, and growth of the soybean looper, *Pseudoplusia includens*. Ann. Entomol. Soc. Amer. 67:66-72.

Lingren, P. D., G. L. Greene, D. R. Davis, A. H. Baumhover, and T. J. Henneberry. 1977. Nocturnal behavior of four lepidopteran pests that attack tobacco and other crops. Ann. Entomol. Soc. Amer. 70:161-167.

Marston, N. L., C. E. Morgan, G. D. Thomas, and C. M. Ignoffo. 1976. Evaluation of four techniques for sampling soybean insects. J. Kans. Entomol. Soc. 49:389-400.

Martin, P. B., P. D. Lingren, and G. L. Greene. 1976. Relative abundance and host preferences of cabbage looper, soybean looper, tobacco budworm, and corn earworm on crops grown in north Florida. Environ. Entomol. 5:878-882.

McDunnough, J. 1944. Revision of the North American genera and species of the phalaenid subfamily Plusiinae. Mem. S. Calif. Acad. Sci. 2:175-232.

Mitchell, E. R. 1967. Life history of *Pseudoplusia includens* (Walker) (Lepidoptera:Noctuidae). J. Ga. Entomol. 2:53-57.

Mitchell, E. R. 1972. Female cabbage loopers inhibit attraction of male soybean loopers. Environ. Entomol. 1:444-446.

Mitchell, E. R. 1973. Nocturnal activity of adults of three species of loopers, based on collections in pheromone traps. Environ. Entomol. 2:1078-1080.

Mitchell, E. R., J. C. Webb, A. H. Baumhover, R. W. Hines, J. W. Stanley, R. G. Endris, D. A. Lindquist, and S. Masudo. 1972. Evaluation of cylindrical electric grids as pheromone traps for loopers and tobacco hornworms. Environ. Entomol. 1:365-368.

Mitchell, E. R., R. B. Chalfant, G. L. Greene, and C. S. Creighton. 1975. Soybean looper: populations in Florida, Georgia, and South Carolina as determined with pheromone-baited BL traps. J. Econ. Entomol. 68:747-750.

Morishita, F. S., R. N. Jefferson, S. T. Besemer, and W. A. Humphrey. 1967. *Pseudoplusia includens* — a pest of floricultural crops in southern California. J. Econ. Entomol. 60:1758.

Morris, A. P. 1958. Summary of insect conditions — 1957. Virginia. USDA Coop. Econ. Insect Rep. 8(9):152-156.

Myers, T. V., and L. P. Pedigo. 1977. Seasonal fluctuation in abundance, reproductive status, sex ratio, and mating of the adult green cloverworm. Environ. Entomol. 6:225-228.

Newman, G. G., and G. R. Carner. 1975. Disease incidence in soybean loopers collected by two sampling methods. Environ. Entomol. 4:231-232.

Newsom, L. D., R. L. Jensen, D. C. Herzog, and J. W. Thomas. 1975. A pest management system for soybeans. La. Agr. 18:10-11.

Onsager, J. A. 1976. The rationale of sequential sampling, with emphasis on its use in pest management. USDA Tech. Bull. 1526:1-19.

Reid, J. C., and G. L. Greene. 1973. The soybean looper: pupal weight, development time, and consumption of soybean foliage. Fla. Entomol. 56:203-206.

Roelofs, W. L., and R. T. Carde. 1977. Responses of Lepidoptera to synthetic sex pheromone chemicals and their analogues. Annu. Rev. Entomol. 22:377-405.

Rudd, W. G., and R. L. Jensen. 1977. Sweep net and ground cloth sampling for insects in soybeans. J. Econ. Entomol. 70:301-304.

Rudd, W. G., W. G. Ruesink, L. D. Newsom, D. C. Herzog, R. L. Jensen, and N. F. Marsolan. 1980. The systems approach to research and decision making for soybean pest control. In C. B. Huffaker, ed. New technology of pest control. John Wiley & Sons, N.Y.

Shepard, M., and G. R. Carner. 1975. Distribution of insects in soybean fields. Can. Entomol. 108:767-771.

Shepard, M., and G. L. Greene. 1980. Sampling cabbage looper larvae. in Greene, G. L., ed. State of the knowledge on the cabbage looper. Florida Monogr. (in press).

Shepard, M., G. R. Carner, and S. G. Turnipseed. 1974. A comparison of three sampling methods for arthropods in soybeans. Environ. Entomol. 3:227-232.
Shorey, H. H., L. K. Gaston, and J. S. Roberts. 1965. Sex pheromones of noctuid moths. VI. Absence of behavioral specificity for the female sex pheromone of *Trichoplusia ni* versus *Autographa californica*, and *Heliothis zea* versus *H. virescense* (Lepidoptera:Noctuidae). Ann. Entomol. Soc. Amer. 58:600-603.
Strayer, J., and G. L. Greene. 1974. Soybean insect management. Florida Coop. Ext. Serv. Circ. 395.
Teetes, G. L., N. M. Randolph, and M. L. Kinman. 1970. Notes on noctuid larvae attacking cultivated sunflowers. J. Econ. Entomol. 63:1031-1032.
Tietz, H. M. 1972. An index to the described life histories, early stages, and hosts of the Macrolepidoptera of the continental United States and Canada. The Allyn Museum of Entomology, Sarasota, Fla. 1041 p.
Tumlinson, J. H., E. R. Mitchell, S. M. Browner, and D. A. Lindquist. 1972. A sex pheromone for the soybean looper. Environ. Entomol. 1:466-468.
Turnipseed, S. G. 1974. Sampling soybean insects by various D-Vac, sweep, and ground cloth methods. Fla. Entomol. 57:217-223.
USDA. 1956. Truck crop insects. USDA. Coop. Econ. Insect Rep. 5:988-989.
USDA. 1963. Truck crop insects. USDA. Coop. Econ. Insect Rep. 13:318-319.
Waters, W. E. 1955. Sequential sampling in forest insect surveys. Forest Sci. 1:68-79.
Wuensche, A. L. 1976. Relative abundance of seven pest species and three predacious genera in three soybean ecosystems in Louisiana. M. S. thesis, Louisiana State University, Baton Rouge. 384 p.

Chapter 8

Sampling Green Cloverworm on Soybean[1]

Larry P. Pedigo

I. Introduction

The green cloverworm, *Plathypena scabra* (F.) (Lepidoptera:Noctuidae), is one of the most widespread soybean insects in North America, occurring predominantly in the eastern half of the United States (Fig. 8-1). Epiphytotics on soybean have been reported since 1919 (Stone and Pedigo 1972a) and have tended to increase in frequency since that time (Pedigo 1974).

A. Host Plants and Nature of Damage

The most common hosts of the green cloverworm are soybean, alfalfa, clovers, field bean, lima bean, and pea, in that order. Twenty-nine other plant species, mostly legumes, are also fed upon and support development (Pedigo et al. 1973). However, the green cloverworm is known best for its damage to soybean.

Damage occurs when larvae (Fig. 8-2) consume leaf tissue between the main veins, giving leaves a tattered appearance (Fig. 8-3). Populations are present in almost every field each year, and during endemic years, peaks in numbers occur mid-to-late August. In outbreak years, the larval population peak occurs two to three weeks earlier, when plants are in the more susceptible flowering and pod-fill (R2-R5) stages (Pedigo et al. 1972a).

---

[1] Journal Paper No. J-8774 of the Iowa Agricultural and Home Economics Experiment Station, Ames, Iowa. Project No. 1956.

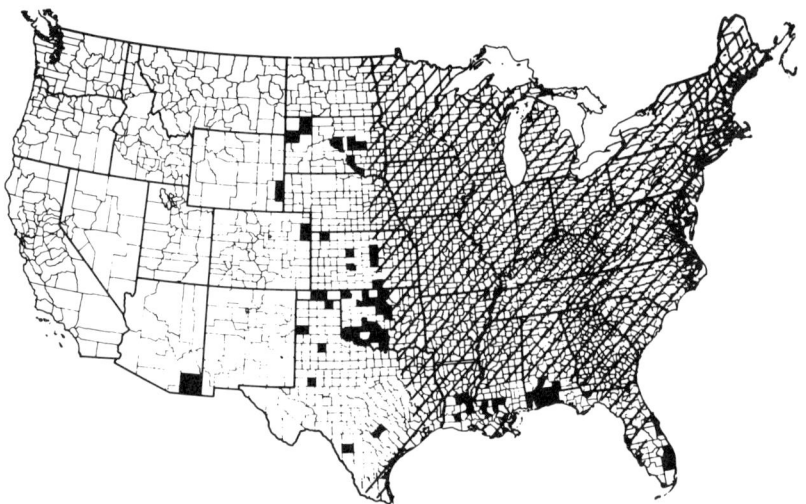

Figure 8-1. Distribution of the green cloverworm in the United States. Cross-hatched area indicates distribution reported by Hill (1925). Blackened areas are new county records that lay along or outside Hill's distribution records (Pedigo et al. 1973).

Figure 8-2. Stage-6 green cloverworm larva.

Figure 8-3. Green cloverworm damage to stage R2 soybean.

B. Life Cycle and Phenology

Green cloverworms overwinter as unmated adults (Fig. 8-4) and possibly pupae (Fig. 8-5) in regions somewhere south of ca. 41°N latitude (Myers and Pedigo 1978). Mating occurs in early spring, and moths disperse northward, laying eggs (Fig. 8-6) singly on available hosts. Eggs hatch in ca. five days, and larvae feed for ca. 19 days, consuming an average of 105 cm^2 of soybean leaf tissue (greenhouse) during development (Stone and Pedigo 1972b). Pupation usually occurs on or just beneath the soil surface, with bits of soil and plant material being incorporated into the cocoon. Adults emerge in ca. 10 days, giving an approximate generation time of 34 days.

Generations of green cloverworms are distinguishable in soybean if samples are taken frequently (at least twice weekly). Probably as many as four generations occur on southern regions (Hill 1925), but two are most common in northern regions.

Peak population levels vary greatly among seasons. During endemic years, population numbers may be <3 larvae/row-m but during epiphytotics, may be >100/row-m. Because of the explosive nature of populations, they should be monitored each growing season for the early detection of outbreaks.

Sampling procedures for monitoring populations have been reported from several studies. Most of these studies emphasized comparison of techniques for

Figure 8-4. Green cloverworm adults: male (left), female (right).

Figure 8-5. Green cloverworm pupa.

Figure 8-6. Green cloverworm egg.

collecting larval data. Only a few studies have been conducted with the adult and egg stage and none with the pupal stage.

II. Sampling Adult Populations

Adult green cloverworm populations have been sampled mostly with techniques that yield relative estimates. Primarily, these are: light traps, waterpan traps, and sticky traps. More recently a flushing technique has been tested for absolute estimates of adult green cloverworm (L. Pedigo, unpublished).

Light traps have been, by far, the most widely used technique for making adult population estimates. Some older studies used lanterns (Forbes 1923a,b) and incandescent lamps in obtaining moth data, and many reported green cloverworm catches as a part of general sampling programs for noctuids and other Lepidoptera (Knutson 1944).

Pedigo et al. (1973) reported samples taken from the 1970 growing season in Iowa with blacklight traps. Traps collected 643 moths, showing peaks in July and cessation of activity in October. Numbers of males captured were far greater than females, producing a ♂:♀ ratio of ca. 7:1. Comparison with sex ratios of emerging adults, indicated that blacklight trap estimates showed sex bias. These results conflicted with those of Knutson (1944) whose catches produced a sex ratio of ca. 1:1.

Myers and Pedigo (1977a) continued studies with blacklight traps in Iowa to assess their potential in forecasting larval population peaks. For these studies a 15-watt blacklight trap (Harding et al. 1966) was used, with a dichlorvos strip (replaced every 21 days) serving as the killing agent. Moth collections from the trap were made every three days. It was found in these studies that peak adult numbers were collected 11-13 days before a peak in larval numbers in the same field. Conclusions from these data were that adult estimates based on blacklight samples had potential in forecasting approaching peaks in larval numbers, with a warning of ca. five to seven days.

Waterpan traps (Fig. 8-7) were used by Myers and Pedigo (1977b) to monitor adult populations and to assess the effects of weather on trap catch. Round waterpans 39 cm in diameter were used, mounted on 10 × 10 cm posts at 1 m above the soil surface. Water level in the trap was maintained at 6 cm and a surfactant (25 ml Chiffon liquid soap/20 ℓ of water) was added to "wet" moths falling into the trap, thereby preventing their escape. The collecting solution was replaced every four days. Using this method, major peaks in moth flight were detected in Iowa, and these were closely associated with nightly minimum temperatures of $21°C$ or higher.

Sticky traps (Fig. 8-8) have also been used to detect peaks of moth flight (Myers and Pedigo 1977b). With this technique, a surface is coated with an adhesive material (e.g., Tack-Trap) and moths touching the surface will adhere to it, unable to escape. Numbers of moths collected in this way were compared with those of waterpan traps in Iowa studies. Results showed that numbers trapped with waterpans greatly exceeded those captured with sticky traps. Be-

Figure 8-7. Waterpan trap for sampling green cloverworm adults. Center cage was used for impounding virgin females to assess potential for attracting males.

cause of the nature of adhesives used on sticky traps, they have the added disadvantage of being difficult to handle and are not recommended for green cloverworm sampling.

A moth-flushing technique was tested (L. Pedigo, unpublished) with one surveyor walking between two rows and brushing the plants on both rows with a rod held in each hand. Three assistants, one in back and one on each side of the surveyor, helped make observations. One of the assistants recorded the data. Sampling was begun on the downwind side of the field, when possible, and proceeded upwind. Moths then tended to fly into the already sampled area when flushed. The flushing technique caused moths to fly, and these could easily be seen and identified by the surveyor and the assistants. Correct identification of green cloverworms in flight was confirmed by frequently capturing flushed moths with a sweep net. Use of this technique in Iowa made it possible to detect peaks of adult moth flight that corresponded to those detected with black-light trapping.

Figure 8-8. Sticky trap for sampling green cloverworm adults. Adhesive was placed on aluminum foil and adhesion surface could be easily removed from holder.

III. Sampling Egg Populations

Little literature exists concerning the sampling of green cloverworm eggs. Only Pedigo et al. (1973) mention taking counts of eggs on leaves from soybean fields. In these studies, they removed 1 080 leaflets (3/plant, 360 plants) over a three day period in 1968 and brought them to the laboratory for counting. In the selection of leaflets, plants were chosen randomly, and a leaflet was removed from each vertical one-third of the plant canopy. A total of 120 eggs and empty choria were counted, giving a mean of 0.11 eggs and empty choria/leaflet for the study period. Analysis showed that eggs were more heavily concentrated in the middle third of the canopy (49.2%), than the upper (29.2%) or lower (21.6%) thirds and that no preference was shown for upper or lower leaflet surface. These in situ counts gave absolute estimates but were very time-consuming to make. Further studies are needed to develop rapid and accurate sampling techniques for this stage.

IV. Sampling Larval Populations

The efficacy of larval sampling procedures is greatly dependent on two major behavioral characteristics of the larvae: violent undulations when disturbed and stratification of certain age groups in the soybean canopy. Larval undulating or flipping behavior occurs when either the substrate on which larvae rest is abruptly jostled or the larvae themselves are touched. This behavior may be an advantage for techniques such as the shake cloth method, which requires dislodging, or it may be a disadvantage for techniques such as the vacuum net, which must retain larvae in an inverted net. Because not all sampling techniques cover the entire soybean plant in the process of accounting for individuals (e.g., sweep netting along the top of the canopy), larval stratification can produce errors in estimating population age structure. Studies have shown that young larvae show no preference for strata but that fifth and sixth stage larvae are most frequently found in the upper third of the soybean canopy (Pedigo et al. 1973). Therefore, samples from only the top of the canopy may strongly bias estimates towards greater proportions of large individuals. Consequently, knowledge of the sampling technique with relation to these behavioral patterns is important in evaluating the accuracy of estimates for specific techniques.

In comparing sampling techniques, a method that produces accurate absolute estimates is needed to serve as a base for the comparison. Methods used to collect data on all the larvae in a given length of row have included: (1) direct observations (Pedigo et al. 1972b), (2) fumigation cage (Kretzschmar 1948, Pedigo et al. 1972a, Hillhouse and Pitre 1974, Marston et al. 1976), and (3) a bag-sampling method (Hammond and Pedigo 1976) (see Chapter 2). Direct observations with in situ counts (Fig. 8-9) are begun by carefully placing white enameled pans on either side of the row where the sampling is to be done to catch

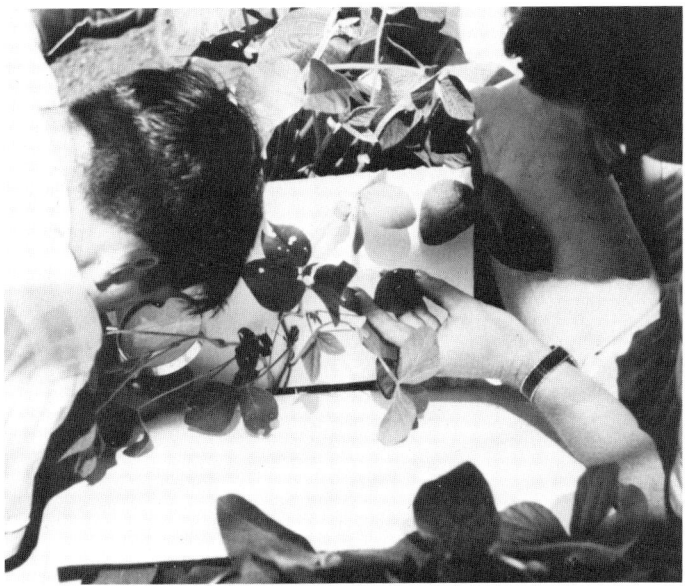

Figure 8-9. Taking in situ counts of green cloverworms on single plants.

any larvae that drop from the plants. White poster-board sheets then are placed on either side of a plant in the row to isolate it, and each leaflet and stem is scanned for larvae. Counts then are recorded for all larvae on the plant, including those that dropped into the pans. Fumigation cages (Fig. 8-10) differ in size and shape but all enclose a length of row and rely on a fumigant (hydrogen cyanide, potassium cyanide, or pyrethrum aerosol) to kill all larvae in the enclosure. Counts are made of dead larvae that drop into a receptacle at the base of the cage or onto the soil surface. Bag samples have been taken by placing a large (120 ℓ) plastic bag over 0.3 row-m of soybeans and cutting the plants at soil level. Bags are tied and taken to the laboratory where larval counts are made.

These absolute-estimate techniques have, at least, two characteristics in common: small sample variability and relatively great cost. For example, mean relative variation, $RV = 100\ S_{\bar{x}}/\bar{x}$, of fumigation-cage samples has been reported as 21% (Pedigo et al. 1972a) and 59% (Marston et al. 1976) lower than those of sweep net samples taken in the same sample area. Greater costs are often incurred because of the time required to remove all larvae from a given area. Fumigation cage samples incur five (Hillhouse and Pitre 1974) to 10 (Pedigo et al. 1972a) times the cost (time) of sweep net samples. Great cost prevents these techniques from being used in extensive survey programs. However, they are indispensable in many intensive sampling programs and serve as a basis for standardization and calibration of relative sampling techniques (see Chapter 2).

Figure 8-10. Fumigation cage for obtaining absolute estimates of green cloverworm larvae in soybean.

Relative sampling techniques are usually less expensive than absolute sampling techniques but often produce greater variation. Probably the most common of these used to sample green cloverworm larvae are the sweep net, ground cloth, and vacuum net. Pitfall traps have also been used to a limited extent.

Several variations of the sweep net approach have been applied. Pedigo et al. (1972a) swept across and down 2 rows in a box-pattern. With this approach, they found that sweep samples gave great relative net precision and produced estimates of great fidelity to the actual larval population. Turnipseed (1974) compared estimates based on sweeping across two rows and sweeping on alternate sides of the same row, finding no significant differences between means.

Hillhouse and Pitre (1974) conducted one of the most comprehensive studies of sweeping techniques in soybeans, comparing four methods (see Chapter 2). They found that sweeping upward against one row produced estimates not significantly different from those of a fumigation cage and required only 20% of the time.

In a study of sample efficiency, Hammond and Pedigo (1976) placed three levels of green cloverworm larvae in 0.66-row-m plots and took one sweep in each plot, upward against the foliage. Data showed that only 14 to 17% of the larvae were removed with the sweep net, but that trends in numbers captured closely followed the densities of plots infested with increasing number of green cloverworm.

It can be concluded from these studies that sweep sampling is an acceptable technique for sampling green cloverworm larvae, when incorporated in an appropriate sampling program. The method is probably most useful in extensive sampling studies, where absolute estimates are not required. Probably the best approach for these studies is to sweep upward against the foliage of one row. In this way, age-class bias will be minimized and acceptable levels of precision may be obtained.

Ground cloths with either the beat or the shake method have been used widely by professional entomologists and soybean growers. No special equipment is needed to take samples, and the method requires little time (see Chapter 2).

In general, most studies comparing various sampling techniques have considered the ground cloth method to give absolute estimates of green cloverworm larvae. However, only Hillhouse and Pitre (1974), and Hammond and Pedigo (1976) presented data on efficiency (percentage of total larvae accounted for in an area) specifically for green cloverworm larvae. Both studies showed that estimates based on ground cloth techniques were not significantly different from those of absolute techniques. In the Hammond and Pedigo study, known numbers of larvae were also placed in plots and then shaken out. Results of these data showed that an average of 90% of the larvae were shaken out of the plant canopy when plants were in growth stages R2 and R4.

Ground cloth samples require about 1.5-9.0 min/row-m to obtain data (depending on whether or not counting time is included), and precision levels generally are adequate for intensive and extensive sampling studies. Because of the high level of efficiency, this technique can be used to obtain absolute estimates. Being able to make these estimates, combined with the advantage of low cost, make the ground cloth one of the best methods of studying green cloverworm larval populations.

The vacuum-insect net (D-Vac) has been included in many studies comparing larval sampling techniques. It is not used widely, however, in extensive survey or grower-based decision programs. Both vertically held and horizontally held procedures have been used to obtain larval estimates with this device (see Chapter 2). Pedigo et al. (1972a), Shepard et al. (1974), and Turnipseed (1974) found that their vertically held technique (900 cm^2 head cone) was not adequate for estimating larval populations. Estimates using this approach were proven strongly biased for smaller larvae and, therefore, are not acceptable for many population studies. Marston et al. (1976) modified the technique by holding a 300 cm^2 distal cone against the side of the row. Estimates obtained were not significantly different from those of an absolute technique and no age-class bias was detected with this approach. However, even though age-class bias can be avoided, the vacuum net still has the disadvantages of being expensive to purchase and cumbersome to use. These disadvantages will probably prevent its widespread adoption, particularly in extensive survey programs and by soybean growers.

Pitfall traps have been used to sample green cloverworm larval populations only in Iowa (Pedigo et al. 1972a). Here, weekly samples were taken with a

Figure 8-11. Tending a pitfall trap used in sampling green cloverworm larvae. Reservoir contains 95% ethanol.

12 cm-diam trap (Fig. 8-11), charged with 95% ethanol. Because larvae do not normally crawl about on the surface, except in search of pupation sites, relative estimates obtained are often erratic and do not have great fidelity to the actual population. Therefore, this technique is not recommended for population studies.

V. Sampling Program

The major considerations that enter into the design of an insect sampling program are: (1) sampling-unit size, (2) number of samples required, (3) pattern of sampling, and (4) time of sampling (Southwood 1978, see also Chapter 3). Sufficient data in these areas are not available to design an entire program for all stages of the green cloverworm. However, most data are available for design of larval programs, particularly with regard to the ground cloth technique.

Sample-unit size for the ground cloth method is largely determined by preference, but data are usually reported in number of larvae per 0.3-row-m. Both large sample units (1.5-row-m) and small units (0.3-row-m) have yielded relatively low RVs, and cost is probably comparable on a row-meter basis (Hillhouse and Pitre 1974, Hammond and Pedigo 1976). Given the number of row-meters to be sampled, a large number of small samples may provide more precise data than a smaller number of large samples. Greater interplot time, however, is a distinct disadvantage with small samples, and should be considered in the selection of a sample-unit size and sample number as related to cost.

Because of cost restrictions, selection of the sample-unit size will partly determine the number of samples that can be taken. However, considering time as a major constraint, the number of samples taken will be mostly determined by the level of precision desired. A mean RV of 13.0 was obtained for estimates in central Iowa, where 20 (0.3-row-m) samples/0.4 ha were taken in stage R3 and R4 soybeans (Hammond and Pedigo 1976). For lepidopterous larvae in general, Hillhouse and Pitre (1974) obtained a mean RV of 8.5 (with transformed data, $\sqrt{\bar{x} + 0.5}$) by taking 36 sample units (18-row-m/sample, 6-row-m/subsample, 1.5-row-m/sampling-unit) in an approximately 0.2-ha area. In the first instance (Iowa) larvae from 0.1% of the total row area were counted, and in the second, larvae from about 3% of the total area were counted. Based on relative net precision ($100/\overline{RV} \, C_s$, where C_s = cost in man h for all samples), these sampling programs were nearly equal (20 0.3-row-m units = 7.7 and 36 1.5-row-m units = 7.8 with transformed means).

A. Sample Number, Pattern, and Timing

To determine the number of samples required for a given level of precision the formula $n = [S_0/(\bar{x}_0 C)]^2$ was used where, n = estimated number of samples required, S_0 = sample standard deviation, $\bar{x}_0$ = sample mean, and C = level of precision ($S_{\bar{x}}/\bar{x}$) desired (e.g., 0.1 for intensive sampling programs) (Karandinos 1976, see also Chapter 3).

Because the dispersion of sampling data (1 0.3-row-m samples) is approximated by a Poisson distribution (Hammond and Pedigo 1976), $s^2 = \bar{x}$ for these samples and the expression $n - 1/\bar{x}_0 C^2$ is applicable.

A family of curves calculated with this equation for 0.3-row-m samples is shown in Fig. 8-12. These curves or calculations using this equation can be used to determine the number of 0.3-row-m samples required per 0.4 hectare (after preliminary samples have been taken) to estimate the mean and obtain a desired level of reliability. As indicated, number of samples required decreases as higher levels of RV are accepted and as the population mean increases. The difficulty in obtaining low RVs when population means are low is evident, with cost becoming a major limiting factor when $\bar{x} < 15$ larvae/row-m and desired RV < 15.

To avoid bias, the pattern of sampling green cloverworm larvae should be based on a random selection of sampling sites. It is often convenient to obtain random coordinates of row by number of steps from one end of the row in determining locations of samples before going to the field. In locating sampling sites for green cloverworms, most researchers have used a stratified-random pat-

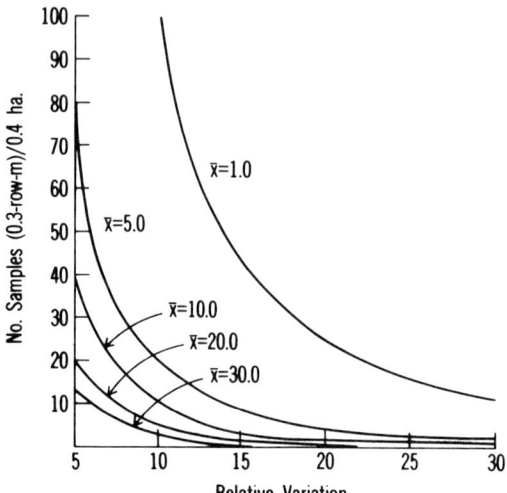

Figure 8-12. Sampling curves for green cloverworm larvae at various population sizes. $\bar{x}$ = mean number per 0.3-row-m.

tern (Pedigo et al. 1972a, Turnipseed 1974, Hillhouse and Pitre 1974, Marston et al. 1976), but Hammond and Pedigo (1976) found that a completely random design was also quite acceptable for obtaining estimates. Estimates made from samples taken in a small area (e.g., 0.4 ha) of a field can subsequently be used to estimate the population of the field if the area is located randomly (nested sample, Bancroft and Brindley 1958) and if it is representative of the field as a whole.

The timing of sampling activities will be greatly determined by the objectives of the sampling program. For detailed population studies, samples will need to be taken at least weekly and perhaps every two to three days (Pedigo et al. 1972b) from the beginning of the soybean growing season until plant maturity occurs. Weekly samples will probably allow the detection of major population peaks but would not give adequate data on dynamics of population age structure. If sampling is to be done simply for the detection of damaging populations, early season activities can be eliminated. Green cloverworm outbreaks most frequently occur during mid-July, especially in the Corn/Soybean Sector (National Academy of Sciences 1975).

Adequate programs of surveillance could, therefore, be initiated weekly from the beginning of July until plants reach the green-bean stage (R6), thereby allowing the detection of impending outbreaks.

B. Surveillance and Sequential Decision Procedure

A sequential decision program has been developed for surveillance of green cloverworm populations (Hammond and Pedigo 1976). The program (Fig. 8-13)

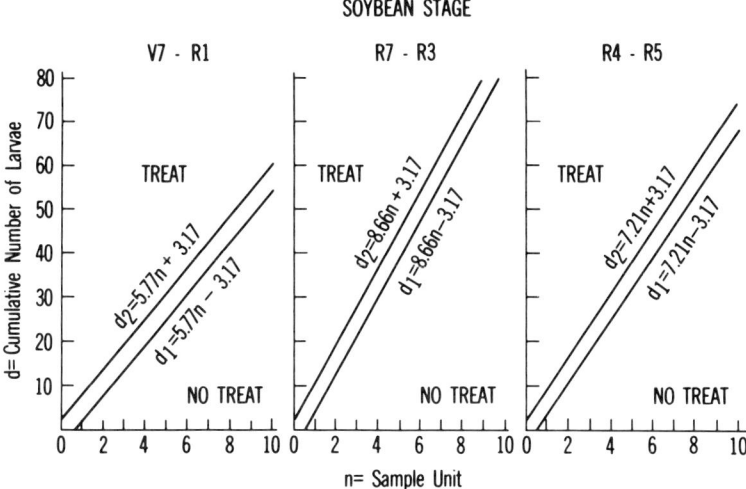

Figure 8-13. Sequential decision programs for green cloverworm larvae at soybean plant-growth stages V7-R1, R2-R3, R4-R5. Based on 0.3-row-m ground cloth samples.

was developed for late vegetative and early reproductive plant-growth stages and is applicable to soybean in the Corn/Soybean Sector. It is useful for farmers, scouts, and survey entomologists who must initiate pest-management activities or make recommendations. Classification of a population level requiring treatment is based on economic thresholds of 24, 36, and 30 larvae/row-m for plant-growth stages V7-R1, R2-R3, and R4-R5, respectively. These thresholds were set at two-thirds of the economic-injury levels developed earlier (Pedigo 1974).

To use the program, the surveyor takes a minimum of three random, 0.3-row-m ground cloth samples in a field and accumulates the number of green cloverworm larvae. The surveyor then determines the plant-growth stage and enters the data into the appropriate graph. If the accumulated number falls above or below the double parallel lines, sampling stops and a decision is made whether to treat or not to treat. If the accumulative number falls between the parallel lines, no decision can be reached, and sampling is continued. If, after taking 10 samples, a decision cannot be reached, sampling is discontinued, and the field is sampled using the same sampling procedure four to five days later. For convenience Tables 8-1 to 8-3 can be used in accumulating sampling data. Using the sequential decision approach to surveillance, farmers, scouts, and survey entomologists may save up to 50% of the time normally spent sampling.

Table 8-1. Sequential decision plan for sampling green cloverworm larval populations at soybean plant-growth stage V7-R1 [a]

Number of samples taken [b]	Cumulative number of larvae collected	Cumulative number of larvae			
			Low level		High level
1	—		—		—
2	—		—		—
3	—	LOW POPULATION	14	CONTINUE SAMPLING	21
4	—		19		27
5	—		25		32
6	—		31		38
7	—		37		44
8	—		42		50
9	—		48		56
10	—		54		61

(HIGH POPULATION column label on right)

[a] Based on a market value of $184.60/ton ($5.04/bu) and a control cost of $7.73/ha ($3.13/a)
[b] Taken with the ground cloth method (0.3-row-m).

Table 8-2. Sequential decision plan for sampling green cloverworm larval populations at soybean plant-growth stage R2-R3 [a]

Number of samples taken [b]	Cumulative number of larvae collected	Cumulative number of larvae			
			Low level		High level
1	—		—		—
2	—		—		—
3	—	LOW POPULATION	22	CONTINUE SAMPLING	30
4	—		31		38
5	—		40		47
6	—		48		56
7	—		57		64
8	—		66		73
9	—		74		81
10	—		83		89

(HIGH POPULATION column label on right)

[a] Based on a market value of $184.60/ton ($5.04/bu) and a control cost of $7.73/ha ($3.13/a)
[b] Taken with the ground cloth method (0.3-row-m).

Table 8-3. Sequential decision plan for sampling green cloverworm larval populations at soybean plant-growth stage R4-R5 [a]

Number of samples taken [b]	Cumulative number of larvae collected	Cumulative number of larvae	
		Low level	High level
1	–	–	–
2	–	–	–
3	–	18	25
4	–	25	33
5	–	32	40
6	–	40	47
7	–	47	54
8	–	54	61
9	–	61	69
10	–	68	76

(Low level column: LOW POPULATION; between columns: CONTINUE SAMPLING; High level column: HIGH POPULATION)

[a] Based on a market value of $184.60/ton ($5.04/bu) and a control cost of $7.73/ha ($3.13/a)

[b] Taken with the ground cloth method (0.3-row-m).

Acknowledgments

Sincere thanks are extended to my past and present graduate students for their many contributions to sampling green cloverworms.

References

Bancroft, T. A., and T. A. Brindley. 1958. Methods for estimation of size of corn borer populations. Proc. Tenth Int. Congr. Entomol. 2:1003-1014.
Forbes, W. T. 1923a. Trap-lantern record at Ithaca, New York (Lepidoptera). Can. Entomol. 55:151-158.
Forbes, W. T. 1923b. Trap-lantern record at Ithaca, New York (Lepidoptera). Can. Entomol. 55:176-184.
Hammond, R. B., and L. P. Pedigo. 1976. Sequential sampling plans for the green cloverworm in Iowa soybeans. J. Econ. Entomol. 69:181-185.
Harding, W. C., J. G. Hartsock, and G. G. Rohwer. 1966. Blacklight trap standards for general insect surveys. Bull. Entomol. Soc. Amer. 12:31-32.
Hill, C. C. 1925. Biological studies on the green cloverworm. USDA Bull. 1336: 17 p.
Hillhouse, T. L., and H. N. Pitre. 1974. Comparison of sampling techniques to obtain measurements of insect populations on soybeans. J. Econ. Entomol. 67:411-414.
Karandinos, M. G. 1976. Optimum sample size and comments on some pub-

lished formulae. Bull. Entomol. Soc. Amer. 22:417-421.

Knutson, H. 1944. Minnesota Phalaenidae (Noctuidae). The seasonal history and economic importance of the more common and destructive species. Minn. Agr. Exp. Sta. Tech. Bull. 165:85.

Kretzschmar, G. P. 1948. Soybean insects in Minnesota with special reference to sampling techniques. J. Econ. Entomol. 41:586-591.

Marston, N. L., C. E. Morgan, G. D. Thomas, and C. M. Ignoffo. 1976. Evaluation of four techniques for sampling soybean insects. J. Kans. Entomol. Soc. 49:389-400.

Myers, T. V., and L. P. Pedigo. 1977a. Forecasting green cloverworm larval population peaks. Iowa State J. Res. 51:363-368.

Myers, T. V., and L. P. Pedigo. 1977b. Factors influencing flight activity of the adult green cloverworm. Proc. N. Centr. Br. Entomol. Soc. Amer. 32:21.

Myers, T. V., and L. P. Pedigo. 1978. Winter survival of the green cloverworm, *Plathypena scabra* (F.) in central Iowa (Lepidoptera:Noctuidae). J. Kans. Entomol. Soc. 51:288-293.

National Academy of Sciences. 1975. Pest Control: an assessment of present and alternative technologies. vol. 2. Corn/Soybeans Pest Control. National Academy of Sciences, Washington, D.C. 169 p.

Pedigo, L. P. 1974. Bioeconomics of Iowa soybean insects. Proc. N. Centr. Br. Entomol. Soc. Amer. 29:56-61.

Pedigo, L. P., G. L. Lentz, J. D. Stone, and D. F. Cox. 1972a. Green cloverworm populations in Iowa soybean with special reference to sampling procedure. J. Econ. Entomol. 65:414-424.

Pedigo, L. P., J. D. Stone, and G. L. Lentz. 1972b. Survivorship of experimental cohorts of the green cloverworm on screenhouse and open-field soybean. Environ. Entomol. 1:180-186.

Pedigo, L. P., J. D. Stone, and G. L. Lentz. 1973. Biological synopsis of the green cloverworm in central Iowa. J. Econ. Entomol. 68:665-673.

Shepard, M., G. R. Carner, and S. G. Turnipseed. 1974. A comparison of three sampling methods for arthropods in soybeans. Environ. Entomol. 3:227-232.

Southwood, T. R. E. 1978. Ecological methods. Halsted Press, N. Y. 524 p.

Stone, J. D., and L. P. Pedigo. 1972a. Selected bibliography of the green cloverworm, *Plathypena scabra* (Lepidoptera:Noctuidae). Bull. Entomol. Soc. Amer. 18:24-26.

Stone, J. D., and L. P. Pedigo. 1972b. Development and economic-injury level of the green cloverworm, *Plathypena scabra* (F.), on soybean in Iowa. J. Econ. Entomol. 65:197-201.

Turnipseed, S. G. 1974. Sampling soybean insects by various D-Vac, sweep, and ground cloth methods. Fla. Entomol. 57:217-223.

SECTION III
Coleopterous Defoliators

Chapter 9

Sampling Mexican Bean Beetle on Soybean

Sam G. Turnipseed and Merle Shepard

I. Introduction

The Mexican bean beetle, *Epilachna varivestis* Mulsant (Coleoptera:Coccinellidae), is among the best known of the insect pests in the United States. It, like the cotton boll weevil, is considered to be a native of Mexico, having invaded the United States in the mid-to-late 1800s.

A. Geographical Distribution

The range of the Mexican bean beetle was limited to the western United States for 75 years after its introduction, the eastern movement presumably limited by the arid plains (Eddy and McAlister 1927, Pallister 1949). After its accidental introduction into the east in the late 1920s (Thomas 1924) it spread rapidly, until it currently inhabits almost all the states east of the Mississippi River. Its current range of distribution extends from Panama to southern Ontario (Kogan 1977) and from Arizona, Utah, and Idaho to the east coast, with a distinct longitudinal gap through the Great Plains (Fig. 9-1.).

B. Host Plants and Nature of Damage

Mexican bean beetle larvae and adults feed on the underside of the leaves of soybean plants and have a peculiar feeding mechanism. The mandibles scrape and crush the lower epidermal and parenchymatous tissues. The cellular contents of

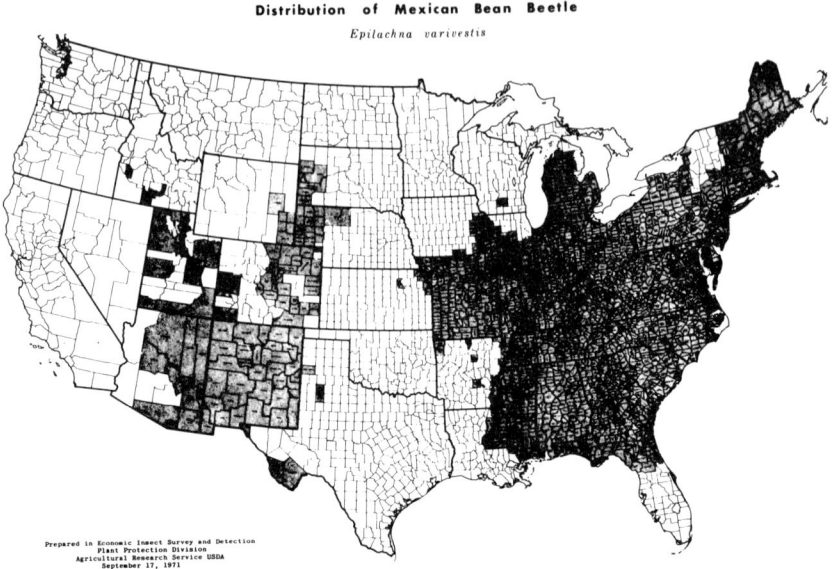

Figure 9-1. Distribution map of *Epilachna varivestis* in the United States. (From USDA, 1971).

these crushed tissues are expressed and sucked into the food channel. The residual tissues, which are not ingested, remain on the leaves as ridges displaying a typical lacework aspect (Fig. 9-2). Adult beetles may open irregular holes, especially when feeding on younger or thinner leaves. The leaves thus injured may not display the characteristic lacework appearance.

Hosts of the Mexican bean beetle are restricted to plants of the family Leguminosae. The preferred hosts are apparently most varieties of snap bean, lima bean, and cowpea (Kogan 1977). *E. varivestis* has long been recognized among the most severe pests of those crops (Howard 1922). Uncultivated plants recorded as hosts include members of the genus *Desmodium*. Early records described only sparse feeding on soybean (Garman 1923, Thomas 1924, Hamilton 1929), but more recently it has become a serious pest of this crop in the coastal plain areas of Delaware, Maryland, Virginia, North and South Carolina, Georgia, and in southern Indiana. In 1977, outbreaks occurred on soybean in Kentucky and Tennessee (C. R. Edwards and G. L. Lentz, personal communication).

C. Life Cycle and Phenology

Eggs, larvae, pupae, and adults occur together on soybean through most of the season (Turnipseed 1973). Only the third and fourth instars and adults consume

Sampling Mexican Bean Beetle on Soybean

Figure 9-2. Typical feeding marks of a fourth instar Mexican bean beetle on a young soybean leaflet. (Courtesy of M. Kogan).

significant amounts of foliage, but populations do not usually reach economic proportions until August because soybean growth during June and July compensates for damage and beetle populations are not usually sufficiently dense to warrant concern until late in the growing season (Carner et al. 1974). Early season predation by beneficial arthropods, such as lady beetles and geocorids, results in high mortality of first generation eggs and larvae.

After overwintering in clumps of grass or other protected areas near soybean fields, adults (Fig. 9-3A) that emerge in the spring usually begin feeding on hosts such as snap or lima beans, completing one or more generations before moving to soybean. However, many beetles move directly from hibernation to early

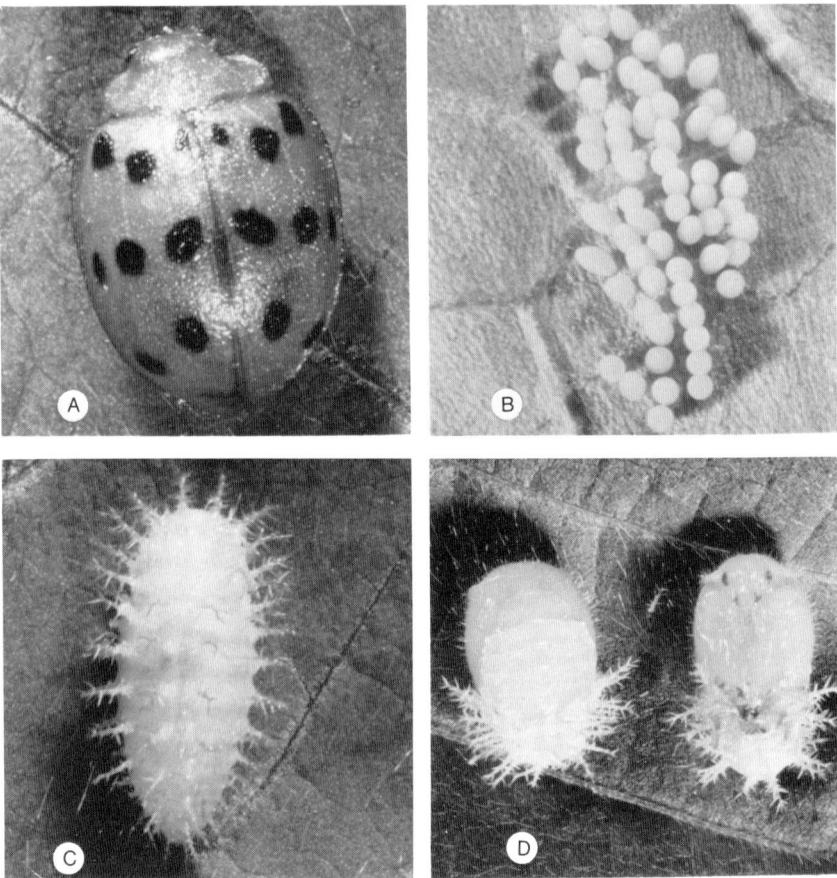

Figure 9-3. Mexican bean beetle life stages: (A) Adult, (B) Egg cluster, (C) Larva, (D) Pupae—dorsal and ventral views. (Courtesy of M. Kogan).

planted soybean where they feed for a few days before depositing egg clusters, each consisting of 40-60 yellow eggs (Fig. 9-3B), on the undersurface of lower leaves. An average of 500 and a maximum of over 1 500 eggs are oviposited per female in ca. 40 days. After a week or longer, depending on temperatures, small yellow larvae hatch from eggs and begin feeding gregariously. There are four larval stages with developmental times for each stage ranging from 3.5 to 7 days, but development is slower when soybean is utilized as food (Bernhardt and Shepard 1978). Fourth stage larvae are approximately 8 mm long, yellow, and covered with several rows of branching yellowish spines (Fig. 9-3C). Late in the season the spines may be completely black. Fully developed fourth instars attach themselves by the caudal end of their abdomens to the underside of lower leaves and change into orange-yellow pupae (Fig. 9-3D). Several pupae often occur on one leaflet. When first emerging from the pupa, adults are soft bodied and light lemon in color. Soon, eight small black spots appear on each elytron, which has turned to dark yellow (Fig. 9-3A). With advancing age a copper hue is attained. Up to four generations may occur during a season, depending on the latitude (Fig. 9-4).

The adults are strong fliers, but dispersal is facilitated by winds. The early northward spread of the species has been calculated at about 300 km per year. Recovery of marked beetles 8 km away from the point of release after two days was reported by Howard (1922).

II. Sampling for Adults and Larvae

Adults and larvae occur together on soybean plants and any sampling technique employed will capture both life forms. Sampling adults and larvae simultaneously is particularly convenient if the objective of the sampling program is to acquire information for management decisions.

Economic losses generally occur from adult and larval feeding in mid-to-late season and most of the sampling for control decisions or research is required during this period. However, adults are sometimes numerous in some areas on soybean in early season, and sampling at this time may be necessary.

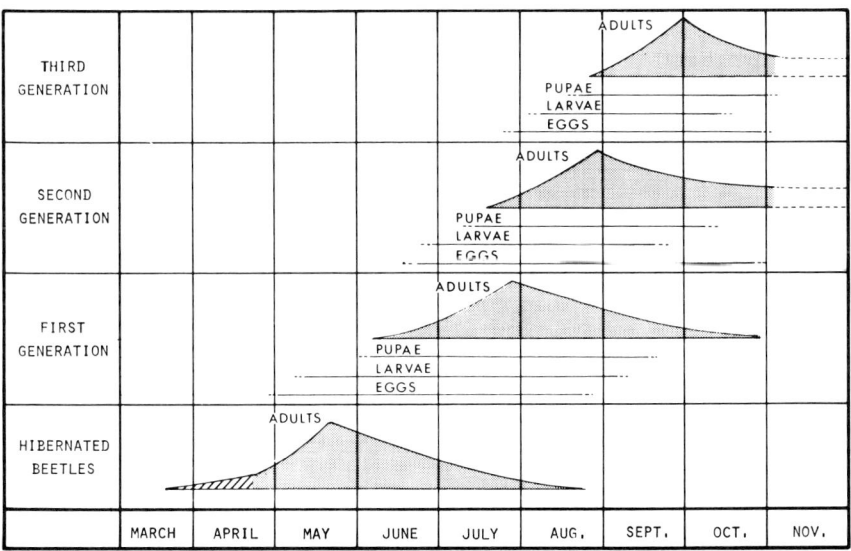

Figure 9-4. Phenology of *Epilachna varivestis* in southern United States. (Courtesy of M. Kogan, unpublished).

A. Sampling Adults from Young Plants

A two-step procedure is efficient for detecting adults on young plants for pest management decisions: first walk through the field and observe damage to foliage (the characteristic brown netting effect is easily detected at this time). If no damage is evident, insufficient numbers of beetles are present for concern. Where damage is present, an estimate of the adult and larval populations can be made by visual observations of a measured length of row (e.g., 4-row-m), by sweep net or absolute sampling methods. Absolute samples may be taken from young plants using the fumigation cage (Kretzschmar 1948) or clamp trap techniques (Mayse et al. 1978, see Chapter 2).

B. Sampling Adults and Larvae from Larger Plants

After the crop attains a height of 30-40 cm, adult and larval populations may be estimated by the standard relative sampling procedure for above-ground soybean arthropods, or measured by some absolute method (see Chapter 2).

C. Comparison and Calibration of Relative Sampling Methods

Relative sampling methods studied for the Mexican bean beetle include variations of the sweep net, ground cloth, and vacuum net methods (see Chapter 2). Shepard et al. (1974) counted larvae and adults in three 1-ha soybean fields at weekly intervals using these three sampling methods: ground cloth with sample-units of 1.2-row-m; 20 sweeps across two rows; and 9-row-m with a 30 cm^2 D-Vac head cone. The data (Fig. 9-5) indicated that ground cloth and sweep net methods were equally effective in sampling Mexican bean beetle larvae and yielded higher means with lower coefficients of variation than did D-Vac. Further, the D-Vac sampling often was unable to detect the presence of larvae and was also ineffective in sampling low populations of adults.

Variations of the ground cloth, sweep net, and D-Vac methods were compared for sampling adults, larvae, and pupae of the Mexican bean beetle (Turnipseed 1974). The results are shown in Tables 9-1 and 9-2. All sweep net and ground cloth methods were equally effective in sampling adults; however, more larvae were sampled with ground cloth compared to sweep net methods. This may be due to vertical stratification of the various life stages, early instar larvae being located primarily in lower levels of the canopy; D-Vac methods proved unsatisfactory for sampling this insect.

More recent research has been conducted in South Carolina on the calibration of the sweep net and ground cloth methods using an absolute sampling procedure. Comparisons were made in a plot of 'Cobb' soybean with a dense canopy of foliage and plants 1.2 m in height. Samples were taken on September 8 and

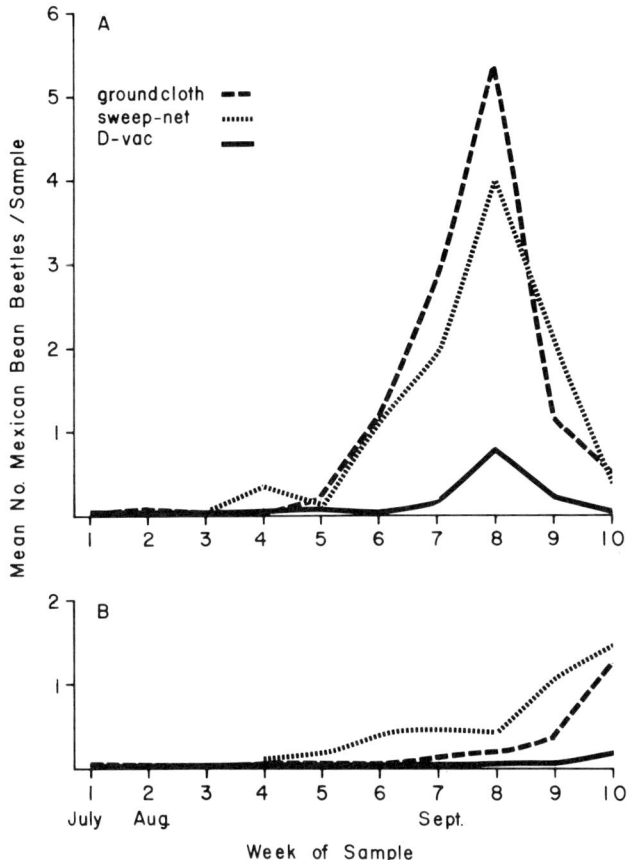

Figure 9-5. Mean numbers of Mexican bean beetle larvae. (A) and adults; (B) collected from soybean using ground cloth, sweep net, and D-Vac sampling methods. (From Shepard et al. 1974, by permission from the Entomological Society of America).

September 28 at growth stages R3 and R6, respectively, at each of six randomly selected sites.

Absolute samples were taken from a 1-m section of row at each site. A total of four samplers were involved—two sampled and counted all life stages, one observed any escaping adults, and one recorded the counts. The sampling proceeded as follows:

> An observer approached the sample site from the side, stopping one row away;
> Two other observers moved along the row from either end and carefully separated the section to be sampled from adjacent plants and rows;

Table 9-1. Mean numbers of the Mexican bean beetle sampled from soybean using D-Vac, sweep, and ground cloth methods

Sampling method* and size	Average number per sample**				
	1968		1969		
	Adults	Larvae	Adults	Larvae	Pupae
DVH 6.1 m	4.5 a	9.3 a	19.5 a	4.3 a	0.8 a
SAC 15 times	16.0 b	49.8 b	110.0 c	19.5 ab	2.3 a
GCB 2.4 m	16.8 b	128.5 c	–	–	–
GCB 1.2 m	–	–	54.5 b	31.3 b	10.3 b

*DVH = D-Vac horizontal; SAC = sweep across, GCB = ground cloth beat.
**Means followed by the same letter are not significantly different at the 5% level (Duncan's multiple range test).

Table 9-2. Mean numbers of the Mexican bean beetle sampled from soybean by two variations each of D-Vac, sweep, and ground cloth methods, 1972

Sampling method* and size	Average number per sample** and coefficients of variability (%)			
	Adults		Larvae	
	$\bar{x}$	C.V.	$\bar{x}$	C.V.
DVV 7.6 m	0.8 a	66.7	8.0 a	42.1
DVH 7.6 m	1.0 a	115.5	3.3 a	72.7
SAC 15 times	8.0 a	91.0	22.5 b	52.7
SAL 15 times	3.8 a	55.0	9.5 ab	56.0
GCB 2.1 m	12.0 a	41.4	48.8 c	41.4
GCS 2.1 m	17.0 a	150.9	56.3 c	45.5

*DVV = D-Vac vertical; DVH = D-Vac horizontal; SAC = sweep across; SAL = sweep along; GCB = ground cloth beat, GCS = ground cloth shake.
**Means followed by the same letter are not significantly different at the 5% level (Duncan's multiple range test).

A 3-m × 3-m sheet of polyethylene was placed at the base of plants along 1 m of row;

Plants were bent over the plastic sheet, the bases were clipped, and the plants and insects were enclosed within the sheet;

The sheet and plants were taken to the edge of the field, and all stages of the Mexican bean beetle from the sheet and plants were counted, as were the few adults that flew away during this process.

A similar process was followed for ground cloth samples; however, after plants were bent over a 1-m cloth, the plants were beaten six to eight times with hands and forearms to dislodge insects. Counts were made in the field directly from the cloth—first of adults and then larvae and pupae.

Fifteen sweeps were taken across and through the upper third of two adjacent rows using a heavy 36 cm diameter sweep net. Contents of the net were taken to the edge of the field, placed on a polyethylene sheet, and the insects collected were counted.

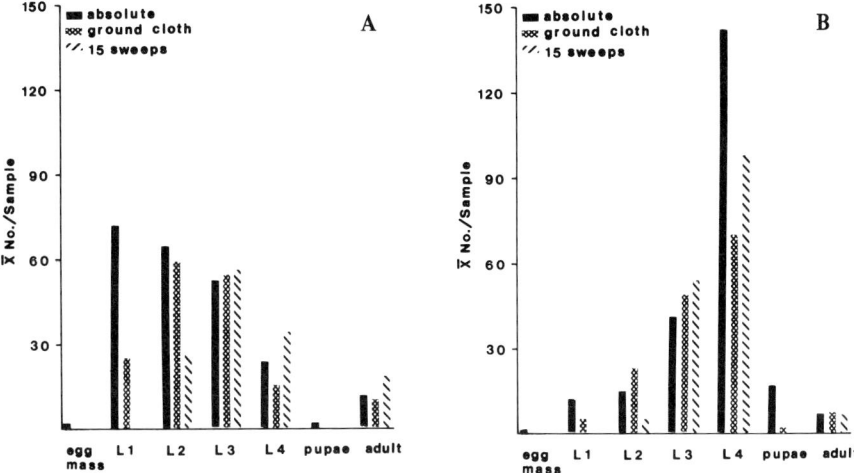

Figure 9-6. Comparison of absolute, ground cloth, and sweep net methods for sampling life stages of the Mexican bean beetle in soybean L_1 - L_4 = larval instars. (A) Samples collected on September 21, 1977; (B) Samples collected on September 28, 1977.

Absolute samples required about 21 minutes for completion of each sample, sweep net samples required 5 minutes each, and ground cloth sampling took an average of 8 minutes. Ground cloth samples required ca. one-third the time of absolute ones, and were efficient in estimating all larval instars and adults (Fig. 9-6). Sweeping 15 times required the least amount of time and efficiently estimated adults and third and fourth instar larvae. Second instars were taken by sweeping, but in much lower numbers compared with ground cloth or absolute methods, and very few first instars were collected by this method. Eggs are normally deposited in the lower one-third of the canopy and only in the second instar do larvae migrate upward to any extent. Because sweeping only involves the upper portion of the canopy, few early stage larvae were taken. Major defoliation is caused largely by late stage larvae and adults. Thus, sweeping for adults and large larvae should produce effective population estimates for these two life stages.

D. Dispersion and Aggregation of Adults and Larvae

The highly aggregated patterns of Mexican bean beetle larval populations in soybean fields has been reported by Shepard and Carner (1976). Neyman's type A and negative binomial distributions described their populations with greatest frequency. Changes over time in the dispersion parameter (k) of the negative binomial describes mathematically how, as the larvae develop, they move away from the egg mass. The k values for the larvae were 0.34 for first instars, 0.49 for the second, 0.63 for the third, and 1.10 for fourth instars. Adults were not so aggregated and numbers obtained from sampling their populations often fitted the Poisson distribution.

III. Sampling Eggs and Pupae

Both the eggs and pupae of the Mexican bean beetle are sessile and are found on the leaves. Sampling these stages is much more laborious than sampling larvae and adults. Estimates of numbers of egg masses or pupae may be made in young beans by examination of randomly selected sections of row. To facilitate this sampling, a 1-m wooden stick can be used to raise lower leaves from the soil surface so that their undersurfaces can be observed for eggs and pupae. In general, accurate population estimates for these forms are difficult. Because these forms are sessile, removal of a given section of row (e.g., 1 m) in several places should allow absolute samples for comparison.

Estimates of egg mass numbers can only be made by direct observation (Fig. 9-6). Pupae, however, may be dislodged and therefore collected by the ground cloth method. Our observations suggest that about 15% of the pupae present may be dislodged while using ground cloth sampling (Table 9-1, Fig. 9-6).

A comparison of a cursory visual examination was made with the absolute method described above. The cursory examinations for eggs and pupae consisted of 3 m of row and were taken placing the forearm under the lower foliage, rolling it over along with a 20-30 cm section of row and quickly counting egg masses and pupae. The results are presented in Table 9-3. Note that cursory examination could detect high numbers of egg masses or pupae but was less effective when numbers were low.

Table 9-3. Comparison of cursory examination with absolute sample method for egg masses and pupae in soybean, 1977

Date of Sample	Egg masses			Pupae		
	Avg number/m		Ratio	Avg number/m		Ratio
	Cursory	Absolute	(Cursory: Absolute)	Cursory*	Absolute	(Cursory: Absolute)
Sept. 8	3.61	8.06	1:2.23	0.72	2.83	1:3.93
Sept. 21	0.28	1.67	1:5.96	0.00	1.0	–
Sept. 28	0.08	0.78	1:9.75	9.00	17.0	1:1.89

*Cursory examination compared with complete examination (leaf by leaf) of standing plants on Sept. 8 and with complete examination of plants that were cut off on Sept. 21 and 28.

IV. Concluding Remarks

These studies do not present an adequate comparison of absolute, ground cloth, and sweep net methods for soybean in different stages of growth. For example, if soybean height was only 60 cm and the canopy was not closed, the sweep net method would probably provide more reliable estimates of first or second instar larvae, and data in Table 9-1 (Turnipseed 1974) indicate that some pupae may be collected. For reliable estimates, absolute samples should be taken as the situation warrants for comparison with estimates by ground cloth or sweep net methods. This can be accomplished easily using the method described below.

In simple scouting programs designed to make treatment decisions, potentially damaging populations can be monitored by the following steps:

1. Check several locations around the perimeter of fields (going in 10-20 rows) for the characteristic feeding injury of adults or larvae. If less than 5-10% defoliation is observed, no further sampling is necessary. If injury approaches economic levels, go to the next step.
2. Sample by sweeping or ground cloth in several places in the field to determine population levels. If numbers of larvae at early developmental stages are considered important, the ground cloth should provide more reliable information.
3. Combine the above steps with cursory examination for egg masses and pupae. Compare with absolute counts, if necessary.

References

Bernhardt, J. L., and M. Shepard. 1978. Validation of a physiological day equation: Development of the Mexican bean beetle on snap beans and soybeans. Environ. Entomol. 7:131-135.

Carner, G. R., M. Shepard, and S. G. Turnipseed. 1974. Seasonal abundance of insect pests of soybeans. J. Econ. Entomol. 57:487-493.

Eddy, C. O., and L. C. McAlister, Jr. 1927. The Mexican bean beetle. S. C. Agr. Exp. Sta. Bull. 236:38 p.

Garman, H. 1923. The Mexican bean beetle in Kentucky. Ky. Agr. Exp. Sta. Circ. 31:16 p.

Hamilton, C. C. 1929. The Mexican bean beetle and how to control it. N. J. Agr. Exp. Sta. Circ. 216:16 p.

Howard, N. F. 1922. The Mexican bean beetle in the southeastern U.S. J. Econ. Entomol. 15:265-271.

Kogan, M. 1977. *Epilachna varivestis* Muls. pp. 391-394. in J. Kranz, H. Schmutterer, and W. Koch, eds. Diseases, pests and weeds in tropical crops. Verlag Paul Parey, Berlin. 666 p.

Kretzschmar, G. P. 1948. Soybean insects in Minnesota with special reference to sampling techniques. J. Econ. Entomol. 41:586-591.

Mayse, M. A., M. Kogan, and P. W. Price. 1978. Sampling abundances of soybean arthropods: Comparison of methods. J. Econ. Entomol. 71:135-141.

Nichols, M. P., and M. Kogan. 1972. A bibliography of the Mexican bean beetle, *Epilachna varivestis* (Mulsant) (Coleoptera:Coccinellidae). Ill. Natur. Hist. Surv. Biol. Notes 77:20 p.

Pallister, J. C. 1949. Mexican bean beetle. Natur. Hist. 58:162-165.

Shepard, M., and G. R. Carner. 1976. Distribution of insects in soybean fields. Can. Entomol. 108:767-771.

Shepard, M., G. R. Carner, and S. G. Turnipseed. 1974. A comparison of three sampling methods for arthropods in soybeans. Environ. Entomol. 3:227-232.

Thomas, F. L. 1924. Life history and control of the Mexican bean beetle. Ala. Agr. Exp. Sta. Bull. 221:99 p.

Turnipseed, S. G. 1973. Insects. pp. 545-572 in B. E. Caldwell, ed. Soybeans: Improvement, production, and uses. Amer. Soc. Agron. Pub., Madison, Wisconsin. Agron. Ser. 16. 681 p.

Turnipseed, S. G. 1974. Sampling soybean insects by various D-Vac, sweep, and ground cloth methods. Fla. Entomol. 57:217-223.

Chapter 10

Sampling Bean Leaf Beetles on Soybean

Marcos Kogan, Gilbert P. Waldbauer, Gilles Boiteau, and Cathy E. Eastman

I. Introduction

The genus *Cerotoma* Chevrolat (Coleoptera:Chrysomelidae) contains ca. 15 species (Wilcox 1972), most (if not all) of which are associated with wild and cultivated leguminous hosts. *C. trifurcata* (Forster) is by far the best known of the species, and much of this chapter is based on information gathered on this species. Bean leaf beetle (Popenoe 1889) is the official common name of *C. trifurcata*, but it has also been used to designate other species of *Cerotoma* (Nichols et al. 1974). Although most information on *C. trifurcata* is probably also applicable to other species of the genus, the reader should be very cautious in making such assumptions. The efficiency of various sampling procedures may differ rather markedly among species because of differences in the species behavior. For example, the alarm response of *C. trifurcata* is usually a drop to the ground (partial thanatosis), which makes the ground cloth a good sampling method for this species. *C. ruficornis* (Olivier), on the other hand, responds to alarm with a quick escape flight; in this case the ground cloth is rather inefficient. Too little is known about other species of *Cerotoma* to permit a detailed comparative analysis.

A. Geographical Distribution

The genus *Cerotoma* is limited in distribution to the Nearctic and Neotropical regions, and one or more species are well established on soybean in most areas where soybean is grown within these regions. There are, however, no published records of *Cerotoma* on soybean in the extreme latitudinal reaches of the crop (south of southern Brazil and north of southern Minnesota).

The given distribution of the species of *Cerotoma* recorded feeding on soybean is shown in Fig. 10-1 (Bouseman and Kogan 1980). *C. trifurcata* is the main species in North America, and its range of occurrence extends from southern Canada to the Gulf States and from the Atlantic coast westward to South Dakota in the north and to Arizona in the south. *C. ruficornis* is widely distributed in the West Indies, and on the mainland it occurs in Florida and Texas and from Mexico to Costa Rica.

B. Host Plants and Nature of Damage

The few species of *Cerotoma* that have well documented host records apparently maintain an oligophagous relationship with herbaceous *Leguminosae*. A list of the valid species with known host associations is presented in Table 10-1.

At least 10 species of *Cerotoma* are known to feed on soybean (Bouseman and Kogan 1980). The host-plant range of *C. trifurcata* is almost certainly restricted to the Leguminosae; reports that it feeds on corn and cotton are probably in error (Herzog 1973). Regarding the wild legumes native to North America, adult *C. trifurcata* have been reported feeding on species of *Amphicarpaea, Desmodium, Lespedeza,* and *Strophostyles*. They also feed on a variety of cultivated legumes in addition to soybean, including cowpea and numerous species and varieties of *Phaseolus* (i.e., string, wax, and lima bean) (Chittenden 1897, Eddy and Nettles 1930, Isely 1930). Adults have been collected also on forage legumes such as clover and alfalfa (Herzog 1973, Waldbauer and Kogan 1976a). The host-plant range of *C. trifurcata* is known largely from adult feeding, but it is generally assumed that host ranges of adults and larvae coincide. Larvae have been found on roots of soybean, cowpea, and *Phaseolus* (Isely 1930), while eggs have been taken from soil near stems of soybean plants (Waldbauer and Kogan 1973, 1975), *Desmodium,* and *Strophostyles* (Waldbauer and Kogan 1976a).

Damage caused by *C. trifurcata* to soybean results from the following: (1) defoliation, damage to pods and seeds, and transmission of virus diseases by adults; and (2) destruction of roots and nodules by larvae feeding below ground level.

In the southern United States—especially in Arkansas, Louisiana, and North Carolina—adults of *C. trifurcata* are frequently economically important defoliators of soybean, but they are not often a threat in the Midwest. In Illinois, for example, damaging populations of adults were recorded in 1953, 1955, 1959, 1964, 1966, 1973, 1977, and 1978; but in 1964 and 1978, years of peak abundance, insecticides had to be applied to only about 0.2-0.4 percent of the hectareage (Kogan et al. 1974, and M. Kogan, G. Waldbauer, C. Helm and C. Eastman unpublished). In North Carolina, *C. trifurcata* problems on soybean are most common in the northeastern section of the coastal plain, where the species appears to prefer heavy organic, poorly drained soils (Deitz et al. 1976). In Brazil, *Cerotoma* spp. have not caused major damage as defoliators, but they

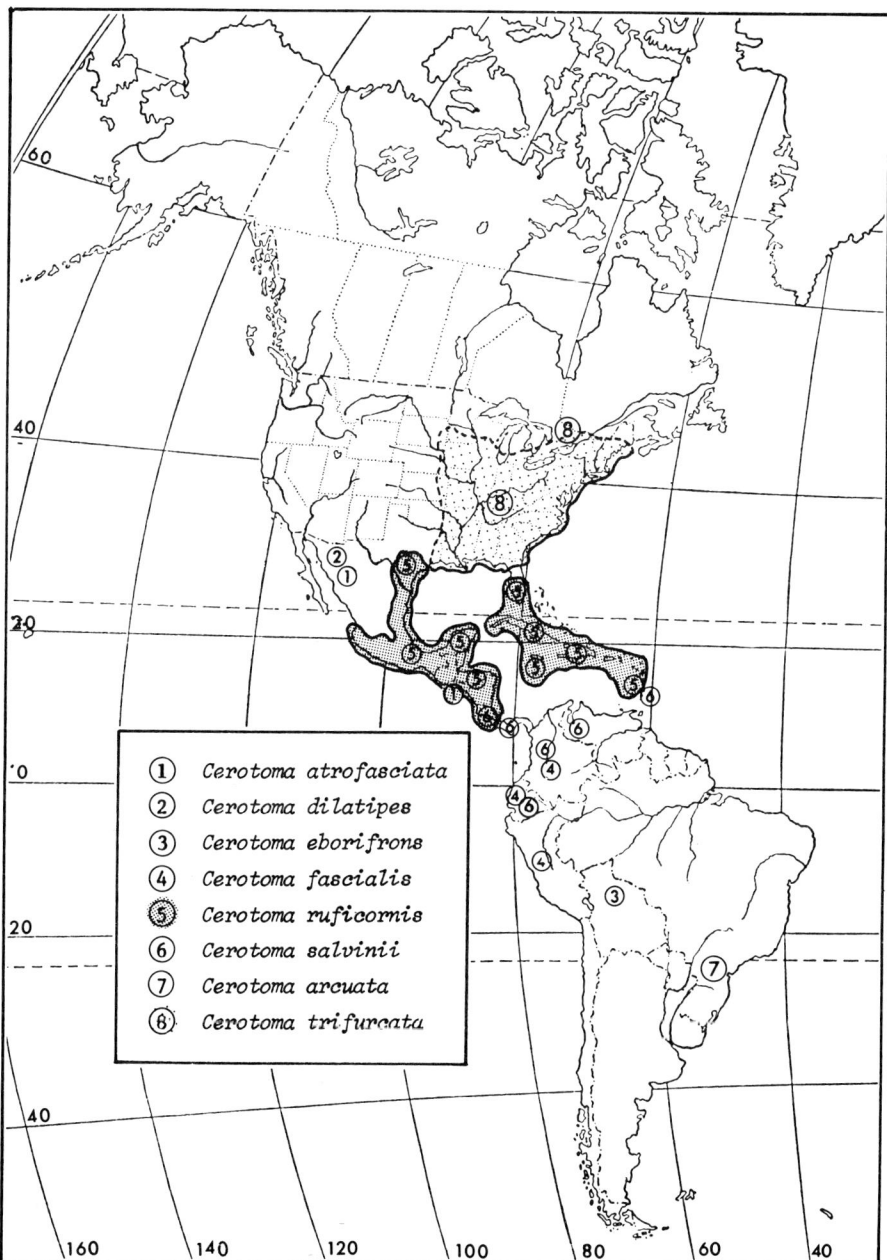

Figure 10-1. Geographical distribution of eight *Cerotoma* species that have been recorded in association with soybean.

Table 10-1. Host records and distribution of species of *Cerotoma* known to feed on soybean

Species	Host plants	Country	Reference[a]
C. arcuata (Olivier)	Leguminosae		
= *C. variegatus* (F.)	*Canavalia ensiformis*	Venezuela	1
= *C. tingomarianus* Bechyné	*Cassia alata*	Venezuela	1
	Dolichos lablab	Venezuela	1
	Glycine max	Bolivia, Brazil, Colombia, Peru, Trinidad	2
		Venezuela	1
	Phaseolus vulgaris	Venezuela	1
	Vigna sinensis	Brazil	3
		Venezuela	1
	Compositae		
	Vernonia scabra	Venezuela	1
	Gramineae		
	Avena sativa	Venezuela	1
	Malvaceae		
	Sida spp.	Venezuela	1
	Solanaceae		
	Capsicum annuum	Brazil	3
	Solanum melongena	Venezuela	1
	(?) "Timbo"	Brazil	3
	Uncertain	Ecuador, Guyana, Surinam, Venezuela	2
C. atrofasciata Jacoby	Leguminosae		
	Glycine max	El Salvador, Costa Rica	2
	Phaseolus vulgaris	Mexico	2
	Uncertain	United States (Arizona, Texas), Guatemala, Honduras, Nicaragua	2
	Leguminosae		
C. dilatipes Jacoby	*Glycine max*	Mexico	2
	Phaseolus vulgaris	Mexico	4
C. eborifrons Ruppel	*Glycine max*	Bolivia	2
		Brazil	9A
C. fascialis Erickson	*Glycine max*	Colombia	2,5
		Ecuador	2
	Phaseolus vulgaris	Ecuador, Peru	2
		Colombia	5
C. ruficornis (Olivier)	*Dolichos lablab*	Costa Rica	11
		Venezuela	1

Table 10-1 (cont.)

Species	Host plants	Country	Reference[a]
	Glycine max	Belize, Mexico, Puerto Rico, West Indies	2
	Phaseolus limensis	United States (Florida, Texas)	2
	P. vulgaris	Costa Rica	5,11
		El Salvador	7
	Vigna sinensis	Puerto Rico	8
	Vigna unguiculata	Costa Rica	6
		Cuba	10
	Uncertain	Guatemala, El Salvador, Nicaragua, Costa Rica, Honduras	2
	Leguminosae		
C. salvinii Baly	*Glycine max*	Colombia, Costa Rica, Ecuador, Panama, Trinidad	2
	Phaseolus vulgaris	Venezuela	1
	Vigna sinensis	Colombia, Venezuela	9,1
C. trifurcata (Forster)	*Amphicarpaea bracteata*	United States	12,13
	Cassia fasciculata		14
	Desmodium cuspidatum		13
	D. illinoense		13
	D. tortuosum		15
	Glycine max		16
	Lespedeza striata		15
	Medicago hispida		14
	M. sativa		13
	Melilotus officinalis		14
	Phaseolus limensis		15
	P. vulgaris		16
	Strophostyles helvola		12,13
	Trifolium dubium		14
	T. incarnatum		14
	T. pratense		17
	T. repens		18
	T. resupinatum		18
	Vicia villosa		17
	V. ludoviciana		17
	Vigna sinensis		16

Table 10-1 (cont.)

Species	Host plants	Country	Reference[a]
	Leguminosae		
C. uncicornis (Germar)	Glycine max	Brazil	2,3
	Phaseolus vulgaris	Brazil	3
	Rutaceae		
	Citrus sp.	Brazil	3

[a]
1. Guagliumi (1966)
2. Bouseman and Kogan (1980)
3. Silva et al. (1968)
4. F. Pacheco in Ruppel (1978)
5. Harries (1975)
6. Risch (1976)
7. Diaz (1972)
8. Perez and Cortes-Mollor (1970)
9. Ruppel (1962)
9A. Ruppel (1978)
10. Kvicala et al. (1970)
11. Gamez (1976)
12. Isely (1930)
13. Waldbauer and Kogan (1976a)
14. D. C. Herzog, University of Florida Agricultural and Education Center, personal communication.
15. Chittenden (1897)
16. Many authors [see bibliography of Nichols et al. (1974)]
17. Davis (1950)
18. Herzog (1973).

may have contributed to the overall level of defoliation caused by other pests (Panizzi et al. 1977).

Adults feeding on leaves make rounded holes that vary considerably in size. This type of leaf injury is generally distinguishable from that caused by lepidopterous larvae (Fig. 10-2A). Colonizing adults often arrive in the field shortly after the plants germinate and may kill very small plants. Entire stands of seedling garden beans have been destroyed in this way (Isely 1930). Seedling soybean can be injured, and some replanting may be necessary if seedlings are killed or if cotyledons are extensively damaged before the unifoliolate leaves open.

Bean leaf beetle adults seem to prefer young plant tissue; when vegetative growth terminates they consume tender pod tissue. Damage is usually limited to the outer layers of the pod (Fig. 10-2B). Developing seeds are not frequently attacked, but seeds within injured pods may suffer from exposure, resulting in inferior quality. Thus, pod damage by *C. trifurcata* is most serious in soybean grown for seed (Deitz et al. 1976).

Adult *C. trifurcata* are known to transmit bean pod mottle virus, cowpea mosaic virus, and southern bean mosaic virus (Walters 1969). The effect of bean pod mottle virus on soybean yields varies with variety and percent of the plants infected, significant yield reductions occurring when 20 to 40% of the plants in a field are infected (Horn et al. 1973). Infection rates exceeding 40% have been reported from fields in Louisiana and Arkansas. In North Carolina, the virus is more damaging to late maturing soybean. Bean pod mottle virus in soybean was not detected in Illinois until 1974 (Milbrath et al. 1975), but no economic

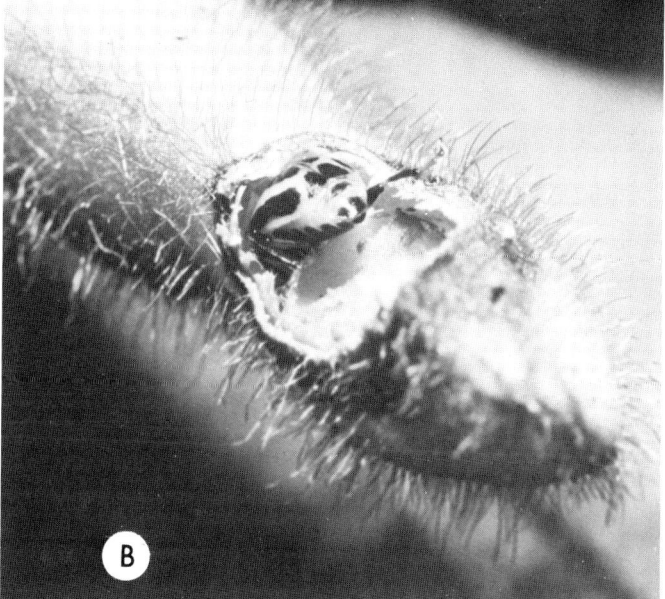

Figure 10-2. Injury of *C. trifurcata* adults to soybean: (A) foliage feeding, (B) pod feeding.

damage has yet been reported. Economic damage due to this same virus occurred in the states of Paraná and Santa Catarina, Brazil (Panizzi et al. 1977).

The economic importance of larval feeding on roots and nodules (Fig. 10-3) is not fully understood although this kind of injury is likely to have some serious impact. Direct injury to soybean nodules has been shown to result in decreased ability of soybean to fix atmospheric nitrogen. The effect on yield, however, has not been fully quantified (Newsom et al. 1978). Such damage is likely to be most serious when coupled with defoliation or stem injury, both of which may also decrease nitrogen fixation (see Chapter 1).

There is no experimental work on economic injury levels for the bean leaf beetle. Computations based on estimated feeding rates of adults (on the order of 2 cm^2 per beetle per day over an average 20-day period) resulted in a probability of economic damage at a population of about 6.5 adult beetles per plant at the R4 stage of development (Kogan 1976).

Figure 10-3. Injury of *C. trifurcata* larvae to soybean N_2-fixing nodules.

II. Life Cycle and Phenology

Most *Cerotoma* species are small (5.0-10.0 mm long), suboval, convex, and highly variable in color pattern. Ground color of the elytra of *C. trifurcata* (Fig. 10-4A-B) may be beige, pink, salmon, orange, or crimson; and the elytra are

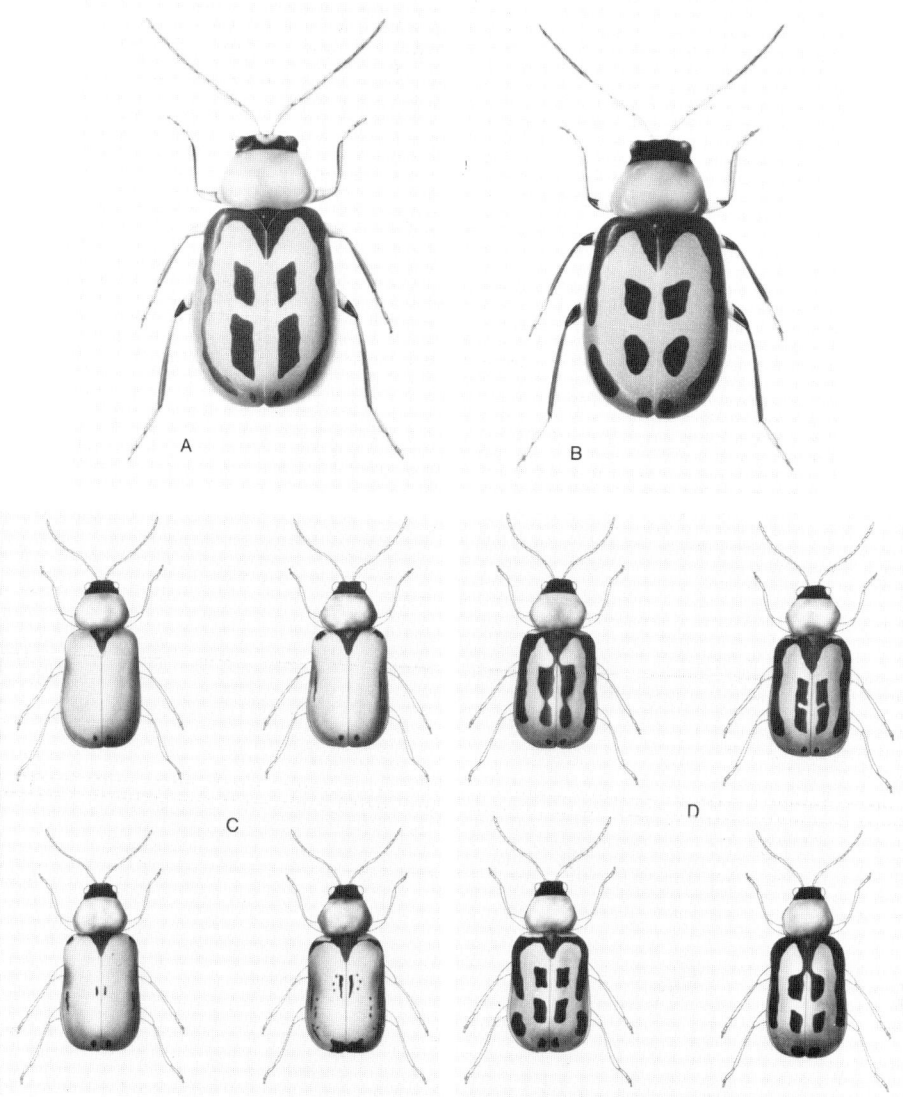

Figure 10-4. Adult *C. trifurcata:* (A) male, (B) female—typical marking pattern; (C-D) elytral markings—extreme variations of patterns. (By courtesy of Bouseman and Kogan 1980, original drawings by J. Sherrod).

variously marked with black spots and vittae ranging from nearly immaculate to heavily patterned (Fig. 10-4C-D). Similar color and pattern variation is observed in *C. ruficornis* (Fig. 10-5A-D). The color of the frons is the best character to

Figure 10-5. Adult *C. ruficornis:* (A) male, (B) female—typical marking pattern; (C-D) elytral markings—extreme variations of patterns. (By courtesy of Bouseman and Kogan 1980, original drawings by J. Sherrod).

separate the sexes; the female frons is black, the male frons is light tan (Horn 1893) (Fig. 10-6A-B).

Some subtle biological differences among color variants of *C. trifurcata* have been detected (Herzog 1973). No differences in adult life history were noted, but beige and pink reproductive females lived much longer than corresponding nonreproductives, while longevity of other color variants was similar whether

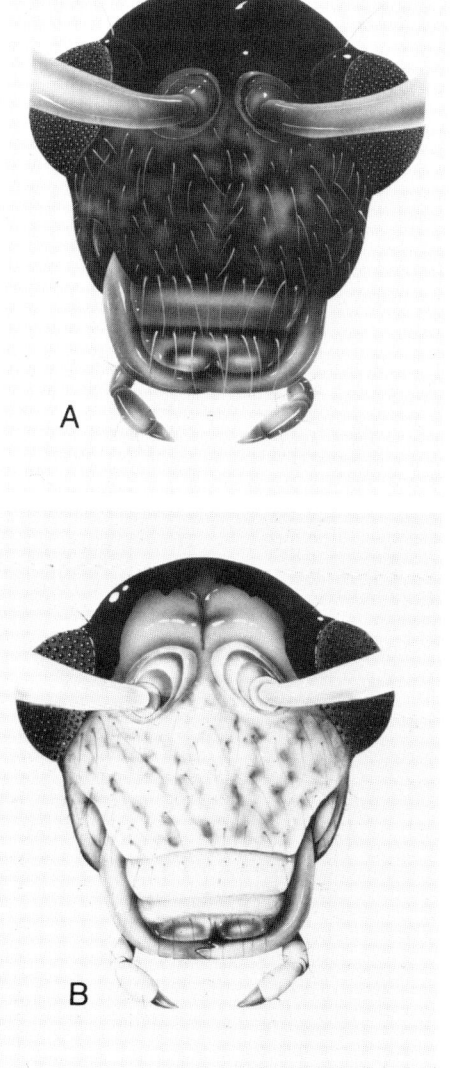

Figure 10-6. Head of *C. trifurcata*, frontal view: (A) female frons, usually black; (B) male frons, usually tan and more coarsely punctuated. (By courtesy of Bouseman and Kogan 1980, original drawings by J. Sherrod).

fecund or not. Females with elytral markings confluent (Herzog 1968) laid more eggs than all others, and significant differences in weight were detected among color variants. Color, however, was not significantly correlated with temperature tolerance; but, in general, marking pattern variants showed greater similarity of response to temperature than did the color forms, suggesting the existence of subtly differing selective values for the many forms under various ecological conditions (Herzog 1973).

C. trifurcata has been found overwintering, as adults, in various habitats available around soybean fields: in clumps of grass (Isely 1930), under a rock (Eddy and Nettles 1930), and in leaf litter, plant debris, and loose top soil (Herzog 1968, Waldbauer and Kogan 1976a, Boiteau 1978). In North Carolina, overwintering bean leaf beetles were found in various sites but were consistently at higher densities in woodlands where they dispersed to at least 520 m. From the edge there was no active selection of sites, and the dispersion of beetles within forested habitats was clumped (Boiteau 1978). Some individuals of the first generation and all of the second generation overwinter. Cold-hardiness experiments showed that beetles of both generations are equally capable of surviving the low temperatures experienced in North Carolina. The highest cold temperature causing significant mortality is near $-5.5°C$ (Boiteau 1978).

Adult beetles are active well before soybean emergence. Beetles were reported emerging from overwintering sites from Mississippi (McConnell 1915) and Louisiana (Herzog 1968) to Illinois (Waldbauer and Kogan 1976a) in early April. In North Carolina, emergence is spread over a period of three months, 60% occurring in May. Emerged beetles were found to disperse over weeds and shrubs growing along ditches, roadsides, and field edges (Boiteau 1978). During early spring in Illinois, adults may be found in alfalfa and sweet clover; but they disappear from these plants when soybean emerges, and no eggs have been found in soil samples from alfalfa fields in which adults had been collected (Waldbauer and Kogan 1976a; L. Turner, M. Kogan and G. Waldbauer unpublished). The nature and importance of suitable wild legume hosts for food, shelter, and oviposition for bean leaf beetle adults prior to availability of soybean have not been determined.

Oviposition in soybean fields starts as soon as plants emerge from the soil. Eggs (Fig. 10-7B) are spindle-shaped, about 0.8 mm long, dark orange, with a coarsely reticulated chorion. Eggs are deposited in the soil at a maximum depth of 3.8 cm, and 92% of the eggs in a field are within a 7.6 cm radius of a soybean stem (Waldbauer and Kogan 1973,1975). Thus, females seem to be attracted to the plant rhizosphere for oviposition. The total oviposition period on soybean averages 38 days in the laboratory, during which time females lay about 350 eggs (L. Turner and M. Kogan unpublished). Total oviposition has been reported to range from 138 to 250 (Eddy and Nettles 1930, Isely 1930, Herzog et al. 1974). Eggs hatch after an incubation period of 5-7 days at $26 ± 1°C$ (Herzog et al. 1974).

Sampling Bean Leaf Beetles on Soybean

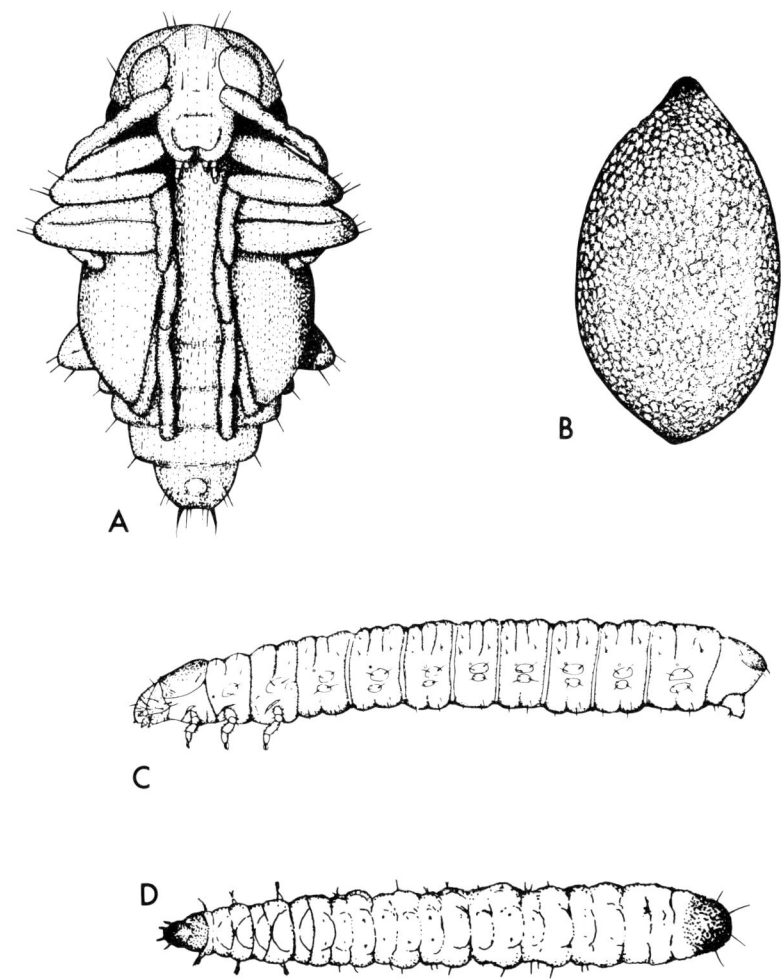

Figure 10-7. Developmental stage of *C. trifurcata*. (A) pupa–ventral view; (B) egg; (C) larva–lateral view; (D) larva–lateral view. (A-C after Isely 1930; D– redrawn from Eddy and Nettles 1973 by L. LaMere).

Larval life is spent in the soil. There are three larval instars, and the older instars seem to disperse away from the tap root, probably in search of nodules, their prime source of nourishment (Levinson et al. 1980). Developmental rates for larvae reared on cowpea cotyledons in the laboratory at 26.2 ± 0.6°C were (in days) 4.55 for the first stage, 3.4 for the second, 7.9 for the third plus prepupal stage, and 7.1 for the pupal stage (L. Turner, M. Kogan and W. Ruesink unpublished). Thus, the total larval development takes 15-16 days at 26.2°C, and it takes about 50 days at 16°C (Isely 1930). Based on laboratory studies, a theoretical threshold for larval development of 10.2°C was computed (L. Turner, M. Kogan and W. Ruesink unpublished) that suggests that soil conditions in the Midwest are adequate for larval development at the onset of the soybean growing season. These temperatures, however, are still rather low, which may account for a field developmental rate about twice to three times as long as that recorded in the laboratory at 26°C. Mature larvae (Fig. 10-7C-D) are ca. 10.0 mm in length. They are whitish, subcylindrical, with three pairs of legs and an anal proleg.

Mature larvae form an earthen cell in the soil where the prepupal and pupal stages are spent. The pupal stage (Fig. 10-7A) takes ca. seven days at 26.2°C (L. Turner, M. Kogan and W. Ruesink unpublished); thus, the complete life cycle from egg to adult emergence takes about 30 days at this temperature in the laboratory. Developmental rates of 47 days at ca. 21°C to 24.8 days at 30°C were reported by Isely (1930).

The seasonal life history of *C. trifurcata* has been studied in Arkansas (Isely 1930), Illinois (Kogan et al. 1974, Waldbauer and Kogan 1976a), South Carolina (Eddy and Nettles 1930), and North Carolina (Boiteau et al. 1979a). In most southern states the species has three complete generations a year, with first generation adult populations peaking about July 1, second generation adults emerging at the end of July and beginning of August, and third generation adults emerging in the middle of September (Eddy and Nettles 1930, Isely 1930). In Illinois and North Carolina, there are two complete generations. Colonizing adults may be present in soybean fields from late May to the end of June, first generation adults from early or mid-July to late August or early September, and the second generation adults from late August or early September until the crop matures (Kogan et al. 1974, Waldbauer and Kogan 1976a, Boiteau 1978). Figure 10-8 shows the phenology of *C. trifurcata* in Illinois. In North Carolina (and perhaps also in Illinois) some individuals of the first generation and all of the second generation overwinter (Boiteau 1978).

Overwintering beetles undergo a reproductive diapause. Diapausing females are generally unmated and have small, immature ovaries (Boiteau et al. 1979b). Few or no eggs are laid by these adults until the following spring. Diapause can be broken in the laboratory by a temperature of 26.7°C and 14 hour light:10 hour dark photoperiod. Under these conditions females kept in overwintering cages started ovipositing within 25 days (L. Turner and M. Kogan unpublished).

In summary, peaks of adult abundance range between four in the southern United States and three in the north, corresponding to the colonizing population that appears as soon as soybean emerges from the soil and two or three generations during the year. Peak egg abundance can be recorded shortly following peak adult abundance (Fig. 10-9). Synchronization of *C. trifurcata* seasonal life history with the soybean crop phenology determines the relative abundance of beetle populations in the Midwest. Late planting has resulted in the virtual disappearance of the beetles for a period of three years, and because of senescence of soybean leaves before peak emergence of second generation adults, the crop

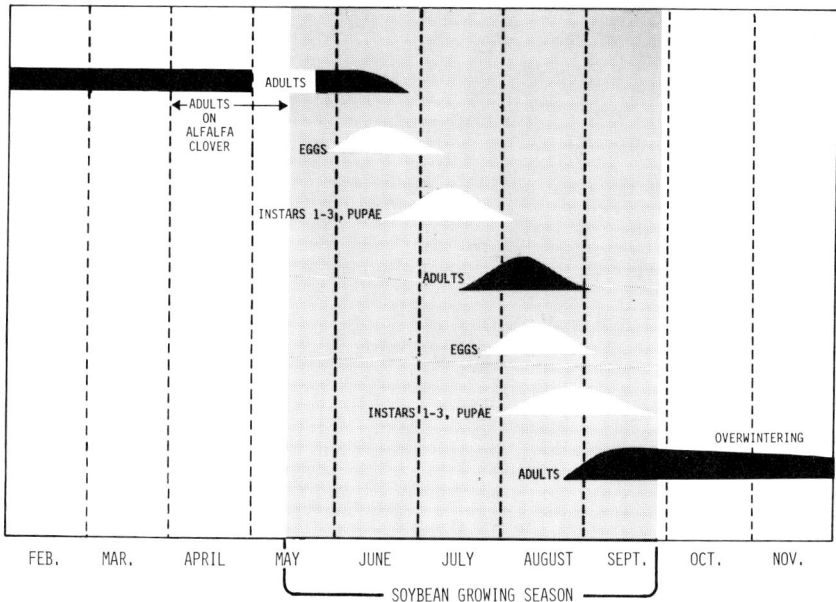

Figure 10-8. Phenology of *C. trifurcata* in Illinois. In North Carolina (and perhaps also in Illinois) a few adults of the first generation also overwinter.

in Illinois generally escapes damage by what is usually the largest population of adults present during the summer (Waldbauer and Kogan 1976b). Peak emergence of second generation adults occurs well before senescence of soybean leaves in the South (Boiteau 1978).

III. Sampling Adult Populations

Adult bean leaf beetle populations have been sampled by the four main procedures used in arthropod population studies on row crops: direct observations, sweep net, ground cloth, and suction net, although most of the work has been done with the first three methods (see Chapter 2). Direct observations and counts over a stretch of row were used very early in the season while plants were in the seedling (VC) stage and until they reached the V4 stage (Kogan et al. 1974, Waldbauer and Kogan 1976b, Boiteau 1978). Direct observations of bean leaf beetles can yield absolute population estimates if carried out with enough care, but the method becomes increasingly inaccurate as plants grow and beetle populations increase. Sampling units of 5-10 m of row are usually considered adequate for direct observation sampling. *C. trifurcata* are very conveniently sampled by this method as they are easily spotted on the sparse foliage or on the

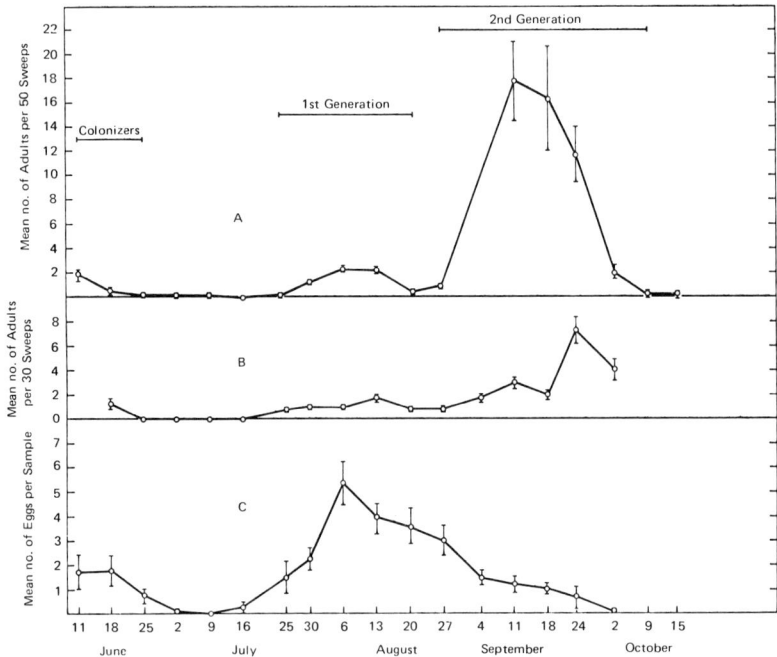

Figure 10-9. Seasonal occurrence of adults and eggs of *C. trifurcata* in central Illinois. Egg samples are 10-cm diameter cores cut 7.6 deep, centered on one plant or on a clump of plants. (From Waldbauer and Kogan 1976a by permission from the Entomological Society of America).

ground (see Chapter 2). Cool early-morning hours are a particularly good time for using direct observations as beetles are not very active then.

Most population studies and scouting for pest management programs have used the sweep net or the ground cloth as a relative sampling technique. In a comparative study of various sweeping procedures, Hillhouse and Pitre (1974) concluded that sweeping upward against the foliage of one row was the most efficient method for sampling adult *C. trifurcata*. However, Boyer and Dumas (1963) recommended use of the ground cloth. One of the main problems with use of the sweep net is the standardization of the sweeping motions (Jensen 1976). On the other hand, the ground cloth method displays higher variability than sweep net and consequently requires larger samples except for high populations. From this, one may conclude that selection of a sampling method for bean leaf beetle adults should not be made a priori. First it must be determined whether the interest is in an estimate of the actual population or in management sampling (Rudd and Jensen 1977). Then various intrinsic and extrinsic factors that may impede one from achieving an expected level of precision with each method must be taken into account.

A. Comparison and Calibration of Relative Sampling Methods

Several variations of sweep net sampling procedures, suction net, and ground cloth sampling have been compared; and attempts to calibrate these various methods have been made by using either the ground cloth (Kogan et al. 1974, Turnipseed 1974, Rudd and Jensen 1977) or fumigation, clam cages, or removal of whole plants as estimates of absolute populations (Hillhouse and Pitre 1974, Marston et al. 1976, M. Kogan, G. Waldbauer, C. Helm and C. Eastman unpublished).

Within-field correlation between sweep net and ground cloth samples was inconsistent (Hillhouse and Pitre 1974—sweeping across one or two rows; Kogan et al. 1974—pendulum sweeps; Rudd and Jensen 1977—sweeping across one row). Rudd and Jensen (1977) obtained a good correlation (r^2 = .803, n = 21) when they regressed numbers of bean leaf beetles caught by the sweep net method on numbers caught by the ground cloth using the mean values per field. They considered the method sufficiently accurate to recommend use of the equation y = (0.387 ± .09)x (where y = the estimated "absolute" population and x = number of beetles in 25 sweeps) to convert sweep net catches to absolute estimates based on ground cloth catches. They recommended the use of the sweep net for sampling bean leaf beetles on the basis that it would take about 50 percent more observations to determine bean leaf beetle populations to a given reliability (measured by $s/\bar{x}$) using the ground cloth method than would be necessary using the sweep net method. Furthermore, the ground cloth method was 2.28 times more costly than a sampling unit of 25 sweeps. They concluded that the sweep net method is economically more efficient particularly when estimating populations of one or two species. However, the ground cloth was better when making comprehensive surveys of many species. There is also evidence that the ground cloth method is as good or better than the sweep net at high populations (Kogan et al. 1974, Rudd and Jensen 1977).

Based on a study conducted on four soybean fields in central and southern Illinois with an exhaustive absolute sampling procedure entailing harvest of caged plants (see Chapter 2), M. Kogan, G. Waldbauer, C. Helm and C. Eastman (unpublished) suggested that conversion of sweep net and ground cloth samples to absolute estimates was too variable to permit use of any general conversion factor. They recommended that sampling programs for research purposes on the bean leaf beetle should be carefully calibrated by the plant caging procedure conducted under the prevailing conditions of the study (row spacing, erectness of the plants, and stage of growth) and taking into account the ability of the surveyors involved in the study. For scouting purposes, however, adequate levels of reliability could be achieved with the use of the ground cloth at moderate to high population levels (number of beetles per row-meter > 10) or sweep net at all population levels.

The suction net (D-Vac) has not yielded good results in sampling bean leaf beetles, and it has not been used to any appreciable extent (Turnipseed 1974).

B. Spatial Patterns and Sampling Program for Adults

Bean leaf beetle adult population spatial patterns in soybean fields are perhaps best described by the negative binomial distribution. Studies conducted in Illinois indicate a tendency to moderate aggregation expressed by "k" values between 6.8 and 7.5 (Kogan et al. 1974; M. Kogan, G. Waldbauer, C. Helm and C. Eastman unpublished). Studies in North Carolina yielded a common k_c = 6.51 for combined 1976-1977 data, although the 1976 data produced k = 2.6 and 1977 produced k = 4.08 (Boiteau 1978).

The measured degree of aggregation has been shown to vary with the seasonal fluctuations in abundance of teneral beetles (Boiteau 1978). The distinctly lower k_c (0.73) obtained for teneral beetles in North Carolina is very similar to the one for eggs (0.64) in Illinois. This indicates that bean leaf beetle aggregation results from a patchy ovipositional pattern. The measured degree of aggregation has also been shown to vary with various sampling factors such as size of the sampling unit and sampling method (see Chapter 3), hence the possibility that results of some studies comparing sampling methods for the bean leaf beetle were affected by inadequate consideration of sampling unit size. Boiteau (1978) has shown that population densities within individual fields were quite uniformly distributed from one area to another except where the homogeneity of the crop was broken by local changes in drainage, soil fertility, and soybean variety or stage. Therefore, homogeneous fields of up to 300 acres need to be sampled only at one location to obtain a satisfactory estimate of *C. trifurcata* abundance. Much greater variability of within-field spatial patterns was, however, reported in one field in Illinois (Waldbauer and Kogan 1976a).

Curves describing the number of ground cloth and sweep net samples necessary to obtain expected levels of precision at various population levels are shown in Figs. 10-10 and 10-11. The curves for ground cloth sampling were computed using k = 6.84 and the equation $n = (k \pm \bar{x})/(C^2 k \bar{x})$ where C = reliability factor defined as a percent of the $S_{\bar{x}}$ to the mean (see Chapter 3). The curves for sweep net sampling were computed using two different equations, one using Morisita's I_δ (Ono 1967) and the other using the regression of the index of mean crowding to *C. trifurcata* mean density (Iwao and Kuno 1971). It is apparent from Fig. 10-11 that 30 to 50 random samples of 20 sweeps each will give estimates of population densities satisfactory for pest management decision making. However, ecological studies would require more intensive sampling (Boiteau 1978). Curves in both figures show that the number of samples required decreases as the required level of precision decreases and population mean increases.

The frequency of the sampling program will depend on the purpose of the program. Since bean leaf beetles colonize soybean fields very early in the season, sampling for both population studies and scouting should start immediately after seedling emergence. In the Midwest, beetles apparently seldom migrate from one field to the next within the same season (Waldbauer and Kogan 1976a,b); and in this region, for scouting purposes, the absence of colonizers

Sampling Bean Leaf Beetles on Soybean

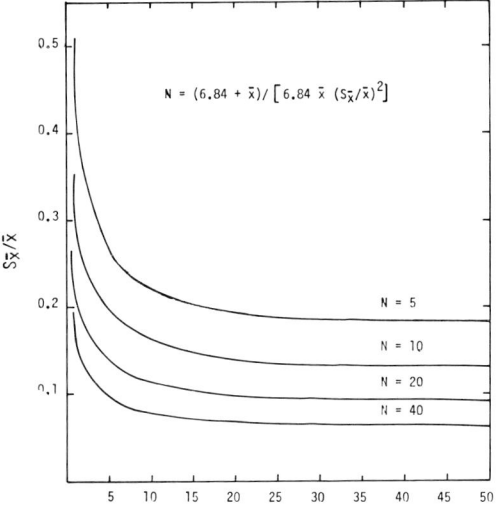

Figure 10-10. Curves describing the number of 1-m ground cloth samples necessary to achieve a certain level of precision according to population density (From M. Kogan, G. Waldbauer, C. Helm and C. Eastman, unpublished).

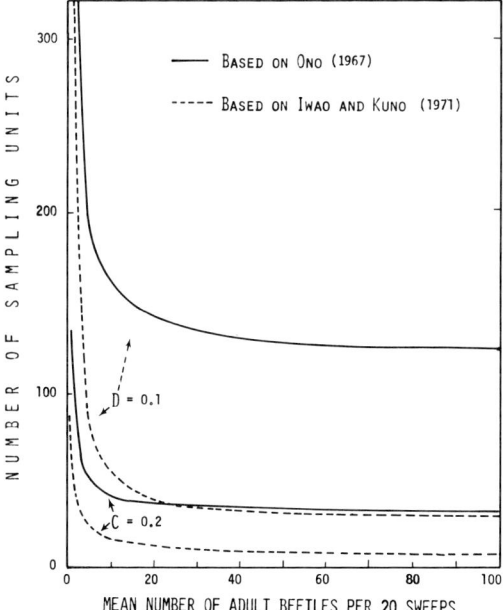

Figure 10-11. Curves describing the number of 20-sweep sampling units necessary to achieve a certain level of precision according to population density (From Boiteau 1978).

from the fields within 1-2 weeks from seedling emergence may very well signal a bean leaf beetle-free season for those fields. In the South it is fairly common for beetles to move to adjacent later-planted soybean fields when plants in the earlier planted, colonized field begin to mature (D. C. Herzog, personal communication). For population studies, sampling *C. trifurcata* at weekly intervals has generally allowed detection of population fluctuations, although at peak abundance closer intervals may be required for more timely detection of phenological phenomena as related to climatic and crop growth conditions.

An additional consideration in timing of bean leaf beetle surveys is the apparent diel beetle population fluctuation. *C. trifurcata* populations have a diurnal cycle and are more intense on soybean plants around 1700 h (Kogan et al. 1974, Boiteau et al. 1979c), but the levels of *C. dilatipes* populations on soybean in Mexico remain constant throughout the day (Pacheco 1976). Since such diel fluctuation, when present, may influence results of the sampling program, it would be advisable to survey the fields at approximately the same time of day throughout the growing season. Bean leaf beetles are daytime fliers with a peak period of flight activity between 1100 and 1500 h (Boiteau et al. 1979d). Therefore, sampling should be avoided if possible during this period because it is characterized by strong fluctuations in the abundance of beetles on the plant.

Sequential sampling programs are presented here for the early (colonizers of plants at seedling stage) and the mid-season populations (first generation in the Midwest, about mid-July, with plants at the R3 stage). The estimated population level at the economic threshold, assumed here to be one-half of the computed economic injury levels (Kogan 1976), is one-half beetle per plant at the seedling stage and three beetles per plant at the R3-R4 stage of plant growth. At a mean plant population of 20 plants per row-m, for 1-m row spacing, these economic thresholds are 10 beetles/row-m and 60 beetles/row-m, respectively. Assuming that the ground cloth estimates absolute populations within 0.4 of the actual population density (M. Kogan, G. Waldbauer, C. Helm and C. Eastman unpublished), the sequential sampling plans for adult beetles on plants at the R3-R4 stage were established for moderate populations between 12 and 18 beetles/row-m (Figs. 10-12 and 10-13). This program will be best used with a ground cloth 0.5-m long and with beat or shake samples taken on the two rows adjacent to the ground cloth (see Chapter 2). Use of the sequential sampling plan is explained in the caption of Fig. 10-12.

A sequential sampling program for southern determinate soybean cultivars is presented in Fig. 10-14. It shows the upper and lower limits obtained for various soybean stages. Acceptance and rejection lines of the sequential plan were determined using an equation (Iwao 1975) that takes into account the index of basic contagion and the density-contagiousness coefficient. The computation of economic injury levels is presented in Boiteau (1978). To use this program the scout must walk between two rows swinging a 38-cm diameter sweep net upward alternatively on each row synchronized with successive steps. Each stroke of the net counts as one sweep and 20 sweeps constitute a sample unit.

Sampling Bean Leaf Beetles on Soybean

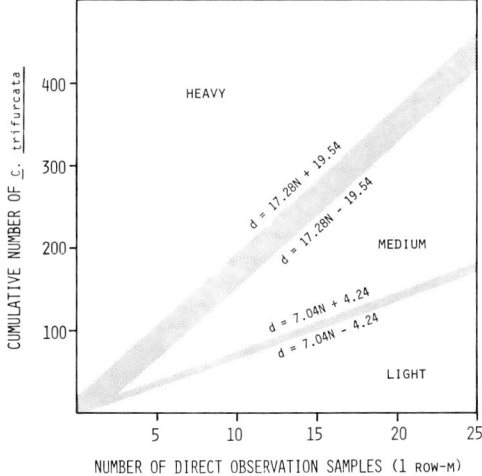

Figure 10-12. Sequential sampling plan for adult *C. trifurcata* on soybean at stage VC to V4 using direct observations over 1 row-m as the sampling unit. Successive samples are taken and the counts are added. After five samples if counts fall in any of the white triangular regions, a decision may be reached based on whether the population is light, medium, or heavy. If cumulative counts fall in one of the shaded zones, additional samples have to be taken (see Chapter 4). (From M. Kogan, G. Waldbauer, C. Helm and C. Eastman, unpublished).

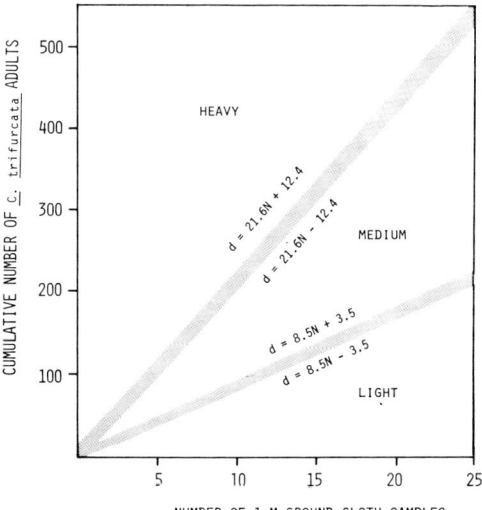

Figure 10-13. Sequential sampling plan for adult *C. trifurcata* on soybean at stage R2-R4, based on ground cloth samples. Use of plan is explained in caption of Fig. 10-12. (From M. Kogan, G. Waldbauer, C. Helm and C. Eastman, unpublished).

Sequential sampling programs for pest management decisions are described in Chapter 4. The programs help reduce the cost of scouting by establishing the minimum number of samples needed to reach a treatment or no-treatment decision at various population levels.

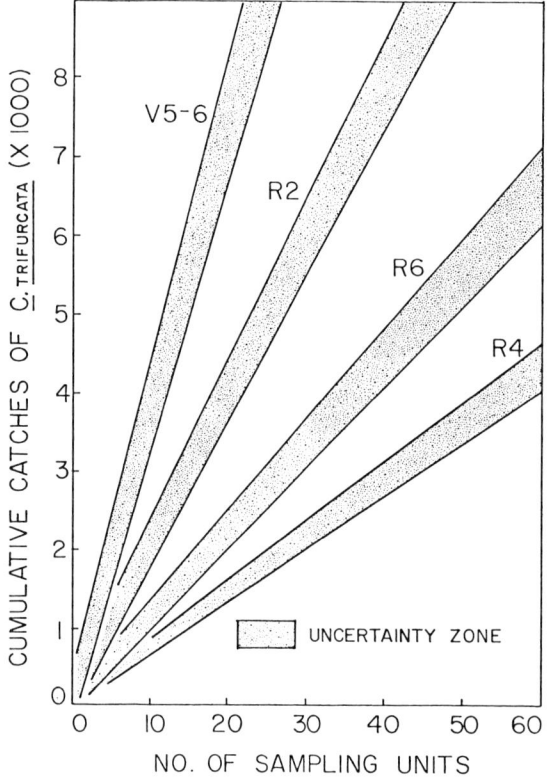

Figure 10-14. Sequential sampling plan to classify *C. trifurcata* populations relative to a critical density at four stages of soybean growth. Treatment and no-treatment areas are above and below the uncertainty zone for each growth stage. (Boiteau 1978).

IV. Sampling Eggs, Larvae, and Pupae

Sampling bean leaf beetle eggs, larvae, and pupae involves two distinct phases: (1) collection of soil samples from the field and (2) extraction of eggs, larvae, and pupae from the soil and separation of the sampled organisms from organic debris. The methods described in this chapter have been tested rather extensively with the bean leaf beetle. Methods used in collection of other soil inhabiting arthropods are discussed in Chapters 16 and 26. The material covered in this section is based mainly on Waldbauer and Kogan (1973, 1975), Anderson and Waldbauer (1977), and Levinson et al. (1980).

A. Collection of Soil Samples

Soil samples are generally collected with long-handled soil corers provided with a close fitting plunger for extruding the cores. Studies conducted in Illinois used three types of corers. Two of the corers were circular in cross section, one 10.2 cm and the other 5.1 cm in diameter. The third corer was square in cross section,

Sampling Bean Leaf Beetles on Soybean

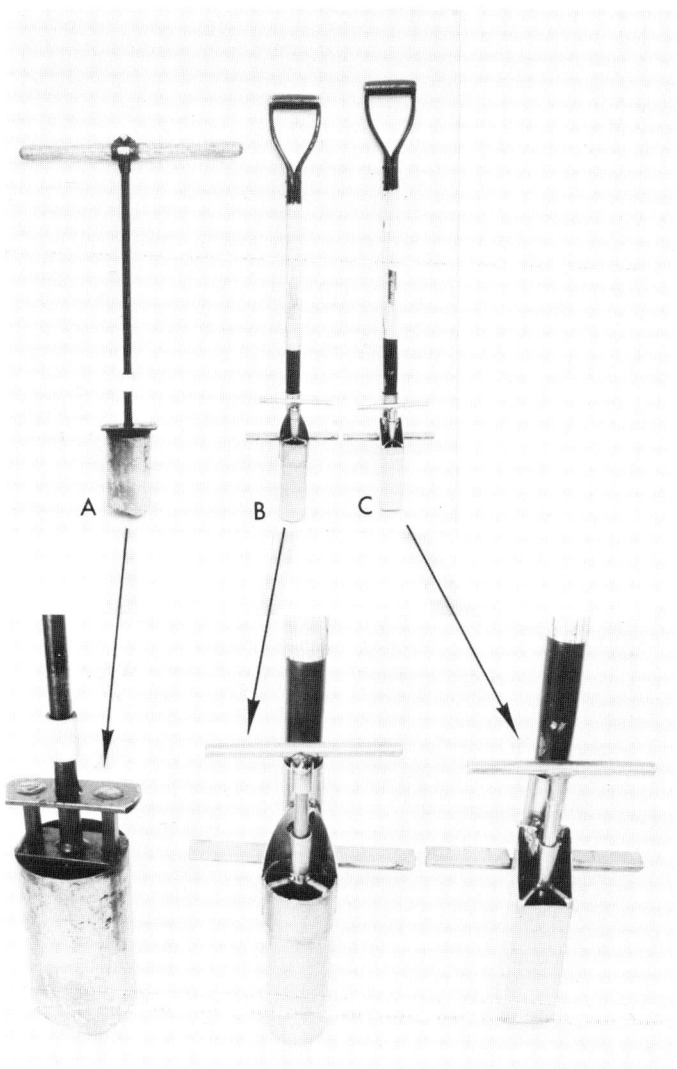

Figure 10-15. Soil corers used for sampling *C. trifurcata* eggs and larvae: (A) 10.2 cm-diameter round corer, (B) 5.1 cm-diameter round corer, (C) 5.1 cm on a side square corer.

measuring 5.1 cm on a side (Fig. 10-15). The circular cores are more easily extruded from the corer; the square corer was used only in determining distance of oviposition from a row of plants (Waldbauer and Kogan 1975). Plants to be sampled are cut off at ground level, and the circular corer is centered on the stump that remains. Individual samples are placed in plastic bags closed with a tie and properly labeled. Samples for eggs can be stored at 4°C for up to four weeks before processing. Samples for larvae and pupae must be frozen if they cannot be processed immediately after collection (Anderson and Waldbauer 1977).

C. trifurcata eggs were found within the upper 3.8 cm layer of soil, and about 77% of the eggs were found within 2.5 cm of a plant; thus, a cylindrical sample 5.1 cm in diameter, 3.8 cm deep (Fig. 10-16A) and centered on a plant or clump of plants is sufficient for determining relative number of eggs. With the use of the cubical corer it was found that 92% of the eggs were located within 7.6 cm of a plant or clump of plants (Fig. 10-16B,C) (Waldbauer and Kogan 1975). Soil samples obtained within these parameters, therefore, are likely to provide absolute density of eggs in a field with better than 90% precision.

C. trifurcata larvae were found most often (92.3%) within 7.6 cm of the soil surface, and no larvae were found below a depth of 15.1 cm, despite the fact that plenty of nitrogen-fixing nodules were found below this depth. Most larvae were found within an area 10 cm wide about the row center (Anderson and Waldbauer 1977). Although older (third instar) larvae showed a tendency to disperse away from row centers, none were found more than 20 cm distant from the plant row (Levinson et al. 1980).

B. Extraction of Eggs, Larvae, and Pupae from Soil

Two techniques have been used to separate bean leaf beetle eggs from soil. Most work has been done with a system of nested sieves; another more expensive although more automated technique uses a rotating screen. The first detailed studies of techniques for the extraction of bean leaf beetle eggs from soil used flotation and a system of concentric screens (Waldbauer and Kogan 1973, 1975). The rotatory screen method has been used extensively in rootworm research on corn.

1. Stationary Nested Screens

The Illinois Egg Separator, originally described by Horsfall (1956), has for many years been used by laboratories and mosquito-abatement districts for routine separation of eggs of flood-water mosquitoes from soil. Processing samples is rapid, and rate of recovery is consistently high. The method consists of using tap water to wash the sample through a series of nested sieves of decreasing mesh. The portion caught on the finest sieve is further cleaned by flotation. The eggs and some of the residual debris float in a saturated aqueous solution of NaCl (brine). Sand and other heavy particles sink and are discarded. When transferred to tap water, the eggs and a small portion of the remaining debris sink while

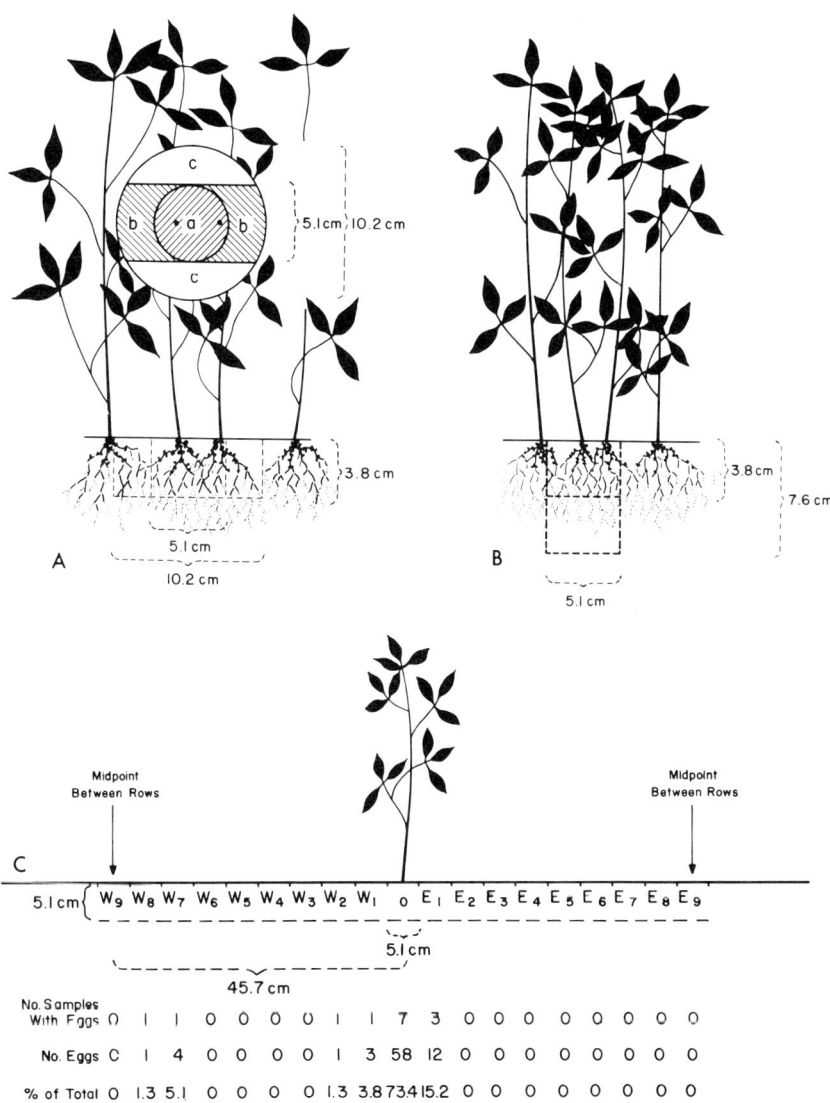

Figure 10-16. Location of cylindrical core samples in relation to soybean plants used to determine stratification of egg populations: (A) core sample divided into inner and outer cylinders—view from above and in longitudinal section; (B) core sample divided into upper and lower halves; (C) location of cubical core samples in relation to rows of soybean plants (each data point is the total of 10 sampling units). (From Waldbauer and Kogan 1975, by permission of the Entomological Society of America).

most of the debris floats and is decanted and discarded. Details of the method and dimensions and plans for building the apparatus are given by Horsfall (1956). See Waldbauer and Kogan (1973, 1975) for the adaptation of this method for use with bean leaf beetles.

For the Illinois Egg Separator, the following equipment is used (Fig. 10-17A-D):

1. plastic pans (30 x 20 x 10 cm)
2. a hose with a nozzle that delivers anything from a wide cone of spray to a forceful jet
3. three large removable superimposed (nested) sieves in a rotating rack (16, 20, and 100-mesh from top to bottom)
4. small transfer sieves (100 mesh) and a tripod to hold them
5. a soil separatory funnel made of 400 ml straight outlet percolator (Corning glassware) that is 46 cm tall and has a 15.5 cm diameter opening on top, the small spout replaced with a 2 cm diameter glass tube closed with a cork stopper
6. a thin stainless steel baffle approximating in size and shape the longitudinal section of the funnel
7. a bubbling stirrer made of a hollow plexiglas rod through which air is passed from an aquarium pump
8. wash bottles filled with tap water or brine
9. white porcelain casseroles.

The extraction proceeds as follows: (1) the sample is put into a plastic pan, covered with tap water containing a small quantity of detergent, and allowed to soak for one or two hours. (2) The sample is rinsed from the pan into the top sieve (16-mesh) of the superimposed set and washed down through the sieves with a narrow and moderately forceful cone of spray. The sieves are shaped and arranged so as to minimize loss from the sample by splashing (Fig. 10-17A). The 16- and 40-mesh sieves are removed when no more material passes through them; the material in the 100-mesh sieve frequently becomes clogged, but it is easily unclogged by directing a strong jet from the hose down through the backed-up water to the surface of the screen. (3) Tap water is used to wash the sample from the 100-mesh sieve into a plastic pan (Fig. 10-17B) and from the pan into a transfer sieve where it is allowed to drain briefly. (4) With a wash bottle filled with salt solution the sample is washed into the soil-separatory funnel which is nearly full of salt solution and in which the bubbling stirrer has been placed (Fig. 10-17C). The bubbles from the stirrer rising from the bottom prevent the rapidly settling debris from trapping eggs. After stirring above the settled portion for a few seconds, the stirrer is withdrawn; a stainless steel baffle that prevents formation of a vortex when the funnel is drained is put in place; and the contents are allowed to settle. Within one or two minutes the eggs and a small amount of debris float to the top, while most of the debris settles to the bottom. (5) The deposit of sand and other heavy debris and all but 6 or 7 cm of the salt

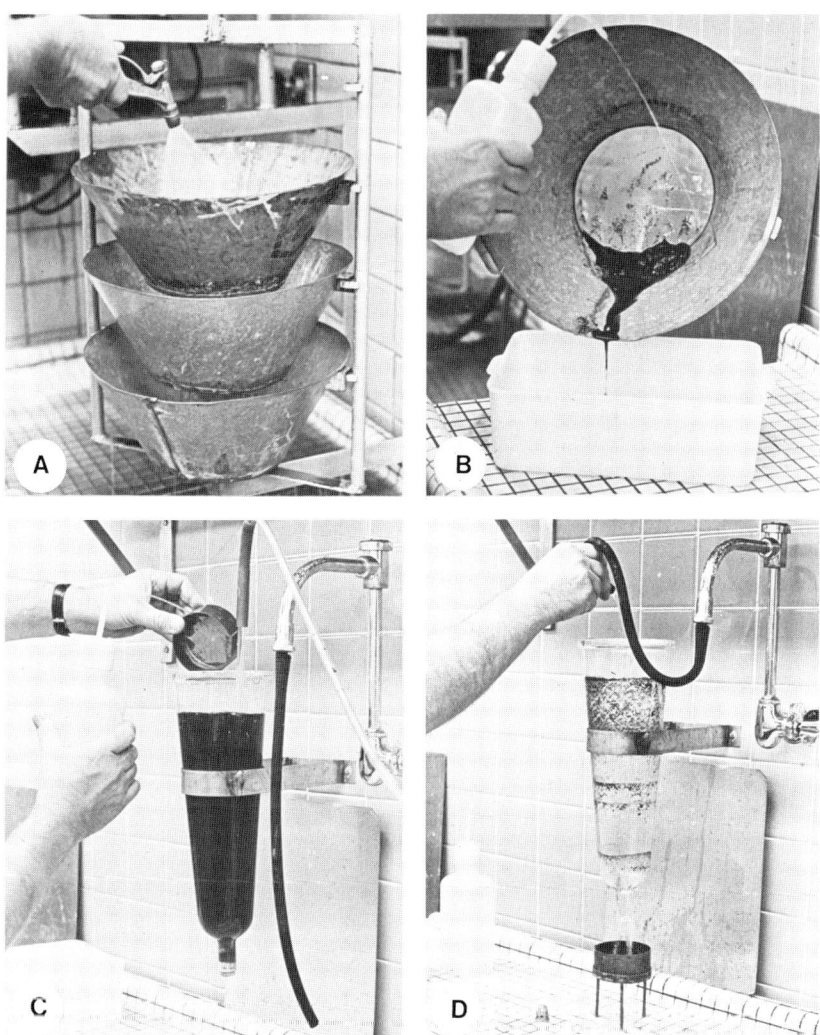

Figure 10-17. Egg separation technique using the system of nested sieves (Waldbauer and Kogan 1973,1975): (A) washing the sample through superimposed sieves in a rotating rack; (B) transferring the material collected on the bottom 100-mesh screen; (C) washing the sample from a transfer sieve into a separatory funnel containing a saturated sodium chloride solution, a bubbling stirrer is kept inside the funnel; (D) washing eggs and debris that floated in the salt solution into a transfer sieve. (See p. 228 for final stages of the procedure).

solution column are drained into a plastic pan. The solution can be reused after being strained through a 100-mesh sieve. The remaining solution is drained through a transfer sieve, and all material caught on the sides of the funnel is washed into the same sieve with tap water (Fig. 10-17D). (6) The material on the transfer sieve is washed into a casserole with tap water. Within a few minutes the eggs and some debris settle to the bottom while a large amount of debris floats to the top. The floating debris and most of the water are decanted a little at a time; sometimes there is so little debris left in the casseroles that there is no need to decant before picking out the eggs. (7) Under a dissecting microscope the eggs are picked out with a bulb pipette. Eggs are searched for both in the bottom of the casserole and in the debris floating at the edge of the meniscus. If there is too much debris, the samples can be stirred and divided among two or more casseroles. The method is rapid, but it still takes two persons about 10-12 minutes per sample.

The accuracy of the extraction method was tested by processing samples containing a known number of eggs obtained from laboratory-maintained cultures. Soil used in these tests was taken from corn fields and was free of bean leaf beetle eggs. This test showed a recovery rate of 92-94%, which was considered adequate for most population studies.

2. Extraction Apparatus with Rotatory Screen

Other extraction apparatuses have been used in surveys of rootworm (*Diabrotica* spp.) eggs. One such machine that was successfully tested for separation of bean leaf beetle eggs (C. Helm and M. Kogan unpublished) was described by Shaw et al. (1976).

In this machine, which is constructed of stainless steel sheeting (Fig. 10-18), a soil sample is placed in a sieve insert inside a 22.8 cm diameter funnel beneath a shower head, and the sample is sprayed with full water pressure. The shower head is controlled by a foot pedal. The funnel-shaped sieve insert, constructed of 30-mesh stainless steel screen, fits inside the funnel with 2.5 cm clearance between the mesh sieve and the funnel. Fifty percent of the sieve's surface is open area, allowing an easy, rapid flow of eggs and soil in suspension.

The internal design is illustrated in Fig. 10-18. A 1/10-horsepower 60-RPM gearmotor, using drive-pulley reduction to 50 RPM, rotates a 50-mesh stainless steel screen cylinder. The cylinder is 12.5 cm in diameter and 75 cm long and has a 10% slope (the slope is very important). The cylinder is mounted on a 1.85 cm diameter shaft. Seven fan TeeJet nozzles, mounted both in front of and behind the cylinder, spray it with water. The water containing eggs and soil spills into a trough at the lower end of the cylinder. This trough is slanted toward the front of the machine, causing eggs and soil to drop into a collecting tray constructed of the same 50-mesh stainless steel screen used in the cylinder. A nozzle sprays the eggs and debris into the collecting tray.

Sampling Bean Leaf Beetles on Soybean

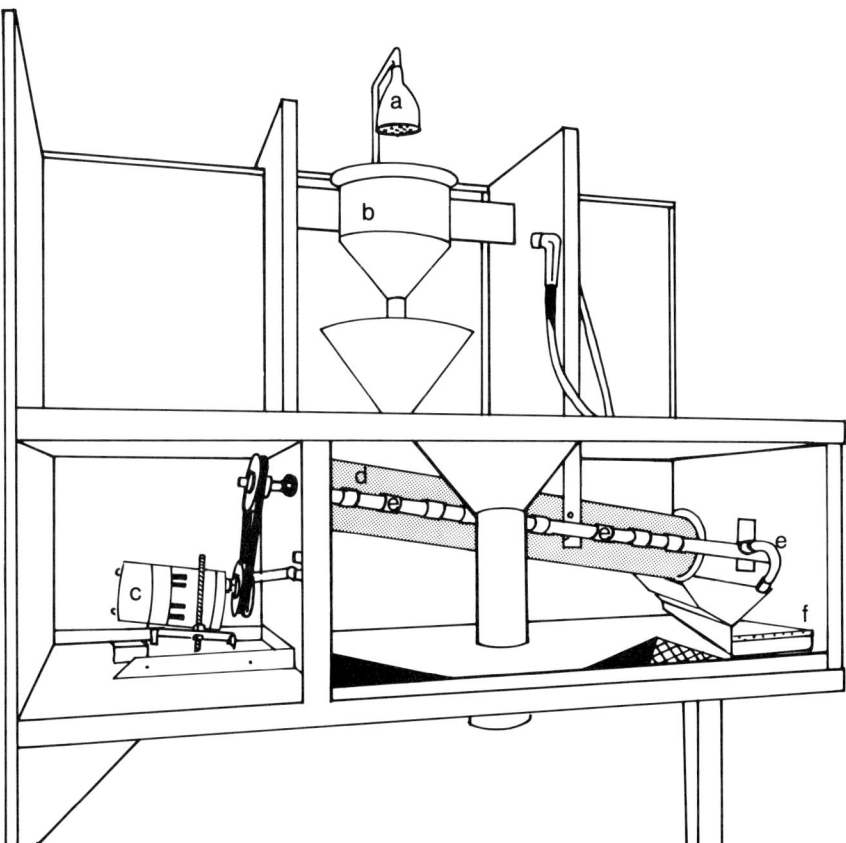

Figure 10-18. Egg separation system using a rotatory screen: (a) shower head, (b) funnel containing 30-mesh screen, (c) electric gearmotor, (d) rotatory 50-mesh stainless steel screen cylinder, (e) TeeJet nozzles, (f) collecting tray.

A plastic garbage can under the machine collects waste water and soil. Residues of soil clinging to the sides of the sieve, the receiving funnel, and the funnel immediately below are flushed down with water sprayed from a common kitchen sink sprayer.

The soil sample is sifted through a 0.62 cm hardware cloth screen and mixed thoroughly. About 0.5 liter of soil is measured and placed in the sieve in the top funnel under the shower head. The sample is sprayed for three minutes, and any remaining soil is flushed down with the hand-operated kitchen sprayer. The sample moves through the screen cylinder, and in 3-4 minutes, almost all of the eggs and remaining debris are captured in the collecting tray. The collecting tray is sprayed thoroughly with the hand-operated sprayer and removed from the machine; the eggs and remaining debris are flushed into a 1000-ml separatory funnel with a stopcock opening of approximately 7 mm. Approximately 500 ml of 2 M magnesium sulphate solution ($MgSO_4$) is used to flush the eggs and debris into the funnel. The funnel is shaken vigorously and returned to the holding rack; in about 30 seconds, the heavier debris sinks to the bottom of the funnel. The stopcock is then opened, and the debris and about 450 ml of the $MgSO_4$ solution are drained from the funnel. Care should be taken not to drain off all of the solution, since the eggs will be floating on its surface. The stopcock is closed, 500 ml of water is added to the funnel, and the funnel is shaken again. The sample must settle for no less than one minute (this time is critical) to permit the eggs to sink to the bottom of the funnel. Next the stopcock is opened and closed quickly to drain the eggs and about 50 ml of water into a petri dish. The eggs are counted under a binocular microscope, and the number of eggs per liter of soil is recorded. The $MgSO_4$ solution can be reused by straining it through a sieve to remove minor debris. This system has been critically calibrated for bean leaf beetle egg separation and yielded better than 90% recovery (C. Helm, M. Jeffords, and M. Kogan unpublished).

The selection of salt solution for the bean leaf beetle eggs is not as critical as that for *Diabrotica* spp. NaCl is perfectly acceptable and much cheaper than $MgSO_4$. *C. trifurcata* eggs are still viable after up to a 10-minute immersion in saturated NaCl solution (Waldbauer and Kogan 1973).

3. Extraction of Larvae and Pupae

A method for extraction of larvae and pupae from soil was developed based on the brine flotation method, but it dispenses with the preliminary sifting of the soil (Anderson and Waldbauer 1977). In this method the soil sample is directly mixed with brine in a plastic pan. The extraction proceeds as follows: (1) The soil sample is placed in a plastic pan and covered with brine. (2) The clods are finely crumbled, with care taken not to destroy larvae. (3) The brine-covered sample is slowly poured into the separatory funnel already containing the bubbling stirrer (material sticking to the pan is washed into the funnel with brine). (4) The contents of the funnel are stirred as the bubbling stirrer is withdrawn (material clinging to the stirrer is washed into the funnel with brine). (5) The baffle is placed in the funnel to prevent a vortex from forming. (6) After the contents settle, all but the top 6 or 7 cm is drained into a plastic pan. (7) The remaining material is drained through a 100-mesh transfer screen (material sticking to the funnel and baffle is washed onto the screen with tap water). (8) The contents of the screen are washed with tap water into a petri dish for examination against a dark background under the dissecting microscope. Some larvae float in tap water and others sink.

Occasionally (in less than 10% of the samples), a larva is trapped by debris and lost in step 6. These larvae usually float on the discarded debris and brine and are easily found by a quick visual check. The larger roots were split, but in only one sample of the more than 100 samples processed was a larva found in a root. The brine can be reused if it is strained through a 100-mesh screen. Freezing the samples or soaking them in a 5% solution of Calgon detergent may be necessary to break up clods of heavy clay soil.

Recovery of a known number of larvae placed in soil from corn fields was between 97 and 99%, irrespective of soil texture (Anderson and Waldbauer 1977). Eggs and larvae can be separated simultaneously using a slightly modified procedure (Levinson et al. 1980). The combined procedure for larval and egg extraction basically follows the method described by Anderson and Waldbauer (1977), except that larvae and pupae are floated and recovered from the saturated salt solution instead of being recovered from tap water after the sample has been flushed from the separatory funnel through the collecting screen.

C. Spatial Patterns and Sampling Program for Eggs and Larvae

Bean leaf beetle egg samples collected in Illinois in July of 1974 (Waldbauer and Kogan 1975) showed, as expected, a considerable degree of aggregation. Aggregated patterns of egg samples are expected if females lay several eggs at a time. Bean leaf beetle females laid 10-15 eggs per day in the laboratory (L. Turner and M. Kogan unpublished). If the full daily egg complement is deposited in one site, one may expect that soil samples either contain zero or more than one egg per sample. In fact, of 175 samples analyzed, 41 samples contained zero eggs, 120 contained more than 1 egg (range 2-34 eggs), and only 14 samples contained one egg. The computed k_c for the egg distribution was 0.64. This k_c was used in estimating the number of samples necessary to achieve a required level of reliability (expressed as $S_{\bar{x}}/\bar{x}$) and a sequential sampling plan for egg populations (Figs. 10-19 and 10-20). Because of the quantitative nature of the soil sampling and egg extraction techniques, these calculations involve a very close approximation of absolute population parameters.

Because of inadequate field data, no attempts were made to provide a similar sampling program for larvae and pupae. Much more research needs to be done in this area.

V. Concluding Remarks

Early in the season (at stage VC-V4) direct counts are the most practical and reliable method for *C. trifurcata* surveys on soybean. After V4 for most research and pest management purposes, the ground cloth and sweep net provide reliable

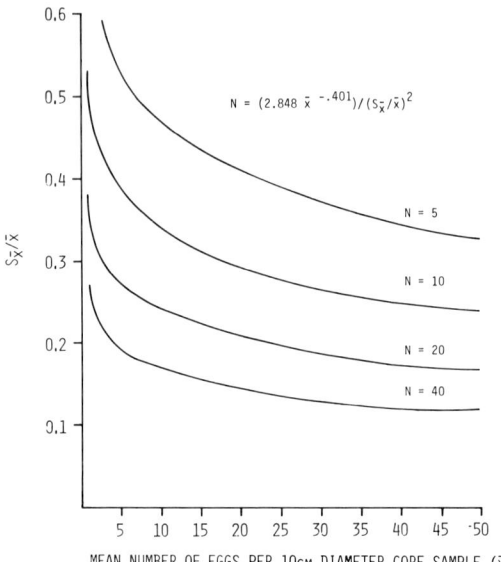

Figure 10-19. Curves describing the number of 10 cm-diameter cylindrical core samples necessary to achieve a certain level of precision according to population density. (Based on data from Waldbauer and Kogan 1973).

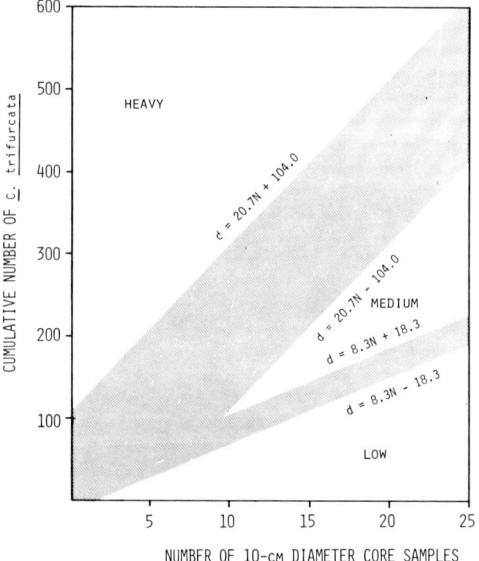

Figure 10-20. Sequential sampling plan for *C. trifurcata* eggs in soil around soybean plants at stage VC-V4. Use of plan is explained in caption of Fig. 10-12. (Based on data from Waldbauer and Kogan 1973).

population estimates. Both sweep net and ground cloth can be converted to absolute populations under given conditions for research purposes. Sweeping is more economical for scouting purposes except when monitoring a complex of species. It is, therefore, suggested that a full season sampling program for adults start with direct counts over a measured length of row (five to ten 10-m row lengths should provide an adequate indication of population levels for scouting purposes).

After V4, the sweep net or the ground cloth provide satisfactory estimates of population densities when following the parameters of the sequential plans. These plans provide a more economic and reliable way of allocating the proper amount of effort in the scouting program. Knowledge of phenological relationships between pests and the crop permits concentrating the greatest amount of effort during the time when the plant is most susceptible to injury and when pest populations are most likely to occur in damaging levels.

Acknowledgments

We are grateful to John Bouseman, Illinois Natural History Survey, for providing host and distribution records, and for critically reviewing the manuscript. Figures 10-4, 10-5, and 10-6 were reproduced from color plates made by Mr. John Sherrod, University of Illinois, for the monograph on soybean *Cerotoma* by Bouseman and Kogan (1980). Much of the information contained in this chapter resulted from research supported in part by the National Science Foundation and the Environmental Protection Agency, through a grant (NSF GB-34718) to the University of California. The findings, opinions, and recommendations expressed herein are those of the authors and not necessarily those of the University of California, National Science Foundation, or Environmental Protection Agency. The original research of Gilles Boiteau was conducted while he was at the North Carolina State University at Raleigh.

References

Anderson, T. E., and G. P. Waldbauer. 1977. Development and field testing of a quantitative technique for extracting bean leaf beetle larvae and pupae from soil. Environ. Entomol. 6:633-636.
Boiteau, G. 1978. Descriptive study of the biology and distribution of the adult bean leaf beetle in eastern North Carolina. Ph.D. diss., North Carolina State University. 182 p.
Boiteau, G., J. R. Bradley, Jr., and J. W. Van Duyn. 1979a. Bean leaf beetle: Emergence patterns of adults from overwintering sites. Environ. Entomol. 8:427-431.
Boiteau, G., J. R. Bradley, Jr., and J. W. Van Duyn. 1979b. Bean leaf beetle: Some seasonal anatomical changes and dormancy. Ann. Entomol. Soc. Amer. 72:303-307.
Boiteau, G., J. R. Bradley, Jr., and J. W. Van Duyn. 1979c. Bean leaf beetle: Diurnal population fluctuations. Environ. Entomol. 8:615-618.
Boiteau, G., J. R. Bradley, Jr., and J. W. Van Duyn. 1979d. Bean leaf beetle: Flight and dispersal behavior. Ann. Entomol. Soc. Amer. 72:298-302.

Bouseman, J. K., and M. Kogan. 1980. Ecology and systematics of bean leaf beetles (genus *Cerotoma*, Coleoptera:Chrysomelidae) associated with soybean. Univ. of Illinois, College of Agriculture, INTSOY ser. (in press).

Boyer, W. P., and W. A. Dumas. 1963. Soybean insect survey as used in Arkansas. USDA Coop. Econ. Insect Rep. **13**:91-92.

Chittenden, F. H. 1897. The bean leaf beetle. USDA Div. Entomol. Bull. (n.s.) **9**:64-71.

Davis, J. J. 1950. Insects of Indiana for 1950. Proc. Indiana Acad. Sci. **60**:178-182.

Deitz, L. L., J. W. Van Duyn, J. R. Bradley, Jr., R. L. Rabb, W. M. Brooks, and R. E. Stinner. 1976. A guide to the identification and biology of soybean arthropods in North Carolina. N. C. Agr. Exp. Sta. Tech. Bull. **238**:246 p.

Diaz, A. J. 1972. Estudio y caracterización de un mosaico del frijol de costa (*Vigna sinensis*) en El Salvador. Phytopathology **62**:754.

Eddy, C. O., and W. C. Nettles. 1930. The bean leaf beetle. S. C. Agr. Exp. Sta. Bull. **265**:25 p.

Gamez, R. 1976. Los virus del frijol en Centroamerica. IV. Algunas propriedades y transmisión por insectos crisomelidos del virus del moteado amarillo del frijol. Turrialba (Costa Rica) **26**:160-166.

Guagliumi, P. 1966. Insetti e aracnidi delle piante comuni del Venezuela segnalati nel periodo 1938-1963. Inst. Agr. per l'Oltremare, Firenze. Rel. Monogr. Agr. Subtrop. Trop. (n.s.) **86**:391 p.

Harries, V. 1975. Zur innerartlichen Variabilitat, Wirtspflanzen-Preferenz und Schadendeutung von Blattkäfern der U.F. Galerucinae (Col. Chrysomelidae) in Feldkulturen des Cauca-Flusstals/Columbien. Z. Angew. Zool. **62**:491-497.

Herzog, D. C. 1968. Seasonal, locational, and sexual variation in the color pattern of the bean leaf beetle, *Cerotoma trifurcata* (Forst.). M.S. thesis, Louisiana State University. 87 p.

Herzog, D. C. 1973. Some biological implications of polymorphism in the bean leaf beetle, *Cerotoma trifurcata* (Forster). Ph.D. diss., Louisiana State University. 176 p.

Herzog, D. C., C. E. Eastman, and L. D. Newsom. 1974. Laboratory rearing of the bean leaf beetle. J. Econ. Entomol. **67**:794-795.

Hillhouse, T. L., and H. N. Pitre. 1974. Comparison of sampling techniques to obtain measurements of insect populations on soybeans. J. Econ. Entomol. **67**:411-414.

Horn, G. H. 1893. The Galerucini of boreal America. Trans. Amer. Entomol. Soc. **20**:57-144.

Horn, N. L., L. D. Newsom, and R. L. Jensen. 1973. Economic injury thresholds of bean pod mottle and tobacco ringspot virus infection of soybeans. Plant Dis. Rep. **57**:811-813.

Horsfall, W. R. 1956. A method for making a survey of floodwater mosquitoes. Mosquito News **16**:66-71.

Isely, D. 1930. The biology of the bean leaf beetle. Ark. Agr. Exp. Sta. Bull. **248**:20 p.

Iwao, S. 1975. A new method of sequential sampling to classify populations relative to a critical density. Res. Popul. Ecol. **16**:281-288.

Iwao, S., and E. Kuno. 1971. An approach to the analysis of aggregation pattern in biological populations. pp. 541-713 in G. P. Patil, E. C. Pielou, and W. E.

Waters, eds. Statistical ecology. vol. 1. Spatial patterns and statistical distributions. Penn. State Univ. Press, Univ. Park, PA. 582 p.

Jensen, R. L. 1976. Are you doing a complete job of sweeping the field? Agri-Fieldman 32:31-32,34.

Kogan, M. 1976. Evaluation of economic injury levels for soybean insect pests. pp. 515-533 in L. D. Hill, ed. World soybean research. Proc. World Soybean Res. Conf., Interstate Print., Danville, Illinois. 1073 p.

Kogan, M., W. G. Ruesink, and K. McDowell. 1974. Spatial and temporal distribution patterns of the bean leaf beetle, *Cerotoma trifurcata* (Forster), on soybeans in Illinois. Environ. Entomol. 3:607-617.

Kvicala, B. A., J. Smrz, and N. Blanco. 1970. Some properties of cowpea mosaic virus isolated in Cuba. Phytopathol. Z. 69:223-235.

Levinson, G. A., G. P. Waldbauer, and M. Kogan. 1980. Distribution of bean leaf beetle eggs, larvae and pupae in relation to soybean plants: Determination by emergence cages and soil sampling techniques. Environ. Entomol. (in press).

McConnell, W. R. 1915. A unique type of insect injury. J. Econ. Entomol. 8:261-266.

Marston, N. L., C. E. Morgan, G. D. Thomas, and C. M. Ignoffo. 1976. Evaluation of four techniques for sampling soybean insects. J. Kans. Entomol. Soc. 49:389-400.

Milbrath, G. M., M. R. McLaughlin, and R. M. Goodman. 1975. Identification of bean pod mottle virus from naturally infected soybeans in Illinois. Plant Dis. Rep. 59:982-983.

Newsom, L. D., E. P. Dunigan, C. E. Eastman, R. L. Hutchinson, and R. M. McPherson. 1978. Insect injury reduces nitrogen fixation in soybeans. La. Agr. 21:15-16.

Nichols, M. P., M. Kogan, and G. P. Waldbauer. 1974. The literature of arthropods associated with soybeans. III. A bibliography of the bean leaf beetles *Cerotoma trifurcata* (Forster) and *C. ruficornis* (Olivier) (Coleoptera: Chrysomelidae). Ill. Natur. Hist. Surv. Biol. Notes 85:16 p.

Ono, Y. 1967. Estimation of animal numbers. pp. 87-107 in Seitaigaky Jisshu-Sho' (Guide for Experiments in Ecology), Tokyo (In Japanese) (Not seen in the original).

Pacheco, F. 1976. Seasonal and daily fluctuation of soybean insect populations in the Yaqui Valley, Sonora, Mexico. pp. 584-593 in L. D. Hill, ed. World soybean research. Proc. World Soybean Res. Conf., Interstate Print., Danville, Illinois. 1073 p.

Panizzi, A. R., B. S. Correa, D. L. Gazzoni, E. B. de Oliveira, G. G. Newman, and S. G. Turnipseed. 1977. Insetos da soja no Brasil. EMBRAPA, Cent. Nac. Pesq. Soja, Bol. Tec. 1:20 p.

Perez, J. E., and A. Cortes-Mollor. 1970. A mosaic virus of cowpea from Puerto Rico. Plant Dis. Rep. 54:212-215.

Popenoe, E. A. 1889. Some insects injurious to the bean. Kans. Agr. Exp. Sta. Annu. Rep. 2:206-212.

Risch, S. 1976. Effect of variety of cowpea (*Vigna unguiculata* L.) on feeding preferences of three chrysomelid beetles, *Cerotoma ruficornis rogersi*, *Diabrotica balteata* and *Diabrotica adelpha*. Turrialba (Costa Rica) 26:327-330.

Rudd, W. G., and R. L. Jensen. 1977. Sweep net and ground cloth sampling for insects in soybeans. J. Econ. Entomol. **70**:301-304.

Ruppel, R. F. 1962. Identification of cowpea mosaic in Colombia. Bull. Entomol. Soc. Amer. **7**:170.

Ruppel, R. F. 1978. A new species of *Cerotoma* from Bolivia (Coleoptera, Chrysomelidae). J. Kans. Entomol. Soc. **51**:28-30.

Shaw, J. T., R. O. Ellis, and W. H. Luckmann. 1976. Apparatus and procedure for extracting corn rootworm eggs from soil. Ill. Natur. Hist. Surv. Biol. Notes **95**:1-4.

Silva, A. G. d'A. e, C. R. Gonçalves, D. M. Galvão, A. J. B. Gonçalves, J. Gomes, M. de N. Silva, and L. de Simoni. 1968. Quarto catálogo dos insetos que vivem nas plantas do Brasil seus parasitos e predadores. Part 1, Vol. 1:1-422, Vol. 2:423-906; Part 2, vol. 1:1-622. Ministério da Agricultura, "Serviço de Defesa Sanitária Vegetal," Rio de Janeiro, Brazil.

Turnipseed, S. G. 1974. Sampling soybean insects by various D-Vac, sweep, and ground cloth methods. Fla. Entomol. **57**:217-223.

Waldbauer, G. P., and M. Kogan. 1973. Sampling for bean leaf beetle eggs: Extraction from the soil and location in relation to soybean plants. Environ. Entomol. **2**:441-446.

Waldbauer, G. P., and M. Kogan. 1975. Position of bean leaf beetle eggs in soil near soybeans determined by a refined sampling procedure. Environ. Entomol. **4**:375-380.

Waldbauer, G. P., and M. Kogan. 1976a. Bean leaf beetle: Phenological relationships with soybean in Illinois. Environ. Entomol. **5**:35-44.

Waldbauer, G. P., and M. Kogan. 1976b. Bean leaf beetles: Binomics and economic role in soybean agroecosystems. pp. 619-628 in L. D. Hill, ed. World soybean research. Proc. World Soybean Res. Conf., Interstate Print., Danville, Illinois. 1073 p.

Walters, H. J. 1969. Beetle transmission of plant viruses. Adv. Virus Res. **15**:339-363.

Wilcox, J. A. 1972. Chrysomelidae: Galerucinae, Luperini: Aulacophorina, Diabroticina. *Coleopterorum catalogus supplementa*, Part 78, No. 2:221-431. W. Junk Pub., Gravenhage, Holland.

SECTION IV
Other Foliage Feeders

Chapter 11

Sampling Aphids in Soybean Fields

Michael E. Irwin

I. Introduction

Everywhere soybean grows, aphids form a part of the community. Aphids are either colonizers of soybean fields or they are transient members of the soybean community. Several species of aphids successfully colonize and reproduce parthenogenetically on soybean in Asia and parts of Africa: *Aphis craccivora* Koch, *Aphis glycines* Matsumura, *Aulacorthum solani* (Kaltenbach). In the western hemisphere, aphids are mostly transient, although on occasion at least two species are known to form small, usually temporary colonies in soybean: *A. craccivora* in an experimental plot in Illinois and in the high jungles of Peru and *Aphis gossypii* Glover in Ecuador. *A. gossypii* established colonies on greenhouse grown soybean (M. E. Irwin unpublished).

Colony forming aphids are actual or potential pests of soybean. Ants that tend aphids are sometimes associated with aphid colonies (Fig. 11-1) (Kogan 1978). Aphids place a considerable strain on growing plants by sucking nutrients and sap from leaves and stems. Yellow patches on leaves that later develop into necrotic areas are typical symptoms of aphid injury.

Regardless of aphid colonization potential, assessment of numbers and kinds of aphids flying over and, more importantly, landing on soybean canopies is of importance because several species of aphids are active vectors of several types of viruses that affect soybean. Yield losses vary depending on many factors, but large reductions resulting from natural infections by soybean mosaic virus have been reported (Muraveva 1973, Ross 1977).

Two major objectives are apparent for sampling aphid populations in soybean fields: (1) experimental field research involving colonization, population dy-

Figure 11-1. *Aphis glycines* colonizing soybean in Korea, tended by ants. (Courtesy of M. Kogan).

namics, effects of pesticides, movement and dispersal, natural enemy−aphid interactions, aphid−disease interactions, and particularly the epidemiology of virus diseases; and (2) sampling for pest management purposes. Each of these objectives involves sampling aphids forming colonies on plants and in the air or landing on the canopy (transient alates). The detail needed to fulfill the first objective is far greater than that needed to satisfy the latter. Thus, it is essential to define the objective of the sampling program for aphids before selecting the actual sampling method(s).

II. Aphid Life Cycle and Flight Behavior

The only species of aphids that seem in any way important to soybean crops belong to the Aphididae: Aphidinae. Many alternate host plants during the year, going from a tree species to an herbaceous species. Others are polyphagous but do not alternate host types during the season. However, a considerable number

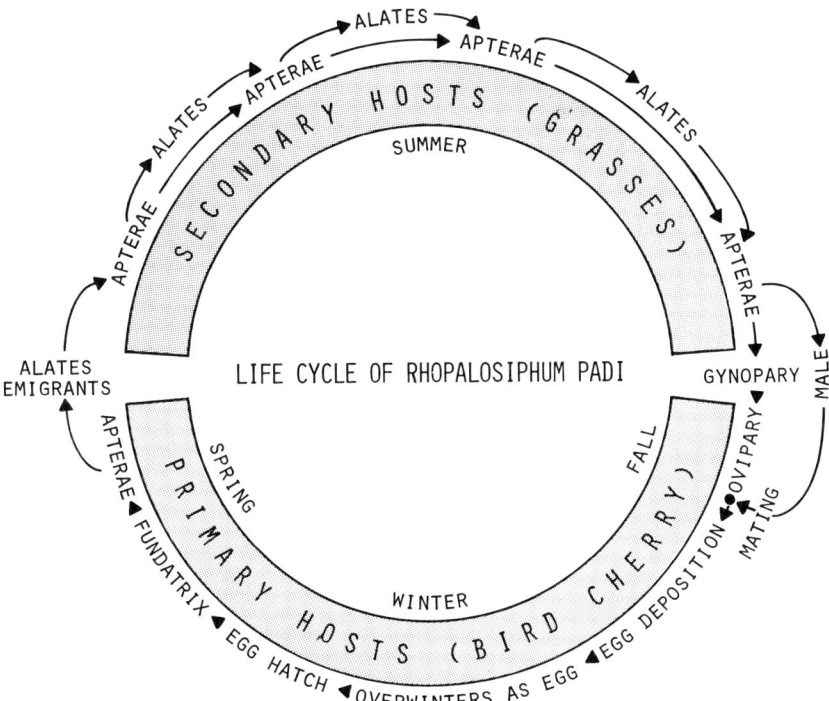

Figure 11-2. Schematic representation of the life cycle of *Rhopalosiphum padi* (After Dixon 1977).

that are transient in soybean fields are monophagous and usually have tree species as their only colonizing hosts. A typical life cycle of a host-alternating species, *Rhopalosiphum padi* (Linnaeus), is shown in Fig. 11-2. In tropical and subtropical areas, the parthenogenic phase of the cycle may continue throughout the year, but in areas where winters are more severe, the sexual stages are also present. Typically there are several structurally different morphs in a species, including sexual and asexual forms, and these are associated with the passage of the season, movement from host to host, and overcrowding (Dixon 1977).

Soybean seems to be a secondary host of host-alternating aphid species. Thus, colonization of soybean by alate aphids is cued by seasonal and behavioral factors during the previous morph stage. Once a colony is formed on soybean, it will increase, mainly as apterae, until cued by overcrowding or short day lengths to other morphs; these morphs could be either parthenogenic alates that could recolonize soybean or other secondary hosts, or sexual alates (gynoparae and males) that search out primary hosts, usually tree species.

Some emigrant and midsummer alates of noncolonizing species are capable of transmitting soybean viruses. Flight behavior, which is the key to understanding aphid colonization potential and virus epidemiology, was reviewed by Kring (1972).

Many aphid species migrate, some long distances. Moericke (1955) divided the flight behavior of migrant alates into four phases: (1) rest, the time prior to flight; (2) flight or distance flight, characterized by strong vertical movement; (3) attack flight, during which alates alight on plants, probe, and then resume flight; and (4) settlement phase, during which the aphid produces viviparous young: this settlement phase can be interrupted by a second attack phase, after which settlement resumes.

Upon initial take-off, migrating alates move upward in a persistent object-avoiding flight into the sky. Nonmigrating alates fly at a low level over open ground or directly toward and among foliage and shadows. Nonmigrating flight can be straight and directional, or it can be spiraling (Kring 1972). Aphids have been recorded flying up to 50 m/min in still air, and some species tend to alight when winds are greater than their flight speed. Others have been recorded to fly into the wind (Kennedy and Booth 1963, Kring 1972). Some aphids may leave wind transportation and alight (Kring 1972). Exhaustion, reduced light, and increased wind speed result in flight termination. Alighting can also be a response to stimuli such as color, odor, or sound.

This cursory explanation of the complexities of aphid life cycles and flight behavior provides minimum background for developing adequate sampling procedures for aphids.

III. Sampling Aphid Populations

The sampling portion of this chapter will be divided into two sections: sampling for aphids within the plant canopy and sampling for alate aphids in the air and alighting on the canopy.

A. Sampling for Aphids within the Plant Canopy

Soybean yield losses or reduced seed quality due to aphid feeding occur only when large populations are present (Kobayashi et al. 1976). Damage resulting from aphid colonization is probably a product of several factors, including climate, other stresses on the crop, age of the aphid colonies, and their distribution on the plants and in the crop. Sampling for aphid spatial patterns and abundance should usually be linked with sampling for ants that tend and protect them (Fig. 11-1) and natural enemies that destroy them. In areas where aphid-borne viruses are a problem, counting aphids is not enough. It may be much more important to separate the count into numbers of alates and alatoid nymphs, and apterae and apteroid nymphs. This information could be important in predicting aphid flights that could spread viruses in and between soybean fields.

Heathcote (1972) divided techniques for sampling aphid populations on plants into five categories: direct observation, plant clipping, sweep net, suction net sampler, and ground cloth (see Chapter 2). Suction samplers are unsuited for extracting aphids from many broadleaf plants (Shorey 1963, Heathcote 1972). Similarly, the sweep net is not efficient for estimating populations of aphids because several factors influence catches (Saugstad et al. 1967) and because samples give very low estimates (Fenton and Howell 1957). Also, specimens are usually severely damaged and not identifiable to the species level. Beating aphids from foliage proved most effective on shrubs and trees but does not seem suited for quantifying aphid populations on soybeans. Cook (1955), for instance, found that populations of *Acyrthosiphum pisum* (Harris) were impossible to correlate when sampling alfalfa with sweep net, tip-sampling, and ground cloth (in his case a sticky board beneath a shaken pea plant).

In essence, aphids colonizing soybean can best be sampled by plant clipping and extracting or by direct observation.

1. Plant Clipping and Extraction

Plant clipping requires considerable effort but it is reliable. The sample unit consists of a certain number of leaves, stems, or other plant parts from various strata randomly collected in a field. Since aphid colonies are patchy in a field, many scattered samples are necessary to estimate numbers of aphids per unit area. This, however, can grossly underestimate the number per infected plant and that could be dangerous since damage to a small portion of the field could occur even though numbers might be well below estimated damage levels on an area-wide basis. Build-up of alates from colonizing clones may also influence the spread of aphid-borne plant viruses to soybean and to crops other than soybean in the surrounding area.

Plant parts selected per sampling should depend to a large extent on the colonizing behavior of the aphid species. In Korea *A. solani*, for example, colonizes soybean leaves in the upper half of the canopy while *A. glycines* often colonizes soybean stems (Kogan 1977, 1978). If aphids prefer certain strata, the sampling procedure should take advantage of this preference to maximize results for expended effort.

When intact stems or leaves containing living aphids are brought to the laboratory, new nymphs may be viviposited by adults and predators may consume existing aphids. By keeping the sample in an ice chest, insect activity can be greatly reduced and population fluctuations can be minimized.

Removing aphids by brushing them from foliage or by washing foliage in cold or hot water proved to be the most accurate methods for collecting and separating large quantities of aphids (Rabasse and Bouchery 1977). Brushing, however, was time consuming, so, for most purposes, using a detergent water bath would be most suitable.

Therefore, the most reliable and easiest method of collecting and transporting samples of aphids from stems and leaves is to place samples individually in containers partially filled with detergent water. The containers are shaken to dislodge aphids from the substrate. The aphids, upon settling to the bottom of the container, are separated from the sample substrate, sieved, taken to the laboratory, and, there, identified and counted (see Chapter 13).

2. Direct Observations

Counting aphids in situ (direct observations) is to be preferred to removing plants or parts of plants, provided the direct counts are reliable or are taken when only relative estimates of the population are required; with large numbers of aphids it may be better to check samples in the laboratory (Heathcote 1972). Since soybean only has a few easily distinguishable colonizing species, direct observations seem to be adequate for most purposes. When few aphids are present, counting by direct observation is less time-consuming than washing aphids and counting them in the laboratory. When many aphids are present, the reverse is probably true. If comparable counts are to be made throughout the growing season, it is advisable to use the same technique for the entire season, so far as possible. Choice of either method should be decided on criteria of reliability (RV) and relative net precision (RNP), which takes into account the cost (in time) required to obtain a sample unit by a given method (see Chapters 2 and 3).

Certain species of aphids readily drop from plants when disturbed while others do not. When sampling, either by clipping or direct observation, the possible defense reaction (dropping) of the aphid, due at least partially to aphid alarm pheromones, should be kept in mind and the sampling approached cautiously.

Direct observation counting can be divided into several categories: (1) unit area counts, (2) timed counts, (3) absolute counts, and (4) infestation scales (see Heathcote 1972). Unit area counts, e.g., numbers of aphids per cm^2 of leaf surface or cm of stem length, could be useful for determining economic injury levels. This concept can be extended to numbers of aphids per m^2 of plants, or it can be expressed as numbers per ha. For population dynamics studies, it is important to define spatial patterns of aphids in the field (see Chapter 3). Crowding can be extremely important in morph determination and production since, in general, viviparous alates are produced on highly crowded plants whereas viviparous apterae are produced on sparsely populated plants.

The number of aphids or aphid colonies observed and recorded in a given amount of time constitute timed counts. Shands and Simpson (1955) counted the number of aphids found per minute of observation and the percentage of pure *Myzus persicae* (Sulzer) colonies. This method required familiarity with locating aphid colonies and in detecting pure colonies (i.e., one species). Uniformity of results depended on maintaining a uniform work speed throughout the day and the season.

Absolute counts involve counts of total aphids per plant or per plant part. Number of aphids per plant are difficult to tally when soybean plants become large or when population densities are high. Number per specified plant part may be more practical. Reliable comparative data can be obtained by consistently sampling the same plant part. Since soybean leaves are trifoliolate, one may standardize the leaflet to be sampled (e.g., the center leaflet). Vertical stratification of aphids on soybean plants has not been determined.

3. Gross Estimates of Infestation Levels

Infestation scales are perhaps the most popular form of counting aphids within a field crop. This is especially true if sampling for management decisions, although it can also be used for relative population estimates. These scales are moderately reliable as long as the surveyors are experienced in their use. This system is fast and requires no laboratory counting, although specimens should be brought into the laboratory on occasion to confirm species identification.

Bandong and Litsinger (1976) and Litsinger et al. (1977) patterned an infestation scale on that of Banks (1954). They rated aphid abundance per soybean plant on the following modified scale:

1 = no aphids
3 = winged aphids to small colony
5 = several colonies
7 = many distinct colonies
9 = many indistinct colonies

The above scale was easy to use and aphid abundance was readily comparable among sites and among dates within a site (J. A. Litsinger, personal communication). This soybean rating scale included two aphid species, *A. craccivora* and *A. glycines,* but no attempts were made to separate the species.

B. Sampling for Alate Aphids in the Air and Alighting on the Canopy

Wherever soybean is grown, transient alate aphids form a part of the fauna. Aphids fly or are carried by winds into soybean fields where they alight, probe, usually discover that soybean is not a suitable host plant, and fly on to other plants and other fields. Some are captured by predators, spider webs, or for other reasons form a more permanent part of the energetics of a soybean ecosystem.

Aerial sampling of alate aphids has been reviewed by Taylor and Palmer (1972). This review catalogs types of trapping techniques and contains sections on limitations of trapping and methods of trapping to meet desired objectives. Their review emphasizes sampling for migration.

Aphid trapping devices can be divided into *filter traps,* those that collect air and filter out the aphids; *impact traps,* those onto which aphids are projected and deposited; or *light traps,* those that attract aphids to a light source at night, a time during which aphid flight activity is very low (Taylor and Palmer 1972). It is preferable to classify the aerial aphid sampling methods not by trapping devices but by trapping principle or measure. In other words, it is the catch constituency that is important to classify: is the catch a near absolute account of whatever one is attempting to measure, is it relative with respect to absolute numbers, or is it skewed with respect to absolute or relative numbers of the species mix?

1. Skewed Measures

Skewed measure traps either attract or repel some aphid species more than they do other species. Yellow pan traps and black or sticky glass plates have this attribute. Yellow attracts certain species of aphids more strongly than it does other species (Eastop 1955, Roach and Agee 1972, Taylor and Palmer 1972), whereas aluminum, black, and even clear glass 'repel' some aphid species more than others. Light traps attract aphids and thus give skewed measures of densities.

Heathcote (1957) found that the size of yellow impact traps influences the number of aphids caught and that the numbers caught are fewer per unit area if the surface area is larger. Costa and Lewis (1967) gave the relationship between trap size and aphid catch as linear when plotted on a square root scale. They also suggested that different species of aphids were differentially attracted to different sizes of the same colored trap. Rabasse et al. (1976) presented evidence suggesting that the linear relationship proposed by Costa and Lewis (1967) proved valid for only a few species and compensation conversions should be approached cautiously. Costa and Lewis (1967) suggested that circular pan traps were more representative of aphid abundance than rectangular or square ones because the area of the trap subtended by flying aphids would be equal, no matter the direction of flight.

In an experiment conducted over a 'Williams' soybean field in central Illinois in 1978, certain species of aphids were found to be more abundant on horizontal, yellow, liquid-filled traps than on ermine lime, liquid-filled traps. Furthermore, these same species, for the most part, were more abundant on ermine lime traps than on other trap surfaces: aluminum, mirror, black, and clear glass. At the same time, no differences existed among species captured in large enough numbers to analyze for black, aluminum, clear glass, or mirror horizontal, liquid-filled traps (M. Irwin, S. Halbert and R. Goodman unpublished).

Traps that differentially attract or repel species of aphids cannot provide a good measure of the relative abundance of these species. If, however, one wishes to know the relative abundance of any one species over the season or among seasons, there seems to be no objection to the use of a skewed measure sampling device.

2. Relative Measures

Aerial nets provide information on the relative densities of aphid species at any given height or stratum above the ground. Aerial nets pivoted or directed into the wind (Fig. 11-3) (Shands et al. 1942, Davis and Landis 1949, Johnson 1950, Taylor 1962a, Ashby 1976) have been used for aphid sampling. Sticky coated screens perform similarly (Shands et al. 1956) in that they strain aphids from a volume of air. Vertical impact traps such as cylindrical sticky traps (Fig. 11-4), if they are not painted yellow or some other differentially attractive or repelling color, operate on the principle of impaction rather than by filtering out specimens. These types of traps have been experimented with and widely used (Johnson 1950, Heathcote 1957, Dickson et al. 1956, Zettler et al. 1967, Roach and Agee 1972, Rogerson 1975).

Figure 11-3. Modified wind directed, vertically oriented trap for capturing small insects on windward and leeward sides of cylindrical, tanglefoot coated polyethylene cups.

While the catches are considered to be relative with respect to aerial densities of aphid species, these sampling devices do not give a relative measure of any given species through time. The greater the volume of air passed through the filter-type trap, and the higher the velocity of the air current (wind) passing by a vertical impaction-type trap, the greater the catch per unit of time, assuming equal aerial densities of aphids. This means that wind speed is an added factor in the makeup of the resulting catch.

A few types of traps combine the skewed trap bias along with bias inherent in relative measuring devices. An example is a vertically oriented cylindrical trap painted yellow (A'Brooke 1968, Heathcote et al. 1969, Adlerz 1976).

3. Absolute Measures

There are very few trap types that approach measuring absolute densities of aphids. Two different traps measure near absolute densities of aphids in a particular stratum of air: Johnson-Taylor suction trap (Fig. 11-5) (Taylor 1955,

Figure 11-4. A cylindrical, vertically oriented sticky trap.

1962b) and rotary traps or whirligig traps (Fig. 11-6) (Chamberlin and Lawson 1945, Taylor 1962b). These trap types are discussed and compared by Taylor and Palmer (1972). They each measure aerial density, and each is extremely useful for aphid migration studies: where did the aphids come from, how long might they have been in the air? These questions can be answered by monitoring migrating populations over long distances at different heights by suction traps or rotary traps. When tagging or labeling aphid populations (Pettersson 1968, Mackauer 1972, Guss and Branson 1972, Taimr et al. 1978) in conjunction with these trapping methods, much can be learned. A predictive capability can possibly evolve (see Section V).

For colonization potential and for potential as virus vectors, other parameters should be monitored: landing rates, staying times, take-off rates. It is important to know how many of which species of aphids are landing in the canopy per unit area over a given time interval. That will describe the colonization rate (when

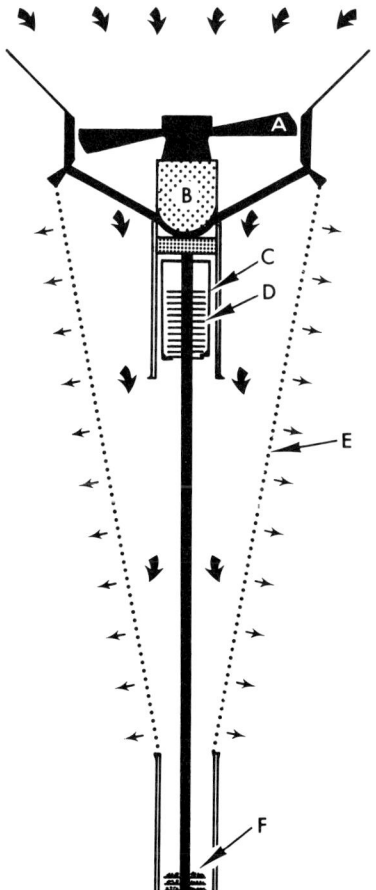

Figure 11-5. The Johnson-Taylor suction trap. Arrows show air flow. Insects are collected between disks at bottom of trap. (A) fan blades, (B) fan motor, (C) release mechnism for brass disks, (D) brass disks, (E) 26 mesh copper gauze cone, (F) brass disks separating hourly catches of aerial plankton within collecting tube.

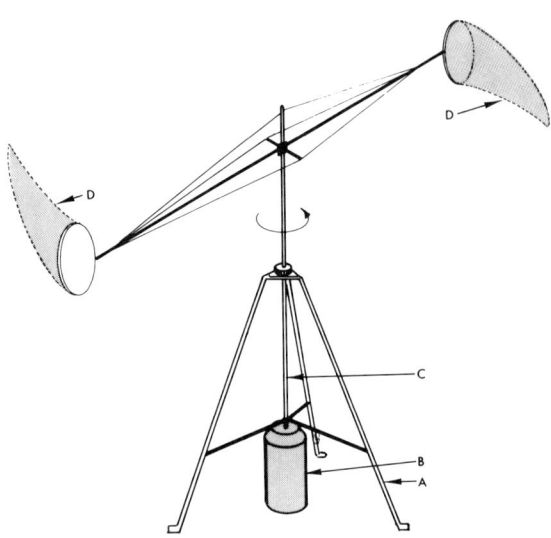

Figure 11-6. A rotary or whirligig trap. (A) stand, (B) motor, (C) central shaft, (D) rotating nets.

take-off rates and deaths are subtracted). It will also give an indication of transmission potential, especially if probing rates and durations can be calculated.

For viruses transmitted in a nonpersistent manner, as are most aphid-borne soybean viruses, a relatively short staying time with frequent take-offs and frequent landings is probably optimal. For viruses transmitted in a persistent manner (e.g., soybean dwarf virus in Japan), longer staying times probably optimize virus transmission. Thus, for either colonization or virus-vector monitoring, what seems desirable in terms of sampling alate aphids in and above a soybean canopy is not to measure aphid density in the air above a field, but to measure landing rates, take-off rates, and staying times. A specific trap to satisfy one of these objectives—landing rates—is being developed (M. Irwin, S. Halbert, G. Schultz, R. Goodman unpublished)

A trap designed to give absolute records of all aphid species alighting on a canopy would need the following properties: (1) horizontal attitude, (2) position over and relatively close to canopy top, (3) neutral reflectivity and color spectrum relative to canopy. The size or shape of the trap would be immaterial if it were neutral with respect to the canopy.

The trapping surface of the horizontal ermine lime trap is a square ceramic tile, 15.2 cm on a side (231 cm^2) of rugose texture and of a color most closely resembling that of soybean leaves. The title is commercially called ermine lime (SE 11, H & R Johnson, Stoke-on-Trent, England). The reflective spectrogram of the ermine lime tile is similar to that of young 'Williams' soybean leaves. Although it less accurately matched readings of a single mature 'Williams' leaf, it most closely matched readings if several mature 'Williams' leaves were stacked together; overlapping leaves occur naturally in a soybean canopy. When the ermine lime tile was emersed in ethylene glycol, the reflective wavelengths were somewhat muted, and most closely approached those of stacked 'Williams'

Sampling Aphids in Soybean Fields

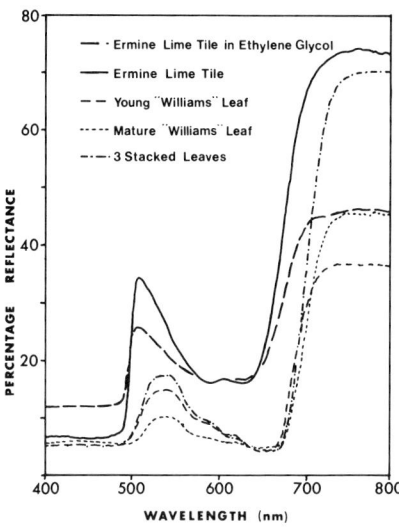

Figure 11-7. Reflective wavelength spectrogram of the horizontal ermine lime trap and various 'Williams' soybean leaves. Large dashed lines give reflective spectrograph of the ermine lime trap emersed in ethylene glycol.

leaves (Fig. 11-7). Soybean leaves are not shiny, and the rugose texture of the tile seemed to scatter the reflected light more than would one of smooth-surface.

The current design of the horizontal ermine lime trap that approaches values closest to actual landing rates on the canopy is figured (Fig. 11-8). The tile is placed in a square plexiglass tray 15.7 cm on a side and 3.5 cm deep (inside measurements). Several small holes are drilled along the top 0.5 cm of all sides of the tray to insure that rainwater can run off without overflowing the tray. A chemistry ring stand with arm and attachment clamp are fitted to the tray. The ring portion is placed under the tray and four square plexiglass pieces, 1 cm on a side, are glued to the bottom of the tray within but adjacent to the ring's perimeter. When the ring stand is in a horizontal position, the tray snugly fits on top of it. The ring stand is connected by its clamp to an iron rod driven vertically into the soil. The stand and tray can be easily adjusted up and down by loosening the clamp, moving the ring stand and tray to the desired height, then tightening the clamp over the rod.

From experiments conducted in 1978 (M. Irwin, S. Halbert unpublished), it appears that as long as the trap is placed within 0.5 m of the canopy, exact height is not critical (Table 11-1). Surface reflectance, whether sticky (coated with Tanglefoot) or liquid-filled (the tray filled with a 70:30 mixture of ethylene glycol:water) made some difference in the catch constituency (Table 11-2). More *Myzocallis asclepiadis* (Monell) and *R. padi* were captured on sticky surfaced tiles, while more *Rhopalosiphum maidis* (Fitch) and *Aphis citricola* (Van der Goot), were caught in liquid-filled traps. Many other species showed no significant preference for one surface over the other.

M. Irwin and S. Halbert (unpublished) have attempted to calibrate the trap by placing it alongside an equal area of canopy and collecting all alighting aphids for given time intervals. Too many variables existed to accurately test

Figure 11-8. The horizontal ermine lime trap in situ in a soybean field, Urbana, Ill.

Table 11-1. Mean number of aphid alates captured per horizontal glass plate (231 cm^2) at canopy level and at 0.5 m above canopy level during the 1978 season, Urbana, Ill. (N = 5)

	Canopy level	0.5 m Above canopy
Rhopalosiphum maidis (Fitch)	61	55
Aphis citricola (van der Goot)	1	2
Myzocallis asclepiadis (Monell)	25	20
Capitophorous hippophaes (Walker)	3	4
Macrosiphum euphorbiae (Thomas)	1	4
Therioaphis trifolii (Monell)	2	6
Rhopalosiphum padi (L.)	7	11*[1]

[1] Horizontally paired numbers followed by an asterisk (*) are significantly different from one another (P < .05), according to the Duncan-Waller test.

Sampling Aphids in Soybean Fields

Table 11-2. Mean number of alates captured per horizontal ermine lime trap (231 cm^2) on sticky (tanglefoot covered) and in liquid-filled containers (ethylene glycol) during the 1978 season, Urbana, Ill. (N = 5)

	Sticky surface	Liquid-filled surface
Rhopalosiphum maidis (Fitch)	53	78*[1]
Aphis citricola (van der Goot)	13	42*
Myzocallis asclepiadis (Monell)	68	45*
Capitophorous hippophaes (Walker)	11	17
Macrosiphum euphorbiae (Thomas)	3	23
Therioaphis trifolii (Monell)	4	5
Rhopalosiphum padi (L.)	7	4*

[1] Horizontally paired numbers are significantly different from one another, according to the Duncan-Waller test (P < .05), if the paired number are starred (*).

one against the other. This trap seems to approach the ideal, but final acceptance must await confirmation of little or no bias toward or away from the tile relative to the canopy.

Number of sample units needed with the horizontal ermine lime trap: Taylor's power law (Taylor 1961, see Chapter 3) was fitted from 1976, 1977, and 1978 data of aphid landings on horizontal ermine lime traps (M. Irwin unpublished). The mean regression line for three consecutive years of aphid catches (Fig. 11-9, 1976,1977,1978; grey or stippled line) conformed to the formula

$$s^2 = 0.8355\bar{x}^{1.3054}, r = 0.87$$

while the formula for the individual years varied from it:

$$s^2 = 0.6233\bar{x}^{1.6126}, r = 0.89 \qquad 1976$$

$$s^2 = 1.1472\bar{x}^{1.0244}, r = 0.85 \qquad 1977$$

$$s^2 = 0.7481\bar{x}^{1.3106}, r = 0.86 \qquad 1978$$

where s^2 = variance and $\bar{x}$ = mean.

By plotting the mean versus the ratio of the standard error over the mean ($S_{\bar{x}}/\bar{x}$ = RV) for the regression line for the mean of all three years, one can calculate the number of sample units (i.e., horizontal ermine lime traps) needed to achieve a desired level of reliability given the approximate mean number of alates per trap (Fig. 11-10). For instance, if one wanted to be within 20% of the mean number of aphids caught per week and the estimate of the mean were 30 aphids per trap per week (a very conservative estimate), one would need two horizontal ermine lime traps.

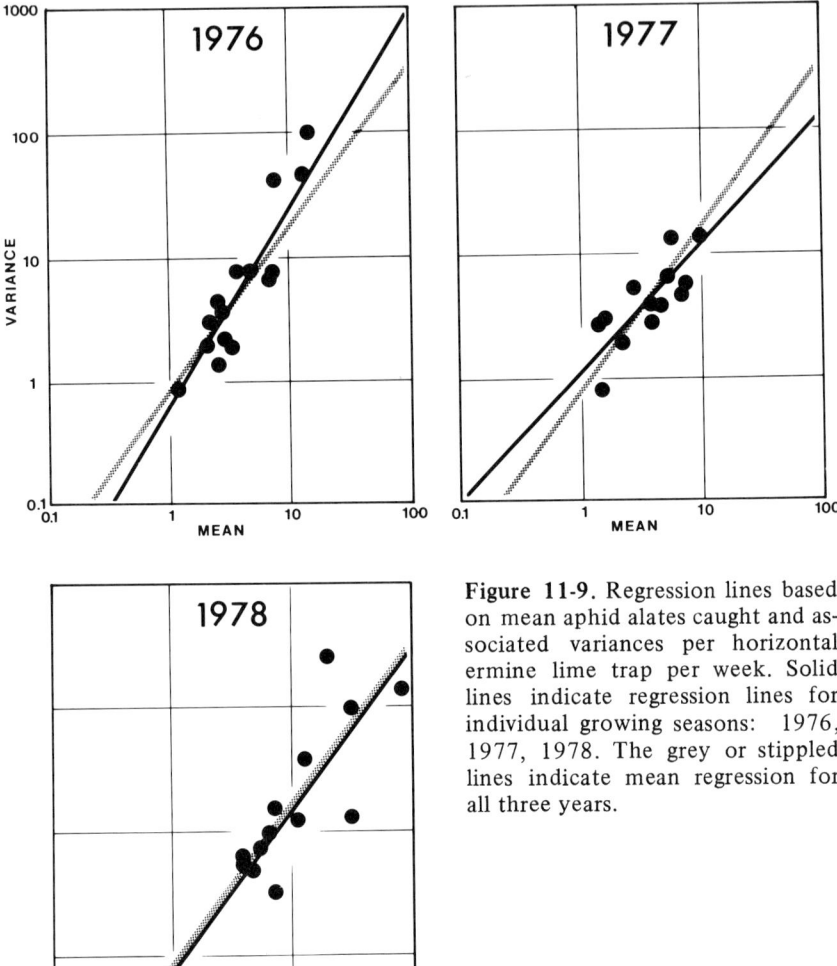

Figure 11-9. Regression lines based on mean aphid alates caught and associated variances per horizontal ermine lime trap per week. Solid lines indicate regression lines for individual growing seasons: 1976, 1977, 1978. The grey or stippled lines indicate mean regression for all three years.

IV. Live Trapping for Determining Field Vectors of Soybean Viruses

An important method for specific use in studies of aphids as vectors of viruses is trapping and subsequent assaying of live specimens. A trap was constructed that intercepts aphids while flying or being carried by wind currents (S. Halbert, M. Irwin and R. Goodman unpublished). It consists of a relatively fine mesh strip of netting, ca. 1.5 m high and 5 m long staked to fence posts on the downwind side

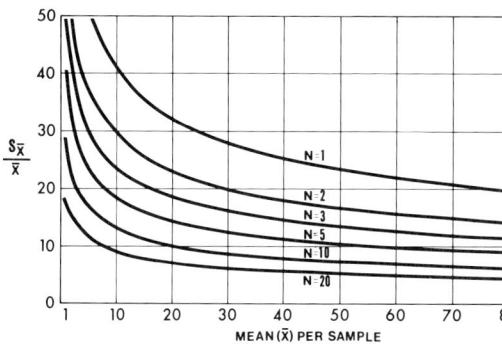

Figure 11-10. Relationship between the number of horizontal ermine line traps (N), aphids per trap week ($\bar{x}$), and the accuracy of the estimate of numbers per horizontal ermine lime trap ($S_{\bar{x}}/\bar{x}$). Graph based on regression of three years' catch of variance and means on a log scale (see Fig. 11-9). Regression line $s^2 = 0.8355\bar{x}^{1.3054}$, where s^2 = variance, $\bar{x}$ = mean; n = 46; r = 0.87.

of a heavily virus-infected field. The net must be mobile so that it can be moved to the downwind side. As aphids are intercepted by the net they are removed; each one is placed on a test plant and the plant caged. After ca. 1 hr, the cage is removed, the aphid preserved and the plant placed in a greenhouse to await possible symptom development. This works well for nonpersistent viruses. For persistent viruses, the cages should be left on overnight or longer, depending on latent period. This sampling device gives relative numbers of aphid species caught, the percentage of each species that transmitted, and the percentage of each species transmission relative to total aphid catches. This method provides a way of determining which species are most important in the field spread of viruses and, while time-consuming, allows one to determine target species for control.

V. Predicting Aphid Abundances in Soybean Fields

Use of the horizontal ermine lime trap in soybean fields in central Illinois seems to provide approximate absolute landing rates for aphids (M. Irwin, R. Goodman, S. Halbert and G. Schultz unpublished). During most seasons from 1976 through 1978, somewhere between 80 and 100 species are trapped.

During the 1976 growing season (17 weeks), a mean of 4 085 aphids/m^2 were captured. In 1977 (18 weeks) only 2 494 aphids/m^2 were caught. In 1978 (16 weeks) 9 698 aphids/m^2 were collected. Thus, the numbers of aphids landing on the soybean canopy in central Illinois were not consistent from year to year. The timing of flights of any given species was also inconsistent. For instance, *R. padi* was most abundant in the fall of 1975 and 1977 and most abundant in the spring of 1976. *A. citricola* was abundant in 1975 and 1978 but not in 1976 or 1977. A few species seemed more predictable. *A. craccivora*, when abundant, was caught primarily in the spring and early summer. *R. maidis*, while varying in numbers among years, was consistently more abundant in the fall.

Way and Cammell (1973) approached the problem of predicting the infestation levels of *Aphis fabae* Scopoli by counting eggs on primary hosts during the winter, and by counting active stages on primary hosts in the spring. The forecasting of infestation levels based on these parameters, especially egg counts, was economical and the predictive value high. Whalon et al. (1978) used pan traps to help adjust rates of computer simulated colonization and population dynamics of *M. persicae*. The model was closely tied to climatic data. Both of the aforementioned forecasting systems were predicated on single colonizing species.

VI. Concluding Remarks

With soybean and noncolonizing aphid species, the problem becomes more complex. In studies of aphid flights in central Illinois, the origin of aphid alates that land in soybean is uncertain. Were they migratory or nonmigratory? In fact, the primary and secondary hosts of many of these species are unknown. Until the population dynamics of the most important species (i.e., the species most responsible for the early to midseason spread of soybean mosaic virus), is understood the ability to predict population trends is very limited. From one set of studies conducted in 1977 (M. Irwin and R. Goodman unpublished), it appears as though some species have flights about the same time over a fairly wide area in any given year. This was documented with *R. maidis*, the most abundant species. If this holds true for other species, continual monitoring of aphid landings at one sight might provide data for a large region. In this sense, some predictive value could be obtained.

Acknowledgments

I acknowledge with gratitude the effort exerted by colleagues toward the goal of accurate sampling of aphids alighting in soybean fields: Drs. Susan E. Halbert, Agricultural Entomology, Gerald A. Schultz, Agricultural Entomology and Plant Pathology, and Robert M. Goodman, Plant Pathology, University of Illinois. This paper is based on work supported in part by the International Soybean Program (INTSOY) through the United States Agency for International Development contracts cm/ta-c-73-19 and TA/C-1294 and through an Illinois Agricultural Experiment Station competitive Hatch grant. The views and interpretations are those of the author and should not be attributed to the Agency for International Development or to any individual acting in their behalf.

References

A'Brook, J. 1968. The effect of plant spacing on the numbers of aphids trapped over the groundnut crop. Ann. Appl. Biol. **61**:289-294.

Adlerz, W. C. 1976. Comparison of aphids trapped on vertical sticky board and cylindrical aphid traps and correlation with watermelon mosaic virus 2 incidence. J. Econ. Entomol. **69**:495-498.

Ashby, J. W. 1976. An aphid trap used in virus-vector studies. New Zealand Entomol. **6**:187.

Bandong, J. P., and J. A. Litsinger. 1976. Insect and disease control of field crops planted after rice. Report of research trials in Manaoag, Pangasinan 1975-76. Cropping Systems Program, International Rice Research Institute, Los Baños, Philippines. 66 p., mimeo.

Banks, C. J. 1954. A method for estimating populations and counting large numbers of *Aphis fabae* Scop. Bull. Entomol. Res. **45**:751-756.

Chamberlin, J. C., and F. R. Lawson. 1945. A mechanical trap for the sampling of aerial insect populations. Mosquito News **5**:4-7.

Cook, W. C. 1955. Survey methods: pea aphid. USDA Coop. Econ. Insect Rep. **1955**:1-28.

Costa, C. L., and T. Lewis. 1967. The relationship between the size of yellow water traps and catches of aphids. Entomol. Exp. Appl. **10**:485-487.

Davis, E. W., and B. J. Landis. 1949. An improved trap for collecting aphids. U.S. Bur. Entomol. Plant Quar. **ET-278**:3 p.

Dickson, R. C., M. M. Johnson, R. A. Flock, and E. F. Laird, Jr. 1956. Flying aphid populations in southern California citrus groves and their relation to the transmission of the tristeza virus. Phytopathology **46**:204-210.

Dixon, A. F. G. 1977. Aphid ecology: Life cycles, polymorphism, and population regulation. Annu. Rev. Ecol. Syst. **8**:329-353.

Eastop, V. F. 1955. Selection of aphid species by different kinds of insect traps. Nature (London) **176**:936.

Fenton, F. A., and D. E. Howell. 1957. A comparison of five methods of sampling alfalfa fields for arthropods. Ann. Entomol. Soc. Amer. **50**:606-611.

Guss, P. L., and T. F. Branson. 1972. The use of ^{75}Se in feeding studies with the corn leaf aphid (Hemiptera (Homoptera) Aphididae). Ann. Entomol. Soc. Amer. **65**:303-306.

Heathcote, G. D. 1957. The optimum size of sticky aphid traps. Plant Pathol. **6**:104-107.

Heathcote, G. D. 1972. Evaluating populations on plants. pp. 105-145 in H. F. van Emden, ed. Aphid technology. Academic Press, London. 454 p.

Heathcote, G. D., J. M. Palmer, and L. R. Taylor. 1969. Sampling for aphids by traps and by crop inspection. Ann. Appl. Biol. **63**:155-166.

Johnson, C. G. 1950. The comparison of suction trap, sticky trap and tow net for quantitative sampling of small airborne insects. Ann. Appl. Biol. **37**:268-285.

Kennedy, J. S., and C. O. Booth. 1963. Free flight of aphids in the laboratory. J. Exp. Biol. **40**:67-85.

Kennedy, J. S., C. O. Booth, and W. J. S. Kershaw. 1959. Host finding by aphids in the field. I. Gynoparae of *Myzus persicae*. Ann. Appl. Biol. **47**:410-423.

Kobayashi, T., T. Oku, Y. Maeta, K. Takahashi, and T. Matsushima. 1976. Studies on the soil application of insecticides. IX. Effects of some insecticides on arthropod pests and virus diseases of the soybean. Bull. Tohoku Nat. Agr. Exp. Sta. **53**:15-28.

Kogan, M. 1977. Soybean entomology in Korea. Consultant report to the Crop Improvement Research Center and Institute of Agricultural Sciences, Office of Rural Development, Suweon, Korea. September 1977. College of Agr., Univ. of Ill., Urbana. 25 p.

Kogan, M. 1978. Soybean entomology in Korea, second report. Consultant report to the Crop Improvement Research Center, Crop Experiment Station, and Institute of Agricultural Sciences, Office of Rural Development, Suweon, Korea. July 24-August 4. College of Agr., Univ. of Ill., Urbana. 14 p.

Kring, J. B. 1972. Flight behavior of aphids. Annu. Rev. Entomol. 17:461-492.

Litsinger, J. A., C. B. Quirino, M. D. Lumaban, and J. P. Bandong. 1977. Grain legume pest complex of three Philippine rice-based cropping systems. Cropping Systems Program, International Rice Research Institute, Los Baños, Philippines. 39 p., mimeo.

Mackauer, M. 1972. Antennal amputation as a method for bio-marking aphids. J. Econ. Entomol. 65:1725-1727.

Moericke, V. 1955. Über die Lebensgewohnheiten der geflügelten Blattläuse (unter besonderer Berücksichtigung des Verhaltens beim Landen). Z. Angew. Entomol. 37:29-91.

Muraveva, M. F. 1973. Soybean mosaic in the Khabarovsk region (in Russian). Tr. Dal'nevost. Kraev. Nauchno-Issled. Inst. Lesnogo Khoz. 13:156-158.

Pettersson, J. 1968. Tagging aphids. Opusc. Entomol. 33:219-229.

Rabasse, J. M., and Y. Bouchery. 1977. Nouvelles données sur les méthodes d'évaluation des populations de pucerons (Homoptères, Aphididae): séparation des insectes de leur plante-hôte et dénombrement. Ann. Zool. Écol. Anim. 9:407-423.

Rabasse, J. M., E. Brunel, R. Delecolle, and J. Rouze-Jouan. 1976. Influence de la dimension de pièges à eau colorés en jaune sur les captures d'aphides dans une culture de carotte. Ann. Zool. Écol. Anim. 8:39-52.

Roach, S. H., and H. R. Agee. 1972. Trap colors: Preference of alate aphids. Environ. Entomol. 1:797-798.

Rogerson, J. P. 1975. Grease-coated traps for catching alate aphids in potato crops. Entomol. Mon. Mag. 111:229-230.

Ross, J. P. 1977. Effect of aphid-transmitted soybean mosaic virus on yields of closely related resistant and susceptible soybean lines. Crop Sci. 17:869-872.

Saugstad, E. S., R. A. Bram, and W. E. Nyquist. 1967. Factors influencing sweep net sampling in alfalfa. J. Econ. Entomol. 60:421-426.

Shands, W. A., and G. W. Simpson. 1955. Survey methods: populations of potato-infesting aphids and of aphid eggs on primary hosts in Maine. USDA Coop. Econ. Insect Rep. 1955:28-30.

Shands, W. A., G. W. Simpson, and J. E. Dudley. 1956. Low-elevation movement of some species of aphids. J. Econ. Entomol. 49:771-776.

Shands, W. A., G. W. Simpson, and F. H. Lathrop. 1942. An aphid trap. U.S. Bur. Entomol. Plant Quar. ET-196:6 p.

Shorey, H. H. 1963. Differential toxicity of insecticides to the cabbage aphid and two associated entomophagous insect species. J. Econ. Entomol. 56:844-847.

Taimr, L., J. Holman, J. Kriz, and A. Kudelova. 1978. Large-scale radiophosphorus marking of the hop aphid (*Phorodon humuli* Schrank) in situ on the primary host. Z. Angew. Entomol. 86:145-160.

Taylor, L. R. 1955. The standardization of air-flow in insect suction traps. Ann. Appl. Biol. 43:390-408.

Taylor, L. R. 1961. Aggregation, variance and the mean. Nature (London) 189:732-735.

Taylor, L. R. 1962a. The efficiency of cylindrical sticky insect traps and suspended nets. Ann. Appl. Biol. 50:681-685.

Taylor, L. R. 1962b. The absolute efficiency of insect suction traps. Ann. Appl. Biol. **50**:402-421.

Taylor, L. R., and J. M. P. Palmer. 1972. Aerial sampling. pp. 189-234 in H. F. van Emden, ed. Aphid technology. Academic Press, London. 454 p.

Way, M. J., and M. E. Cammell. 1973. The problem of pest and disease forecasting—possibilities and limitations as exemplified by work on the bean aphid, *Aphis fabae*. Proc. 7th British Insecticide and Fungicide Conference. **3**:933-954.

Whalon, M. E., B. A. Bajusz, and Z. Smilowitz. 1978. Updating the green peach aphid forecast system with a monitoring procedure. J. N. Y. Entomol. Soc. **86**:327-328.

Zettler, F. W., R. Louie, and A. M. Olson. 1967. Collections of winged aphids from black sticky traps compared with collections from bean leaves and water-pan traps. J. Econ. Entomol. **60**:242-244.

Chapter 12

Sampling Leafhoppers on Soybean

Charles G. Helm, Marcos Kogan, and Bob G. Hill

I. Introduction

Numerous species of leafhoppers (Homoptera: Cicadellidae) have been recorded in association with soybean in the United States. A survey conducted in Arkansas detected 37 species of leafhoppers (Tugwell et al. 1973). Forty-seven species have been recorded in Illinois based on specimens in the International Reference Collection of Soybean-Associated Arthropods (IRCSA unpublished). Sixteen species were listed from Mississippi (Kincade et al. 1970), and 30 species were listed from Missouri (Blickenstaff and Huggans 1974). A simple key for the identification of the most common genera of Cicadellidae is presented in an appendix to this chapter.

Not all species collected are abundant enough to be considered permanent inhabitants of soybean fields and are most likely transients or weed-feeders. Several other species are very common although their ability to colonize soybean is not always known. Among the most abundant and annually recurring species are: *Empoasca fabae* (Harris), *Scaphytopius acutus* (Say), *Aceratagallia sanguinolenta* (Prov.), *Paraphlepsius irroratus* (Say), *Agallia constricta* Van Duzee, *Graminella nigrifrons* (Forbes), and *Graphocephala versuta* (Say) (Kincade et al. 1970, IRCSA unpublished). In all four surveys, the most abundant leafhopper species was *Empoasca fabae*, the potato leafhopper. Therefore, many of the sampling procedures discussed in this chapter were developed in potato leafhopper research. Although little information exists regarding their effectiveness for other species, these procedures can serve at least as a starting point in research on populations of those other species.

A. Geographical Distribution

E. fabae is generally distributed throughout the eastern half of the United States, extending westward to the Rockies and throughout the north central states (Fig. 12-1). A few specimens have been recorded from the lower elevations of California, Arizona, and New Mexico, but higher populations are most likely to occur in the eastern, midwest, and northern plain states. In general, climate and

Figure 12-1. Geographical distribution of *Empoasca fabae*. (Based on DeLong 1938).

Table 12-1. Host records for the seven most abundant Cicadellidae species associated with soybean in North America

Species name	Common name	Host-records	Reference
Empoasca fabae	Potato leafhopper	Nearly all leguminous plants (especially alfalfa, clover, beans), potatoes, apple.	Poos and Wheeler 1943
Aceratagallia sanguinolenta	Clover leafhopper	Legumes (especially clover, alfalfa); abundant on many economic crops; grass feeders, in meadows and pastures.	DeLong 1948 DeLong 1965
Agallia constricta	--	Large numbers on economic crops; grass feeders, in meadows and pastures.	DeLong 1948 DeLong 1965
Graminella nigrifrons	Blackfaced leafhopper	Grass feeders; common on lawns, pastures, meadows, and most cultivated crops.	DeLong 1948
Graphocephala versuta	--	Grassy habitats, herbaceous vegetation.	DeLong 1948
Paraphlepsius irroratus	--	Pastures, meadows, wet prairies; on almost every cultivated crop.	DeLong 1948
Scaphytopius acutus	--	Grass feeders, pastures and meadows; genus is highly polyphagous.	DeLong 1965

altitude appear to be most important in determining the distribution and abundance of this species. It is most abundant at elevations below 1000 m and in relatively humid regions receiving 75-100 mm of rainfall monthly during the growing season (DeLong 1938). *E. fabae* has been recorded in South America (Poos and Wheeler 1943), but there is a possibility that other small green *Empoasca* species have been misidentified as *E. fabae*.

B. Host Plants and Nature of Damage

Information regarding the host range of leafhoppers is both confusing and inadequate. While the literature is filled with references concerning leafhoppers and their supposed hosts, this information is not always totally accurate and quite often too vague. Such records are often merely adult feeding records or incidental visitation records and are not necessarily indicative of a given species' ability to oviposit and reproduce on a specific host (Oman 1949, Beirne 1956).

Some leafhopper species have a rather restricted host range, and some adult leafhoppers may feed on plants not used for oviposition (Oman 1949). The more common species found in soybean are highly polyphagous. Poos and Wheeler (1943) listed over 100 species of host plants from which nymphs of the potato leafhopper were collected and reared to the adult stage.

E. fabae occurs on nearly all leguminous plants, including alfalfa, clover, and bean. This species is one of the most abundant insects found on soybean in the Midwest (Balduf 1923, Kretzschmar 1948, Tugwell et al. 1973, Blickenstaff and Huggans 1974, IRCSA unpublished). It commonly reaches damaging population levels on the aforementioned forage legumes, as well as bean, cowpea, potato, apple, and occasionally various ornamentals (DeLong 1938).

The six remaining most abundant soybean-associated species are similarly polyphagous. Host records for these species are summarized in Table 12-1 based mainly on DeLong (1948, 1965).

In spite of this relatively high abundance, the damage potential of leafhoppers to soybean has not been fully explored. Leafhoppers can damage agricultural crops either directly by their feeding or oviposition on plant tissues or indirectly as vectors of plant viruses (Oman 1949).

Direct damage caused by leafhopper feeding is difficult to measure. Most leafhoppers feed on the under-surface of leaves, sucking plant juices from the mesophyll. Although occasional localized mottling or stippling may result, in most instances in soybean such feeding apparently has little direct effect on yields or overall seed quality. However, *E. fabae* feeds by puncturing the phloem and xylem tubes. The result of this type of feeding is a disease-like injury known as hopperburn, which is clearly observed in heavy infestations on glabrous or sparse pubescent soybean lines (Fig. 12-2). The exact nature of hopperburn is not fully known, but hopperburn symptoms are distorted leaf veins, leaf margins curled upward and inward, and leaves gradually changing color from green to yellow to brown as they dry out and die (Smith and Poos 1931).

Figure 12-2. Glabrous 'Clark' soybean stunted by heavy *Empoasca fabae* infestations. Plants in the background and on the left-hand side are 'Clark' soybean with normal pubescence. Both lines were planted at the same time.

These symptoms in common bean and in susceptible soybean most likely result from repeated feeding and oviposition along the midribs of the leaves. As the cells are disrupted and growth inhibited along the under side of the leaves, the upper surface grows faster, and the bending and curling follows (DeLong 1965).

Leafhopper populations and susceptibility of soybean to hopperburn are affected by the degree of plant pubescence (Poos and Smith 1931, Johnson and Hollowell 1935, Wolfenbarger and Sleesman 1963). Robbins and Daugherty (1969) found glabrous varieties of soybean to have both the highest numbers of leafhoppers and highest ovipositional rates, while the dense pubescent varieties had the lowest numbers and lowest incidence of oviposition. More specifically, it appears that length and orientation of hairs rather than density alone (Broersma et al. 1972, Turnipseed 1977) are major factors in protecting most commercial cultivars from serious damage by this insect.

Cage-tests have shown that a heavy infestation of potato leafhopper is capable of producing the stunting, curling, necrosis, and drying characteristics of hopperburn in commercially grown varieties (Ogunlana and Pedigo 1974a). However, normal population levels in soybean in any given year are apparently below the economic threshold, as there is little evidence of damage or any apparent effect on yields due to leafhopper feeding (Kincade et al. 1970, Ogunlana and Pedigo 1974b, Turnipseed and Kogan 1976, C. Helm and M. Kogan unpublished).

Leafhoppers have been implicated in the transmission of disease organisms to various crop plants. The only positive association with a disease of soybean has been established when *Scaphytopius fuliginosus* Osborn was shown to transmit the causal agent of the disease known as "machismo" in Colombia (Granada 1979).

C. Life Cycle and Phenology

Leafhopper species may overwinter as either eggs, nymphs, or adults (Oman 1949). *E. fabae* is known to overwinter in fairly permanent breeding areas in the Gulf states, primarily on alfalfa and other legumes. Each spring adults migrate northward on warm air currents (Medler 1957, Pienkowski and Medler 1964). *E. fabae*, which does not overwinter in the north central states (Poos 1932, DeLong and Caldwell 1935), appears rather abruptly and abundantly on cultivated hosts each spring. These early colonizers are predominantly fertile females (Pienkowski and Medler 1964).

Colonizing females begin laying eggs within tissues of adequate host plants. Eggs (Fig. 12-3A) are laid singly beneath the epidermis, primarily in the main veins on the underside of the leaves with a few eggs placed in the petioles or young, tender stems. Eggs may be produced for a month or more at the rate of two or three eggs a day. The incubation period averages 10 days, although cool temperatures may extend it beyond 20 days. Nymphs (Fig. 12-3B-F), similar in overall appearance to adults although paler and lacking wings, gradually increase in size and greenness through five molts. Development during the summer may require only seven to eight days but averages two weeks under fluctuating field conditions. Cool temperature will extend nymphal development to as long as 26 days. Adults (Fig. 12-3G) are small (3.5 mm long) wedge-shaped leafhoppers, pale green, with rather inconspicuous white spots on the head and pronotum. They may begin mating within 24 hr of emergence with oviposition beginning in 3 to 10 days (DeLong 1938).

A season of high humidity and high precipitation is favorable to large leafhopper populations (DeLong 1938), but dry weather and high temperatures for as little as a week are likely to cause drastic reductions in the populations. While higher humidity will attenuate the effects of high temperatures, as temperature approaches 32°C mortality is high regardless of humidity (Decker and Cunningham 1967). Similarly, unseasonably low temperatures can be a factor in greatly reducing populations. Under average climatic conditions in central Illinois, there are two complete generations and a partial third or even fourth, with great overlap among generations (Fig. 12-4). In Illinois the immigrating adults ordinarily first appear on alfalfa or clover, before the emergence of soybean seedlings.

Thus, potato leafhoppers have immediate access to soybean seedlings for feeding and oviposition. Populations generally peak between mid July and early August (Fig. 12-4), during the middle of the soybean growing season, with adults present on plants until harvest maturity (Tugwell et al. 1973, Blickenstaff and Huggans 1974, Ogunlana and Pedigo 1974b, C. Helm and M. Kogan unpublished). Early season populations may actually hold the greatest threat to soybean. Field populations at that time mostly closely approach an economic injury level deter-

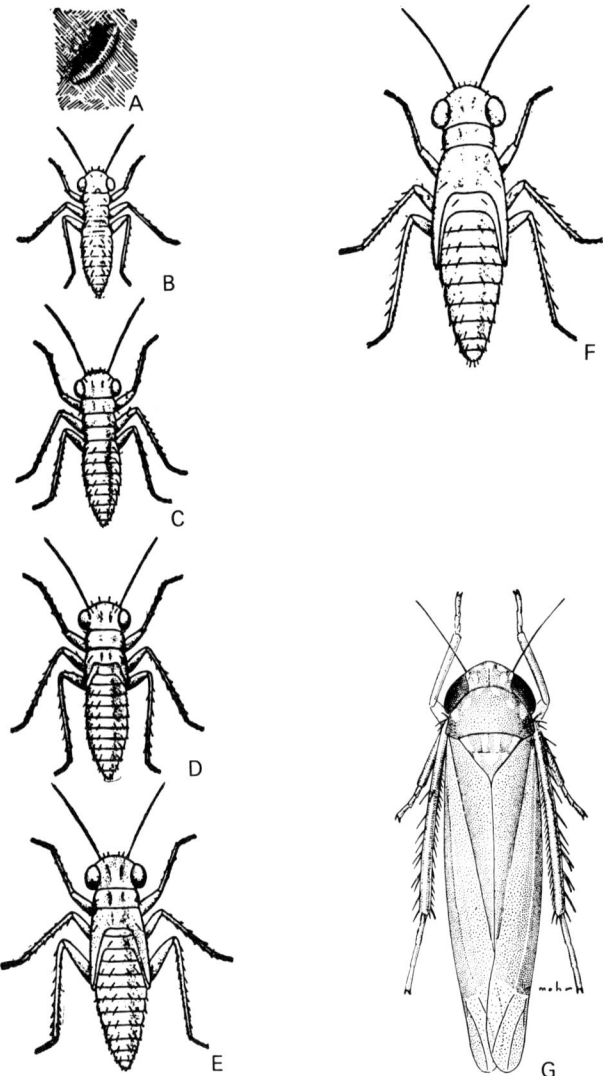

Figure 12-3. *Empoasca fabae* life cycle: (A) eggs; (B-F) nymphs; (G) adult. (From DeLong 1938 and 1948).

mined by Ogunlana and Pedigo (1974a,b). Furthermore, soybean adjoining alfalfa may be particularly threatened early in the season following the cutting of alfalfa, as leafhoppers have been shown to move from freshly cut alfalfa into neighboring bean fields (Poston and Pedigo 1975).

Sampling Leafhoppers on Soybean

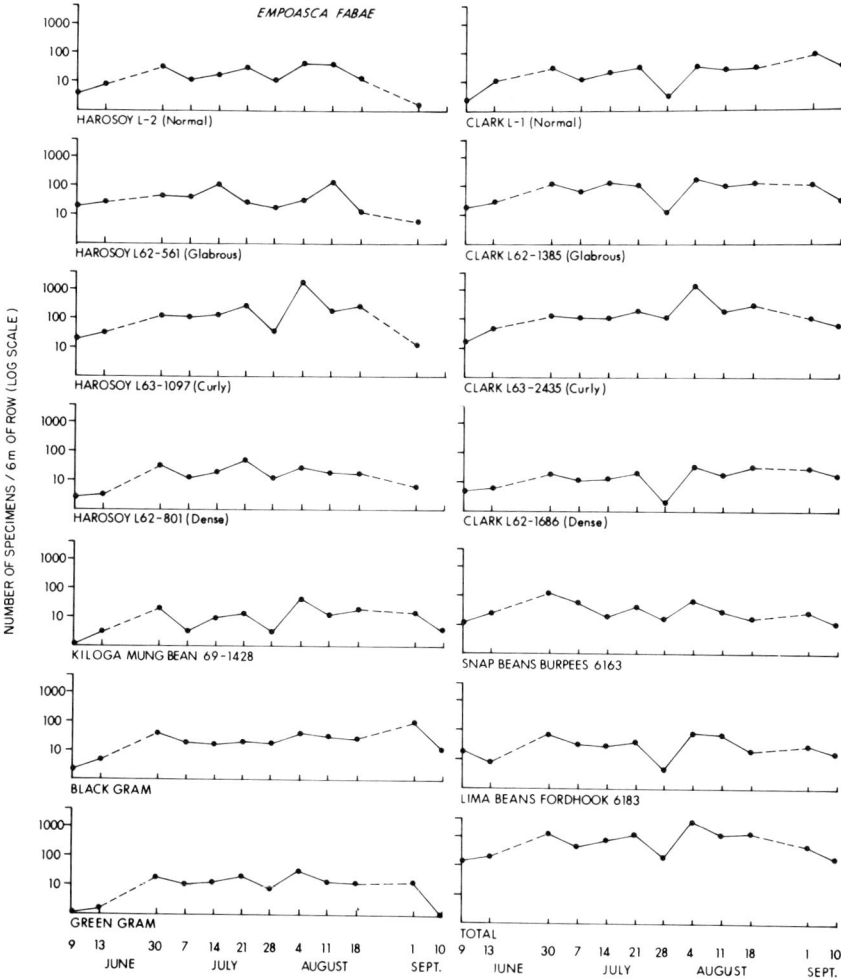

Figure 12-4. Seasonal occurrence of adult *Empoasca fabae* on 13 leguminous hosts in central Illinois. (M. Kogan unpublished).

II. Sampling Adult Leafhoppers

The most commonly used relative methods for sampling adult leafhoppers in soybean fields are the sweep net, direct observations, and the suction net (D-Vac). The main absolute sampling methods are the bagging of whole plants (clam traps) and the phototactic emergence cage.

Many trapping techniques have been used to monitor migrating adult populations. Some of the most complete research on leafhopper migration used nets mounted on automobiles, rotary nets (Cook 1967, see Chapter 11), or airplane collections (Glick 1960).

A. Relative Sampling Methods

The direct observation technique of counts over a measured length of row (see Chapter 2) is recommended as an adequate sampling procedure early in the growing season. As plant size and population increase, this method becomes less effective, but it has been useful for making *E. fabae* population estimates up to the V4 plant stage (C. Helm and M. Kogan unpublished). A variant of this "flushing and counting" technique, useful in activating sluggish forms of leafhoppers, has been to lightly spray a strip of vegetation ahead of the sampler with a pyrethrum spray. Leafhoppers are counted as they jump ahead of the sampler and may be recorded as individuals per unit of area as they fall to the ground (Cook 1967). This technique cannot be used after the plant canopy closes.

Single plant examination for adults of *E. fabae* has similar disadvantages to counts over a measured length of row in that as the season progresses and the plant size and population of leafhoppers increase, it becomes increasingly difficult to spot the quick-moving adults. In this case the sampler must make an initial overall scan of the plant before a detailed examination of leaves, petioles, and stems is begun. Early morning with its cooler temperatures and moister plant conditions is more favorable for sampling *E. fabae* adults, as these conditions should help arrest their activity (Mayse et al. 1978b). Further, as only one plant is being sampled, numbers of leafhoppers are generally fewer, and those that do move off the plant are usually detected.

Perhaps the most widely used method for sampling populations of adult leafhoppers is the sweep net. Factors bearing on this choice as a general sampling tool are also applicable to its selection for leafhopper sampling (see Chapter 2). Sweeping techniques that have proven useful for sampling the quick-moving *E. fabae* adults include a variant of sweeps across one row (DeLong 1932) and the pendulum technique that Boyer (1967) describes as effective in studying populations of Homoptera (see Chapter 19).

Disadvantages to the use of the sweep net in leafhopper sampling are the extreme variations in catch under varying environmental conditions. Wind velocity and temperature are the most important variables affecting sweep net catches of adult leafhoppers. An increase in the wind velocity results in a decreased leafhopper catch, while an increase in temperature results in an increase in the number of leafhoppers (Romney 1945, Cherry et al. 1977). Furthermore, the sweep net samples only the upper one-third to one-half of the soybean canopy later in the season. Although there are no data on stratification of leafhopper adult populations on soybean, the upper canopy sampling may lead to inaccurate estimates of leafhopper populations. Sweep net sampling seems to grossly underestimate the most abundant species, particularly *E. fabae* (Kretzschmar 1948). In spite of these shortcomings, the sweep net is a common choice for a rapid index of species and individuals of leafhoppers present in soybean fields.

The D-Vac suction insect sampler (Dietrick 1961) provides another means of sampling leafhoppers that, according to some researchers, may approach an absolute method (Ogunlana and Pedigo 1974b, Marston et al. 1976).

Details of D-Vac sampling techniques are discussed in Chapter 2. The same factors affecting reliability of sweep net samples are likely to affect D-Vac samples, and care should be taken in converting D-Vac catches directly into absolute densities. Moist field conditions, wind, and especially plant conditions (density of foliage, degree of lodging) may affect D-Vac reliability. The mechanical condition of the machine is also very important for reliable sampling (see Chapter 2). The D-Vac is perhaps the best relative method for sampling populations of adult leafhoppers, although critical studies are still lacking as it applies to populations on soybean.

B. Absolute Sampling Methods

Most attempts at measuring absolute leafhopper populations have dealt with cage techniques: fumigation cages (Romney 1945, Kretzschmar 1948), clam traps (Mayse et al. 1978a,b) or emergence traps (Cherry et al 1977, C. Helm and M. Kogan unpublished) (see Chapter 2). Most if not all leafhoppers on soybean are captured using some trapping procedure (Romney 1945, Mayse et al. 1978a), suggesting this as the technique of choice for studies requiring accurate population estimates or for calibration of the relative sampling methods.

Fumigation cages used in leafhopper population studies have been cylinders with base plates to collect the fallen individuals following fumigation (Romney 1945, Kretzschmar 1948).

The clam trap is a destructive method originally developed for absolute determination of cotton insect populations (Leigh et al. 1970). The clam trap consists of a hinged wooden frame supporting an open organdy collecting bag that is rapidly brought over a row of soybean and clamped shut, trapping insects inside (Mayse et al. 1978b) (Fig. 12-5). Plants are cut off below the bag-frame at ground level, and the entire contents are placed in large plastic bags. After freezing the bag to kill the insects, the contents are washed in a soap-and-water solution, and the liquid is strained through a fine mesh screen. The residue is then examined for leafhoppers. This technique proved superior to either direct observation or sweep net samples for adult *E. fabae* (Mayse et al. 1978a).

Emergence traps have performed with efficiency in sampling alfalfa for adult *E. fabae* (Cherry et al. 1977). The alfalfa emergence trap consists of a large black plastic garbage can with a hole on top to fit an adhesive lined collecting jar and to allow the entrance of light. The trap is carefully placed over alfalfa with use of a handle and loop holder to keep the sampler a distance away from the site so as not to disturb leafhoppers in the area. Positively phototactic insects, including leafhoppers, are supposed to fly into the jar where they are trapped.

While this purportedly gave good estimates of absolute population of leafhoppers in alfalfa, the same trap used in soybean did not (C. Helm and M. Kogan unpublished). Sheet-metal traps fitted with handles and gradually tapering to a collecting jar from a base size corresponding to row width, designed for soybean sampling, were very inefficient (C. Helm and M. Kogan unpublished). As solar intensity is a key to the performance of these traps (Cherry et al. 1977), it may

Figure 12-5. Clam trap—a method for absolute sampling of leafhopper populations. (A) trap, lateral view; (B) trap, front view; (C) positioning trap over sample site; (D) trap clamped in place, plants are clipped at ground level and removed for extraction of arthropods. (By courtesy of M. Mayse, and the Canadian Entomological Society).

be that the soybean canopy affects the effectiveness of the method. Further work is necessary before this technique can be recommended for sampling leafhoppers on soybean.

Limitations to the use of cage techniques in absolute leafhopper population estimates center on the nature of the soybean crop itself and sampling time. As the season progresses, rows become less discrete, and it is difficult to position traps without disturbing the area to be sampled. The clam trap method is very time-consuming, as it requires washing and sorting of specimens after collecting samples in the field. Young, small soybean plants may be sampled as accurately and much more easily by direct observations (Mayse et al. 1978b). Still, for accurate estimates of absolute density with readily accessible or easily made equipment, consideration should be given to the caging technique.

C. Spatial Patterns and Sampling Program for Adult Leafhoppers

Direct counts over 10 m of row early in the season through stage V4 and 0.5-m D-Vac spot sampling thereafter were used in research on economic injury levels for *E. fabae* on curly pubescent 'Clark' soybean (C. Helm and M. Kogan unpublished). Results of weekly samples obtained over a two-year period were analyzed and revealed that leafhopper spatial patterns follow a Poisson distribution with an excellent fit tested by χ^2 ($p < .002$). Applying Taylor's power law (see Chapter 3) to the set of means and variances for 20 samples collected on 'Clark' soybean in east-central Illinois, the relationship $s^2 = 1.3752\bar{x}(\exp 1.5524)$ was obtained from which the number of sample units (0.5 m D-Vac spot sampling) to achieve a prescribed level of reliability was calculated by the equation: $n = [1.3752\bar{x}(\exp -0.4476)] c^2$ where $c = S_{\bar{x}}/\bar{x}$, or the desired reliability factor (see Chapter 3). The curves obtained for $c = 0.1, 0.15, 0.20,$ and 0.25 are shown in Fig. 12-6.

III. Sampling Leafhopper Eggs

As eggs of the potato leafhopper are inserted into the veins and leaf tissues, they are invisible not only because of their small size but also because of this placement. Thus, most techniques for egg sampling must clear the leaf tissue and stain the eggs in a contrasting color. Ordinarily this is accomplished by boiling the leaves in a lacto-phenol solution to clear the leaves and coagulate the protein in the leafhopper eggs (Carlson and Hibbs 1962, Gaddoura and Venkatraman 1967, Robbins and Daughterty 1969). As seen through a dissecting microscope with sub-lighting, the eggs are visible along the veins as outlines or elongate brownish dots (Fig. 12-7). A method proposed by Curtis (1942), although giving satisfactory results, is too time-consuming and complex in that it requires monitoring an acid/base balance for proper staining of eggs.

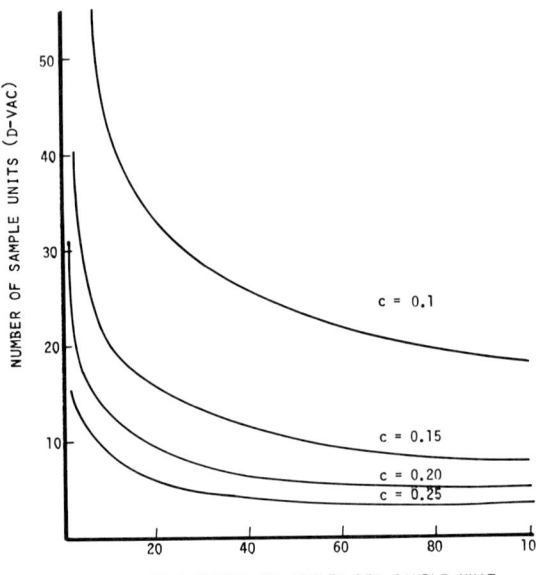

Figure 12-6. Number of sample units necessary to achieve a desired level of reliability to measure *E. fabae* adult populations. Samples obtained with 0.5-m D-Vac spot sample. (C. Helm and M. Kogan unpublished).

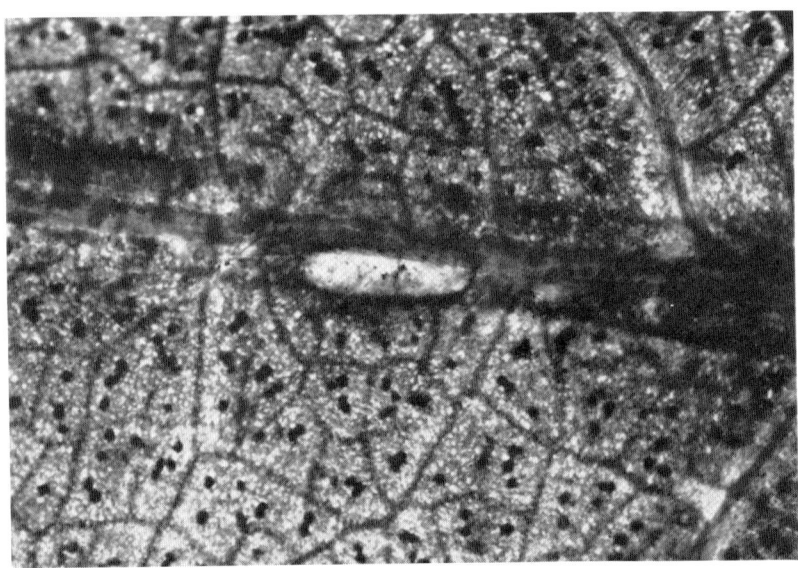

Figure 12-7. *Empoasca fabae* eggs as shown in leaf tissue treated by the lactophenol method. (By courtesy of O. V. Carlson and E. I. Hibbs and the Entomological Society of America).

IV. Sampling Leafhopper Nymphs

Sampling methods for *E. fabae* nymphs take advantage of their typical behavior. Nymphs move fairly freely over the plant, particularly upward, but they seldom leave an individual plant (DeLong 1938). The small size of early instars and their soft, fragile bodies are further factors that may influence the efficiency of any particular method.

A. Relative Sampling Methods

The sweep net is inadequate for sampling nymphal leafhopper populations (DeLong 1932, Mayse et al. 1978a), perhaps due to their small size and tendency to stay attached to the plant.

Counts over a measured length of row are also not applicable to nymphal sampling since nymphs cannot be flushed from the plant for counting. As the dislodged nymphs are small and pale and would be difficult to see on the ground, some collection method must be used. A beating and sticky board technique has been proposed for sampling the brown planthopper (*Nilaparvata lugens* (Stål)) on rice hills (Nagato and Masuda 1978). An adhesive-coated plywood board is held alongside a rice hill while the plants are beaten to dislodge the nymphs onto the sticky surface. A number of boards can be carried to the field in a stacking container for successive samples and later returned to the lab for sorting and counting. The suitability of this method to sampling soybean leafhoppers has not been explored. In adapting this method to soybean, one must consider behavioral differences between the leafhopper species and structural differences between the host plants.

Simonet et al. (1978) used a pan-shake method to sample leafhopper nymphs in alfalfa. Plants are shaken over white enamel pans, and the nymphs are counted as they fall into the pans. This technique has not been tested with soybean. In alfalfa it was reportedly more reliable than direct observations, D-Vac, or sweep net sampling.

Direct observation of single plants is particularly useful for nymphal leafhopper sampling (Mayse et al. 1978a). Although extremely high populations of early instars would make this method very tedious, if not impossible to use, the "stay-at-home" behavior of the nymphs is in the sampler's favor. Disturbed nymphs can be seen running sideways across the leaf surface and up and down the petioles and stems, but seldom do they jump from the plant. However, care must be taken as the early instars are very small and are easily overlooked by all but the most acute observer.

An alternative to total plant examination is leaf unit sampling (DeLong 1938, Kretzschmar 1948, Wolfenbarger and Sleesman 1963, Broersma et al. 1972). Individual trifoliolates are removed from randomly selected plants and examined for the presence of leafhoppers. Leaf samples should be equally distributed over the plant height to account for the vertical stratification of nymphal populations (Table 12-2).

Table 12-2. Stratification of *E. fabae* nymphal populations within the soybean canopy. Nymphs sampled by a leaf washing technique. (Based on Y. Lee, M. Kogan and C. Helm, unpublished)

Dates	Plant growth stage	Bottom 1/3	Middle 1/3	Top 1/3
July 11	V8-9	24.8 ± 15.9	130.6 ± 46.0	127.7 ± 53.5
July 18	R1	12.7 ± 4.8	76.6 ± 40.8	90.7 ± 27.5
July 25	R2	19.7 ± 21.8	68.1 ± 31.4	126.0 ± 66.4
August 1	R3	8.0 ± 4.1	121.1 ± 64.2	312.7 ± 110.4
August 8	R4	14.3 ± 16.0	105.9 ± 105.6	257.9 ± 172.6
August 15	R5	0.2 ± 0.4	0	9.3 ± 14.7
August 22	R5	1.8 ± 2.5	5.9 ± 5.2	18.1 ± 17.7

B. Absolute Sampling Methods

Considerations regarding caging techniques for sampling adult populations also apply to nymphs. The clam trap cage gave adequate results for nymphal *E. fabae* samples, although it detected lower numbers of nymphs than did direct observations (Mayse et al. 1978a). The processing stage of this technique may be a source of this discrepancy as the freezing, thawing, and washing may result in the destruction of the smaller, softer nymphs.

Another method using a leaf washing procedure consists of cutting plants or portions of plants into plastic bags. Plants are cut into top, middle, and bottom sections to provide information on stratification. Detergent water is added to the bagged plants, the bags are vigorously shaken, and the liquid is strained through a 100 mesh stainless steel sieve. Individual leaves, stems, and petioles are washed into the sieve with a light stream of water to remove any remaining detergent that may contain nymphs. The contents of the sieve are then transferred to a petri dish for microscopic examination. The entire process is time-consuming and may also result in crushing nymphs, although it seems to be accurate for absolute population studies (Y. Lee, M. Kogan and C. Helm unpublished). Table 12-2 shows preliminary results of a study on the stratification of leafhopper nymphs during the soybean growing season.

A variation of the bagging techniques and fumigation cage has been used in sampling alfalfa for *E. fabae* nymphs (Jarvis and Kehr 1966). In this method plants are cut, quickly placed in paper bags, and fumigated with a pyrethrum spray. Although time-consuming, it was felt that the resulting counts gave an accurate estimate of total nymphs present. Simonet et al. (1978) used ca. 1-liter paper cartons with a square cut from a dichlorvos strip attached to the lid. Plant samples were clipped and quickly placed inside the cartons. After 24 hr, cartons were shaken in an upside down position, dislodging the dead nymphs into the lid. The technique was reportedly highly efficient for sampling the potato leafhopper on alfalfa. This technique may be adapted for soybean research perhaps by using 3.5-liter containers and by taking leaf samples instead of whole plants.

Little information is available regarding the use of the D-Vac for sampling nymphal populations of leafhoppers on soybean, but its efficiency seems to be reduced as nymphs remain on plants that have been thoroughly sampled with the D-Vac (Ogunlana and Pedigo 1974b, C. Helm and M. Kogan unpublished). The same limitations were reported in sampling leafhopper nymphs on alfalfa (Simonet et al. 1978).

C. Spatial Patterns and Sampling Program for *E. fabae* Nymphs

Absolute samples of *E. fabae* nymphs on curly pubescent 'Clark' soybean were obtained with a plant washing technique (Y. Lee, M. Kogan, and C. Helm unpublished). These samples were used in a preliminary analysis of nymphal spatial patterns within soybean plant canopy strata and within fields (Table 12-2). About 92% of the nymphs are found in the upper two-thirds of the plant canopy, but the middle one-third has a slightly higher proportion of nymphs than the upper one-third. Nymphs at the bottom one-third are more aggregated than those in the upper two-thirds.

Spatial patterns of nymphal populations in soybean are better described by a negative binomial distribution. A common k_c value of 2.32 was computed for the overall population (all three strata combined) at seven stages of plant growth. This k_c was used to compute the number of sample units necessary to achieve a given level of reliability. The equation was: $n = (2.32 + \bar{x})/(2.32 \, c^2 \bar{x})$ (see Chapter 3). The curves computed with $c = 0.1, 0.15, 0.20$, and 0.25 are shown in Fig. 12-8.

Spatial patterns of *E. fabae* nymphs tend to be more random (Poisson) at higher population densities. As populations increase during the season an increased randomness in dispersion is also associated with plant growth until populations finally decline and patterns become more patchy. It may be necessary in view of this seasonal variation to adjust sampling programs, particularly the number of sample units, to conform with these variable patterns.

V. Concluding Remarks

The potato leafhopper is one of the first colonizers of soybean fields in the Midwestern United States. A sampling program for population dynamics studies should start as soon as seedlings emerge. Visual observations and counts of adults over a measured length of row should be used up to stage V4 and the D-Vac thereafter. Detection of eggs should be based on leaflet samples, with eggs visu-

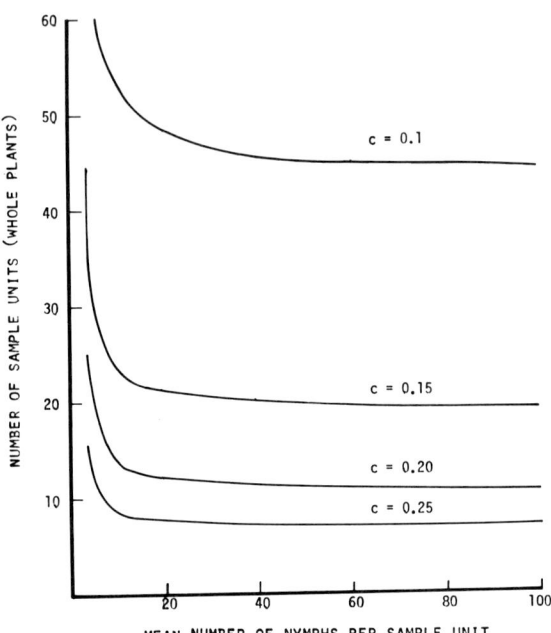

Figure 12-8. Number of sample units necessary to achieve a desired level of reliability to measure *E. fabae* nymph populations on soybean. The sample unit is a single whole plant. Nymphs are separated by the washing technique. Population patterns following the negative binomial distribution. (Y. Lee, M. Kogan and C. Helm unpublished).

alized by the lacto-phenol method. The most reliable sampling procedure for nymphs is clipping of plants or leaves into bags and washing in the laboratory. Nymphs are too small and fragile to be adequately sampled by any of the available relative methods.

There are no generally accepted economic injury levels for leafhoppers on soybean. Therefore, no sequential sampling plans have been proposed although some preliminary plans are under investigation (C. Helm and M. Kogan unpublished).

APPENDIX 12-1 Key to the Common Adult Genera of Leafhoppers Associated with Soybean

(General external morphology Fig. 12-9; heads of common genera Fig. 12-10 and Fig. 12-11)

```
1.      Length 5 mm or less...........................................10
1'.     Length greater than 5 mm......................................2
2(1').  Length 9.4 mm or greater......................................3
2'.     Length less than 9.4 mm.......................................5
3(2).   Submarginal vein of hind wing not evanescent along costal margin
        apically............................................... Aulacizes
3'.     Submarginal vein of hind wing evanescent for a short distance along
        costal margin apically........................................4
```

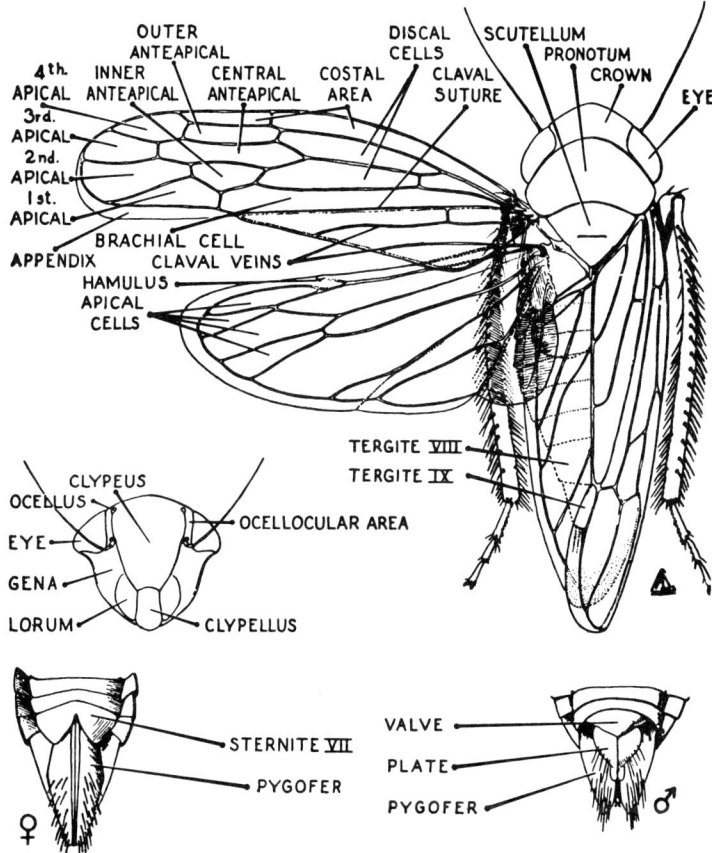

Figure 12-9. External morphology illustrating terminology of body parts and male and female genitalia (see key). (By courtesy of P. W. Oman and the Entomological Society of Washington).

4(3').	Crown long, anterior margin angulate; claval veins usually confluent for a short distance on disk of clavus . . . . *Homalodisca*
4'.	Crown short, anterior margin convex; claval veins not confluent. *Cuerna*
5(2').	Clypellus swollen; ocelli on crown usually near disk 6
5'.	Clypellus not swollen; if so, then ocelli on face and close to the eyes. . 7
6(5).	Ocellocular area produced into a distinct and thick-margined ledge above antennal pit; macropterous, the forewing narrow and not covering the lateral margin of the abdomen . . . . *Cuerna*
6'.	Ocellocular area not forming a distinct ledge above antennal pit; macropterous forms with forewing covering or nearly covering lateral margin of abdomen . . . . *Graphocephala*
7(5').	Genae visible from above . . . . *Scaphytopius*
7'.	Genae not visible from above 8

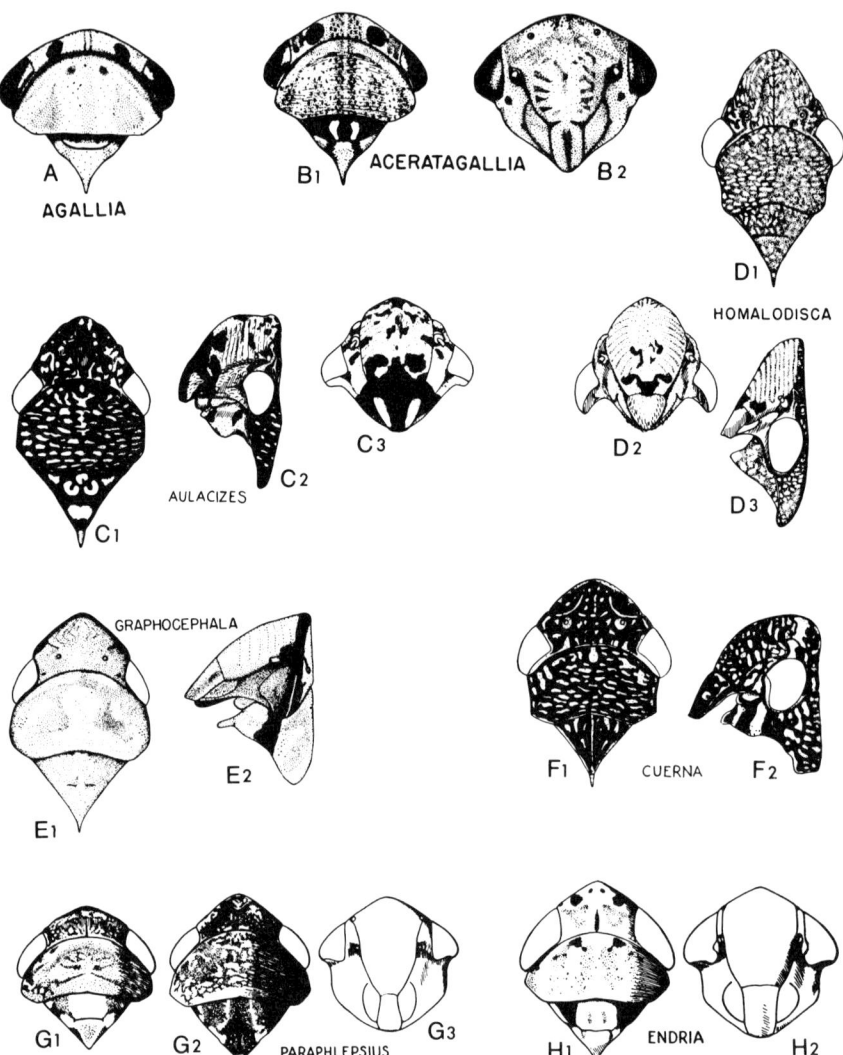

Figure 12-10. Heads of genera of Cicadellidae that include the most common soybean-associated species. (A) *Agallia* dorsal; (B) *Aceratagallia*, 1-dorsal, 2-frontal; (C) *Aulacizes*, 1-dorsal, 2-lateral, 3-frontal; (D) *Homalodisca*, 1-dorsal, 2-frontal, 3-lateral; (E) *Graphocephala*, 1-dorsal, 2-lateral; (F) *Cuerna*, 1-dorsal, 2-lateral; (G) *Paraphlepsius*, 1 and 2-dorsal, 3-frontal; (H) *Endria*, 1-dorsal, 2-frontal (see key). (By courtesy of P. W. Oman and the Entomological Society of Washington).

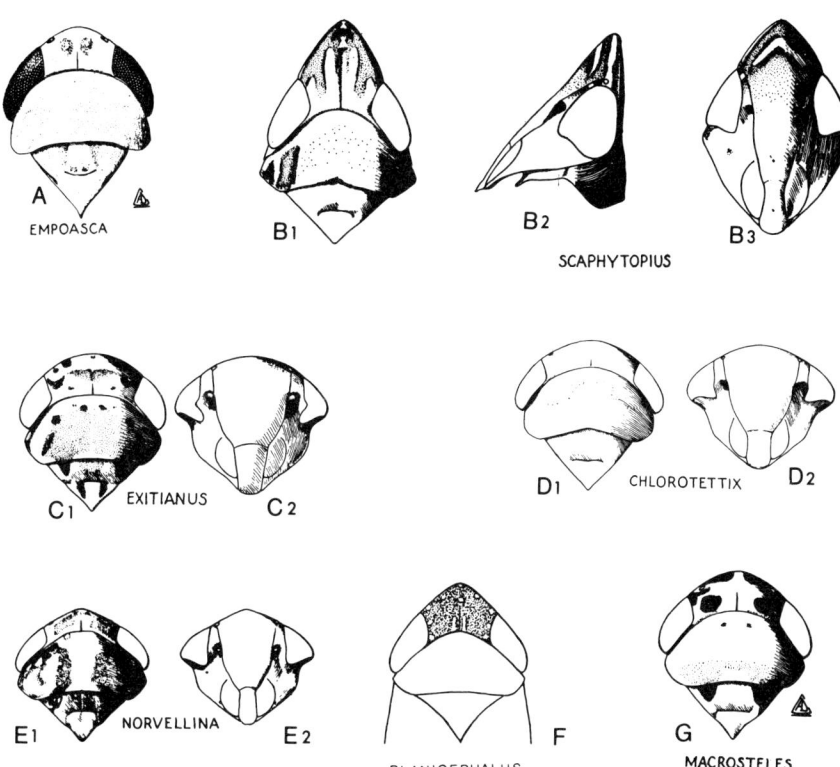

Figure 12-11. Heads of general of Cicadellidae that include the most common soybean-associated species. (A) *Empoasca,* dorsal; (B) *Scaphytopius,* 1-dorsal, 2-lateral, 3-frontal; (C) *Exitianus,* 1-dorsal, 2-frontal; (D) *Chlorotettix,* 1-dorsal, 2-frontal; (E) *Norvellina,* 1-dorsal, 2-frontal; (F) *Planicephalus,* dorsal; (G) *Macrosteles,* dorsal (see key). (A-E and G by courtesy of P. W. Oman and the Entomological Society of Washington; F by courtesy of D. M. DeLong and Illinois Natural History Survey).

8(7').	Head at least as wide as pronotum	*Paraphlepsius*
8'.	Head narrower than pronotum	9
9(8').	Length 5.5 mm or greater	*Chlorotettix*
9'.	Length less than 5.5 mm	*Exitianus*
10(1).	Forewings without crossveins subapically; frail insects greenish in color	*Empoasca*
10'.	Forewings with crossveins subapically; usually robust insects variable in color	11
11(10').	Clypellus swollen; ocelli on crown	*Graphocephala*
11'.	Clypellus not swollen; ocelli on face or, if on crown, much nearer eyes than center of crown	12
12(11').	Lateral frontal sutures terminating at antennal pits; distance between ocelli less than distance between antennal pits	13

12'. Lateral frontal sutures extending beyond antennal pits to or near ocelli; distance between ocelli equal to or greater than distance between antennal pits. 14
13(12). Surface of pronotum finely granulose; vertex of nearly uniform length throughout its width .... *Agallia*
13'. Surface of pronotum transversely rugulose; vertex distinctly longer at middle than near eyes. .... *Aceratagallia*
14(12'). Genae broadly expanded, extending dorsally behind the eyes and visible from above. .... *Scaphytopius*
14'. Genae not visible from above. 15
15(14'). Forewing milky white with broad brown band across posterior half; vertex with impressed transverse furrow just back of anterior margin .... *Norvellina*
15'. Forewing not milky white or without broad brown band or without impressed transverse furrow just back of anterior margin. 16
16(15'). Forewing long and slender with outer anteapical cell absent, without extra cross veins, inner anteapical cell open basally and appendix well developed .... *Macrosteles*
16'. Venation of forewing not as above 17
17(16'). Length 3.5 mm or less; dark brown to black; costal margins of forewings yellowish; face black .... *Planicephalus*
17'. Length greater than 3.5 mm or color and markings not as described above 18
18(17'). Appendix extending around the entire apex of forewing. . . . *Exitianus*
18'. Appendix not extending around entire apex of forewing. ..... *Endria*

Acknowledgments

Much of the original research reported herein was supported in part by the National Science Foundation and the Environmental Protection Agency through a grant (NSF GB-34718) to the University of California. The findings, opinions, and recommendations expressed herein are those of the authors and not necessarily those of the University of California, the National Science Foundation, or the Environmental Protection Agency.

References

Balduf, W. V. 1923. The insects of soybean in Ohio. Ohio Agr. Exp. Sta. Bull. **366**:147-181.

Beirne, B. P. 1956. Leafhoppers (Homoptera: Cicadellidae) of Canada and Alaska. Can. Entomol. **88**(suppl. 2):180 p.

Boyer, W. P. 1967. Survey method for the three-cornered alfalfa hopper (*Spississtilus festinus*) in soybeans in Arkansas. USDA. Coop. Econ. Insect Rep. **17**:324-325.

Blickenstaff, C. C., and J. L. Huggans. 1974. Soybean insects and related arthropods in Missouri. Mo. Agr. Exp. Sta. Res. Bull. **803**(rev.):47 p.

Broersma, D. B., R. L. Bernard, and W. H. Luckmann. 1972. Some effects of soybean pubescence on populations of the potato leafhopper. J. Econ. Entomol. **65**:78-82.

Carlson, O. V., and E. T. Hibbs. 1962. Direct counts of potato leafhopper, *Empoasca fabae*, eggs in *Solanum* leaves. Ann. Entomol. Soc. Amer. **55**:512-515.

Cherry, R. H., K. A. Wood, and W. G. Ruesink. 1977. Emergence traps and sweep net sampling for adults of the potato leafhopper from alfalfa. J. Econ. Entomol. **70**:279-282.

Cook, W. C. 1967. Life history, host plants, and migration of the beet leafhopper in the western United States. USDA. Tech. Bull. **1365**:122 p.

Curtis, W. E. 1942. A method of locating insect eggs in plant tissue. J. Econ. Entomol. **35**:286.

Decker, C. G., and H. B. Cunningham. 1967. The mortality rate of the potato leafhopper and some related species when subjected to prolonged exposure at various temperatures. J. Econ. Entomol. **60**:373-379.

DeLong, D. M. 1932. Some problems encountered in the estimation of insect populations by the sweeping method. Ann. Entomol. Soc. Amer. **25**:13-17.

DeLong, D. M. 1938. Biological studies on the leafhopper, *Empoasca fabae* as a bean pest. USDA Tech. Bull. **618**:60 p.

DeLong, D. M. 1948. The leafhoppers, or Cicadellidae, of Illinois. Ill. Natur. Hist. Surv. Bull. **24**:97-376.

DeLong, D. M. 1965. Ecological aspects of North American leafhoppers and their role in agriculture. Bull. Entomol. Soc. Amer. **11**:9-26.

DeLong, D. M., and J. S. Caldwell. 1935. Hibernation studies of the potato leafhopper (*Empoasca fabae* Harris) and related species of *Empoasca* occurring in Ohio. J. Econ. Entomol. **28**:442-444.

Dietrick, E. J. 1961. An improved backpack motor fan for suction sampling of insect populations. J. Econ. Entomol. **54**:394-395.

Gaddoura, W. M., and T. V. Venkatraman. 1967. A simple method for counting leafhopper eggs inserted in plant tissue. Curr. Sci. (India) **36**:619.

Glick, P. A. 1960. Collecting insects by airplane, with special reference to dispersal of the potato leafhopper. USDA Tech. Bull. **1222**:16 p.

Granada, G. A. 1979. Machismo disease of soybeans: 1. Symptomatology and transmission. Plant Dis. Rep. **63**:47-50.

Jarvis, J. L., and W. R. Kehr. 1966. Population counts versus nymphs per gram of plant material in determining degree of alfalfa resistance to the potato leafhopper. J. Econ. Entomol. **59**:427-430.

Johnson, H. W., and E. A. Hollowell. 1935. Pubescent and glabrous characters of soybeans as related to resistance to injury by the potato leafhopper. J. Agr. Res. **51**:371-381.

Kincade, R. T., L. W. Hepner, and M. L. Laster. 1970. A survey of leafhoppers in soybean fields in Mississippi. J. Econ. Entomol. **63**:1991-1993.

Kretzschmar, G. P. 1948. Soybean insects in Minnesota with special reference to sampling techniques. J. Econ. Entomol. **41**:586-591.

Leigh, T. F., D. Gonzalez, and R. van den Bosch. 1970. A sampling device for estimating absolute insect populations on cotton. J. Econ. Entomol. **63**:1704-1706.

Marston, N. L., C. E. Morgan, G. D. Thomas, and C. M. Ignoffo. 1976. Evaluation of four techniques for sampling soybean insects. J. Kans. Entomol. Soc. 49:389-400.

Mayse, M. A., M. Kogan, and P. W. Price. 1978a. Sampling abundance of soybean arthropods: Comparison of methods. J. Econ. Entomol. 71:135-141.

Mayse, M. A., P. W. Price, and M. Kogan. 1978b. Sampling methods for arthropod colonization studies in soybean. Can. Entomol. 110:265-274.

Medler, J. T. 1957. Migration of the potato leafhopper—a report on a cooperative study. J. Econ. Entomol. 50:493-497.

Nagata, T., and T. Masuda. 1978. Efficiency of sticky boards for population estimation of the brown planthopper, *Nilaparvata lugens* (Stål) (Hemiptera: Delphasidae) on rice hills. Appl. Entomol. Zool. 13:55-62.

Ogunlana, M. O., and L. P. Pedigo. 1974a. Economic injury level of the potato leafhopper on soybeans in Iowa. J. Econ. Entomol. 67:29-32.

Ogunlana, M. O., and L. P. Pedigo. 1974b. Pest status of the potato leafhopper on soybeans in central Iowa. J. Econ. Entomol. 67:201-202.

Oman, P. W. 1949. The Nearctic leafhoppers (Homoptera: Cicadellidae). A generic classification and check list. Mem. Entomol. Soc. Wash. 3:253 p.

Pienkowski, R. L., and J. T. Medler. 1964. Synoptic weather conditions associated with long range movements of the potato leafhopper, *Empoasca fabae*, into Wisconsin. Ann. Entomol. Soc. Amer. 57:588-591.

Poos, F. W. 1932. Biology of the potato leafhopper, *Empoasca fabae* (Harris) and some closely related species of *Empoasca*. J. Econ. Entomol. 25:639-646.

Poos, F. W., and F. F. Smith. 1931. A comparison of oviposition and nymphal development of *Empoasca fabae* (Harris) on different host plants. J. Econ. Entomol. 24:361-371.

Poos, F. W., and N. H. Wheeler. 1943. Studies of host plants of the leafhoppers of the genus *Empoasca*. USDA. Tech. Bull. 850:51 p.

Poston, F. C., and L. P. Pedigo. 1975. Migration of plant bugs and the potato leafhopper in a soybean-alfalfa complex. Environ. Entomol. 4:8-10.

Robbins, J. C., and D. M. Daugherty. 1969. Incidence and oviposition of potato leafhopper on soybeans of different pubescent types. Proc. N. Cent. Br. Entomol. Soc. Amer. 24:35-36.

Romney, V. E. 1945. The effect of physical factors upon catch of the beet leafhopper (*Eutettix tenellus* Baker) by a cylinder and two sweep-net methods. Ecology 26:135-147.

Simonet, D. E., R. L. Pienkowski, D. G. Martinez, and R. D. Blakeslee. 1978. Laboratory and field evaluation of sampling techniques for the nymphal stage of the potato leafhopper on alfalfa. J. Econ. Entomol. 71:840-842.

Smith, F. F., and F. W. Poos. 1931. The feeding habit of some leafhoppers of the genus *Empoasca*. J. Agr. Res. 43:267-285.

Tugwell, P., E. P. Rouse, and R. G. Thompson. 1973. Insects in soybeans and a weed host (*Desmodium* sp.). Ark. Agr. Exp. Sta. Rep. Ser. 214:18 p.

Turnipseed, S. G. 1977. Influence of trichome variations on populations of small phytophagous insects in soybeans. Environ. Entomol. 6:815-817.

Turnipseed, S. G., and M. Kogan. 1976. Soybean entomology. Annu. Rev. Entomol. 21:247-282.

Wolfenbarger, D., and J. P. Sleesman. 1963. Variation in susceptibility of soybean pubescent types, broad beans and runner bean varieties and plant introductions to the potato leafhopper. J. Econ. Entomol. 56:895-897.

Chapter 13

Sampling Phytophagous Thrips on Soybean

Michael E. Irwin and Kenneth V. Yeargan

I. Introduction

Sampling phytophagous thrips populations in soybean fields is seldom undertaken. There are at least five reasons for this: (1) no report has directly linked soybean yield loss or lowered seed quality to thrips abundance or damage caused by thrips abundance or damage caused by thrips, (2) conventional sampling methods used in soybean fields (e.g., sweep net, ground cloth) are inadequate for estimating population levels, (3) it is difficult to contain thrips in and extract them from conventional sampling devices once they are collected, (4) phytophagous thrips are relatively small and may appear inconsequential compared to larger pests, and (5) they are somewhat difficult to identify to species, especially the immature stages.

Phytophagous thrips should be incorporated in a soybean pest management sampling plan because (1) they form a substantial portion of the soybean-associated fauna, (2) when abundant, their feeding stresses soybean plants to the point that, when coupled with other stress factors, yields can be depressed, (3) they are active field vectors of at least one virus that causes yield loss and lowered seed quality of soybean, and (4) they attract large populations of certain species of predators that might later act as buffers to potential increases of other pests.

The procedures adopted for sampling thrips on soybean are specific for measuring changes in thrips population densities, for monitoring thrips flight activity and for studying thrips as vectors of soybean viruses.

This chapter is restricted to sampling thrips in soybean fields and, because of the paucity of published information on the subject, concentrates largely on those species associated with soybean in the midwestern and southern parts of the United States. A more generalized account of thrips sampling is found in Lewis (1973).

A. Thrips Abundance in Soybean Fields

Phytophagous thrips constitute one of the most abundant complexes of arthropods in soybean fields. *Sericothrips variabilis* (Beach), the soybean thrips, was seven times more abundant than the second most numerous species of arthropod, *Empoasca fabae* (Harris), in soybean fields in Missouri (Blickenstaff and Huggans 1962). This plus two additional species of phytophagous thrips, *Frankliniella tritici* (Fitch) and *Frankliniella fusca* (Hinds), accounted for more than half the arthropod specimens caught during the Missouri survey. In an insect survey of soybean fields in the Yaqui Valley, Sonora, Mexico, 75% of all insect specimens collected belonged to a single species, *Caliothrips phaseoli* (Hood) (Pacheco 1976). Clearly, where surveys have been conducted and small arthropods tabulated, phytophagous thrips have formed a substantial and often numerically dominant portion of the above-ground soybean fauna.

B. Thrips Species Composition in Soybean Fields

Adults of ten species of thrips were collected on soybean foliage in Illinois: *Leptothrips mali* (Fitch), *Dendrothrips ornatus* (Jablonowski), *Thrips physapus* L., *Anaphothrips obscurus* (Muller), *Thrips tabaci* Lindeman, *Frankliniella fusca*, *F. tritici*, *S. variabilis*, and *Aeolothrips bicolor* Hinds and *A. fasciatus* (L.) (Irwin et al. 1979). The latter two species were predaceous; all others were phytophagous. Only the five latter species were consistently encountered, and immatures of only the four latter species were collected from soybean foliage. Of those, only *S. variabilis* and *F. tritici* were very abundant. Therefore, in the midwestern United States—and probably in the southern United States as well—only these two species of phytophagous thrips commonly colonize soybean.

Several additional species were collected on horizontal ermine lime traps (see Fig. 11-8) at canopy level in the same Illinois field (Irwin unpublished). Again, the predominant species were *S. variabilis* and *F. tritici.* In addition to the species collected from foliage, the following species were identified: *Frankliniella tenuicornis* (Uzel), *Glyptothrips arkansanus* Hood, *Microcephalothrips abdominalis* (Crawford, D. L.), *Haplothrips mali* (Fitch), *Chriothrips mexicanus* Crawford, D. L., *Plesiothrips perplexus* (Beach), *Caliothrips fasciapennis* (Hinds), and *Frankliniella runneri* (Morgan).

In Missouri several other species of thrips were encountered while sweeping soybean foliage (Blickenstaff and Huggans 1962): *Stomatothrips flavus* Hood* (predaceous), *Chriothrips crassus* Hinds, *C. manicatus* Haliday, *Echinothrips*

---

*This might represent a more recently described species, *S. crawfordi* Stannard (Stannard 1968).

americanus Morgan, *Limothrips cerealium* (Haliday) and *Eurythrips* sp. The complex of Thysanoptera in soybean fields is, therefore, diverse, but a large portion of the specimens collected belong to a very few species.

C. Host Plants and Nature of Damage

Host plants have not been tabulated for *S. variabilis*. According to Stannard (1968), this species prefers leaves of herbs, particularly legumes and is abundant on soybean and alfalfa. *F. tritici* has a very wide host range, occurring abundantly in flowers of most plants.

Yield losses or economic injury levels have not been established for any species of thrips attacking soybean. Small plot insecticide experiments in Arkansas demonstrated plant recovery from thrips damage; no yield reduction was expected (Mueller and Luttrell 1977). Infestations of more than 50 thrips per leaflet caused no subsequent yield loss in South Carolina (Turnipseed 1973). In 1976 a heavy infestation of *S. variabilis* on young soybean plants had little impact on small plot yields in southern Illinois (Irwin and Kuhlman 1979). However, when coupled with other stress factors, e.g., herbicide injury, seedlings began to drop leaves and die (Wedberg and Cooley 1976, Wedberg and Kuhlman 1976).

1. Thrips as Vectors of Soybean Viruses

Thrips are suspected of being the major vector of at least one important soybean virus: tobacco ringspot virus (TRSV) (Bergeson et al. 1964, Goodman and Nene 1976). TRSV is one of the causal agents of bud blight disease of soybean and can substantially reduce yields (Athow and Laviolette 1961, Crittenden et al. 1966). Since TRSV is seed transmitted (up to 100%), seed quality can be severely affected (Desjardins et al. 1954, Athow and Bancroft 1959, Crittenden et al. 1966, Iizuka 1973). It is also known to delay plant maturity (Crane and Crittenden 1962). The incidence of TRSV has been correlated with type of adjacent crops, and the pattern of spread in a soybean field suggests an influx from outside the field by aerial vectors (Athow and Bancroft 1959, Goodman and Nene 1976).

Thrips tabaci larvae have transmitted TRSV from soybean plant to soybean plant in the laboratory, but adults of *T. tabaci* failed to transmit the virus (Messieha 1969). *S. variabilis* has not been shown to transmit TRSV under laboratory conditions (Messieha 1969, M. Irwin and S. Halbert unpublished). *F. tritici* was reported not to transmit TRSV (Messieha 1969), but L. D. Newsom (personal communication) has demonstrated very low efficiency transmission with this species. Very few species and stadia of thrips have been tested as vectors of TRSV, but if, as is suspected, thrips prove to be the more important field vectors of TRSV in soybean, sampling thrips influxes into soybean fields could be extremely important in pest management programs.

2. Impact on Predator Abundance

Moderate infestations of certain species of thrips early in the growing season may be economically beneficial. Preliminary results strongly suggest that the density of *Orius insidiosus* (Say), a thrips predator, in soybean fields is directly correlated with buildup of *S. variabilis* (Mueller and Luttrell 1977, Irwin and Kuhlman 1979). After thrips population levels are suppressed, *O. insidiosus* may act as a buffer to the buildup of other species because it also known to feed on eggs and small larvae of many pests (Barber 1936, Fletcher and Thomas 1943, Watve and Clower 1976). In the long run, therefore, moderate population levels of certain species of thrips in a soybean field could result in the lowering of population levels of potentially more serious pest species. Thus, early thrips infestations might eventually take on a positive predictive significance in a pest management program.

D. Life Cycle and Phenology

For those phytophagous species of thrips that colonize soybean in northern Mexico and the United States, six developmental stages exist: egg, first instar larva, second instar larvae, prepupa, pupa, (Fig. 13-1) and adult (Fig. 13-2). The egg is inserted in leaf or other plant tissue. The second instar larva, just prior to transforming into a prepupa, drops to the soil litter where the prepupa and pupa remain until adult emergence (Watts 1936, Vance 1974). Adults and larvae actively feed on soybean plants.

Eggs are deposited singly, a few per day, in incisions cut in plant tissue by the ovipositor. A single female can lay as many as 300 eggs in her lifetime (Lewis 1973), but in a laboratory test, *F. tritici* laid an average of 30 eggs per female on cotton (range 1-111) (Watts 1936). A short preoviposition period usually precedes a longer oviposition period. A 6-day preoviposition was followed by a 16-day mean oviposition period in *F. tritici* (Watts 1936). Mean developmental times (in days) of several thousand specimens of *F. tritici* reared on young cotton leaves for 40 generations under ambient laboratory conditions follow: 3.3 (egg), 2.2 (first instar), 2.8 (second instar), 1.1 (prepupa), 2.5 (pupa), 25.3 (female longevity), 14.8 (male longevity) (Watts 1936). Laboratory reared *S. variabilis* at $22°C$ and 9 h L : 16 h D photoperiod had the following mean developmental times (in days): 3.1 (first instar), 3.8 (second instar), 0.9 (prepupa), and 3.1 (pupa) (Vance 1974). Because egg laying is extended over several weeks or months, generation overlap tends to obscure discrete population cycles; thus, generation turnover is difficult to determine in the field.

S. variabilis and *F. tritici* apparently overwinter as adults in the southern United States and parts of Mexico. Adults of *Caliothrips fasciatus* (Pergande), the bean thrips, were found to overwinter in central California, remaining dormant for five months (Bailey 1933). Within the United States, the northern limit at which *S. variabilis* and *F. tritici* can successfully overwinter depends to a

Sampling Phytophagous Thrips on Soybean

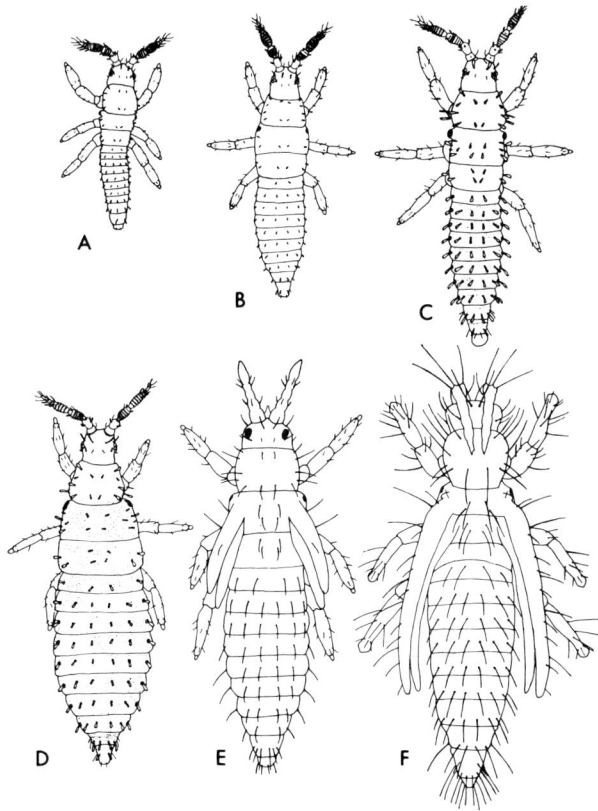

Figure 13-1. Immature stages of *Sericothrips variabilis*.
(A) Early first-instar larva; (B) late first-instar larva;
(C) early second-instar larva; (D) late second-instar larva;
(E) prepupa; (F) pupa. (From Vance 1974).

large extent on the severity of the winter (L. J. Stannard, Jr., personal communication). *F. tritici* has never been found to overwinter in Illinois (Stannard 1968).

In Illinois *S. variabilis* and *F. tritici* colonize soybean early in the growing season, at growth stage V1 (Fehr et al. 1971). They apparently immigrate from the southern states on frontal winds (Stannard 1968, Lewis 1973).

Data collected at Urbana, Illinois and Lexington, Kentucky during the 1976 soybean growing season suggest that populations of *F. tritici* colonize earlier, reach peak levels earlier, and decline sooner than populations of *S. variabilis* (Irwin et al. 1979). A generalized account of the population fluctuation of these species in Illinois is shown in Fig. 13-3. The populations of these two species were considerably denser at Lexington than at Urbana or Columbia, Missouri (Fig. 13-4) (Irwin et al. 1979).

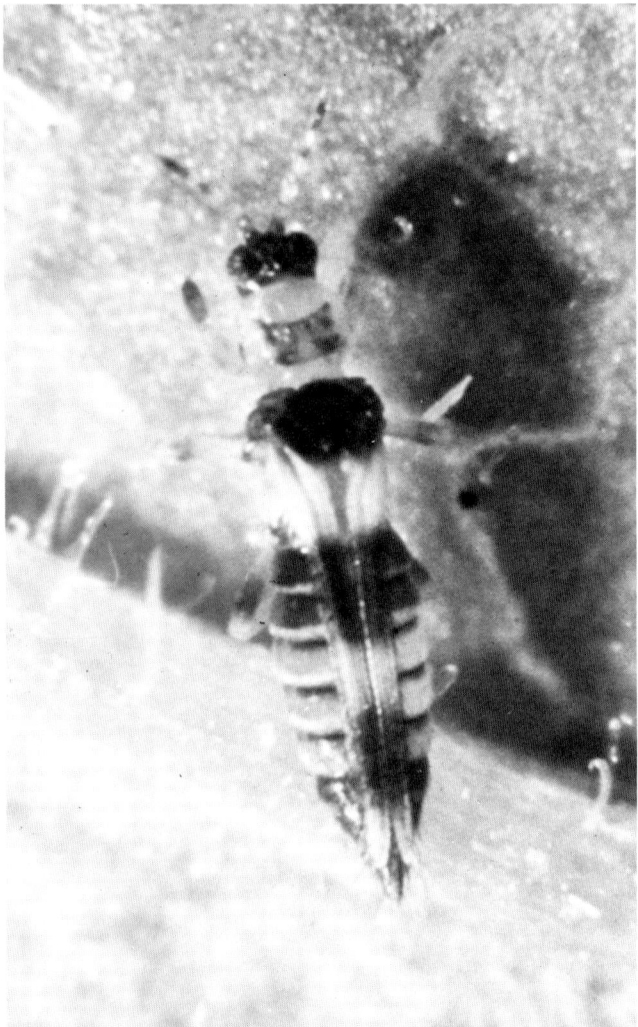

Figure 13-2. Adult of *Sericothrips variabilis* on a soybean leaf. (Photograph courtesy of L. J. Stannard, Jr.).

The species composition and relative abundance of thrips in soybean fields change throughout the growing season. Of the five most commonly collected species in Missouri, only *S. variabilis, F. tritici,* and *F. fusca* were abundant. The physiological growth stage of the soybean crop (and thus, chronological timing) that each species was most abundant differed. During early vegetative stages *F. fusca* was most commonly collected, while during the reproductive stages *S. variabilis* was most abundant (Table 13-1) (Irwin et al. 1979).

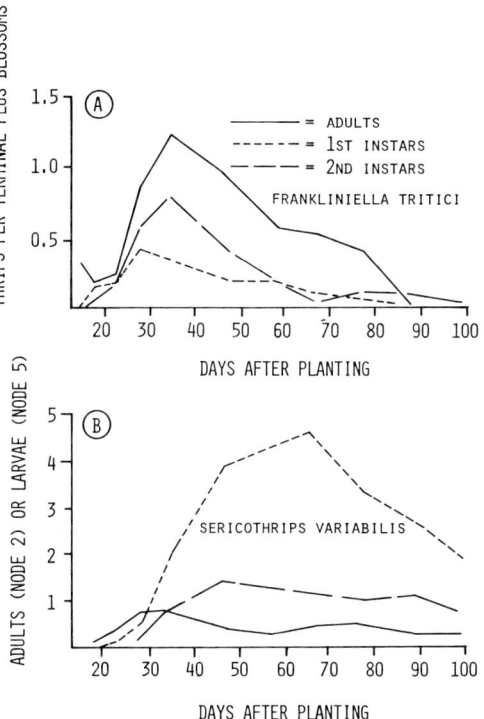

Figure 13-3. Seasonal distribution of (A) *Frankliniella tritici* and (B) *Sericothrips variabilis* in a 'Williams' soybean field, Urbana, Ill., 1976.

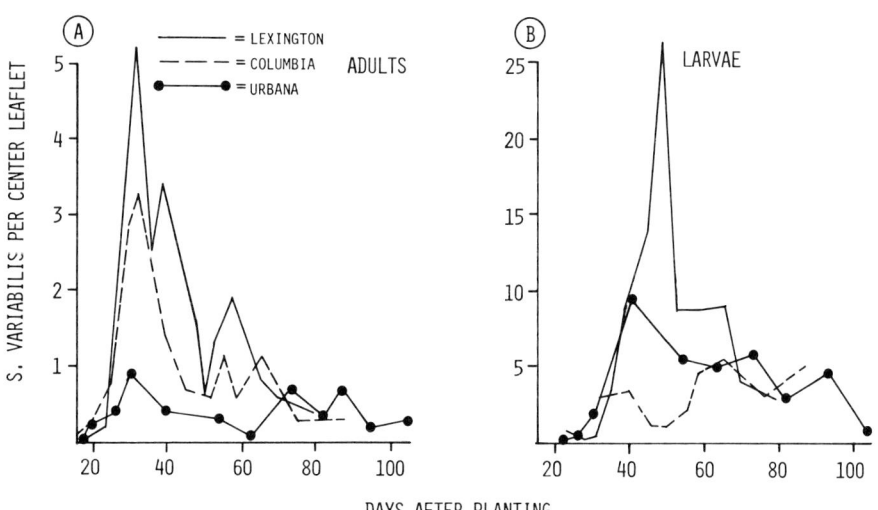

Figure 13-4. Seasonal abundance of *S. variabilis* at Lexington, Columbia, and Urbana at latitudes 38°, 39°, and 40° N., respectively. (A) Mean numbers of adults from center leaflet of uppermost expanded trifoliolate per sampling date. (B) Mean numbers of 1st plus 2nd instars per center leaflet from the 3rd, 4th, 5th, or 6th trifoliolate below the terminal, whichever had the higher number at any given sampling date. (From Irwin et al. 1979).

Table 13-1. Percentage species composition of adult thrips per soybean plant per sampling date at given growth stages at Columbia, Mo., 1976 (adapted from Irwin et al. 1979)

	V1-V3	V4-V6	R1-R3	R4-R5
S. variabilis	29	86	85	95
F. tritici	16	2	11	2
F. fusca	54	11	1	1
A. bicolor	0	1	2	0
A. fasciatus	0	0	1	2
Total adult thrips per plant	7.8	19.6	16.5	26.2

II. Sampling for Adults and Larvae

Specific techniques for sampling thrips have been developed to monitor changes in population densities and to monitor thrips flight activity.

A. Sampling Adults and Larvae on Plants

The literature abounds in methodologies for extracting, concentrating, and tabulating catches of thrips (Lewis 1973). These authors have developed techniques that may improve the standardization of thrips recovery from soybean plants. Four of these techniques will be discussed here. These can be divided into delayed counting and direct counting methods.

1. Delayed Counting Methods

Delayed counting methods are those in which a specific part is removed from the plant and brought into the laboratory for processing, i.e., removal of thrips and tabulation of thrips species and stadia. Processing may also be completed in the field. These methods were devised to maximize sample gathering in the field; they obviously require more time once the samples have been brought to the laboratory. Processing in the laboratory increases the accuracy of species identification and stadia determination since specimens can be viewed under dissecting microscopes and doubtful specimens can be sent for expert determination.

Delayed counting methods involve carefully picking a soybean plant part, placing it in a container, and processing the material to separate thrips from plant parts. These methods can be subdivided into wet-count and dry-count techniques. The dry-count technique was employed in Illinois. The plant parts were placed in 76 mm diameter x 12.5 mm high tins with tight-fitting lids. The plant parts and tin were examined in the laboratory under a dissecting microscope and all thrips were removed from each plant part and container. These were stored in 1-ml vials containing 70% ethanol and later identified.

The wet-count methods were employed in Kentucky (Lexington technique) and Missouri (Columbia technique). A dilute detergent water mixture was used in the "Lexington technique". Plant parts were clipped and allowed to fall singly in 90 cl widemouth polyethylene jars, and the containers were set aside for subsequent processing. Aqueous 95% ethanol in a Zip-loc plastic bag was used in the "Columbia technique". The selected plant part was placed in a bag and brought to the laboratory for processing. For the "Lexington technique", the container was shaken vigorously and the liquid poured through a 100 mesh stainless steel sieve. Each container and plant part was thoroughly rinsed with a jet of water from a wash bottle and the resultant liquid passed through the sieve. Thrips, plus a small amount of debris, remained on the sieve and were washed with 70% ethanol into a Coors 210 ml casserole; after settling, excess alcohol was decanted and the thrips and remaining alcohol were poured into vials. The thrips were later identified to species and stadia and counted. For the "Columbia technique", plant parts were agitated in a bath of 95% ethanol to remove the arthropods. The bag was rinsed and the combined solution was passed through a 200 mesh sieve. The arthropods and debris were then washed into a beaker with 50 ml of water, and transferred by vacuum filtration in a Buchner funnel onto a medium filter paper with a pencilled grid to facilitate counting (Irwin et al. 1979).

The dry-count and Lexington techniques were compared and found to be equally efficient for larvae of *S. variabilis* ($p > 0.1$, Student's t-test), but the dry-count technique proved slightly more efficient for adults of *S. variabilis* ($0.1 > p > 0.05$, Student's t-test) (Table 13-2). The "Columbia technique" recovered $91.5 \pm 1.5\%$ of the adults and $99.3 \pm 0.4\%$ of the larvae found on the third trifoliolate below the terminal bud (Irwin et al. 1979).

Each of these delayed counting techniques has some advantages and disadvantages. All can allow one to put off final counting and determination, thus saving time during summer months. The "Columbia technique" is probably best for minimizing handling of the sample until actual processing. The dry-count technique demands immediate (same day) separation of thrips from plant parts while the "Lexington technique" could allow a day or two between collection and processing. Field processing allows one to resample should accidental spillage occur during processing. During inclement weather, samples could be collected rapidly in the field and processed indoors. It is less time-consuming in the laboratory to process dry-count samples than wet-count samples. Any delayed count method used should be tailored to the needs of the program.

2. Direct Count Method

An alternative to the delayed count methods previously described is the direct count method (see Chapter 2). This method is efficient for *S. variabilis* but does not appear to be adequate for assessing populations of *F. tritici*. When using the technique, one locates the plant part to be sampled and counts the thrips there-

Table 13-2. Comparison of *Sericothrips variabilis* stadia recovered by the Lexington and dry-count techniques from soybean at growth stage R3 using "Student's t-test". Leaf area comparison from both techniques included. Urbana, Ill., July 23, 1976. N = 12

	$\bar{x} \pm S_{\bar{x}}$/ 5th from uppermost center leaflet			
	1st Instar	2nd Instar	Adult	Leaf Area
Lexington technique	5.3 ± 1.2	1.1 ± 0.5	0.2 ± 0.1	88.7 ± 2.2
Dry-count technique	7.7 ± 1.2	1.5 ± 0.3	0.8 ± 0.3	86.8 ± 2.5
P - Value	>.10	>.10	<.10	>.10

on while in the field. This works reasonably well for both adults and larvae, but the more active adults should be counted first. Specimens are not saved, and determination of species and stadia is less accurate than with the delayed count methods. However, results are immediately at hand for pest management decisions. Unfortunately, no data exist about the possibility of knocking *F. tritici* from terminal buds or blossoms onto a contrasting surface and counting them in situ, but this might prove valuable for rapid estimates of population density of that species in the field.

B. Sampling Adults from the Air

Although thrips are weak fliers, their finely fringed wings enable them to remain aloft long enough to be blown to great heights and for long distances (Lewis 1973, Glick 1939). Both *S. variabilis* and *F. tritici* colonize soybean in Illinois and migrate northward each spring in low level jet winds (Stannard 1968, personal communication), usually arriving prior to the emergence of soybean in the fields. While *F. tritici* has numerous host plants—many of which are abundant in early spring—*S. variabilis* has relatively few host plants and population increases are less rapid. Soybean is probably of much less importance to the successful population increase of *F. tritici* than it is to *S. variabilis*, even though Illinois grows over 9 million acres of soybean each year.

The following is based on studies of the flight activity of thrips in soybean fields conducted in Illinois from 1976-1978 (M. Irwin unpublished).

The main sampling device used in these studies is a square lime-green glazed ceramic tile 15.2 cm on a side, covered with a thin film of Tanglefoot. The trap is maintained horizontally at canopy level in soybean fields and the trapped thrips are removed at daily or weekly intervals by washing the tile with acetone. Details of this sampling device are presented in Chapter 11.

Flight with respect to wind direction was monitored with a 30.5 cm diameter x 38 cm high, clear, vertically positioned, cylindrical impact trap positioned at canopy level in a 'Williams' soybean field in Urbana. The trap was vertically divided into four equal quadrants, one of these facing windward, another leeward by action of a wind vane attached to the trap. Two additional portions were 90° to windward and 270° to windward. For present purposes only windward and leeward catches were tabulated.

The results of this experiment indicated that *S. variabilis* impacted on the windward side about nine times out of ten and that *F. tritici* impacted on the leeward side about seven times out of ten. *T. tabaci*, a species known to transmit tobacco ringspot virus, was also found to impact on the leeward side of the trap, but sample numbers were very low (Irwin and Halbert unpublished).

During 1977 a modified version of the original wind-directed trap was placed in a soybean field (see Fig. 11-3). Results substantiated the 1975 *S. variabilis* findings but no significant differences ($p > .05$) were detected for leeward and windward catches of *F. tritici*.

Whether *F. tritici* flies into the wind, flies during calmer times, or is carried by eddies to the leeward side is not known. *S. variabilis* obviously flies or drifts with the wind and is impacted on the windward side. The trap was able to maintain its position relative to wind direction at wind speeds as low as 2 km/hr. The most reasonable explanation for *F. tritici* impacting on the leeward side of the trap is that this species tends to fly in low speed wind (< 2 km/hr) and generally flies against the wind.

III. Sampling to Monitor Colonization Patterns

Techniques to measure flight activity have provided an adequate basis for studies on colonization of soybean fields by thrips.

In Illinois, Kentucky, and Missouri both *S. variabilis* and *F. tritici* colonize soybean early in the growing season, but *F. tritici* is more abundant earlier. Experiments were conducted in Illinois using eight horizontal ermine lime traps placed 21 m apart along a row in a 100 x 200 m 'Williams' soybean field with two additional tile traps in a grassy strip, one outside of each edge of the field (M. E. Irwin unpublished). Early season records indicated that the influx of *F. tritici* was from outside the field and that no correlation was apparent between numbers of thrips caught in various areas of the field and wind direction. Tile trap catches two weeks after planting showed a large concentration of *F. tritici* in the grassy borders (Fig. 13-5A,a), whereas later weeks showed a reversal of this trend, with fewer *F. tritici* in traps in the grassy borders (Fig. 13-5A,bcd). These studies, unlike those of Lewis (1973), showed no correlation between wind pattern and concentration of thrips species.

S. variabilis was too scarce early in the season to determine the effect of grass borders on its pattern of dispersion. The pattern in the transect was erratic between five and nine weeks after planting (Fig. 13-5B,ab), but thereafter became more stable with fewer catches in border traps (Fig. 13-5B,cd). Monthly mean catches (as a percent of total catch) further illustrate that *S. variabilis* becomes more uniformly dispersed about the time of canopy closure (Fig. 13-5C,abcd).

Only three species of thrips were abundant enough to allow generalizations,

294   M. E. Irwin and K. V. Yeargan

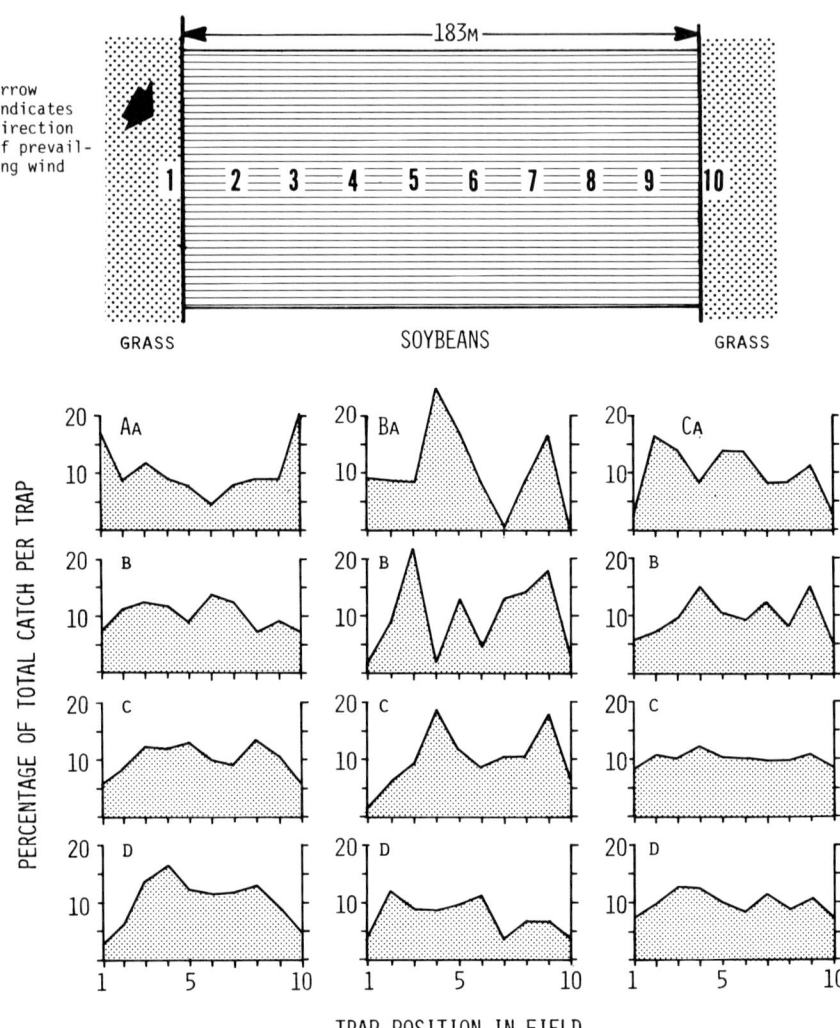

Figure 13-5. Map of 'Williams' soybean field near Urbana, Ill. during the 1976 growing season showing grassy borders (dotted area) and soybean (horizontal lines) with positions of ten horizontal ermine lime traps (1-10). Trap position 1 to the south, 10 to the north; 1 and 10 in the grassy area adjacent to the soybean field. (A) *F. tritici*, percent weekly catch at each trap position; (A,a) 2nd week after planting, (A,b) 4th week, (A,c) 7th week, (A,d) 10th week. (B) *S. variabilis*, percent weekly catch at each trap position; (B,a) 5th week after planting, (B,b) 9th week, (B,c) 10th week, (B,d) 11th week. (C) *S. variabilis*, percent monthly catch at each trap position throughout the growing season; (C,a) June, (C,b) July, (C,c) August, (C,d) September.

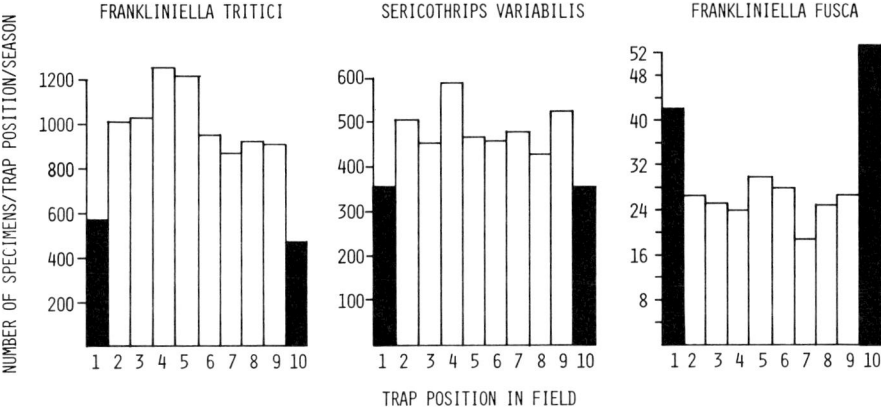

Figure 13-6. Distribution of thrips spp. in a south (S) to north (N) transect of a 'Williams' soybean field near Urbana, Ill., 1976. Total specimens captured during the growing season. Shaded area represents trap locations in the grass borders. White area represents trap locations within the soybean field.

and it is interesting to note that of them, the two species that colonize soybean (*S. variabilis* and *F. tritici*) were caught more often in traps within the field, while *F. fusca,* a species that apparently does not colonize soybean, was caught more frequently in border traps (Fig. 13-6).

Flight activity, however, seems to vary from year to year depending on yet unidentified factors. Observations in Illinois on *F. tritici* revealed double flight activity peaks in both 1976 and 1977 (Fig. 13-7A); however, both peaks were earlier and lower in 1977. Similarly, *S. variabilis* had a single flight activity peak in both 1976 and 1977, but the peak was earlier and lower in 1977 (Fig. 13-7B,8). A trend seems apparent: a double peak for *F. tritici,* a single peak for *S. variabilis;* in both years the first peak of *F. tritici* occurred earlier than that of *S. variabilis;* both *F. tritici* peaks preceded that of *S. variabilis* in both years (M. Irwin unpublished).

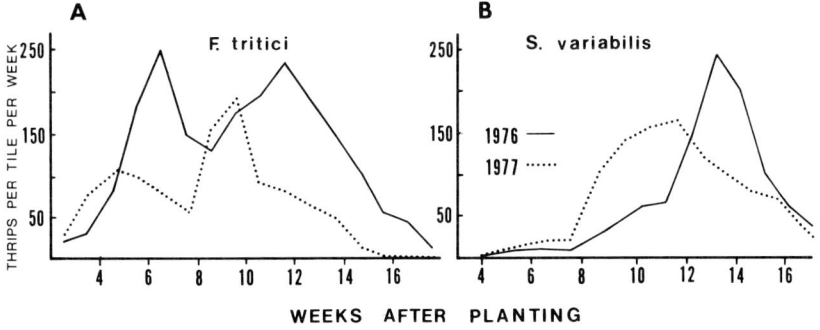

Figure 13-7. Flight activity curves for the 1976 and 1977 seasons, near Urbana, Ill. (A) *F. tritici,* (B) *S. variabilis.*

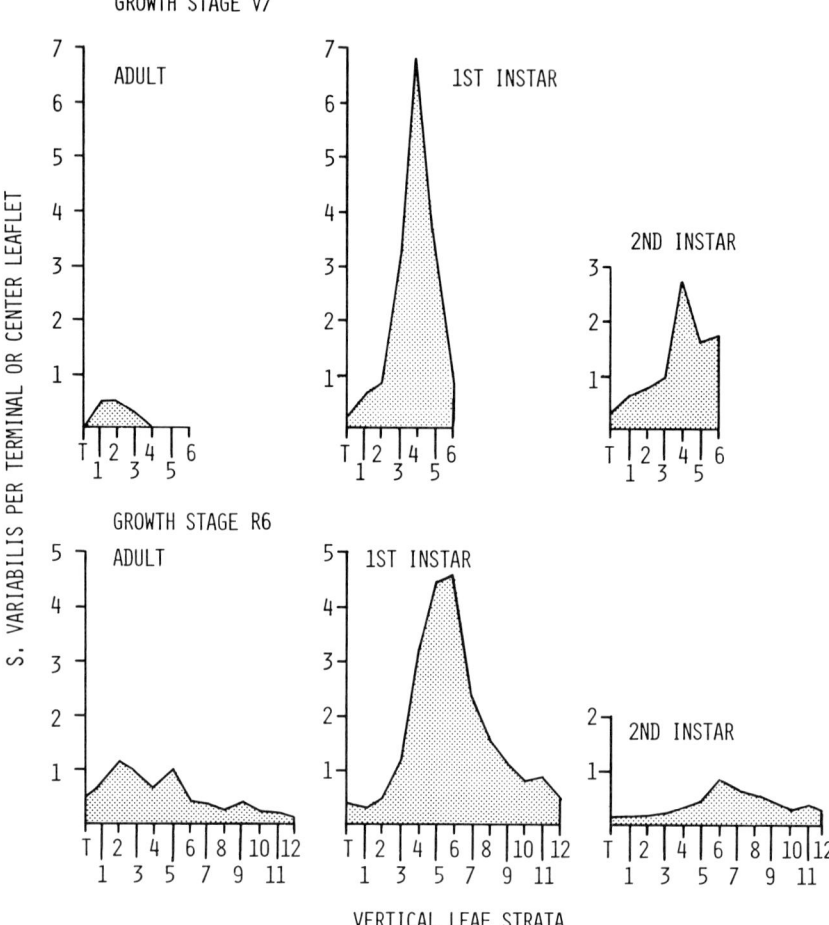

Figure 13-8. Within-plant stratification of *Sericothrips variabilis* by life stage on 'Williams' soybean at growth stages V7 and R6, Urbana, Ill., 1976. T = terminal; 1-12 = center leaflets at nodes from uppermost expanded trifoliolate (1) downward on the main stem of the soybean plant.

IV. Sampling to Measure Population Densities and Spatial Patterns

Techniques developed to extract thrips from plant parts were used to determine, within plant and within field, spatial patterns of larval and adult thrips.

Phytophagous thrips are very small and differentially occupy various parts of a soybean plant. *S. variabilis,* for instance, inhabits leaves while *F. tritici* is largely found in growth buds and blossoms (Irwin et al. 1979). Conventional sampling techniques are adequate for estimating absolute or relative thrips abundances in the field, and, randomly selected plant parts may or, more frequently, may not provide accurate population density estimates. For maximum efficiency, sampling those plant parts that most sensitively reflect population changes is recommended. If one were to sample an entire plant, extraction of thrips would become cumbersome and time consuming, and the number of samples processed per unit time would be few. If, however, only those plant parts that most sensitively reflect population fluctuations were sampled, extraction problems would be minimized and handling time per sample unit would be reduced, allowing time for processing more sample units when needed. Microhabitats most frequently occupied by various *F. tritici* and *S. variabilis* stadia in soybean fields have been identified (Irwin et al. 1979).

S. variabilis is not randomly dispersed over the soybean plant but is normally found on opened leaves. Adults are most abundant on upper leaf strata (nodes 1 and 2 from top) and larvae are concentrated on slightly lower leaf strata; on nodes 3-4 (from top) during vegetative plant growth stages and on 4-6 (from top) during the reproductive plant growth stages of soybean (R2-R6) (Fig. 13-8) (Irwin et al. 1979).

F. tritici is usually found in the terminal buds and blossoms of soybeans. All actively-feeding stages are concentrated in the same area. This species colonizes terminals of soybeans early in the growing season and remains in growth tips and terminal buds until blossoms are formed. At that point, there appears to be a shift in population density from growth tips to blossoms. After bloom has ceased, a shift back to growth tips occurs (Fig. 13-9) (Irwin et al. 1979).

Spatial patterns of both *S. variabilis* and *F. tritici* within soybean fields ranged from uniform to aggregated, and no relationship between pattern and plant growth stage or thrips density could be detected from studies in soybean fields in Illinois (Irwin et al. 1979). However, all sampling was performed well within the borders of the field, and field edges might influence the density of thrips.

According to Lewis (1973), the spatial disposition of individuals on plants is usually associated with food and shelter, but can be influenced by other factors, e.g., patchiness in dispersion of host plant, changes in microweather and shelter, feeding, mating, and oviposition behavior, and possibly by an innate urge to aggregate. Within a soybean field, the plant spatial pattern is uniform. At any one time in a given field, the plant growth stage is relatively consistent, as is the plant height and density.

The initial pattern of colonization may be the major factor influencing within-field patterns of thrips in soybeans. If the pattern is aggregated (patchy), taking many small sample units from several distinct areas of the field will reduce sampling error.

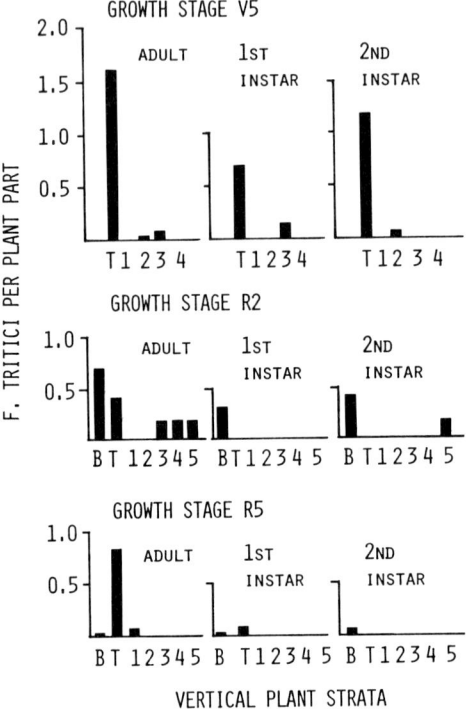

Figure 13-9. Within-plant stratification of *Frankliniella tritici* by life stages on 'Williams' soybean at growth stages V5, R2 and R5, Urbana, Ill., 1976. B = blossoms; T = terminal bud; 1-5 = center leaflets at nodes from uppermost expanded trifoliolate (1) downward on the main stem of the soybean plant.

V. Sampling Program

A. Sample Time

The occurrence of thrips in a soybean field is most critical during the prereproductive developmental stages of the crop (V1-R1). This seems true if one considers the potential of thrips damage lowering seed quality and quantity, the potential of thrips to transmit tobacco ringspot virus, or the associated buildup of *O. insidiosus* as a subsequent buffer to the population increases of other pest species. Therefore, in a pest management program, sampling for thrips densities might be restricted to the early part of the growing season.

Sampling should be conducted about the same time of day for comparisons among sampling dates to be most valid. Preliminary experiments conducted in Illinois suggest a slight, though significant ($P < .05$) difference in catches of *S. variabilis* on the same leaf strata at different times of the day (M. Irwin unpublished).

B. Sample Unit

The sample units selected depend on the objective of the study, but if plant parts are chosen, those plant parts that most sensitively reflect changes in abundance should be sampled: terminal bud for *F. tritici*, center (or lateral) leaflet of the uppermost fully expanded trifoliolate for *S. variabilis* adults, and center (or lateral) leaflet of the 4th (before bloom) or 5th (during and after bloom) from uppermost fully expanded trifoliolate for *S. variabilis* larvae. There is no significant difference in numbers collected from central or lateral leaflets of the same node (Irwin et al. 1979). For sampling *S. variabilis* larvae, if less than four fully expanded trifoliolates are present on the plant, the lowest one should be selected (Irwin et al. 1979).

C. Sampling Method

For studies of flight activity horizontal ermine lime traps at canopy level should be used (see Chapter 11). This sampling device seems to give a near absolute account of what is actually landing on soybean leaves at canopy level; for most purposes—e.g., colonization potential, transmission of viruses—that is the type of data needed. Most of these authors' studies were based on samples collected at weekly intervals. Studies of spatial patterns and population densities, as well as sampling for pest management should use direct observation in the field or any of the methods based on extraction of individuals (dry or wet method) from plant parts.

D. Number of Sample Units

Regression lines computed from observed means and variances plotted on a log-log scale (Fig. 13-10A,B) gave formulas for calculating the numbers of sample units needed, given the approximate number of thrips per sample unit and the required reliability of the sample (Taylor 1961). The same regression line fit the plant part samples for both *F. tritici* and *S. variabilis*: $s^2 = 1.729\bar{x}^{1.3127}$, whereas, for the flight activity samples, one regression line fit *F. tritici* plus early season to mid season, *S. variabilis*: $s^2 = 0.7714\bar{x}^{1.6387}$ and a second line fit late season *S. variabilis*: $s^2 = 0.2449\bar{x}^{1.5847}$. The difference in regression lines for early and late season *S. variabilis* indicates a change in spatial pattern from more patchy (early season) to more uniform (late season) (W. G. Ruesink and M. E. Irwin unpublished). This coincides with canopy closure and a decided change in sex ratio, from an apparent surplus of males to an approximately equal number of males and females.

By plotting the mean ($\bar{x}$) vs. the ratio of the $S_{\bar{x}}/\bar{x}$ (i.e., RV) for these three linear regression lines, one can calculate the number of sample units needed to achieve a desired level of accuracy given the approximate mean number of individuals per sample unit (Fig. 13-11A,B,C).

VI. Concluding Remarks

During the 1976 soybean growing season, thrips were monitored in Illinois for flight activity and for population density measurements. Flight activity was monitored by horizontal ermine lime (sticky tile) traps and population density

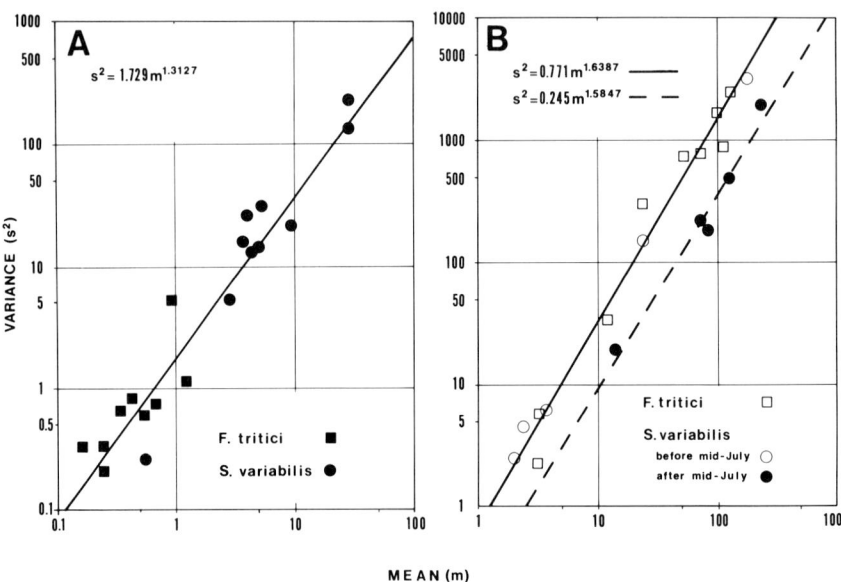

Figure 13-10. Regression lines computed from observed means ($\bar{x}$) and variances (s^2) plotted on a log-log scale. (A) *F. tritici* and *S. variabilis* by plant part sampling. (B) *F. tritici* and *S. variabilis* by flight activity sampling. Curve $s^2 = 0.7714\bar{x}^{1.6387}$ serves for *F. tritici* plus early season *S. variabilis* (to canopy closure). Curve $s^2 = 0.2449\bar{x}^{1.5847}$ serves for late season *S. variabilis*. (After canopy closure).

was assessed by leaf and terminal sample. Both sampling techniques resulted in very high catches of adult *S. variabilis* and *F. tritici* when compared with all other species. However, a comparison of the four most commonly collected species illustrates the difference in proportion of species over the entire season (Fig. 13-12, note particularly *S. variabilis* and *F. tritici*). The leaf and terminal sample technique, used to determine the relative abundance of these two species within the canopy, showed that *S. variabilis* was four times more abundant overall. The sticky tile technique, however, indicated that *F. tritici* was twice as abundant overall (Fig. 13-12). This apparent paradox can be interpreted.

If the phenologies of adult thrips are plotted using the two sampling techniques, the picture becomes clearer (Fig. 13-13A-D). For *F. tritici* there are apparently two major peaks of flight activity (Fig. 13-13A), although from leaf samples there is but one population density peak in soybean (Fig. 13-13B). The entire second peak of flight activity is most likely a product of population buildup on other host plants, thus accounting for the greater proportion of *F. tritici* in horizontal ermine lime trap samples than in terminal samples.

The patterns of *S. variabilis* can also be explained. It is our contention that fewer adults immigrate, but these reproduce more successfully on soybean than *F. tritici*, and adults become active only late in the season as soybean plants

Sampling Phytophagous Thrips on Soybean

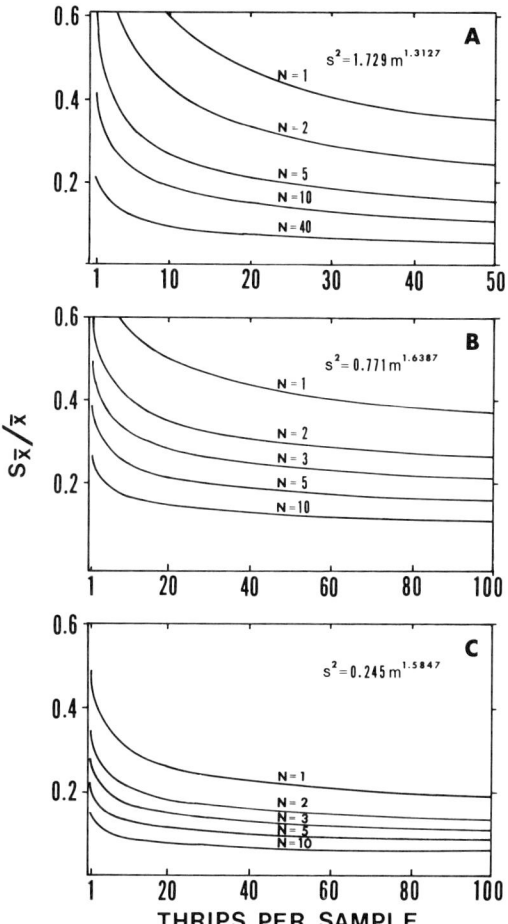

Figure 13-11. Relationship between number of samples (N), thrips per sample, and accuracy of the estimate of numbers per sample. (A) Sampling plant parts for *F. tritici* and *S. variabilis*. (B) Sampling flight activity of *F. tritici* and early to mid-season *S. variabilis* (to canopy closure). (C) Sampling flight activity of late season *S. variabilis*. (After canopy closure).

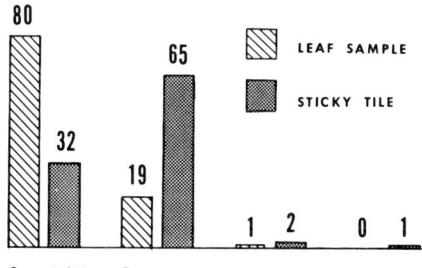

Figure 13-12. Four most commonly collected thrips spp.; proportions captured by horizontal ermine lime trap and leaf sampling.

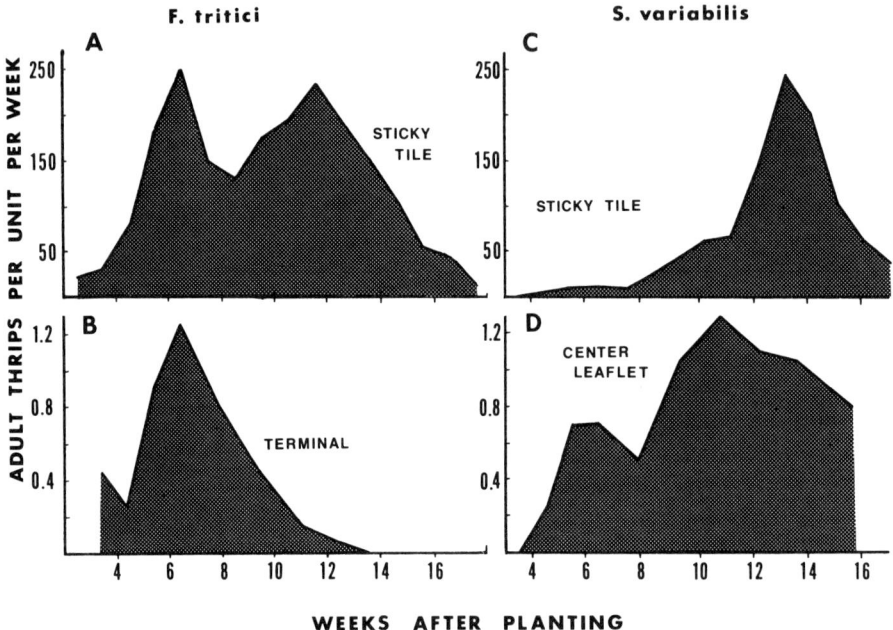

Figure 13-13. Seasonal phenologies of adult *F. tritici* (A,B) and *S. variabilis* (C, D), comparing abundances between flight activity patterns (A,C) and population fluctuations within the soybean canopy (B,D).

begin to senesce (Fig. 13-13C). Thus, horizontal ermine lime trap samples suggest a flight peak of *S. variabilis* after the adult population within the soybean canopy begins to decline (Fig. 13-13D).

If one did not have comparative data from horizontal ermine lime trap and plant part catches, it would be possible to wrongly conclude that *F. tritici* were more abundant in soybean. This example points out the dangers of using a sampling technique to accomplish an objective for which it was not designed.

References

Athow, K. L., and J. B. Bancroft. 1959. Development and transmission of tobacco ringspot virus in soybean. Phytopathology **49**:697-701.

Athow, K. L., and F. A. Laviolette. 1961. The relation of seed-transmitted tobacco ring spot virus to soybean yield. Phytopathology **51**:341-342.

Bailey, S. F. 1933. The biology of the bean thrips. Hilgardia **7**:467-522.

Barber, G. W. 1936. *Orius insidiosus* (Say), an important natural enemy of the corn earworm. USDA Tech. Bull. **504**:24 p.

Bergeson, G. B., K. L. Athow, F. A. Laviolette, and M. Thomasine. 1964. Transmission, movement, and vector relationships of tobacco ring spot virus of soybean. Phytopathology **54**:723-728.

Blickenstaff, C. C., and J. L. Huggans. 1962. Soybean insects and related arthropods in Missouri. Mo. Agr. Exp. Sta. Res. Bull. 803:51 p.

Crane, J. L., and H. W. Crittenden. 1962. Effect of deflowering and tobacco ring spot virus on the occurrence of *Cercospora kikuchii* on soybean. Phytopathology 52:1217.

Crittenden, H. W., K. M. Hastings, and D. M. Moore. 1966. Soybean losses caused by tobacco ring spot virus. Plant Dis. Rep. 50:910-913.

Desjardins, P. R., R. J. Latterell, and J. E. Mitchell. 1954. Seed transmission of tobacco ring spot virus in Lincoln variety of soybean. Phytopathology 44:86.

Fehr, W. R., C. E. Caviness, D. T. Burmwood, and J. W. Pennington. 1971. Stage development descriptions for soybeans, *Glycine max* (L.) Merrill. Crop Sci. 11:929-930.

Fletcher, R. W., and F. L. Thomas. 1943. Natural control of eggs and first instar larvae of *Heliothis armigera*. J. Econ. Entomol. 36:557-560.

Glick, P. A. 1939. The distribution of insects, spiders and mites in the air. USDA Tech. Bull. 673:150 p.

Goodman, R. M., and Y. L. Nene. 1976. Virus diseases of soybeans. pp. 91-96 in R. M. Goodman, ed. Expanding the use of soybeans. Proc. of a Conference for Asia and Oceania. Univ. of Illinois, College of Agriculture. INTSOY ser. 10. 261 p.

Iizuka, N. 1973. Seed transmission of viruses in soybean. Tohoku Nat. Agr. Exp. Sta. Bull. 36:131-141.

Irwin, M. E., and D. E. Kuhlman. 1979. Relationships among *Sericothrips variabilis* (Beach), systemic insecticides and soybean yield. J. Ga. Entomol. Soc. 14:148-154.

Irwin, M. E., K. V. Yeargan, and N. L. Marston. 1979. Spatial and seasonal patterns of phytophagous thrips in soybean fields with comments on sampling techniques. Environ. Entomol. 8:131-140.

Lewis, T. 1973. Thrips, their biology, ecology and economic importance. Academic Press, London. 323 p.

Messieha, M. 1969. Transmission of tobacco ringspot virus by thrips. Phytopathology 59:943-945.

Mueller, A. J., and R. G. Luttrell. 1977. Thrips on soybeans. Ark. Farm Res. 1977(July August):7.

Pacheco, F. 1976. Seasonal and daily fluctuation of soybean insect populations in the Yaqui Valley, Sonora, Mexico. pp. 584-593 in L. D. Hill, ed. World soybean research. Proc. World Soybean Res. Conf., Interstate Print., Danville, Illinois. 1073 p.

Stannard, L. J. 1968. The thrips, or Thysanoptera, of Illinois. Ill. Natur. Hist. Surv. Bull. 29:215-552.

Taylor, L. R. 1961. Aggregation, variance and the mean. Nature (London). 189:732-735.

Turnipseed, S. G. 1973. Insects. pp. 545-572 in B. E. Caldwell, ed. Soybeans: Improvement, production, and uses. Amer. Soc. Agron. Pub., Madison, Wisconsin. Agron. ser. 16. 681 p.

Vance, T. C. 1974. Larvae of the Sericothripini (Thysanoptera: Thripidae), with reference to other larvae of the Terebrantia, of Illinois. Ill. Natur. Hist. Surv. Bull. 31:145-208.

Watts, J. G. 1936. A study of the biology of the flower thrips *Frankliniella tritici* (Fitch) with special reference to cotton. S. C. Agr. Exp. Sta. Bull. **306**:1-46.

Watve, C. M., and C. F. Clower. 1976. Natural enemies of the banded-wing whitefly in Louisiana. Environ. Entomol. 5:1075-1078.

Wedberg, J. L., and T. A. Cooley. 1976. Insect situation and outlook. pp. 111-130 in Twenty-eighth Illinois custom spray operators training school. Ill. Agr. Coop. Ext. Serv., Ill. Natur. Hist. Surv., Urbana, Illinois, 433 p.

Wedberg, J. L., and D. E. Kuhlman. 1976. Thrips problems in soybeans. pp. 17-20 in Twenty-eighth Illinois custom spray operators training school. Ill. Agr. Coop. Ext. Serv., Ill. Natur. Hist. Surv., Urbana, Illinois. 433 p.

Chapter 14

Sampling Whiteflies on Soybean

Sharad M. Vaishampayan and Marcos Kogan

I. Introduction

Whiteflies (Homoptera: Aleyrodidae) are small delicate insects with sucking mouthparts. Ecologically, aleyrodids are the tropical equivalent of aphids (see Chapter 11). Whiteflies are opportunistic insects with transient populations (Mound and Halsey 1978).

There are virtually no detailed studies of whiteflies on soybean, and only two species have been recorded colonizing soybean in the field: the sweetpotato whitefly, *Bemisia tabaci* (Gennadius), and the bandedwinged whitefly, *Trialeurodes abutiloneus* (Haldeman). The greenhouse whitefly, *Trialeurodes vaporariorum* (Westwood), readily establishes damaging colonies on soybean in greenhouses. This species is a considerable hinderance to research, particularly breeding that uses extensive greenhouse plantings of soybean.

Economic infestations of whiteflies on soybean have been recorded from Brazil (Panizzi et al. 1977), India (Gangrade 1974), and Japan (Kuwayama 1953).

A. Geographical Distribution

The greenhouse whitefly is nearly cosmopolitan as a pest of greenhouse plants and occurs out of doors in warmer climates. *Bemisia tabaci* has a widespread distribution throughout the tropical and subtropical regions of all continents (Fig. 14-1). *Trialeurodes abutiloneus* seems to be restricted to North America, occurring in Mexico and in most states in the United States (Fig. 14-1).

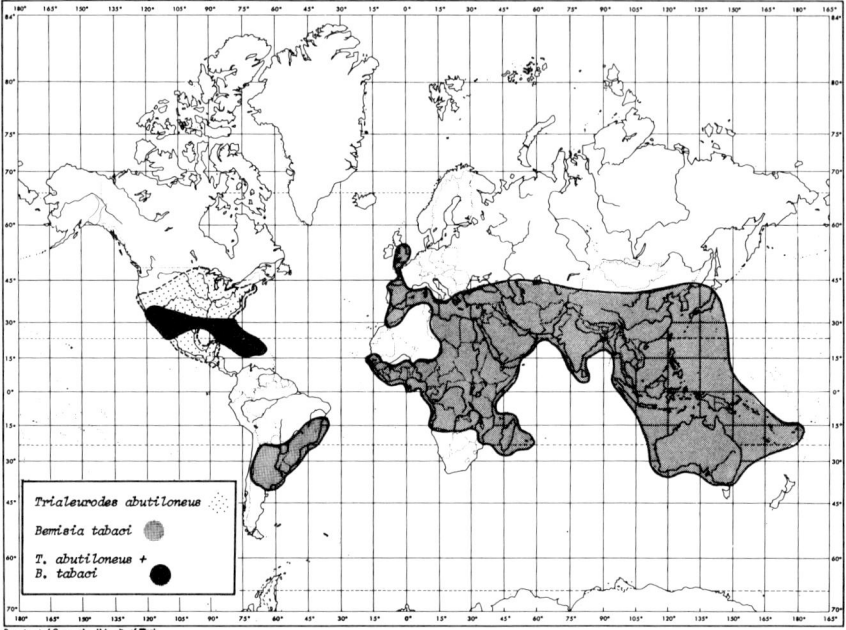

Figure 14-1. Distribution map of *Bemisia tabaci* and *Trialeurodes abutiloneus*. The dark shaded area in North America shows the zone in which the two species overlap. (Based on distribution records in Mound and Halsey 1978).

B. Host Plants and Nature of Injury

Most whiteflies are restricted to dicotyledonous plants. The few records of monocotyledonous hosts are due mainly to feeding by one of the three highly polyphagous species that are associated with soybean.

The host ranges of *B. tabaci, T. abutiloneus,* and *T. vaporariorum* include plants in 63, 33, and 82 botanical families, respectively (Russell 1963, Mound and Halsey 1978). Leguminous host species of *B. tabaci* belong to 32 genera, those of *T. abutiloneus* belong to 15 genera, and the hosts of *T. vaporariorum* belong to 8 genera. The *B. tabaci* host range within the Leguminosae includes trees (e.g., the genera *Caesalpinia, Dalbergia, Erythrina*), shrubs and forbs (e.g., the genera *Bauhinia, Desmodium*). Cultivated forage and grain legumes are also part of this broad host range.

Injury to host plants is caused both by nymphs and adults sucking sap from leaves. Under heavy infestations direct sap feeding may represent a considerable stress factor. Whiteflies also secrete abundant honeydew on which grows a sooty mold. The black mold interferes with normal photosynthetic processes. The honeydew of whiteflies does not seem to attract ants and wasps. In addition to injury due to feeding, whiteflies are potentially damaging as vectors of several important soybean viruses. The best known vector of these diseases is *B. tabaci* (Costa 1976). According to Costa (1975) soybean in Brazil may be infected by agents that cause the soybean crinkle mosaic and the soybean dwarf mosaic. These agents are the *Abutilon* and the *Euphorbia* mosaic viruses respectively. Incidence of these diseases is limited in fields and depends on the presence of weedy reservoirs. Yellow mosaic diseases not caused by the bean yellow mosaic virus, but by a whitefly-transmitted virus-like pathogen, have been reported in India, Brazil, and Puerto Rico (Costa 1975, Goodman and Nene 1976) and Venezuela (Debrot and Ordosgoitti 1975). This disease seems to be a limiting factor to soybean production in certain humid areas in India (Goodman and Nene 1976).

C. Life Cycle and Phenology

Whiteflies lay light-yellow, stalked eggs mostly on the underside of leaves. Nymphs are oval and depressed, pale to greenish yellow. Only the first instars have functional legs. Older instars become attached to the substrate throughout the entire growth period. Adults are small insects with yellow body and hialine wings covered with a white powdery wax (Fig. 14-2). Both sexes are alate and the wing expanse is ca. 3 mm. They are not very active fliers.

The developmental time of *B. tabaci* on *Phaseolus aureus* Roxb. was studied by Murugesan and Chelliah (1978) who have shown that feeding on virus infected plants caused a significant reduction of egg and first instar developmental time. The complete life cycle includes egg, three nymphal stages, puparium, and adult. The duration of the stages is shown in Table 14-1. The adults also displayed a remarkable preference for oviposition on yellow mosaic infected leaves. The mean number of eggs laid on infected leaves was 77, as opposed to 14 eggs on healthy leaves (Murugesan and Chelliah 1978).

There are no phenological studies of whiteflies on soybean. As opportunistic components of the fauna, they seem to explode into huge populations if environmental conditions are adequate. The outbreaks of *B. tabaci* that occurred in Brazil in late 1972 and early 1973 in the states of Paraná and São Paulo, were ascribed to one or more of these causes: (1) the occurrence of highly favorable environmental conditions for whitefly development, (2) expansion of the area planted to soybean providing abundant host material, (3) resurgence due to the elimination of natural enemies caused by the use of nonselective insecticides, (4) occurrence of a new biotype of *B. tabaci* with greater biotic potential (Costa et al. 1973). The nonrecurrence of such outbreaks in succeeding years (M. Kogan unpublished) suggests that favorable environmental conditions were mostly responsible for the 1972-1973 outbreaks.

Adult whiteflies exhibit a significant predilection for feeding and ovipositing on the lower surface of top, young leaves. This habit is pronounced in the greenhouse whitefly. With growth of the host plant there is a definite stratification.

Figure 14-2. Under-side of soybean leaf showing a heavy infestation of *T. vaporariorum* adults and eggs.

Adults and eggs are most abundant on the top leaves, nymphs on the middle leaves, and pupae on the lower leaves.

Table 14-1. Mean developmental time of *Bemisia tabaci* on healthy and on yellow mosaic infected *Phaseolus aureus* plants (Murugesan and Chelliah 1978)

Developmental stage	Number of days	
	Healthy host	Infected host[a]
Egg	6.5	4.4**
1st instar	3.8	3.3*
2nd instar	3.4	3.3
3rd instar	3.0	3.4
Puparium	8.0	7.1
Total developmental time from egg to adult	24.7	21.5

[a] Differences statistically significant by F test
* = ($p < 0.05$)
** = ($p < 0.01$).

II. Sampling Adult Whiteflies

Adult populations of whiteflies can be monitored by taking advantage of their strong specific response to surfaces reflecting within the 520-610 nm region of the spectrum (Vaishampayan et al. 1975). Traps based on this color attraction would bias populations to an overestimate (see Chapter 11). No detailed studies have been conducted on the use of traps to monitor adult populations on soybean.

III. Sampling Eggs, Nymphs, and Pupae

All developmental forms (except the first instars) are sessile; therefore, the measurement of whitefly populations is based on counts of individuals on plants or plant parts. A good study of absolute and stratified random sampling for *T. vaporariorum* to measure parasitization by *Encarsia formosa* Gahan, was published by Mansveld et al. (1978). This study was conducted in the greenhouse but the principles used in the stratified random sampling program may be applicable to field conditions.

A. Absolute Counts

The simplest absolute method is the removal of single plants from a measured length of row or plants from a quadrat and examination of all leaflets. The method is similar to that described for measuring spider mite populations (see Chapter 15).

Samples of plant parts may also serve to provide an adequate populations estimate. In this case, it is necessary to establish vertical stratification patterns if attempts are made to extrapolate population estimates based on leaflets to numbers per plant or numbers per unit of area (see Chapter 13).

Since field infestations by whiteflies can be detected through the symptoms of feeding it is possible to establish a sampling procedure based on the initial detection of these foci. In some cases direct counts in the field may be sufficient to provide population estimates. Counts of adults and nymphs may be done with unaided eye. In other cases removal of plants may be necessary for an accurate evaluation of populations of all developmental stages.

B. Infestation Index

An index of infestation is provided by scoring symptoms of injury: yellowing due to feeding, amount of honeydew on leaves, development of sooty mold, and reduction of plant growth. A grading system on a scale of 1 to 4 may be established as follows: grade 1—healthy plants, free of infestation; grade 2—slightly injured plants, few insects and slight feeding injury present; grade 3—moderate injury, leaves with sticky honeydew on the surface, leaf edges turned down and plant growth slightly arrested; grade 4—development of profuse sooty mold on most leaves, plant growth significantly arrested, large number of whiteflies present on the foliage.

An infestation index (I) for the field is then computed using the formula:

$$I = \frac{(G_1 \times P_1) + (G_2 \times P_2) + (G_3 \times P_3) + (G_4 \times P_4)}{P_1 + P_2 + P_3 + P_4} \qquad (14\text{-}1)$$

where G = grade of infestation, and P = number of plants in that infestation category.

Infestation indices are applicable in programs for screening for plant resistance and in pest management decisions related to the start of insecticide sprays. There is no information on the relationship of infestation levels on soybean and probable reduction in yield.

IV. Concluding Remarks

Whiteflies have been carefully studied as vectors of plant disease. Population studies on soybean, however, are virtually nil. Turnipseed (1977) provided some information on *T. abutiloneus* populations in South Carolina on pubescent and glabrous soybean during several growing seasons. The D-Vac was used as the sampling procedure and one would expect that mostly adult whiteflies were collected by this method. There was no indication of an effect of pubescence on whitefly populations.

It is apparent that advances in epidemiological studies of soybean viruses will require development of more efficient methods for sampling whitefly populations on soybean. Sampling methods described for other similar organisms may be adopted as a starting point (see Chapters 11, 12, and 15).

Acknowledgment

The authors are grateful to Dr. Zile Singh, Professor and Head, Department of Entomology, Haryana Agricultural University, Hissar, India for comments on an early version of the manuscript.

References

Costa, A. S. 1975. Increase in the population density of *Bemisia tabaci,* a threat of widespread virus infection of legume crops in Brazil. pp. 27-50 in J. Bird and K. Maramorosch, eds. Tropical diseases of legumes. Academic Press, N.Y. 171 p.

Costa, A. S. 1976. Whitefly-transmitted plant diseases. Annu. Rev. Phytopathol. **16**:429-449.

Costa, A. S., C. L. Costa, and H. F. G. Sauer. 1973. Surto de mosca-branca em culturas do Paraná e São Paulo. An. Soc. Entomol. Brasil **2**:20-30.

Debrot, C., E. A., and A. Ordosgoitti F. 1975. Estudios sobre un mosaico amarillo de la soya en Venezuela. Agron. Trop. **25**:435-449.

Gangrade, G. A. 1974. Insects of soybean. Jawaharlal Nehru Krishi Vishwa Vidyalaya Tech. Bull. **24**:88 p.

Goodman, R. M., and Y. L. Nene. 1976. Virus diseases of soybeans. pp. 91-96 in R. M. Goodman, ed. Expanding the use of soybeans. Proc. of a Conference for Asia and Oceania. Univ. of Illinois, College of Agriculture, INTSOY Ser. 10. 261 p.

Kuwayama, S., ed. 1953. Survey on the fauna of soybean insect-pests in Japan. Yokendo Pub., Tokio. 129 p.

Mansveld, M. H. E.-R., J. M. Ellenbroek, J. C. van Lenteren, and J. Woets. 1978. The parasite-host relationship between *Encarsia formosa* Gah. (Hym., Aphelinidae) and *Trialeurodes vaporariorum* (Westw.) (Homoptera, Aleyrodidae). Z. Angew. Entomol. **85**:133-140.

Mound, L. A., and S. H. Halsey. 1978. Whitefly of the world. A systematic catalogue of the Aleyrodidae (Homoptera) with host plant and natural enemy data. British Museum (Natural History) and John Wiley and Sons, Chichester. 340 p.

Murugesan, S., and S. Chelliah. 1978. Effect of yellow mosaic infection of the host green gram on the biology of *Bemisia tabaci* (Genn.). Entomon. **3**:41-43.

Panizzi, A. R., B. S. Correa, P. L. Gazzoni, E. B. de Oliveira, G. G. Newman, and S. G. Turnipseed. 1977. Insetos da soja no Brasil. EMBRAPA, Cent. Nac. Pesq. Soja, Bol. Tec. **1**:20 p.

Russell, L. M. 1963. Hosts and distribution of five species of *Trialeurodes* (Homoptera: Aleyrodidae). Ann. Entomol. Soc. Amer. **56**:149-153.

Turnipseed, S. G. 1977. Influence of trichome variations on populations of small phytophagous insects in soybean. Environ. Entomol. **6**:815-817.

Vaishampayan, S. M., M. Kogan, G. P. Waldbauer, and J. T. Wooley. 1975. Visual behavior and host finding by the greenhouse whitefly. 1. Spectral specific responses. Entomol. Exp. Appl. **18**:344-356.

Chapter 15

Sampling Mites on Soybean

Sidney L. Poe

I. Introduction

Among the important arthropod pests of soybean are several species of plant-feeding mites. The Tetranychidae, a family of the most damaging species, spin copious webbing on the plant and, therefore, are commonly called "spiders," red spiders, or spider mites. The species of phytophagous mites that have been observed to injure soybean are listed in Table 15-1. In California two species *Tetranychus urticae* Koch, the twospotted spider mite, and *T. pacificus* McGregor, the Pacific spider mite, caused partial to complete defoliation of soybean and reduced seed yield (Carlson 1969). In Maryland *T. turkestani* Ugarov and Nikolski, the strawberry spider mite, and *T. urticae* were reported as the principal injurious species (Ratcliffe et al. 1960). *T. turkestani* and *T. yusti* (McGregor) were chiefly responsible for mite injury observed in Delaware. Whereas up to 16% of all leaves collected were infested with spider mites, only 2% were infested with *T. yusti;* hence the strawberry spider mite appears to be the most abundant mite pest. *Panonychus* sp. was collected only once (Baker and Connell 1961).

In Florida, *T. urticae, T. yusti* and *Panonychus citri* (McGregor), the citrus red mite, were found in two years of soybean survey. *T. urticae* was found with greatest recurring frequency, while two collections were made of *T. yusti* and one of *P. citri.* No damage by these species was recorded (F. Reid and S. Poe unpublished). *Monomychellus planki* (McGregor) has been reported infesting soybean in South America (Jeppson et al. 1975). *T. tumidus* Banks, the tumid spider mite, is recorded from many hosts and is a common pest in tropical areas.

Common predaceous mite families collected from soybean include the Phyto-

Table 15-1. Phytophagous mites that attack soybean

Species	Location	Reference
Tetranychus urticae Koch	California, Florida, Illinois Maryland	Carlson (1969) F. Reid (unpublished) M. Kogan (unpublished) Ratcliffe et al. (1960)
T. pacificus McGregor	California	Carlson (1969)
T. turkestani Ugarov and Nikolski	Maryland Delaware	Ratcliffe et al. (1960), Baker and Connell (1961)
T. yusti (McGregor)	Delaware, Florida	Baker and Connell (1961), F. Reid (unpublished)
T. tumidus Banks	Florida	S. Poe (unpublished)
Panonychus sp.	Delaware	Baker and Connell (1961)
P. citri (McGregor)	Florida	F. Reid (unpublished)
Monomychellus planki (McGregor)	Brazil, Puerto Rico, Trinidad, Colombia	Jeppson et al. (1975)

seiidae, Ascaidae, Cunaxidae, Bdellidae, Stigmaeidae, and Trombidiidae. Others of saprophytic, fungivorous, or unknown roles that have been observed include Ereynetidae, Pyemotidae, Sejidae, Tydeidae, Rhodacaridae, Tarsonemidae, Acaridae, and Eupodidae (Baker and Connell 1961, Whitcomb 1974, S. Poe unpublished, D. C. Herzog, personal communication). The reader is referred to Baker and Wharton (1964) for aid in identification.

A. Host Plants and Nature of Injury

The Tetranychidae are highly polyphagous, with each species recorded from a large number of hosts (Metcalf et al. 1962).

These pests injure soybean by feeding on the green foliage and pods. Phytophagous mites possess needle-like chelicerate mouthparts that are used to puncture individual plant tissue cells and consume the entire cytoplasmic contents, leaving an empty and irreversibly damaged cell. The presence of numerous empty cells at a site results in the yellow or brown specks or white stipples characteristic of mite-injured leaves. Extensive feeding by large numbers of mites causes the leaves to appear yellow or brown. Those leaves severely injured shrivel, die, and eventually drop from the plant.

Complete defoliation due to mite feeding can reduce pod set and seed yield, the severity depending on timing, duration, and magnitude of the infestation (Carlson 1969). Spider mite feeding on soybean in Delaware has resulted in as much as 50% foliage drop (Milliron 1958). In a study of varietal response to mite injury, total plant weight at harvest, plant height, yield of clean seed, weight of 100 seeds, number of seeds per plant, and length of pods were decreased (Cadapan 1976).

B. Life Cycle and Phenology

Mites are small (0.05 mm) eight-legged arthropods, more closely related to spiders (Araneae) than to insects (Fig. 15-1). Phytophagous species, those that feed and reproduce on plants, are usually characterized by color hues ranging from carmine red to pale yellow to green. Consumption of chlorophyll results in an accumulation in green gut contents that, when digested, often cause the body to appear dark or black.

Stages of growth and development include the egg, larva, protonymph, deutonymph, and adult (Fig. 15-2). Acquiescent period (chrysalis), during which molting to the successive stage is achieved, precedes the adult and each of the nymphal stages. Mites undergoing molts are inactive and anchored to the leaf surface on webbing (Fig. 15-3A-D). Consequently, the forms that may be present at any

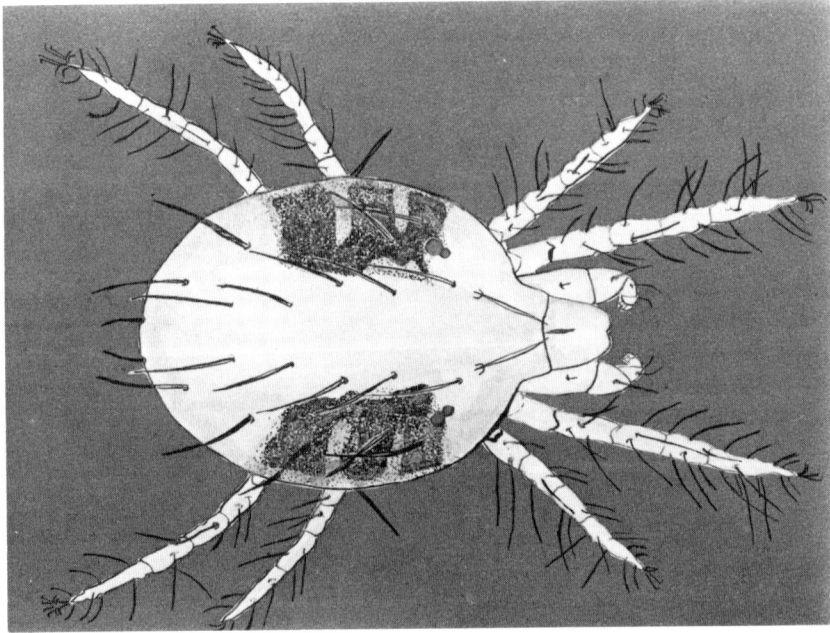

Figure 15.1 Typical tetranychid mite. (Illustration by Ms. Gail H. Childs, Univ. of Florida).

given time include the egg, six-legged larva, protochrysalis, protonymph, deutochrysalis, deutonymph, teliochrysalis, female, and male.

The time required to complete the life cycle varies with environmental conditions and species but usually ranges from 10 to 20 days. Reproduction is of the haploid, diploid type. Unfertilized females lay haploid eggs that develop into males. Sex ratio varies with the age structure of a population but usually favors the females

In the laboratory the most common pest species, *T. urticae,* completes its life cycle in 8-20 days and has a generation interval of 13-16 days. Females lay as many as eight eggs per day over a life span of ca. 30 days (Shih et al. 1976). Individuals develop at a range of temperatures from ca. 12-40°C. In temperate climes spider mites enter a winter diapause that is absent in the mites of tropical areas.

The detection and estimation of spider mite populations must be considered in relation to hosts other than soybean and to population dynamics, dispersal, and dispersion characteristics of these pests. Spider mites, especially *T. urticae* and *T. turkestani,* infest a wide range of acceptable host plants. Generally, mites invade or are dispersed into a field passively by animals (mammals and birds), wind, water, agricultural workers, or machinery. Mites also actively crawl from adjacent weed or other crop hosts into soybean fields. Consequently many

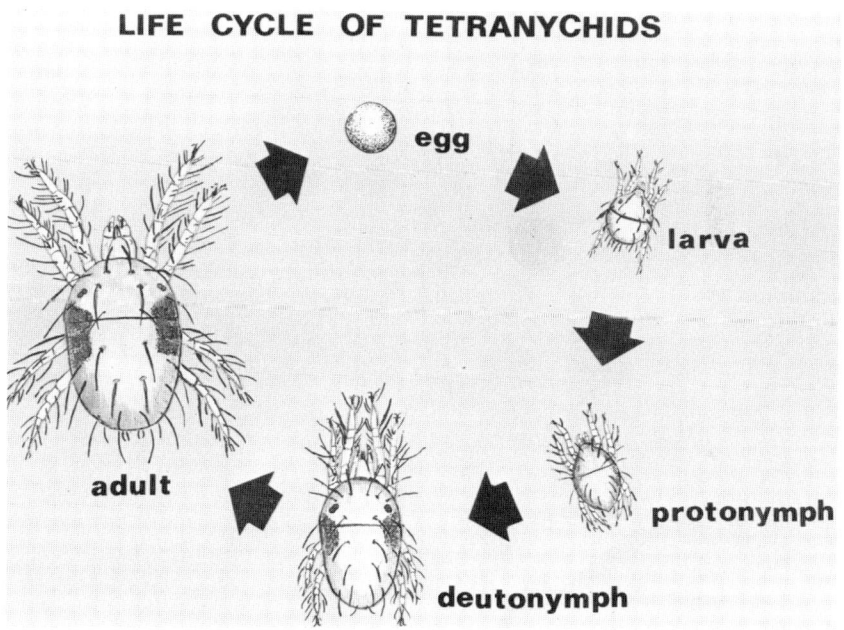

Figure 15-2. Developmental stages in the life cycle of a typical spider mite—family Tetranychidae. (Illustration by Ms. Gail H. Childs, Univ. of Florida).

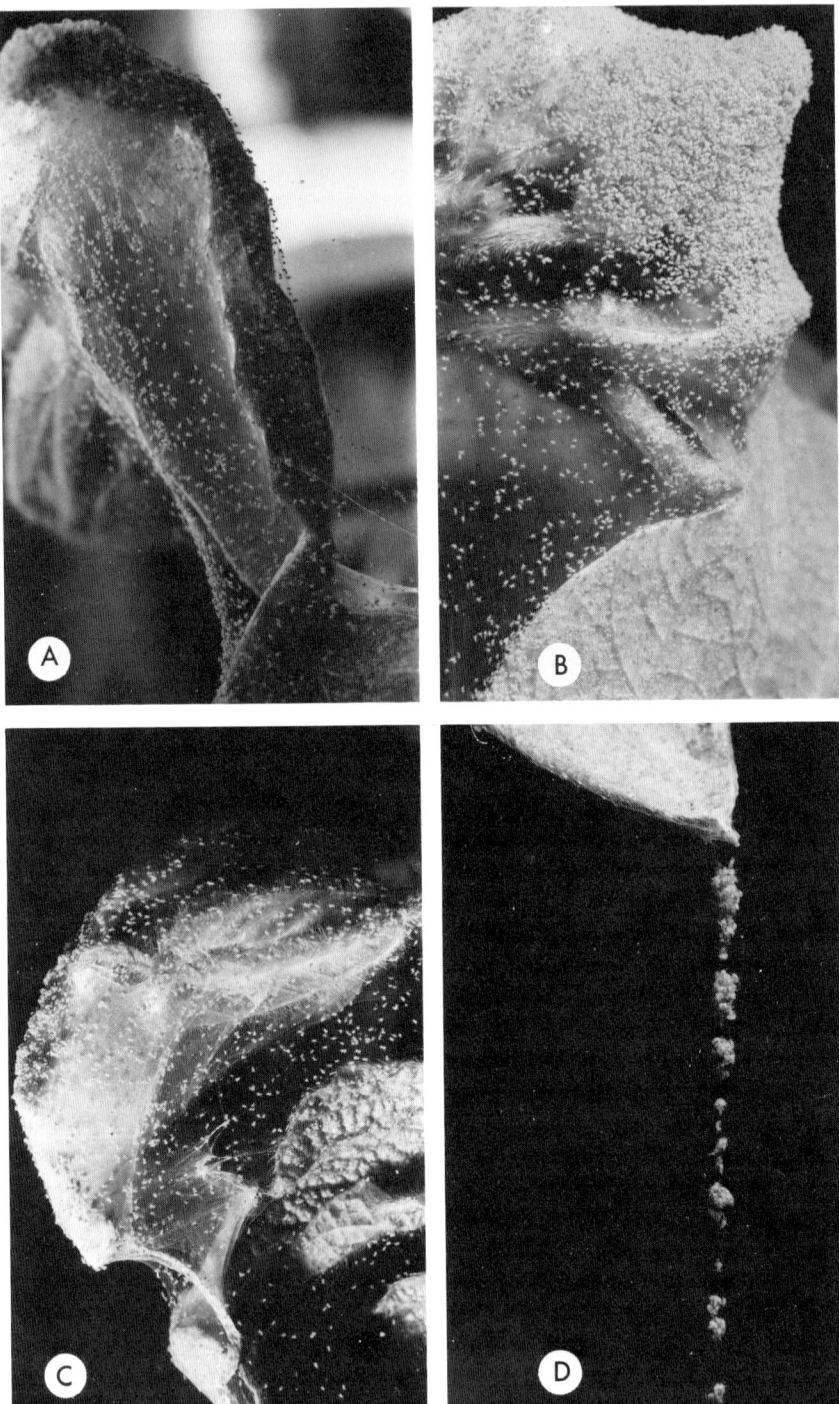

Figure 15-3. Severe spider mite infestation showing heavily webbed leaves and "swarming" behavior of two-spotted spider mite on soybean in the greenhouse. (Courtesy M. Kogan; H. Jeffords photograph).

infestations originate in field edges where plants in border areas provide mite reservoirs.

When a young female (the most common dispersing form) occupies a suitable host plant, she settles and begins to feed and to lay a web over the leaf surface (Fig. 15-3A,C). Oviposition occurs within a few hours, the population begins a rapid increase, and individuals move to other portions of the host plant. Although undersurfaces of lower leaves are preferred, under crowded conditions the entire plant becomes infested. After prolonged infestation and with increasing populations, young females disperse from the parent host plant. Dispersing forms are generally pink to orange, are positively phototactic and seek out points of elevation from which to disperse. Copious webbing is characteristically produced, and a "swarming" phenomenon results in bridging the distance between adjacent plants with webs (Fig. 15-3D). Individuals may crawl along the web to new sites, be blown by wind, transported by mobile objects or drop to the ground in a mass of mites and webbing.

Commonly a field infestation will show defoliated or injured plants at some localized point or focus, with injury extending in a widening arc into the field. As the distance from the focus of infestation increases, the injury becomes less evident. Ratcliffe et al. (1960) described such an infestation that covered ca. 2.4 ha.

Secondary infestations become common when populations reach large proportions and when individuals are dispersed by wind to begin new foci or "hot spots." Secondary foci occur along the path of equipment or workers as their movement serves to spread the population. Hot spots may also appear near or downwind of weedy sites in the field.

Although mite infestations can occur at any time in the cropping season, mid- to late-season populations are more common. Cadapan (1976) noted that soybean is apparently most susceptible to mite injury during the pod and seed development periods (R3-R5). Mite populations increase most rapidly during this period, attaining population levels of over 1000 individuals per leaf (Cadapan 1976).

The most important factor contributing to spider mite outbreaks is the availability of a physiologically suitable host plant. Not all plants in all stages of growth are nutritionally adequate to support the high fecundity rate necessary for an outbreak. Outbreaks are most often associated with the blossoming and fruiting stages of the host plant, rather than with the vegetative growth stages (Huffaker et al. 1969). The phenomenon has often been related directly to plant nutrition (Rodriguez 1958; Henneberry 1962a, 1962b, 1963; Cannon and Connell 1965). It appears, however, that the mineral nutrition of the plant is associated only indirectly as a causal agent. Since mites do not feed directly on elemental nutrients but feed instead on the photosynthates produced in the leaves, the positive correlation between mites and mineral nutrition is oversimplified. The relationship between host, mite, and mite fecundity represents a complex biochemical association that is influenced by grower practices (Hacskaylo 1957, Poe 1971, Poe et al. 1976).

Given a suitable host plant, mite populations will develop only under favorable environmental conditions. Mite injury and outbreaks are associated with hot, dry weather on numerous field crops, including soybean. Relative humidity, field moisture, and temperature may adversely affect spider mite populations (Boudreaux 1958, Nickel 1960, Simpson and Connell 1973). The relationship between field moisture and temperature and the occurrence and density of mite populations on soybean was demonstrated by Simpson and Connell (1973). Using data from eight fields over a seven year period, they showed that much of the variation in infestation levels of *T. turkestani* in soybean could be explained by weather phenomena.

II. Sampling for Adults and Juveniles

Several procedures are available for determining the frequency or relative abundance of spider mites in soybean and other row crops. Most methods are designed to determine the relative number of individuals per plant or leaflet in small plot experiments. Therefore, they must be adapted for use in large fields. Knowledge of the pattern of occurrence in the field, of margin or border plants, of dispersal habits, and of dynamics of populations should be used to aid in initial detection and in implementation of population estimation procedure. Two general techniques have been used to deal with mites on soybean: field surveys and plant or site samples.

A. Field Surveys

Surveys have been used to describe as major or minor the pests of numerous crops (Milliron 1958). Surveys are based on weekly visits to as many fields as possible in a given region. Crops are closely inspected and presence of beneficial species, mites and other organisms noted. Both faunistic (lists or frequency of organisms) and economic (extent of injury) aspects are recorded; however, the latter is highly subjective and not often provided in reports. Although density estimates were not provided in the survey, Milliron (1958) noted infestations as "severe" and estimated as much as 50% foliage loss in local areas of certain fields.

The frequency and magnitude of injury by mites have been surveyed in Maryland and Delaware (Ratcliffe et al. 1960, Baker and Connell 1961) utilizing "approved" methods described in a USDA Cooperative Economic Insect Report (USDA 1955). Weekly visits from July to September were made to ten fields in different locales with one leaf per plant removed from various sites on six plants. Plants were sampled at random along the field borders and leaves were placed in plastic bags for transport and counting under magnification in the laboratory. Quantative estimates were thus available from 60 leaves in ten collections from each field.

Surveys are useful to ascertain the species of mites present, the frequency of their occurrence and, if leaf counts are made, an estimate of relative density. However, in all cases, observations limited to an individual leaf or plant represent only a portion of the habitat available to mites in a soybean field. Foliage inspection, brushing, recovery from duff and litter on the soil, and pitfall traps have been used to collect mites in Florida soybean fields (F. Reid and S. Poe unpublished). Representatives of 15 families were discovered among the 32 species collected. Eleven species were recovered from foliage during both years; however, 16 and 9 species were recovered, respectively, from duff and soil litter for the two years. Thus, while the foliage survey revealed only about half the species richness of the fields, these were the pests. A few free-living, fungivorous, and predaceous species were also recovered from foliage, but by far the greatest proportion of these forms were from the soil litter.

B. Estimation of Relative Density

Various methods of leaf sample site selection and sample sizes have been utilized for sampling spider mites on soybean (Carlson 1969, Simpson and Connell 1973, Bailey and Furr 1975). Within a 50-row border, Simpson and Connell (1973) collected two leaflets weekly from each of five randomly-selected rows. In a small plot experiment, Carlson (1969) selected 20 "infested" leaflets from plants in each test plot. The samples were placed in pint jars and kept refrigerated until processed. The population estimate was made by clamping each leaflet between the halves of a flattened 15.2 cm strap hinge with two 1.27 cm diameter holes drilled, one near the base and one near the center. These "windows" exposed two circular areas on the underside of each leaflet from which numbers of mites and eggs could be counted directly under magnification. Adults, nymphs, and eggs were counted with the aid of a hand magnifier (Bailey and Furr 1975).

Jeppson et al. (1975) summarized several methods of estimating mite populations and pointed out advantages and disadvantages of each technique. Choice of a technique to estimate mite populations or their damage will depend on the specific objectives of the study and the use that is to be made of the data.

The most precise method is direct counting of all stages on leaflets using the aid of a stereomicroscope. This method is time-consuming and must be completed soon after sampling. It requires proper laboratory facilities and loses precision as mite density increases. More commonly, subsamples of leaves are used to obtain direct counts (Carlson 1969).

Imprinting, a technique of crushing mites on leaves placed between the flat surface of absorbent paper, is an indirect estimate of numbers. Although each individual crushed results in a tiny speck or stain on the paper, under high densities stains may coalesce. Counts of spots can be made at a later date; however, stage of development or specific identification can be provided with certainty only at the time of sampling.

Direct observations and counts are made from leaves in the field either through a hand lens or with the unaided eye. Without magnification accuracy is sacrificed, and this method is impossible at high mite densities. The method has merit when implemented by a discard-count system, where only leaves observed to be infested are retained for laboratory counts, all others are discarded. A further improvement is realized when relative numbers are adequate and estimates or categories of infestation are assigned. Levels of mite infestation then fall into a category containing 1-5, 6-10, 11-25, 26-50, > 50 or some similarly defined densities of individuals per leaf. Although precision is sacrificed, this rapid method allows coverage of greater numbers of samples or fields.

Dislodgement of mites by beating foliage over a wide-mouth funnel attached to a collecting apparatus is a useful method to obtain estimates of relative abundance of mites. However, the number of collecting vials required, labels, and portion of foliage sampled are disadvantages when precise measurement is desired. Plant shakes over a ground cloth or white paper may be used to detect presence and even relative estimates of numbers. Movement of individuals on a white background is essential if mites are to be counted with the unaided eye. Advantages of this method are its nondestructive nature and speed. However, the method may be little better than a detection method and should be utilized with caution, realizing that an undetermined portion of the population will not be dislodged from the plants and that the age distribution of the population may vary considerably from that measured.

A machine that brushes mites and eggs from leaflets onto a glass plate coated with adhesive is commercially available (Fig. 15-4). This method has obvious advantages since many more samples can be collected at random, bagged, transported, and quickly processed. Subsamples of the plate counted, using a grid, provide accurate estimation of numbers of each stage and species present. The greatest disadvantages of the brushing machine is that tender or large leaves are either folded, crumpled, or shredded as they pass between the opposing roller brushes, resulting in a messy if not inaccurate sample. Young soybean foliage is generally too tender and older foliage too large for precise measurements using the currently available machine.

C. Sampling Program

The collection of soybean leaves to estimate mite populations should be made with the same considerations given to the sampling of other plants. Removal of foliage should not substantially reduce the plant area, plant vigor, or the mite population. To be practical, collections should be made at random or systematically with a random start in a restricted area where mites are most likely to occur. Knowledge of their contagious pattern and invasion from the field periphery aid in selecting sites for sampling.

The time of year, plant phenology, or growth and general weather conditions conducive to mite problems provide a common sense base from which sampling of soybean for mites can be realistically approached. Dense spider mite popu-

 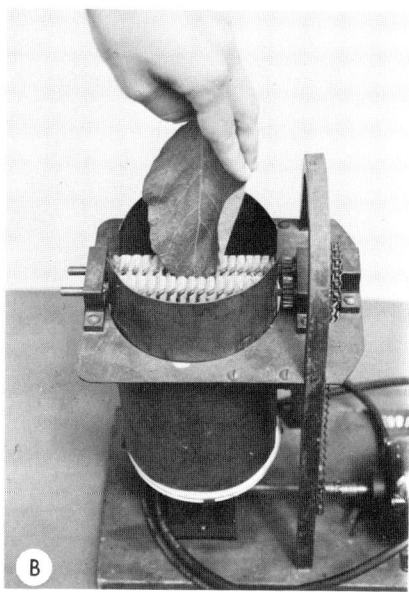

Figure 15-4. (A) Mite brushing machine; (B) Soybean leaf being inserted between brushes. Mites are dislodged onto the collecting glass beneath the cylinder. (Courtesy M. Kogan; M. Jeffords photograph).

lations on row crops are generally encountered on more mature, fruiting or senescent plants; however, younger plants often harbor these mites that become a problem only when conditions approach the optimum. Taking large numbers of samples and numbers of leaves per sample increase the precision and probability of detecting mites at low densities and before injury becomes visible. Statistical treatment of the data from numerous samples to determine the coefficient of variation will aid in selecting an appropriate sample size and number (see Chapter 3).

Just as mites are not randomly dispersed within fields, populations do not generally occur at random on the individual plant. The success of the sampling program is enhanced when leaves most likely to be infested are selected by some method of stratification. Distribution on the plant might depend on species, but early infestations of spider mites are more frequently observed on older, more mature foliage than on younger more tender foliage.

III. Concluding Remarks

Several species of spider mites can be periodic problems in soybean at various locations. Infestations at high densities early in reproductive growth can result in yield loss and plant mortality. Infestations generally begin at a field margin and proceed inward in a widening area. Total density and impact are greatly dependent on host phenology, favorable environmental conditions, natural enemies, and cultural practices.

Population estimation may be approached from either of two objectives: survey sampling for detection and intense sampling for absolute numbers. Surveys should be made with attention to likely location in the field, preferred site on the plant, stage of plant growth, and weather conditions. Intensive sampling should be from a likely site on the host, considering weather, plant age, and controlled variables. Faunistic surveys, to be comprehensive, should consider litter and duff inhabiting species as well as those on foliage.

Estimates can be made from individual plant unit observation, imprint, machine brushing onto a plate, beat cloth, paper, or funnel techniques. Reliability of a sampling method should be estimated by calculating the necessary parameters (see Chapter 3). Reliability is also likely to change with species, population age structure, and density.

The objective of future research should be a comparison of these methods under a range of plant ages and mite population densities. The most appropriate technique to detect and estimate numbers of mites in soybean could then be recommended.

References

Bailey, J. C., and R. G. Furr. 1975. Reaction of 12 soybean varieties to the two-spotted spider mite. Environ. Entomol. 4:733-734.

Baker, J. E., and W. A. Connell. 1961. Mites on soybeans in Delaware. J. Econ. Entomol. 54:1024-1026.

Baker, E. W., and G. W. Wharton. 1964. An introduction to acarology. The MacMillan Co., N.Y. 465 p.

Boudreaux, H. B. 1958. The effect of relative humidity on egg-laying, hatching, and survival in various spider mites. J. Insect Physiol. 2:65-72.

Cadapan, E. P. 1976. The effect of the two-spotted spider mite, *Tetranychus urticae* Koch and several insects on the yield of soybeans. Ph.D. diss., University of California, Berkeley. 111 p.

Cannon, W. N., and W. A. Connell. 1965. Populations of *Tetranychus atlanticus* McG. (Acarina:Tetranychidae) on soybean supplied with various levels of nitrogen, phosphorus and potassium. Entomol. Exp. Appl. 8:153-161.

Carlson, E. L. 1969. Spider mites on soybeans; injury and control. Calif. Agr. 23:16-18.

Hacskaylo, J. P. 1957. Growth and fruiting properties and carbohydrate, nitrogen and phosphorus levels of cotton plants as influenced by Thimet. J. Econ. Entomol. 51:280.

Henneberry, T. J. 1962a. The effect of plant nutrition on the fecundity of two strains of two-spotted spider mites. J. Econ. Entomol. 55:134-137.

Henneberry, T. J. 1962b. The effect of host-plant nitrogen supply and age of leaf tissue on the fecundity of the two-spotted spider mite. J. Econ. Entomol. 55:799-800.

Henneberry, T. J. 1963. Effect of host plant condition and fertilization on two-spotted spider mite fecundity. J. Econ. Entomol. 56:503-505.

Huffaker, C. B., M. van de Vrie, and J. A. McMurtry. 1969. The ecology of tetramychid mites and their natural control. Annu. Rev. Entomol. 14:125-174.

Jeppson, L. R., H. H. Keifer, and E. W. Baker. 1975. Mites injurious to economic plants. University of California Press, Berkeley. 614 p.

Metcalf, C. L., W. P. Flint, and R. L. Metcalf. 1962. Destructive and useful insects. McGraw-Hill Book Co., Inc., N. Y. 1087 p.

Milliron, H. E. 1958. Economic insect and allied pests of Delaware. Del. Agri. Exp. Sta. Bull. 321:55-57, 71.

Nickel, J. L. 1960. Temperature and humidity relationships of *Tetranychus desertorum* Banks with special reference to distribution. Hilgardia 30:41-100.

Poe, S. L. 1971. Influence of host plant physiology on populations of *Tetranychus urticae* (Acarina:Tetranychidae) infesting strawberry plants in Peninsular Florida. Fla. Entomol. 54:183-186.

Poe, S. L., L. Green, R. C. Littell, and C. I. Shih. 1976. Cultural management of pest populations in Saran house-grown Chrysanthemums. pp. 29-40 in F. F. Smith and R. E. Webb, eds. Pest management in protected culture crops. USDA Agr. Res. Serv. NE-85:96 p.

Ratcliffe, R. H., T. H. Bissell, and W. E. Bickley. 1960. Observations on soybean insects in Maryland. J. Econ. Entomol. 53:131-133.

Rodriguez, J. G. 1958. The comparative NPK nutrition of *Panonychus ulmi* Koch and *Tetranychus telarius* (L) on apple trees. J. Econ. Entomol. 51:369-373.

Shih, C. I., S. L. Poe, and H. L. Cromroy. 1976. Biology, life table and intrinsic rate of increase of *Tetranychus urticae*. Ann. Entomol. Soc. Amer. 69:362-364.

Simpson, K. V., and W. A. Connell. 1973. Mites on soybeans: Moisture and temperature relations. Environ. Entomol. 2:319-323.

USDA. 1955. Survey methods. USDA Coop. Econ. Insect Rep., pp. 20-35.

Whitcomb, W. H. 1974. Natural populations of entomophagous arthropods and their effect on the agrosystem. pp. 150-169 in F. G. Maxwell and F. A. Harris, eds. Proceedings of the Summer Institute on Biological Control of Plant Insects and Diseases. University Press of Mississippi, Jackson. 647 p.

SECTION V
Underground Feeders

Chapter 16

Sampling Phytophagous Underground Soybean Arthropods

Cathy E. Eastman

I. Introduction

A great deal of information has been accumulated in recent years on insects that attack above-ground parts of soybean, but relatively little research has been done on those species that spend at least a portion of their life cycle below ground (Turnipseed 1973, Turnipseed and Kogan 1976). This chapter discusses sampling procedures for species whose larvae are injurious to germinating seed, seedlings, roots, and nodules; species whose larvae and adults are predatory on soil- or surface-inhabiting insects are discussed in Chapter 26. Most phytophagous ground-level or underground species are considered to be minor pests, and references to damaging populations in soybean are infrequent. Nodule feeders have been almost totally disregarded, but a recent report (Newsom et al. 1978) indicates that the effects of insect injury on nitrogen fixation directly through nodule destruction, or indirectly through defoliation or stem girdling, may be of greater importance than previously realized.

Pest status notwithstanding, it is increasingly necessary that soil-inhabiting insects in soybean receive greater attention. The information generated will be required for development of economic thresholds for pest complexes and for expansion of integrated pest management systems.

This chapter is intended to serve as a sampling guide for researchers dealing directly with soil-inhabiting pests and for those investigating the role of soil insects in soybean ecosystems. Only those insects most commonly associated with soybean are included, and these are divided into groups depending on the plant parts attacked.

II. Underground Arthropod Species Associated with Soybean

A. Insects Attacking Germinating Seed and Seedlings

1. Seedcorn Maggot

The seedcorn maggot, *Hylemya platura* (Meigen) (Diptera:Anthomyiidae), attacks germinating seed and seedlings of a wide variety of vegetable and field crops. Damage to soybean seed has been reported in the United States (Turnipseed 1973, Turnipseed and Kogan 1976), Canada (Miller and McClanahan 1960), India (Gujrati et al. 1971, Chaudhary et al. 1976), and Japan (Kuwayama et al. 1960).

The white, elongated eggs are laid in soil or on seed. Larvae burrow into the seed, resulting in complete destruction of the germ or in the production of weakened seedlings (Fig. 16-1). When fully grown, the larvae are yellowish-

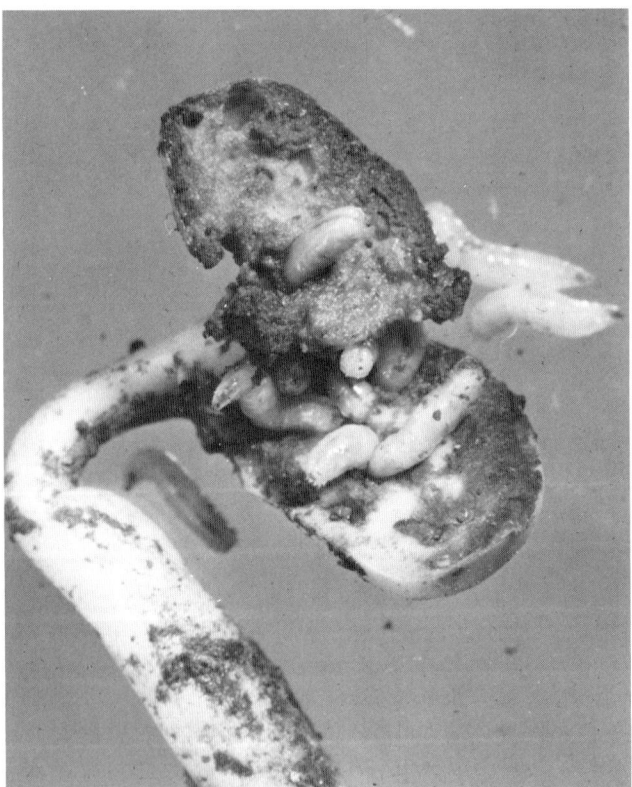

Figure 16-1. Injury by seedcorn maggots to germinated bean seed. (University of Illinois Cooperative Extension Service).

white, legless, and about 6 mm long (Fig. 16-2D). Pupation occurs in reddish-brown puparia in the soil (Fig. 16-2C). The adult flies are grayish-brown and about 5 mm long (Fig. 16-2A,B). Development from egg to adult is completed in 3-4 weeks during the spring and summer months with several generations occurring annually. The species overwinters as larvae inside puparia (Reid 1940, Metcalf et al. 1962).

Infestations of seedcorn maggots are often associated with cool, wet springs and poor field drainage. Freshly cultivated soil with a high organic content is quite attractive to ovipositing flies, and eggs may be laid in disturbed soil prior to planting (Miller and McClanahan 1960). Newly-emerged soybean seedlings appear to be more attractive for oviposition than older plants (Ibrahim and Hower 1979). The patchiness of infested areas within fields may be due to variations in moisture- and heat-holding capacities of soil, resulting in differing degrees of attractiveness for oviposition (Miller and McClanahan 1960).

2. Cutworms

Cutworms (Lepidoptera: Noctuidae) are widely known as pests of a variety of field crops and garden vegetables. Members of the genus *Agrotis,* especially *A. ipsilon* (Hufnagel), have been linked occasionally with injury to soybean seedlings in South America (Van Dinther 1956, Guagliumi 1965, Ferreira 1970, Rizzo 1972), India (Gangrade 1974), Japan (Kuwayama et al. 1960), and the United States. The granulate cutworm, *Feltia subterranea* (F.), produces similar damage to soybean seedlings in South Carolina (Turnipseed 1973) and Florida (D. C. Herzog, personal communication).

Moths of the black cutworm, *A. ipsilon* (Fig. 16-3A), seem to be attracted early in the spring to low-lying field areas containing moderate-to-dense low

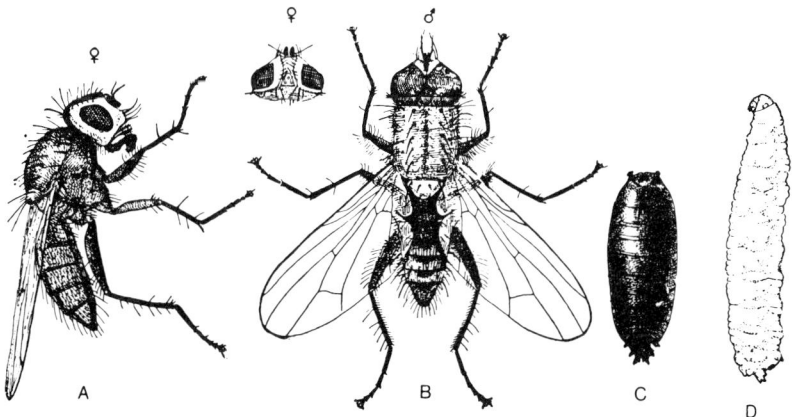

Figure 16-2. Seedcorn maggot, *Mylemya platura*: (A) adult female; (B) adult male; (C) pupa; (D) larva. (Illinois Natural History Survey).

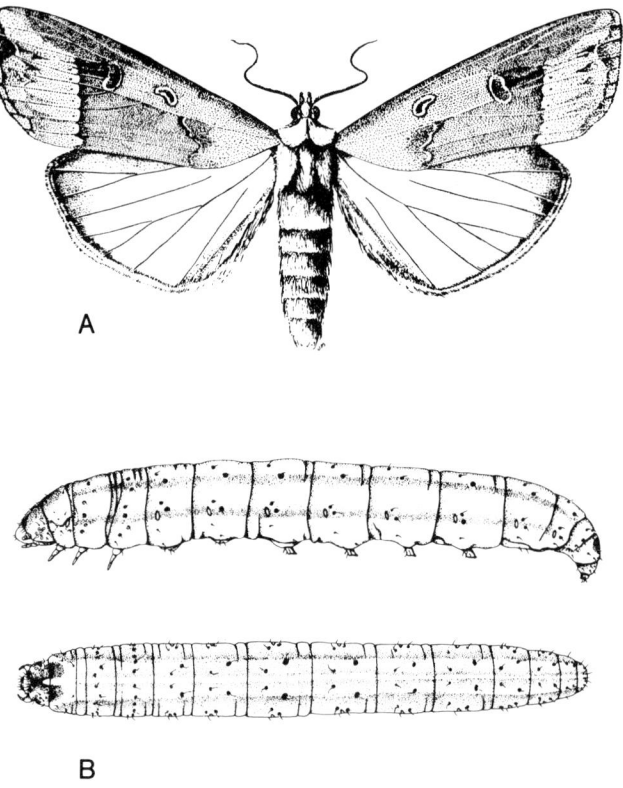

Figure 16-3. Black cutworm, *Agrotis ipsilon*: (A) adult (after Howard 1898, redrawn by J. P. Sherrod); (B) larva. (After Forbes 1890; redrawn by J. P. Sherrod).

weed growth, where the females deposit white, circular eggs singly or in clusters on vegetation or the soil surface. When these areas are cultivated and planted, cutworm larvae are already present and may attack young seedlings (Busching and Turpin 1976,1977). The larva (Fig. 16-3B) is greasy-brown to grayish-brown dorsally with a lighter narrow stripe down the midline. As is the case with other surface cutworms, the larvae destroy seedlings by cutting them off at ground level (Fig. 16-4). Sometimes the larvae drag the seedlings into burrows in the soil at night. Many seedlings may be cut per larva, but the seedlings are not always totally consumed (Metcalf et al. 1962).

The life cycle from egg to adult averages 45.5 days at 25.6°C (Harris et al. 1962). Larvae and pupae are the overwintering stages, and there are several generations per year (Metcalf et al. 1962). Illustrated field keys useful in identification of cutworm moths (Rings 1977) and larvae (Rings and Musick 1976) common in the north-central United States are available. An annotated world bibliography of the black cutworm has been published (Rings et al. 1974, Rings and Arnold 1976, Rings et al. 1978).

Sampling Phytophagous Underground Soybean Arthropods 331

Figure 16-4. Black cutworm larva and injury to soybean seedlings. (Courtesy M. Kogan).

3. Wireworms

Wireworms (Coleoptera: Elateridae) are another group of seed, seedling, and root feeders that are widely known as pests of corn, grasses, root crops, and small grains (Metcalf et al. 1962). Injury to soybean has been attributed to species of *Limonius* and *Melanotus* in the United States (Petty 1966, Edwards and Hallman 1975), *Agriotes* in Japan (Kuwayama et al. 1960), and to undetermined species in the People's Republic of China (Chiang 1977) and Korea (Kogan 1977). Several species of *Aeolus* and *Conoderus* among others have been collected as adults from soybean in the United States, Colombia, and Brazil (International Reference Collection of Soybean-Associated Arthropods, University of Illinois).

In general the larvae are cream to dark brown, elongated and cylindrical, 3 to 4 cm long (Fig. 16-5A). The adults ("click beetles," Fig. 16-5B) oviposit in spring in the soil mainly around roots of grasses. Larvae feed on germinating seed and the roots and below-ground stem portions of young seedlings. This injury kills the seedling or results in slower growth and increased susceptibility to disease (Fig. 16-6). Replanting of portions of the crop is sometimes necessary. The larvae spend 2-6 years feeding on seedlings or roots of older plants prior to pupation (Metcalf et al. 1962). Damage is most likely to occur if soybean is planted after grasses, clover, or alfalfa (Bigger and Petty 1965) and within low-lying, poorly drained field areas (Scott and Aldrich 1970).

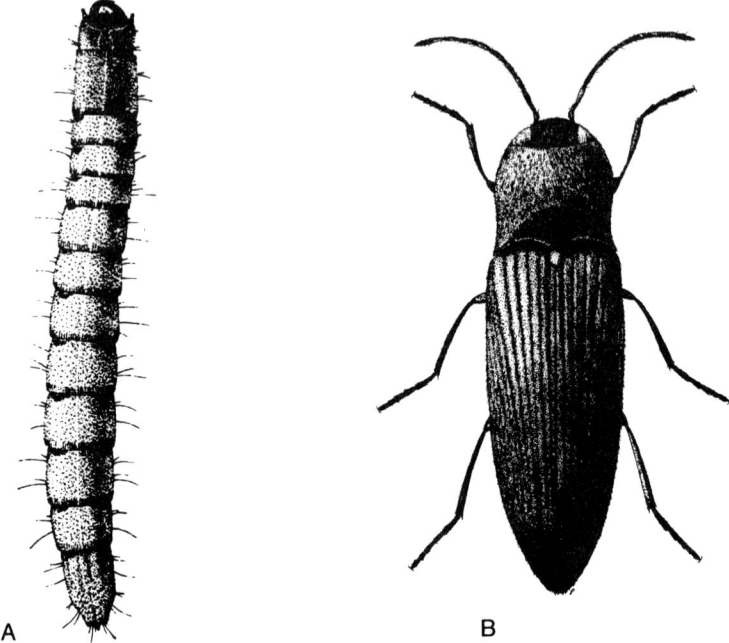

Figure 16-5. Wireworms: (A) unidentified larva; (B) adult of *Melanotus fissilis*. (Illinois Natural History Survey, redrawn by J. P. Sherrod).

Figure 16-6. Wireworm larva attacking germinated soybean seed. (University of Illinois Cooperative Extension Service).

4. Lesser Cornstalk Borer

The larva of the lesser cornstalk borer, *Elasmopalpus lignosellus* (Zeller) (Lepidoptera: Pyralidae), is a semisubterranean pest of grasses and legumes in the central and southern United States through Mexico to southern Brazil and Argentina (Luginbill and Ainslie 1917, Ferreira 1970, Correa 1975). The moths (Fig. 16-7A) are brownish-gray with the forewings of the female almost black. Females are active in the early spring and lay the greenish-white eggs singly on leaves or stems. Johnsongrass and corn are usually the early season hosts; damage to soybean is produced most often by second generation larvae (Motsinger et al. 1967). The mature larva is 16 mm long and is bluish-green with brown transverse bands (Fig. 16-7B).

Larvae feed first on leaves and possibly roots but soon bore into the stem close to or below ground level (Metcalf et al. 1962). When not actually feeding, the larvae retreat into the surrounding soil in tunnels connected to the stem entry

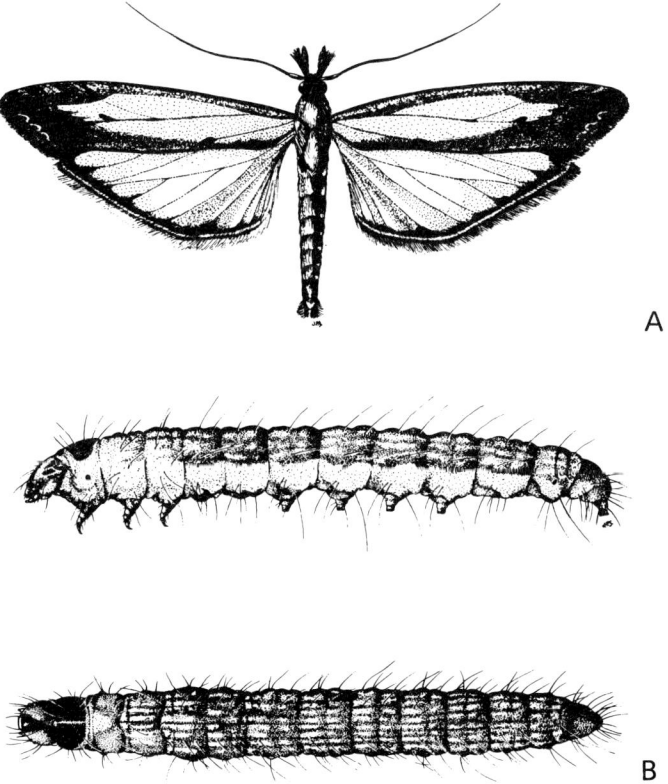

Figure 16-7. Lesser cornstalk borer, *Elasmopalpus lignosellus*: (A) adult male; (B) larva. (After Luginbill and Ainslie 1917, redrawn by J. P. Sherrod).

holes and constructed of sand, soil, and excrement woven together with silk into tube-like structures. Pupation occurs in a case of sand and silk at the end of one of the tunnels, and the pupa is the predominate overwintering stage. Development from egg to adult takes 33-36 days with three full generations and a partial fourth possible in the southern coastal United States (Luginbill and Ainslie 1917, Leuck 1966).

Both seedling and mature soybean are subject to attack, especially seedlings in late-planted fields. Seedlings may be chewed in half, while tunneling of the larva in the lower stem area causes young plants to wilt or lodge easily (Fig. 16-8). Damage to mature plants is usually confined to girdling of the stem at ground level. A larva may destroy several plants. Heavy infestations may completely destroy fields of seedlings necessitating total or spot replanting. Damage is most likely to occur and is most severe in sandy soil during drought conditions or when hot, dry periods follow late plantings (Leuck 1966); seasons with normal or near normal rainfall usually discourage development of economically significant populations of *E. lignosellus* on soybean (M. Linker, University of Florida, personal communication).

Figure 16-8. Lodging of soybean resulting from *Elasmopalpus* larval injury to stems. Cruz Alta, Brazil. (Courtesy of M. E. Irwin).

B. Insects Injurious to Soybean Roots

1. Grape Colaspis

Adult grape colaspis, *Colaspis brunnea* (F.) (Coleoptera: Chrysomelidae), feed on foliage of legumes, corn, grape, strawberries, and various vegetables while the larvae attack mainly the roots of rice, corn, other grasses, clovers, and soybean (Davidson and Peairs 1966). Damage to soybean roots has been reported primarily from north central and midsouthern regions of the United States (Packard 1951, Anonymous 1960, Turnipseed 1973, Deitz et al. 1976). Adults of *C. louisianae* Blake and *C. crinicornis chittendeni* Blake have been collected on soybean in Louisiana (Blake 1974, Chapin 1979).

Grape colaspis overwinter as larvae in cells in the soil. When soil temperatures increase in the spring, the larvae move upward and feed on small roots and the exterior portion of the main root and lower stem (Fig. 16-9). The mature larva is about 7 mm long, pale gray, with a brown head capsule and cervical area (Fig. 16-10B). Pupation occurs in the soil, and adults emerge in early summer. The adult is 5 mm long, light tan to yellowish-brown with elytra containing longitudinal rows of tiny punctures (Fig. 16-10A). There is one generation per year with a partial second generation occurring in the southern states (Deitz et al. 1976).

Damage is most likely to occur in soybean grown for two-to-three successive years in poorly drained or organic soils or when soybean follows clover, lespedeza, or a heavy stand of smartweed (Anonymous 1960, Petty 1966, Deitz et al. 1976). In areas of the field infested with larvae, plants usually display yellowed foliage and are often stunted. Such areas are usually limited in size, however.

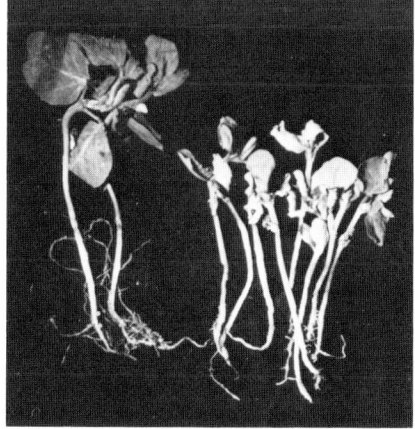

Figure 16-9. Normal soybean seedlings (left) and seedlings showing root injury from *Colaspis* larvae (right). (University of Illinois Cooperative Extension Service).

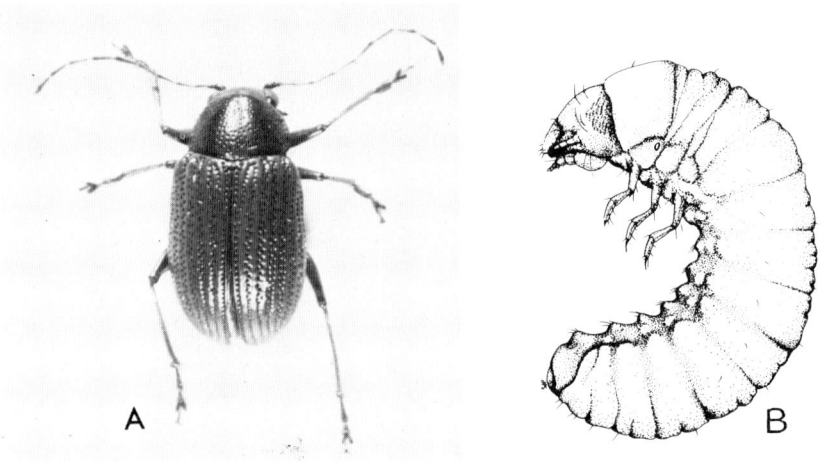

Figure 16-10. Grape colaspis, *Colaspis brunnea*: (A) adult (Illinois Natural History Survey); (B) larva. (Drawing by J. P. Sherrod).

2. White Grubs

The term "white grub" is most frequently applied to larvae of *Phyllophaga* spp. (Coleoptera: Scarabaeidae). These insects attack a wide variety of grasses and grain crops as well as potatoes, strawberries, roses, and nursery stock (Metcalf et al. 1962). Damage to soybean roots has been reported primarily in the midwestern United States (Bigger 1953, Anonymous 1960, Petty 1966, Turnipseed 1973), although S. S. Nilakhe (Louisiana State University, personal communication) collected fairly large numbers of grubs of *P. ephilidae* (S.) from soil samples in soybean fields in southern Louisiana in 1978. Gangrade (1974) described damage to soybean roots by larvae of *P. serrata* (F.) in India. Species of *Holotrichia*, *Anomala*, and *Maladera* are reported to occur in soybean fields in People's Republic of China (Chiang 1977), and unidentified species (probably *Anomala*) have been found in Korea (Kogan 1977). A world bibliography of the genus *Phyllophaga* is available (Pike et al. 1976).

Adults are about 2.5 cm long with dark brown or black bodies and long spiny legs (Fig. 16-11A). Eggs are deposited early in the summer primarily in grassy soil. Mature larvae are 1.3 to 2.5 cm long, often C-shaped, white to creamy or grayish white, with the abdominal tip transparent and with a distinct setal pattern on the underside of the last segment (Fig. 16-11B). Development into adults may take one to four years, during which time the larvae feed on upper roots during spring and summer and then overwinter deep in the soil. Extensive pruning of soybean roots by the larvae results in stunting or even plant mortality (Fig. 16-12), while plants less severely injured may lodge easily because of a

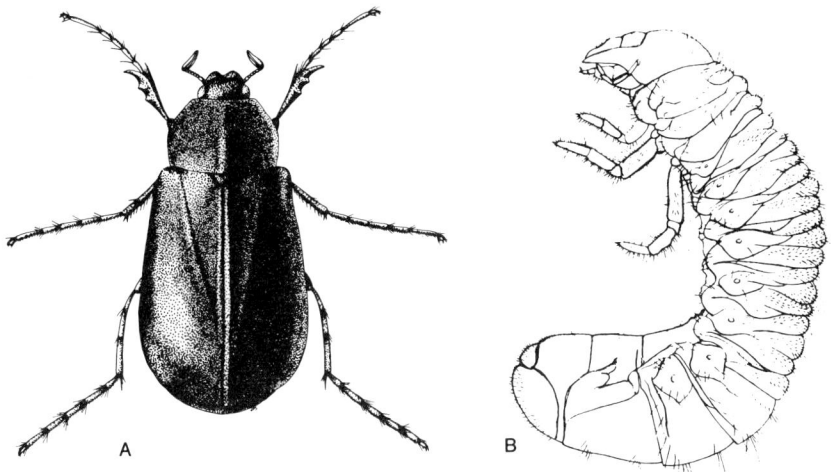

Figure 16-11. White grubs: (A) representative adult (after Forbes 1900, redrawn by J. P. Sherrod); (B) representative larva. (Illinois Natural History Survey).

weakened root system (Metcalf et al. 1962, Anonymous 1968). Damage is usually patchy and is most severe in fields having a two-year corn/two-year soybean rotation (Petty 1966).

C. Insects Attacking Nodules of Soybean

1. Bean Leaf Beetle

The bean leaf beetle, *Cerotoma trifurcata* (Forster) (Coleoptera: Chrysomelidae), is the best known of the *Cerotoma* species which, as adults, defoliate soybean and other legumes. Larvae of *Cerotoma* feed primarily on legume nodules. A thorough discussion of the bean leaf beetle, including the nature of larval damage to soybean and procedures for sampling eggs and larvae in soil, is included in Chapter 10.

2. Soybean Nodule Fly

Adults of the soybean nodule fly, *Rivellia quadrifasciata* (Macquart) (Diptera: Platystomatidae), are among the most common Diptera collected from soybean fields in the United States, but larval damage to soybean nodules was not determined until 1975 (Eastman and Wuensche 1977). This species occurs in southern Canada and in the United States from Florida to New York and west to Montana, Colorado, and Texas (Namba 1956). Larvae of *R. apicalis* Hendel and *R. basilaris*

Figure 16-12. Destruction of soybean roots by white grubs. (University of Illinois Cooperative Extension Service).

Wiedemann have been reported to produce similar injury to nodules of soybean in Japan (Koizumi 1957) and India (Bhattacharjee 1977). Adults of *R. boscii* Robineau Desvoidy have been collected from soybean fields in Missouri (Blickenstaff and Huggans 1974).

Details of the life cycle of *R. quadrifasciata* have not yet been determined. Eggs are similar to rice grains in shape, 1 mm long, and a chalky to creamy white. The full-grown larva (Fig. 16-13A) is about 8 mm long, creamy to tannish white with the posterior spiracles located on russet-colored, sclerotized stigmatic plates containing two thorn-like protuberances. Pupation occurs in russet to reddish-yellow puparia in the soil (Eastman and Wuensche 1977). The adult (Fig. 16-13B) is about 5 mm long with a russet head, black thorax, and reddish-yellow abdomen. The wings are marked with four black bands; the coxae and femora are yellow (Namba 1956).

Sampling Phytophagous Underground Soybean Arthropods

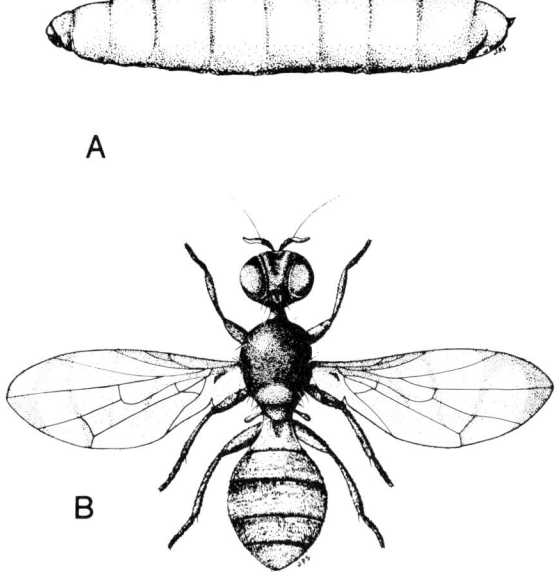

Figure 16-13. Soybean nodule fly, *Rivellia quadrifasciata*: (A) larva; (B) adult. (Drawings by J. P. Sherrod).

Larvae of the soybean nodule fly, like those of the bean leaf beetle, damage soybean nodules by boring a hole into the nodule and consuming the contents (see Fig. 10-3, Chapter 10). While the effects of such injury are still being investigated, it is known that the larvae can destroy a high percentage of a plant's nodules; the nodule damage in turn can reduce nitrogen fixation, plant growth, and seed yield (Newsom et al. 1978). From preliminary observations it appears that early-planted soybean attract and support the largest populations of *R. quadrifasciata* (Eastman and Wuensche 1977). Plants growing in sandy or sandy-loam soils also seem to attract more oviposition, and larval survival is greater in these soils (Eaton 1978).

III. Generalized Techniques for Sampling Soil Arthropods

Several reviews or books of research papers dealing with collection and extraction methods for obtaining soil arthropods are available. Among the most helpful are Murphy (1962a), Edwards and Fletcher (1971), and Southwood (1978). Methods for collecting arthropods from soil samples will be reviewed in the following pages. Pitfall traps, which are often used to catch ground beetles and various larvae crawling on the soil surface, are described in Chapter 26.

Figure 16-14. Collection of soil samples with a commercial golf cup-cutter: (A) The core cylinder is centered over a seedling or seedling cluster; (B) the cylinder is pushed into the soil through twisting motions of the handle; (C) the core cylinder is placed inside a plastic collection bag; (D) the sample is pushed from the cylinder into the bag through upward movement of the lever. (Courtesy C. Helm and M. Kogan).

A. Sample Collection

Soil arthropod samples are collected by removing measured areas of soil with a spade or with a specially designed core sampler. The commercially-available golf-hole borer shown in use in Fig. 16-14A-D and the specially constructed models shown in Chapter 10 are typical of the basic design.

Size and depth of individual soil samples are important considerations and are species and population dependent. Of the insects discussed here, the bean leaf beetle is the only one for which a sampling procedure in soybean has been systematically developed based on distribution of eggs and larvae. Requirements for individual soil sample units for other insects in soybean will necessitate addditional research.

A sequential sampling plan has been developed for sampling eggs of the bean leaf beetle in soybean fields. No programs are available for the other underground and ground-level arthropods in soybean. Plans are available, however, for the sequential sampling of wireworms on potato (Onsager 1974, Onsager et al. 1975) and for white grubs in cut-over land (Ives and Warren 1965) and grain sorghum (Teetes and Sterling 1976).

B. Extraction of Arthropods from Soil Samples

1. Active Methods

Active methods (also called behavioral or dynamic methods) involve forcing or luring living arthropods out of soil or litter samples. Usually the organisms are driven out by generating temperature and moisture gradients within the sample. In a modification of Berlese's funnel extraction technique, Tullgren (1918) placed soil or litter on a wire mesh inside a funnel and subjected the sample to heat from a light bulb suspended over it. This technique has provided the framework for most present-day dry funnel extraction procedures. Indeed, 74% of the soil zoologists and ecologists questioned by Edwards and Fletcher (1971) used some modification of the Tullgren funnel apparatus. Arthropods exiting the sample are collected below the funnels in vials containing ethyl alcohol, picric acid, water, or other fluid. The extraction procedure utilizing this method requires several days. Tullgren-type funnels are shown in use in Fig. 16-15.

Chemicals such as formalin, dimethyl phthalate, and 2-cyclohexyl-4,6-dinitrophenol have also been used, with varying success, to drive arthropods from soil. Complete descriptions of several active extraction methods may be found in Murphy (1962a) and Southwood (1978).

Several factors influence extraction efficiency. A small space should separate the sample from the funnel sides so that the arthropods will not be trapped in water condensing on the sides (Haarlov 1947). The funnels should have steep sloping sides, and every effort should be made to gradually increase the temperature-humidity gradient in the sample during the extraction process

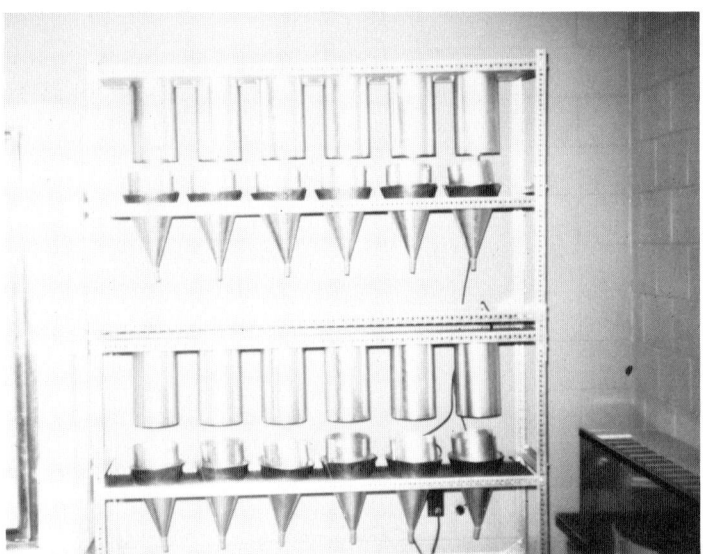

Figure 16-15. Series arrangement of Rothamsted-type Tullgren funnels. With each unit the upper cylinder houses the heating element, a 60-watt light bulb, which is controlled by a variable rheostat. The sample is supported in the middle cylinder by a fine-mesh screen. Arthropods are forced out of the sample and directed by the funnel into an attached collection vial. (Courtesy L. P. Pedigo, Iowa State University).

(Murphy 1962b). Leaving the sample intact reduces the chance of arthropod mortality from too rapid drying of the soil. Inverting the sample in the funnel shortens the distance the organisms have to travel to leave the sample (Edwards and Fletcher (1971). Flotation of the vial contents in saturated cold sugar solutions (Loureiro 1976) or other saturated solutions may be helpful in separating arthropods from soil falling into the vial during the extraction process.

The air-conditioned funnel extractor developed by Macfadyen (1962) may be one of the best of the active or behavioral methods. The apparatus consists of a tightly enclosed box containing a battery of funnels, a water bath and cooling coil, and an attached fan to circulate the cool moist air around the lower part of the samples and collecting funnels. A separate heater and chimney assembly fits onto the top of the box. The contrast of hot, dry air toward the top of the samples and cool, moist air near the bottom creates a steep temperature-moisture gradient within the samples, aiding in expulsion of organisms from the soil into the funnels and collecting vials. In their comparison of extraction methods for terrestrial arthropods, Edwards and Fletcher (1971) found that this air-conditioned apparatus was the most efficient general method—active or passive—for almost all groups of soil arthropods, even the larger insect larvae.

2. Passive Methods

Passive (or mechanical) methods physically separate the arthropods from soil or litter. Murphy (1962a) and Southwood (1978) contain thorough treatments of the basic methods. Of these, handsorting is simplest but also the most inefficient, inaccurate, and laborious. Most passive methods in use today involve washing the soil sample through a series of sieves to remove stones and larger plant matter and then floating the residual material in concentrated solutions of brine, sugar, or magnesium sulphate with a specific gravity of about 1.2. This causes most of the arthropods to float to the surface because of their lower specific gravity. Bubbling air through the solution during the flotation stage helps to free organisms that might be entrapped among sinking soil particles (Ladell 1936). Organic debris also floats in these solutions, and Salt and Hollick (1944) added a final stage to the procedure, namely the shaking of the floating material in a mixture of xylene or benzene and water. Because of the hydrophobic nature of the cuticle, arthropods float on the surface in the organic solvent layer and can be removed, while most of the plant debris remains in the aqueous layer underneath. Murphy (1962c), however, cautioned that insect eggs and dipterous larvae are not separated adequately from plant material by the use of these solvents.

Soil type can greatly affect extraction efficiency. Freezing the samples and/or soaking them in solutions of Calgon or other dispersing agents may be necessary to break up heavy clay soils. Mechanical extraction of arthropods from soil with high organic matter content is often difficult. Reduction of air pressure in the flask containing the arthropod-plant mixture prior to introduction of solvent was recommended by Salt and Hollick (1944) to break up adhering air bubbles that might keep the plant matter afloat. Placing the collecting sieve in boiling water for 30 min might help to sink plant material (Cockbill et al. 1945). In the extraction of arthropods from peat, Hale (1964) found that this material would sink in a solution of magnesium sulphate after the sample had been boiled under reduced pressure for 30 sec. Although interested primarily in Collembola, Hale was able with this technique to extract over 95% of the coleopterous and dipterous larvae present in the peat samples.

Sieve/flotation procedures for extraction of eggs, larvae, and pupae of the bean leaf beetle as described in Chapter 10 are probably adequate for most of the other root- and nodule-feeding species unless the soils have a high organic matter content. With adjustment of water pressure and substitution or addition of larger mesh screens in the sieve and cylinder, the rotary screen extractor (Shaw et al. 1976) described in Chapter 10 could also be adapted for recovery of insect larvae and pupae, thereby permitting the processing of greater numbers of soil samples in less time. Also promising is the washing-flotation method of Montgomery et al. (1979) for recovery of eggs and larvae. The technique involves the piping of water up into a wash tank containing the soil sample and then the overflow of water and floating material into a series of graduated sieves. Debris collecting in the bottom sieve is floated in water and then magnesium sulphate solution for recovery of eggs. Samples ranging in size from < 1 to 5 liters can be processed depending on insect species desired and on soil type. Recovery of first-instar larvae is inefficient with any flotation technique and will be the least accurate.

3. Comparison of Active and Passive Methods

No single method removes all the arthropods or groups of arthropods from soil or litter samples. Some generalizations can be made, however. While a survey of soil zoologists and ecologists revealed that a great majority prefer to use an active extraction method such as modified Tullgren funnels, passive techniques modified from the Salt and Hollick procedure are the methods most commonly cited in literature dealing with recovery of eggs and larvae of arthropods of "economic importance." Flotation usually is more efficient for the larger insects (such as coleopterous and dipterous larvae), but active methods such as the Macfadyen air-conditioned funnels are also quite good. Although they are for the most part inadequate in soils with high organic matter content, passive extraction methods are suitable for processing samples from arable land, which usually contains relatively little organic matter compared to woodland or pasture soils (Edwards and Fletcher 1971).

Convenience and ease of extraction must also be considered. Passive methods demand considerably more "operator time" per sample than do the active methods, which are essentially automatic. Because the arthropods must remain alive and active for successful funnel extraction, soil samples can be refrigerated safely only for about a week without significant increase or decrease in arthropod numbers. Passive methods do not have this restriction, and samples can be stored frozen for several months without decrease in extraction efficiency. Passive methods are also the only means of recovering eggs and pupae. The type of extraction method and its degree of complexity are determined by the organism(s) to be sampled and the type of soil; the number and depth of samples in most instances must be adjusted to fit a single arthropod species or a relatively small group of species of similar behavior and structure.

IV. Sampling for Soil Insects in Soybean

A. Insects Attacking Germinating Seed and Seedlings

Seedcorn maggots and cutworm species have been associated exclusively with early-season damage. Wireworms, although serious full-season pests of other crops, have been recorded primarily as seed and seedling pests in soybean. The lesser cornstalk borer also seems to be most damaging to seedlings. Sampling for these insects in soybean has been done largely by determining the number of infested seed, seedlings, or cut seedlings per unit of row (Table 16-1). For example, *Elasmopalpus* larvae have been sampled in Florida by the surveyor examining the stem base area of 20 plants in a designated length of row in several areas of a field; insecticide applications are made only when plant losses greater than 10 to 15% are anticipated (M. Linker, University of Florida, personal communication). With the seedcorn maggot, however, once the effect on stand count has been observed, the damage has already been done. Conventional sampling for maggots at this point can do little but determine the extent of damage within the field. At present there are no post-emergence insecticide treatments available to control seedcorn maggots, and replanting of portions of the field may be necessary. Damage by maggots rarely occurs to seed in replantings or on older plants (Retan and Fisher 1978).

Cutworms individually can damage several seedlings, and timely sampling may prevent further injury. At present the best scouting method, once injury has been observed, is probably the determination of infested cut plants per unit of row at several locations throughout the field. Insecticide baits are available commercially for early-season control of black cutworm larvae.

Preseason use of baits (Table 16-1) in low-lying fields or in areas with a past history of seedcorn maggot, cutworm, or wireworm problems may be a useful forecasting device, as seed and at-planting pesticide treatments for control of these insects are available. Efficiency of baits for sampling black cutworms in very weedy areas, however, may be decreased because of the greater attractiveness of the weeds (J. Shaw, Illinois Natural History Survey, personal communication). Pitfall traps may also be used for early-season sampling of cutworms.

Because the larval development of wireworms may require several years, soil sampling and extraction methods for wireworms as presented in Table 16-1 and discussed in Sections III B2 and IV B of this chapter may be of benefit for more accurate determination of larval populations. Centrifugation/flotation techniques used by Doane (1969, 1977) to extract wireworm eggs from soil in wheat fields may be helpful. Further work on incidence and abundance of seedcorn maggots, cutworms, wireworms, and lesser constalk borers in soybean must be done before definitive sampling procedures for these insects can be developed.

B. Insects Attacking Roots and Nodules

Of the insects known to damage soybean roots and nodules, published sampling plans are available only for determining incidence of bean leaf beetle on soybean and scarabaeid grubs in cut-over land, permanent meadow, and grain sorghum (Table 16-1). Size and depth of individual soil samples from soybean fields have not been determined for immatures of grape colaspis, white grubs, or soybean nodule flies.

Sifting of dry soil samples through screens has sometimes been used as the sole procedure for extracting larvae, primarily for collection of later instars of the larger, less fragile insects such as wireworms and white grubs. Flotation in brine has also been used as a single extraction method. Most procedures, however,

Table 16-1. Review of selected references on sampling for soil insect groups

Insect	Stage	Type field sampled	Sampling/extraction method	Selected references[a]
Seedcorn maggot	Larva	Soybean	Number of infested seeds per row unit	(4)
	Ovipositing adult	Preplant	Bait	(17), (22), (23), (34)
Black cutworm	Larva	Corn	Number of cut, infested plants per row unit	(24)
	Larva	Corn	Pitfall trap	(3)
	Larva	Corn	Bait	(3), (27)
Wireworm sp.	Larva	Soybean	Number of larvae, infested plants per row unit	(10)
	Larva	Corn	Bait	(32)
	Larva	Laboratory test	Bait	(2)
	Larva	Unspecified	Sieving/sifting soil	(14)
	Larva	Grass; arable land	Sieving/sifting soil	(33)
	Larva	Grass; arable land	Flotation	(5), (26)
	Larva	Potato	Sifting?; sampling	(19), (20), (21)
	Egg, larva	Wheat	Centrifugation/flotation	(6), (7)
Lesser cornstalk borer	Larva	Peanuts	Number of infested plants	(28)
Grape colaspis	Larva	Soybean	Sieving/flotation	(9)
	Larva, pupa	Soybean, lespedeza, rice	Hand sorting?	(25)
White grub sp.	Egg, larva, teneral adult	Pasture	Sieving/flotation	(15)
	Larva	Cut-over ground	Sifting; sampling plan	(13)
	All immatures	Permanent meadow	Sifting; sampling plan	(11), (12)
	Larva	Grain sorghum	Sampling plan	(29)

White grub	Larva	Soybean	Sifting of soil	(18)
Bean leaf beetle	Egg	Soybean	Sieving/flotation	(30)
	Larva, pupa	Soybean	Flotation	(1)
	Larva, pupa	Soybean	Sieving/flotation	(8)
Soybean nodule fly	Adults	Soybean	Emergence cone traps; pitfall traps	(16)

[a] References: (1) Anderson and Waldbauer 1977; (2) Apablaza et al. 1977; (3) Archer and Musick 1977; (4) Bhattacharya and Rathore 1977; (5) Cockbill et al. 1945; (6) Doane 1969; (7) Doane 1977; (8) Eastman and Wuensche 1977; (9) Eaton 1978; (10) Edwards and Hallman 1975; (11) Guppy and Harcourt 1970; (12) Guppy and Harcourt 1973; (13) Ives and Warren 1965; (14) Jones 1937; (15) Kain and Atkinson 1976; (16) R. Koethe, North Carolina State University, personal communication; (17) Miller and McClanahan 1960; (18) S. Nilakhe, Louisiana State University, personal communication; (19) Onsager 1969; (20) Onsager 1974; (21) Onsager et al. 1975; (22) Peterson 1924; (23) Reid 1940; (24) Rings and Musick 1976; (25) Rolston and Rouse 1965; (26) Salt and Hollick 1944; (27) Sechriest and Sherrod 1977; (28) Smith et al. 1975; (29) Teetes and Sterling 1976; (30) Waldbauer and Kogan 1973; (31) Waldbauer and Kogan 1975; (32) Ward and Keaster 1977; (33) Yates and Finney 1942; (34) Yu et al. 1975.

have combined a preliminary soil washing through sieves with some method of flotation. Wet sieving is also necessary if counts of total nodules per sample are to be made, as nodules are easily dislodged from the plant in the sampling and soil washing process.

For the root and nodule-feeding insects listed in this chapter, one of the sieve/flotation techniques rather than dry funnel extraction is probably the method of choice. These techniques are efficient enough to extract the larger insects and permit the storage and processing of a larger number of soil samples than can be processed adequately by funnel techniques. Sieve/flotation methods are less efficient for processing samples of highly organic soils, although boiling the sample under reduced pressure prior to flotation may be helpful. Active methods such as the air-conditioned funnels might be considered if flotation techniques cannot be employed.

V. Concluding Remarks

There has been very little research done to define the critical root and nodule mass needed for soybean growth and yield or to determine the relationship between insect injury to soybean roots or nodules and reduction in yield. Information is available on the ability of the soybean plant to compensate for erratic germination (Pendleton and Hartwig 1973). These findings can be used to project the degree of seed and seedling destruction a field can tolerate without an effect on expected yield. For example, preliminary findings of M. Kogan, C. Helm, and S. Han (unpublished) indicate that 30, 45, and 60 cm gaps in seedling stands made to simulate damage by a moderate cutworm infestation have no effect on plot yield; these data are in agreement with those in Pendleton and Hartwig's report (1973). However, literature on effects of low level-to-severe injury to roots and nodules is scant with the exception of preliminary work by Newsom et al. (1978) and Newsom (unpublished) on the consequences of insect injury to nodules, stems, and leaves as affecting nitrogen fixation.

Moderate infestations of ground-level and underground arthropods usually produce cryptic symptoms of damage or even no observable signs at all. Unlike injury by defoliators and other above-ground pests, damage by phytophagous soil-inhabiting populations often remains undetected by cursory observation unless unusually severe. It is, therefore, understandable that most entomological research in soybean has concentrated on foliage- and pod-feeding pests. The holistic pest management approach, however, implies a broad comprehension of all components of the arthropod community in soybean fields. Furthermore, injuries to soybean roots and nodules when combined with defoliation and pod or stem damage may produce reductions in yield that would have been below an economic threshold if taken singly. Expanded research with soil arthropods is needed as part of an overall study of pest complexes. The influence of conservation tillage practices and the effects of insecticides on soil arthropods are but two other areas needing additional work.

Adequate progress in research on the ecology and economics of soil-inhabiting phytophagous species will depend on the availability of reliable sampling procedures. Much can be extracted from the existing literature on sampling soil arthropods under natural or crop situations. Adaptation of these methods to the soybean fauna is an obvious first step in expanding research programs in soybean soil ecology.

Acknowledgments

Sincere appreciation is expressed to John Sherrod (Illinois Natural History Survey) for the drawings and to Bonnie Irwin (Illinois Natural History Survey) and Jenny Kogan (Soybean Insect Research Information Center) for collection of the literature.

References

Anderson, T. E., and G. P. Waldbauer. 1977. Development and field testing of a quantitative technique for extracting bean leaf beetle larvae and pupae from soil. Environ. Entomol. 6:633-636.
Anonymous. 1960. These are some pests that damage soybeans. Soybean Dig. 20:16-19.
Anonymous. 1968. White grubs. Ill. Coop. Ext. Serv. Entomol. Fact Sheet NHE-23:1-3.
Apablaza, J. U., A. J. Keaster, and R. H. Ward. 1977. Orientation of corn-infesting species of wireworms toward baits in the laboratory. Environ. Entomol. 6:715-718.
Archer, T. L., and G. J. Musick. 1977. Evaluation of sampling methods for black cutworm larvae in field corn. J. Econ. Entomol. 70:447-449.
Bhattacharjee, N. S. 1977. Preliminary studies on the effect of some soil insecticides on soybean nodulation. Pesticides 11:38.
Bhattacharya, A. K., and Y. S. Rathore. 1977. Survey and study of the bionomics of major soybean insects and their control. G. B. Pant University of Agriculture and Technology Res. Bull. 107:324 p.
Bigger, J. H. 1953. Biology and control of *Phyllophaga rugosa* on soybeans. Proc. N. Cent. Br. Entomol. Soc. Amer. 8:29-30.
Bigger, J. H., and H. B. Petty. 1965. Insect infestation of corn roots in Illinois. Ill. Agr. Exp. Sta. Bull. 704:1-8.
Blake, D. H. 1974. The costate species of *Colaspis* in the United States (Coleoptera: Chrysomelidae). Smithson. Contrib. Zool. 181:1-24.
Blickenstaff, C. C., and J. L. Huggans. 1974. Soybean insects and related arthropods in Missouri. Mo. Agr. Exp. Sta. Res. Bull. 803:51 p.
Busching, M. K., and F. T. Turpin. 1976. Oviposition preferences of black cutworm moths among various crop plants, weeds, and plant debris. J. Econ. Entomol. 69:587-590.

Busching, M. K., and F. T. Turpin. 1977. Survival and development of black cutworm (*Agrotis ipsilon*) larvae on various species of crop plants and weeds. Environ. Entomol. **6**:63-65.

Chaudhary, R. R. P., A. K. Bhattacharya, and R. R. S. Rathore. 1976. Field tests for the control of seed maggot, *Delia platura* Mg. attacking soybean. Sci. Cult. **42**:422-425.

Chapin, J. B. 1980. A review of the Louisiana species of *Colaspis* (Coleoptera: Chrysomelidae). Col. Bull. (in press).

Chiang, H. C. 1977. The potential of integrated pest control for grain legumes in the People's Republic of China. Seventh Session of the FAO Panel of Experts on Integrated Pest Control. pp. 1-4 mimeo.

Cockbill, G. F., V. E. Henderson, D. M. Ross, and J. H. Stapley. 1945. Wireworm populations in relation to crop production. 1. A large-scale flotation method for extracting wireworms from soil samples and results from a survey of 600 fields. Ann. Appl. Biol. **32**:148-163.

Correa, B. S. 1975. Levantamento dos lepidopteros pragas e danos. M. S. thesis, Universidade Federal do Parana, Curitiba, Brazil. 120 p.

Davidson, R. H., and L. M. Peairs. 1966. Insect pests of farm, garden, and orchard. 6th ed. John Wiley & Sons, Inc., N. Y. 675 p.

Deitz, L. L., J. W. Van Duyn, J. R. Bradley, Jr., R. L. Rabb, W. M. Brooks, and R. E. Stinner. 1976. A guide to the identification and biology of soybean arthropods in North Carolina. N. C. Agr. Exp. Sta. Tech. Bull. **238**:264 p.

Doane, J. F. 1969. A method for separating the eggs of the prairie grain wireworm, *Ctenicera destructor*, from soil. Can. Entomol. **101**:1002-1004.

Doane, J. F. 1977. Spatial pattern and density of *Ctenicera destructor* and *Hypolithus bicolor* (Coleoptera: Elateridae) in soil in spring wheat. Can. Entomol. **109**:807-822.

Eastman, C. E., and A. L. Wuensche. 1977. A new insect damaging nodules of soybean: *Rivellia quadrifasciata* (Macquart). J. Ga. Entomol. Soc. **12**:190-199.

Eaton, A. 1978. Studies on distribution patterns, ovipositional preference, and egg and larval survival of *Colaspis brunnea* (Fab.) in North Carolina coastal plain soybean fields. Ph.D. thesis, North Carolina State University. 86 p.

Edwards, C. A., and K. E. Fletcher. 1971. A comparison of extraction methods for terrestrial arthropods. pp. 150-185 in J. Phillipson, ed. Methods of study in quantitative soil ecology. IBP Handbook No. 18. Blackwell Sci. Pub., London. 297 p.

Edwards, C. R., and G. J. Hallman. 1975. Insecticides evaluated in field test against wireworms in soybeans. Indiana Agr. Exp. Sta. Bull. **106**:1-7.

Ferreira, E. 1970. Pragas da soja no Rio Grande do Sul. Simp. Brasileiro Soja, 1st. 11 p. mimeo.

Forbes, S. A. 1890. Notes on cutworms. Ill. State Entomol. Rep. **16**:93, plate IV Fig. 2.

Forbes, S. A. 1900. The economic entomology of the sugar beet. Ill. State Entomol. Rep. **21**:163-165, Fig. 97.

Gangrade, G. A. 1974. Insects of soybean. Jawaharlal Nehru Krishi Vishwa Vidyalaya. Jabalpur, M. P., India. 88 p.

Guagliumi, P. 1965. Contributo alla conoscenza dell'entomofauna nociva del Venezuela (Continuagione e fine). Rev. di Agr. Subtrop. e Trop. **59**(10-12): 447-472.

Gujrati, J. P., K. N. Kapoor, and G. A. Gangrade. 1971. Incidence and control of seed-corn maggot, *Hylemya cilicrura* (Rond.) on soybean. Indian J. Entomol. **33**:366-368.

Guppy, J. C., and D. G. Harcourt. 1970. Spatial pattern of the immature stages and teneral adults of *Phyllophaga* spp. (Coleoptera: Scarabaeidae) in a permanent meadow. Can. Entomol. **102**:1354-1359.

Guppy, J. C., and D. G. Harcourt. 1973. A sampling plan for studies on the population dynamics of white grubs, *Phyllophaga* spp. (Coleoptera: Scarabaeidae). Can. Entomol. **105**:479-483.

Haarlov, N. 1947. A new modification of the Tullgren apparatus. J. Anim. Ecol. **16**:115-121.

Hale, W. G. 1964. A flotation method for extracting Collembola from organic soils. J. Anim. Ecol. **33**:363-369.

Harris, C. R., H. J. Mazurek, and G. V. White. 1962. The life history of the black cutworm, *Agrotis ipsilon* (Hufnagel), under controlled conditions. Can. Entomol. **94**:1183-1187.

Howard, L. O. 1898. The principal insects affecting the tobacco plant. USDA Yearb. **1898**:141, Fig. 23.

Ibrahim, Y. B., and A. H. Hower, Jr. 1979. Oviposition preference of the seed-corn maggot for various developmental stages of soybeans. J. Econ. Entomol. **72**:64-66.

Ives, W. G. H., and G. L. Warren. 1965. Sequential sampling for white grubs. Can. Entomol. **97**:596-604.

Jones, E. W. 1937. Practical field methods of sampling soil for wireworms. J. Agr. Res. **54**:123-134.

Kain, W. M., and D. S. Atkinson. 1976. Population studies of *Costelytra zealandica* (White). II. A rapid mechanical extraction method suitable for intensive sampling of *Costelytra zealandica* and other scarabaeids. N. Z. J. Exp. Agr. **4**:391-397.

Kogan, M. 1977. Soybean entomology in Korea. Consultant report to the Crop Improvement Research Center and the Institute of Agricultural Sciences, Office of Rural Development, Suweon, Korea. September 1977, College of Agr., Univ. of Ill., Urbana. 25 p.

Koizumi, K. 1957. Notes on some dipterous pests of economic plants in Japan. Botyu Kagaku **22**:223-227.

Kuwayama, S., K. Sakurai, and K. Endo. 1960. Soil insects in Hokkaido, Japan, with special reference to the effects of some chlorinated hydrocarbons. J. Econ. Entomol. **53**:1015-1018.

Ladell, W. R. S. 1936. A new apparatus for separating insects and other arthropods from the soil. Ann. Appl. Biol. **23**:862-879.

Lueck, D. B. 1966. Biology of the lesser cornstalk borer in south Georgia. J. Econ. Entomol. **59**:797-801.

Loureiro, M. C. 1976. Synecology of edaphic arthropoda in Iowa agro-ecosystems. Ph.D. thesis, Iowa State University. 130 p.

Luginbill, P., and G. G. Ainslie. 1917. The lesser corn stalk-borer. USDA Bull. **539**:1-27, Figs. 3,6.

Macfadyen, A. 1962. Control of humidity in three funnel-type extractors for soil arthropods. pp. 158-168 in P. W. Murphy, ed. Progress in soil zoology. Butterworth's, London. 398 p.

Metcalf, C. L., W. P. Flint, and R. L. Metcalf. 1962. Destructive and useful insects. McGraw-Hill Book Co., N. Y. 1087 p.

Miller, L. A., and R. J. McClanahan. 1960. Life-history of the seed-corn maggot, *Hylemya cilicrura* (Rond.) and *H. liturata* (Mg.) in southwestern Ontario. Can. Entomol. 92:210-221.

Montgomery, M. E., G. J. Musick, J. P. Polivka, and D. G. Nielsen. 1979. Modifiable washing-flotation method for separation of insect eggs and larvae from soil. J. Econ. Entomol. 72:67-69.

Motsinger, R. E., J. L. Bagent, S. D. Hensley, N. L. Horn, and L. D. Newsom. 1967. Soybean diseases and insects of Louisiana. La. Coop. Ext. Serv. Publ. 1558:23 p.

Murphy, P. W. 1962a. Progress in soil zoology. Butterworth's, London. 398 p.

Murphy, P. W. 1962b. Extraction methods for soil animals. I. Dynamic methods with particular reference to funnel processes. pp. 75-114 in P. W. Murphy, ed. Progress in soil zoology. Butterworth's, London. 398 p.

Murphy, P. W. 1962c. Extraction methods for soil animals. II. Mechanical methods. pp. 115-155 in P. W. Murphy, ed. Progress in soil zoology. Butterworth's, London. 398 p.

Namba, R. 1956. A revision of the flies of the genus *Rivellia* (Otitidae, Diptera) of America north of Mexico. Proc. U.S. Nat. Mus. 106:21-84.

Newsom, L. D., E. P. Dunigan, C. E. Eastman, R. L. Hutchinson, and R. M. McPherson. 1978. Insect injury reduces nitrogen fixation in soybeans. La. Agr. 21:15-16.

Onsager, J. A. 1969. Sampling to detect economic infestations of *Limonius* spp. J. Econ. Entomol. 62:183-189.

Onsager, J. A. 1974. A sequential sampling plan for classifying infestations of southern potato wireworm. Amer. Potato J. 51:313-317.

Onsager, J. A., B. J. Landis, and L. Fox. 1975. Efficacy of fonofos band treatments and a sampling plan for estimating wireworm populations on potatoes. J. Econ. Entomol. 68:199-202.

Packard, C. M. 1951. Insect pests of soybean and their control. Soybean Dig. 11:14-18.

Pendleton, J. W., and E. E. Hartwig. 1973. Management. pp. 211-237 in B. E. Caldwell, ed. Soybeans: Improvement, production, and uses. Amer. Soc. Agron. Pub., Madison, Wisconsin. Agron. Ser. 16. 681 p.

Peterson, A. 1924. Some chemicals attractive to adults of the onion maggot, (*Hylemyia antiqua* Meig.) and the seed corn maggot (*Hylemyia cilicrura* Rond.). J. Econ. Entomol. 17:87-94.

Petty, H. B. 1966. Soybean insect problems in Illinois. Proc. N. Cent. Br. Entomol. Soc. Amer. 21:52-54.

Pike, K. S., R. L. Rivers, C. Y. Oseto, and Z. B. Mayo. 1976. A world bibliography of the genus *Phyllophaga*. Nebr. Agr. Exp. Sta. Misc. Publ. 31:21 p.

Reid, W. J., Jr. 1940. Biology of the seed-corn maggot in the coastal plain of the South Atlantic States. USDA Tech. Bull. 723:1-41.

Retan, A. H., and J. Fisher. 1978. Seedcorn maggot. Wash. Coop. Ext. Serv. No. EM4292:1-2.

Rings, R. W. 1977. An illustrated field key to common cutworm, armyworm, and looper moths in the North Central States. Ohio Agr. Res. Dev. Cent. Res. Circ. 227:1-60.

Rings, R. W., and F. J. Arnold. 1976. Supplemental annotated bibliographies of the black cutworm, glassy cutworm, bronzed cutworm, bristly cutworm, and dingy cutworm. Supplement I. Ohio Agr. Res. Dev. Cent. Res. Circ. **212**:1-44.

Rings, R. W., F. J. Arnold, and B. A. Johnson. 1978. Supplemental annotated bibliographies of the black cutworm, glassy cutworm, bronzed cutworm, bristly cutworm, dingy cutworm, dark-sided cutworm, clay-backed cutworm, dusky cutworm, and variegated cutworm. Supplement II. Ohio Agr. Res. Dev. Cent. Res. Cir. **238**:1-59.

Rings, R. W., F. J. Arnold, A. J. Keaster, and G. J. Musick. 1974. A worldwide, annotated bibliography of the black cutworm *Agrotis ipsilon* (Hufnagel). Ohio Agr. Res. Dev. Cent. Res. Circ. **198**:1-106.

Rings, R. W., and G. J. Musick. 1976. A pictorial field key to the armyworms and cutworms attacking corn in the North Central States. Ohio Agr. Res. Dev. Cent. Res. Circ. **221**:1-36.

Rizzo, H. F. 1972. Enemigos animales del cultivo de la soja. Rev. Inst. Bolsa Cereales. **2851**:6 p.

Rolston, L. H., and P. Rouse. 1965. The biology and ecology of the grape colaspis, *Colaspis flavida* Say, in relation to rice production in the Arkansas Grand Prairie. Ark. Agr. Exp. Sta. Bull. **694**:1-31.

Salt, G., and F. S. J. Hollick. 1944. Studies of wireworm populations. 1. A census of wireworms in pasture. Ann. Appl. Biol. **31**:52-64.

Scott, W. O., and S. R. Aldrich. 1970. Modern soybean production. S & A Pub., Champaign, IL. 192 p.

Sechriest, R. E., and D. W. Sherrod. 1977. Pelleted bait for control of the black cutworm in corn. J. Econ. Entomol. **70**:699-700.

Shaw, J. T., R. O. Ellis, and W. H. Luckmann. 1976. Apparatus and procedure for extracting corn rootworm eggs from soil. Ill. Natur. Hist. Surv. Biol. Notes. **96**:1-4.

Smith, J. W., Jr., R. W. Jackson, R. L. Holloway, and C. E. Hoelscher. 1975. Evaluation of selected insecticides for control of the lesser cornstalk borer on Texas peanuts. Texas Agr. Exp. Sta. Progr. Rep. PR-**3303**:16 p.

Southwood, T. R. E. 1978. Ecological methods. Halsted Press, N. Y. 524 p.

Teetes, G. L., and W. L. Sterling. 1976. A sequential sampling plan for a white grub in grain sorghum. Southwest. Entomol. **1**:118-121.

Tullgren, A. 1918. Ein sehr einfacher Ausleseapparat für terricole Tierformen. Z. Angew. Entomol. **4**:149-150.

Turnipseed, S. G. 1973. Insects. pp. 545-527 in B. E. Caldwell, ed. Soybeans: Improvement, production, and uses. Amer. Soc. Agron. Pub., Madison, Wisconsin. Agron. ser. 16. 681 p.

Turnipseed, S. G., and M. Kogan. 1976. Soybean entomology. Annu. Rev. Entomol. **21**:247-282.

Van Dinther, J. B. M. 1956. Soybean insects. Entomol. Ber. Amsterdam **16**:104-109.

Waldbauer, G. P., and M. Kogan. 1973. Sampling for bean leaf beetle eggs: Extraction from the soil and location in relation to soybean plants. Environ. Entomol. **2**:441-446.

Waldbauer, G. P., and M. Kogan. 1975. Position of bean leaf beetle eggs in soil near soybeans determined by a refined sampling procedure. Environ. Entomol. 4:375-380.

Ward, R. H., and A. J. Keaster. 1977. Wireworm baiting: Use of solar energy to enhance early detection of *Melanotus depressus, M. verberans,* and *Aeolus mellilus* in Midwest cornfields. J. Econ. Entomol. 70:403-406.

Yates, F., and D. J. Finney. 1942. Statistical problems in field sampling for wireworms. Ann. Appl. Biol. 29:156-167.

Yu, C., D. R. Webb, R. J. Kuhr, and C. J. Eckenrode. 1975. Attraction and oviposition stimulation of seedcorn maggot adults to germinating seeds. Environ. Entomol. 4:545-548.

SECTION VI
Stem and Axil Feeders

Chapter 17
Sampling Coleopterous Stem Borers in Soybean[1]

William V. Campbell

I. Introduction

In many parts of the world where soybean is grown, a few species of Coleoptera have adapted to feeding, during the larval stage, within the stems of soybean plants. Most of these species are in the Cerambycidae, but at least one species, in southern Brazil, is in the Curculionidae. The following cerambycid species have been recorded in association with soybean: in the United States, *Hippopsis lemniscata* (F.) (Genung and Green 1965) and *Dectes texanus texanus* LeConte; in Peru, *Grammopsoides rufipes* Breuning (M. E. Irwin, personal communication); in India, *Oberea brevis* Swederus (Gangrade 1974); and in Australia, *Zygrita diva* Thomson (Jarvis and Smith 1946). The curculionid recorded as a stem borer in southern Brazil is *Sternechus subsignatus* Boheman (Corseuil et al. 1974).

Considerable research has been conducted only on *D. texanus* in the United States and *O. brevis* in India. Although the species differ in details of their biology and behavior, much of the sampling procedures are probably applicable to both. This chapter is based on research conducted on *D. texanus*, but references to some biological features of *O. brevis* are included based mainly on Bhattacharya and Rathore (1980).

---

[1] Paper No. 5901 of the North Carolina Agricultural Research Service, Raleigh, North Carolina 27650.

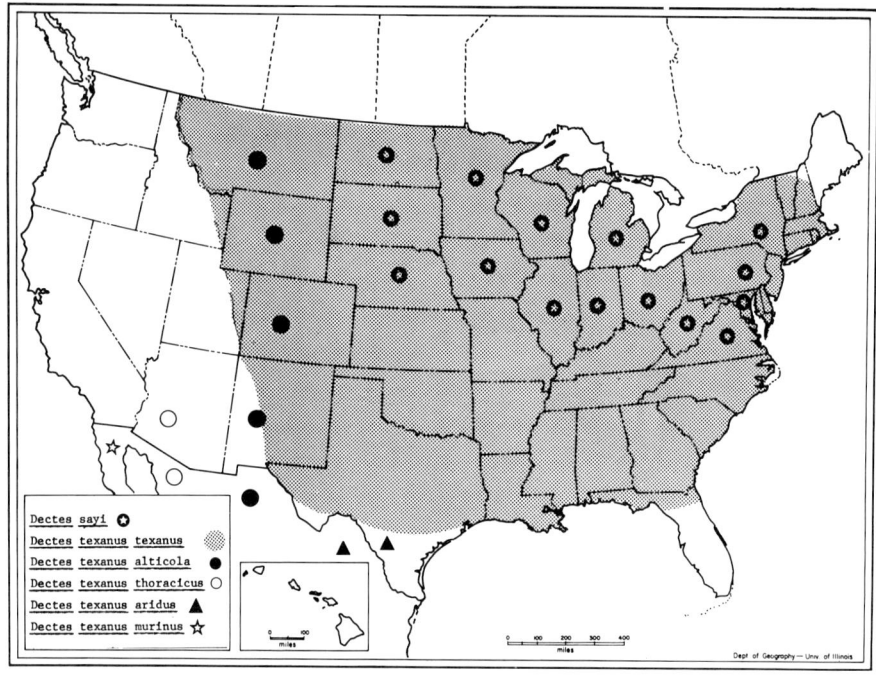

Figure 17-1. General distribution of *Dectes sayi* and *Dectes texanus* in the United States and Mexico.

A. Geographical Distribution

There are only two species in the genus *Dectes*: *D. sayi* Dillon and Dillon and *D. texanus* LeConte. *D. texanus* has several subspecies: *D. t. texanus* LeConte, *D. t. alticola* Casey, *D. t. murinus* Dillon, *D. t. thoracicus* Casey, and *D. t. aridus* Case (Dillon 1956). The distribution of *Dectes* is shown in Figure 17-1. *D. sayi* is limited to the area from New England south to West Virginia and west and north to North Dakota. *D. t. texanus* in its northern range overlaps *D. sayi*. *D. texanus* LeConte is a relatively new pest of soybean in Missouri, North Carolina, Arkansas, Louisiana, and Tennessee (Daugherty and Jackson 1969, Falter 1969, Patrick 1973, Hatchett et al. 1975, Campbell and Van Duyn 1977). *O. brevis* occurs throughout the Indian subcontinent being most serious on soybean in the Madhya Pradesh Province and in the hills of Utter Pradesh and Delhi (Bhattacharya and Rathore 1980).

B. Host Plants and Nature of Damage

D. texanus infests native wild host plants such as common ragweed (*Ambrosia artemisifolia* L.) and cocklebur (*Xanthium pennsylvanicum* Wallr.) in North

Sampling Coleopterous Stem Borers in Soybean 359

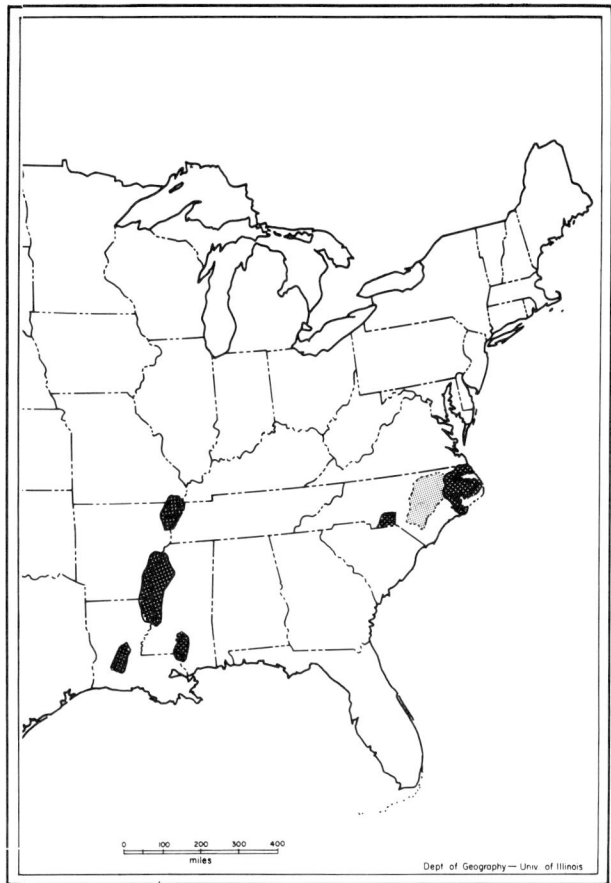

Figure 17-2. Recorded areas with damaging infestations of *Dectes* on soybean in the United States. Dark shaded areas —general infestation. Light shaded area—light infestations.

Carolina and giant ragweed (*Ambrosia trifida* L.) in Missouri (Daugherty and Jackson 1969). As more land was cleared for expanding soybean production and as the native weed hosts were reduced, the feeding behavior of certain *D. texanus* populations seems to have changed toward acceptance of soybean as a host.

Damaging infestations of *D. t. texanus* occur in North Carolina primarily in the northeastern coastal plains, and in restricted areas of Arkansas, Louisiana, Mississippi, and Missouri (Fig. 17-2). In some of these areas, large areas are planted as a soybean monoculture. In North Carolina more frequently a soybean-corn rotation is practiced. The area infested by *Dectes* is atypical of North Carolina farms in several aspects: soils have high organic matter, fields are large and farms contain hundreds of hectares of soybean.

Figure 17-3. Typical *Dectes* girdled soybean stems showing the smooth break at the point girdled.

D. texanus larvae tunnel in the petiole and then into the main stem. Once in the main stem the larvae may move up or down the stem. Mature larvae move down the stem, girdle its inside and pack it with frass. During high wind or heavy rain the stem will break off at the girdled point (Fig. 17-3). This results in a direct loss of soybeans. Fortunately not all the infested stems are girdled prior to harvest. The percentage of infested stems that are girdled averages less than 50% and ranges from 15% to 75%. There is, however, a 10% loss in seed weight for a soybean plant that is infested but not girdled (W. Campbell unpublished).

Adult damage to soybean is negligible and consists of light foliage feeding. Adults oviposit in petioles and may also feed on petiole tissues.

O. brevis has been recorded damaging up to 30% of the plants in Indian fields (Gangrade 1974) with a reduction of seed weight of the order of 50% (Gangrade 1976). *O. brevis* is polyphagous and also infests cowpea [*Vigna sinensis* (L.)], hyacinth bean (*Dolichos lablab* L.), chilies [*Capsicum frutescens* (L.)], and bitter-gourd [*Luffa cylindrica* (L.)] (Gangrade 1974).

Females of *O. brevis* after mating make two parallel girdles on the petiole or, less frequently, on the main stem or side branches. One plant may have up to four pairs of girdles. During initial infestations girdles are green but later turn brown. Leaflets above a girdled petiole slowly dry up starting from the margins. All leaves above a girdled branch or main stem also dry, and these plants with dried leaves are typical of *O. brevis* infestations. Splitting open an infested stem or petiole reveals a dark brown tunnel (Bhattacharya and Rathore 1980).

C. Life Cycle and Phenology

D. texanus is univoltine. Eggs are deposited primarily in the petiole from July to early September. Hatchett et al. (1975) found 83.6% of the eggs in the petiole, 8.6% in the secondary stems, and 7.8% in the primary stem. Larvae feed first in the petiole pith but soon enter the main stem. At plant maturity larvae girdle the stems about 3-6 cm above the ground (Fig. 17-4) and overwinter in the girdled stubble in the ground (Fig. 17-5).

Yellow, legless larvae, ca. 11 mm in length at maturity, commence pupation in the stubble in mid June, and the pupal stage lasts ca. two weeks. Pupae are approximately 9 mm in length and are present from mid June until late August.

The gray colored adult (Fig. 17-6) is approximately 6-10 mm long and 2.5 mm wide. Adult *D. texanus* are distinguished from D. sayi by the second antennal segment never more than one-eighth the length of the first (ca. one-fourth in *D. sayi*), and third segment shorter than first in both sexes (third segment longer than first in male, and slightly shorter in female *D. sayi*); front distinctly transverse and strongly convex; and tarsi and femora fuscous annulate apically, except in subspecies *alticola* (Dillon 1956).

Adults emerge by chewing a hole in the stem (Fig. 17-7) just above the soil line and below the frass plug that seals the girdled end of the stem. Emergence begins in late June and continues until September.

Figure 17-4. Stems are usually girdled 3-6 cm above the soil.

Detailed descriptions of the egg, larva, pupa, and adult stages, the biology, and the seasonal history of *Dectes texanus* are found in Hatchett et al. (1975).

The seasonal history of *Dectes texanus texanus* in North Carolina (Fig. 17-8) coincides with the published records for southern Missouri (Hatchett et al. 1975). The seasonal abundance for overwintered larvae, pupae, and adults in North Carolina is shown in Fig. 17-9. From observations, all larvae pupate prior to August 15, and development of all immature stages of the overwintered population is complete by August 25.

In an experiment conducted in North Carolina, soybean stubble containing larvae was collected in the fall and held in outdoor screen cages for determination of adult emergence. Twenty 20-stem samples were observed weekly for adult emergence (W. Campbell unpublished). The first adults were collected on July 13. Peak emergence was July 20 to August 3, and adult emergence was completed by August 10 (Fig. 17-10).

Sampling Coleopterous Stem Borers in Soybean

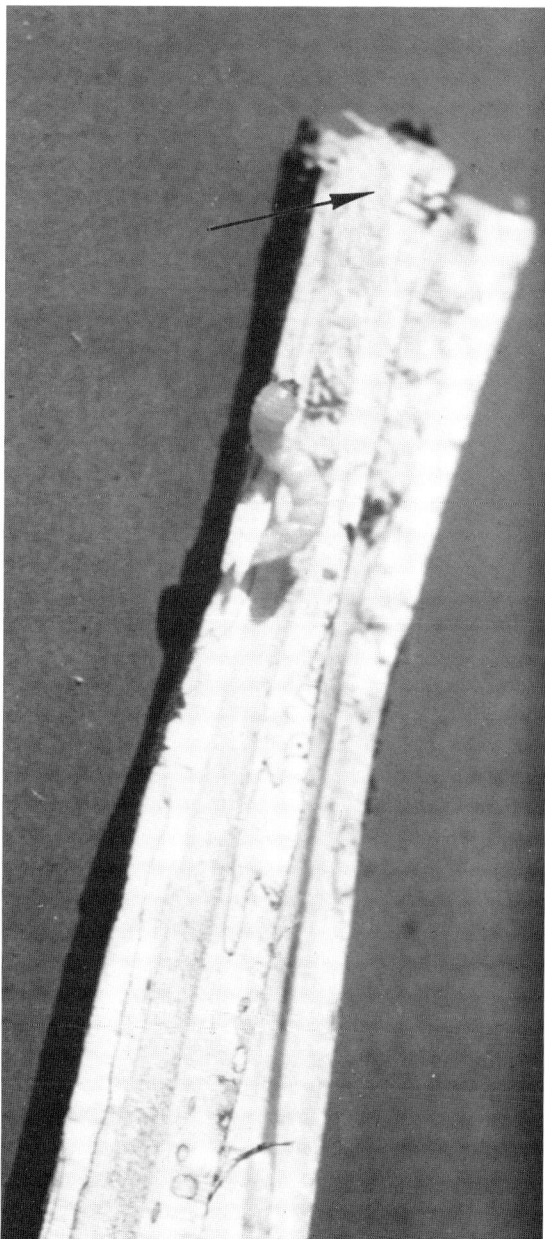

Figure 17-5. Stem split lengthwise to show overwintering larva in the girdled stubble. The frass plug (arrow pointing also at the girdled end) protects the overwintering larva.

Adult *O. brevis* are dark brown with the posterior half of the elytra black. Females lay 8-72 yellowish eggs, ca. 1.5 mm long, which hatch in 3-8 days. The newly emerged larvae are whitish, while the mature larvae are deep yellow and

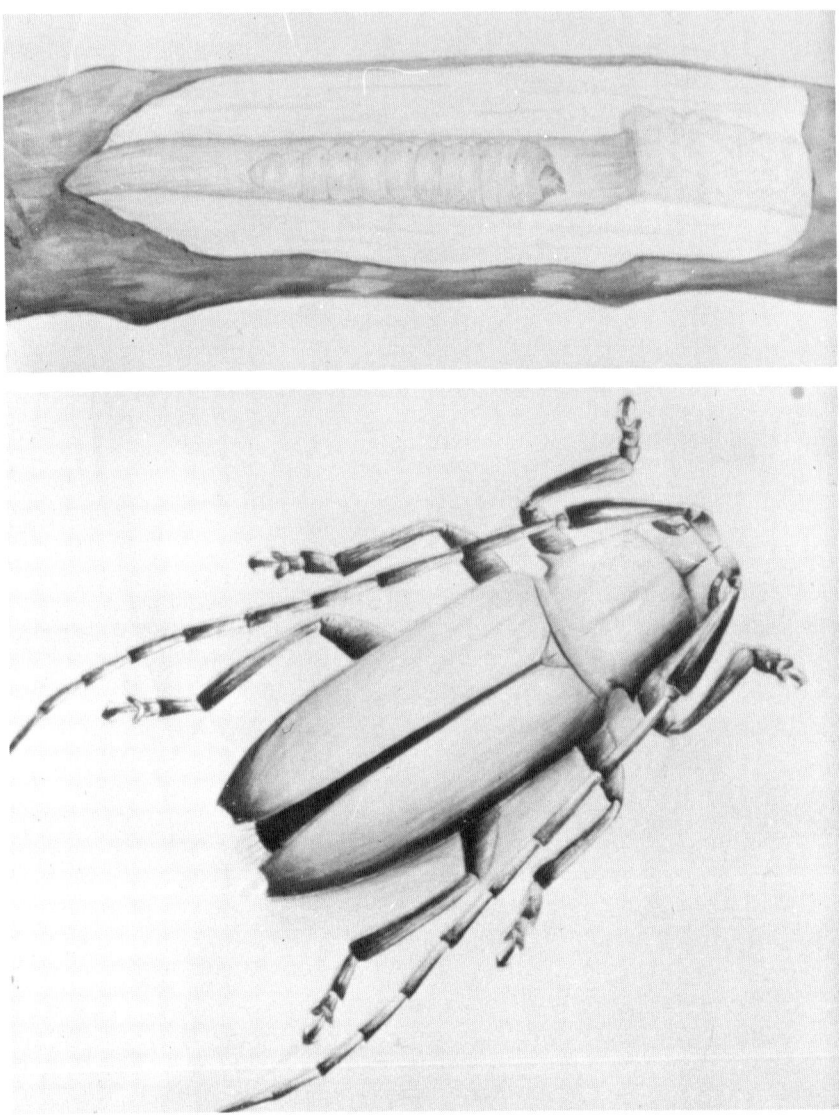

Figure 17-6. *Dectes texanus texanus* LeConte adult and larva.

about 1.8-2.0 cm long. The larvae feed inside the stems for 40-60 days during June and July. The total life cycle is completed in 40-70 days (Bhattacharya and Rathore 1980).

Sampling Coleopterous Stem Borers in Soybean 365

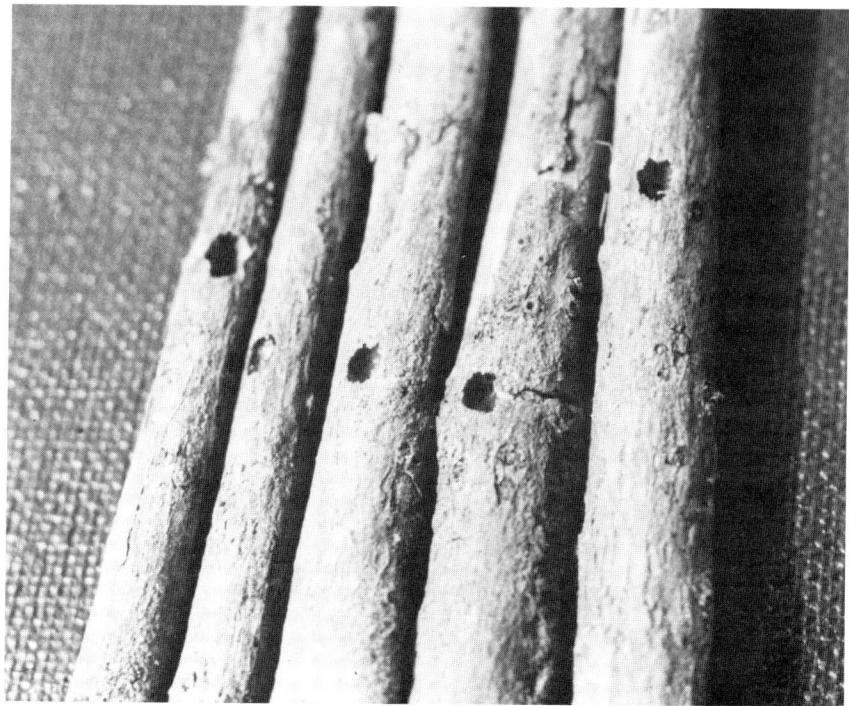

Figure 17-7. Emergence holes of adult *Dectes texanus* cut in the stubble below the frass plug but above the ground.

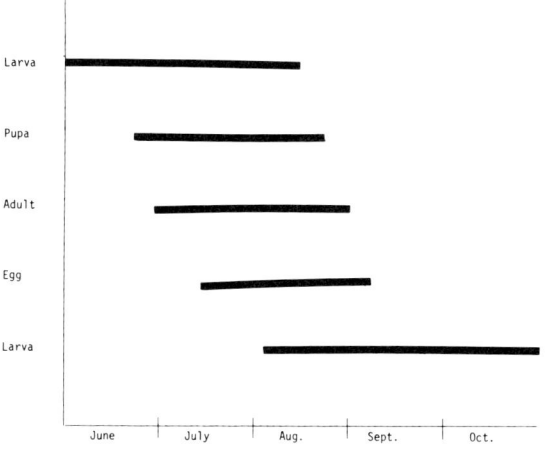

Figure 17-8. Seasonal history of *Dectes texanus texanus* in North Carolina.

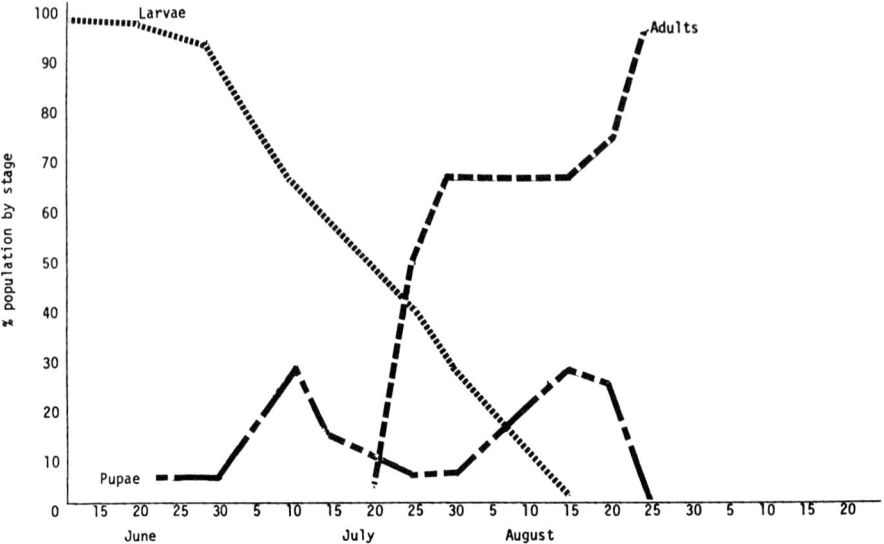

Figure 17-9. Seasonal abundance of *Dectes texanus texanus* overwintering larvae, pupae, and adults. North Carolina. 1977.

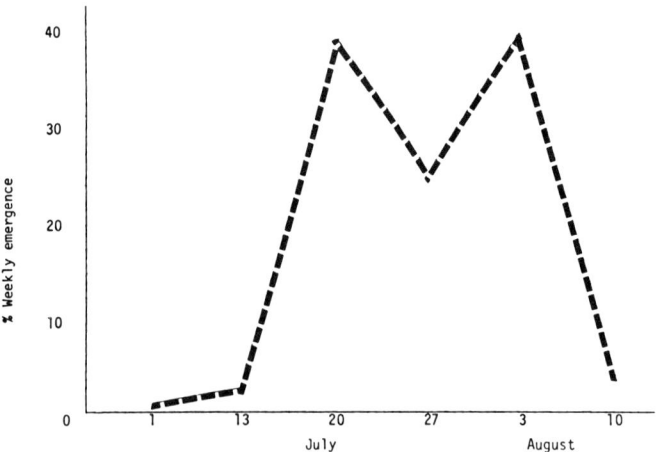

Figure 17-10. Weekly adult emergence of *Dectes texanus texanus* from soybean stubble. North Carolina. 1973.

II. Sampling for Adults

All sampling procedures discussed hereinafter are based on studies on *D. texanus* in North Carolina. Since *Dectes* develops on common ragweed as well as soybean, qualitative sampling to determine the presence of the stem borer may include a biased method. Even though *D. t. texanus* is a strong flier, adults are usually more abundant on the edges of the field adjacent to weeds or adjacent to last year's soybean crop. Adult beetles collected from ragweed numbered two to four times the number collected from adjacent soybean (W. Campbell unpublished).

Careful observation, especially of the upper canopy, will reveal the presence of stem borers in the field. *D. t. texanus* flies if disturbed; therefore, care must be exercised when using a direct observation method for sampling adults. If the beetle does not fly when disturbed, it will drop to the closest foliage. Starting from an upside down position they will land right side up on their feet. Direct observations should start from the top of the plant downward. The beetle prefers to rest on the upper one-third of the plant. Observations should start at the sides or the ends of the field next to weed hosts. Usually it is recommended to examine 1-m lengths of row for 20 sites at the ends, sides, and center of the field. The number of *Dectes* adults per hectare or acre may be calculated according to formulas 2-1 or 2-2 (see Chapter 2). If one is only interested in ascertaining the presence of the stem borer in the field, then surveying plants closest to alternate weed hosts will suffice.

The ground cloth method is not a satisfactory sampling technique because *D. texanus* is an active adult and readily flies or drops to lower foliage when disturbed.

Sweep net collections were made with a 38-cm diameter net by taking 10 sweeps across the top of two adjacent rows at each of 10 stops in the field. The sweep net sampling method was more efficient than the ground cloth; however, numbers of adults collected were low. Approximately 60% of all sweep net sample units contained adult *Dectes* (W. Campbell unpublished). No quantitative studies exist on minimum number of sweep net sample units necessary to achieve a desired level of reliability. Neither has the sweep net method been calibrated by comparison to an absolute sampling procedure.
parison to an absolute sampling procedure.

Sticky traps may be used to determine the presence of adults and their seasonal abundance. Traps used in North Carolina were built of a 30 cm square wood frame with a 20 x 20 mesh screen stapled to the frame. The screen was coated with Stickem. Traps were nailed to a wooden post secured in the ground. Traps were positioned at various heights and compass directions.

Data obtained from sticky traps (Table 17-1) suggest that this method of sampling is effective for determining the seasonal distribution of adults. Peak numbers were collected between August 3 and August 10.

The height of the sticky trap from the ground is more important than its compass direction orientation (W. Campbell unpublished). Traps placed 1.5 m above the ground were ineffective for sampling adults, whereas traps placed at 1 m

Table 17-1. Weekly abundance of adult *Dectes texanus texanus* collected by three sampling methods. Washington County, North Carolina. 1977

Data	Ground Cloth[a]	Sweep Net[b]	Sticky Trap[c]	Total
July 6	0	0	2	2
July 14	0	3	12	15
July 20	3	6	19	28
July 28	1	13	26	40
Aug. 3	1	5	79	85
Aug. 10	1	5	167	173
Aug. 17	0	0	43	43
Aug. 24	0	0	8	8
Aug. 31	0	0	1	1
Total	6	32	357	395

[a] Total adults collected on each date from a meter length of 10 paired rows (total 20 meters)
[b] Total adults collected on each date from 10 sweeps across the top of two adjacent rows in 10 locations (total 100 net sweeps)
[c] Total adults collected from four traps (30 cm^2) at 10 locations (total 40 traps).

above the ground collected 99% of the recorded adults (Table 17-2). This marked effect of trap height on sampling efficiency indicates the adults fly just at soybean canopy height. The general pattern of adults based on compass direction collection indicates that Dectes seems to be randomly distributed within a soybean field.

III. Sampling for Eggs

Eggs are laid in well developed petioles primarily in the upper half of the plant. The yellow eggs may be seen with the naked eye but some magnification is helpful. The egg contrasts slightly with the white pith in the petiole.

Petioles should be split lengthwise with a scalpel or razor blade. The thin chorion makes the eggs fragile, and care must be taken in splitting the petiole.

Table 17-2. Effect of height from ground and compass direction of sticky trap on the collection of adult *Dectes texanus texanus* in a soybean field. Washington County, North Carolina. 1977

Trap height	Number of adults collected/5 traps 30 cm^2				
	North	South	East	West	Total
1 m	63	84	104	103	354
1.5 m	3	0	0	0	3
Total	66	84	104	103	357

Sampling for eggs in North Carolina should start about the last week in July and continue until September. Sample units have been based on splitting four petioles per plant from the upper half of 25 plants at each site. To determine distribution and potential damage as well as presence, it would be desirable to sample the ends, sides, and middle of the fields.

IV. Sampling for Larvae

Dectes larvae consume the pith area of the petiole and then chew into the main stem. The infested soybean leaf wilts and later drops from the plant. A rust colored abscission scar with a nearly round-to-oblong hole in the center is indicative of *Dectes* infestation.

To use the leaf abscission scar as an index for stem borer populations, four rows are selected at random on each end, side and middle of the field and 25 plants are examined in each row (100 plants per 4 rows constitute one sample). More than one petiole per plant may be infested, but only one larva will survive to girdle the main stem and overwinter. A biased sample for presence of *Dectes* would involve selecting the first row of soybean adjacent to ragweed or other weed hosts. Experiments using the 25-plant sample showed that the first row (adjacent to ragweed) had 73% of the plants with *Dectes*-related abscission scars while rows 16 and 32 had 19% and 27% abscission scars, respectively (W. Campbell unpublished).

After the larva enters the stem it may chew its way through nodes up and down the stem to finally girdle the stem just before harvest. The simplest qualitative survey is to wait until near harvest. At that time fields are inspected for girdled plants—those that display the characteristic smooth break just above the soil line.

An adequate sample for determining percent girdled stems consists of counting the number of girdled stems in 25 consecutive stems in four locations at each site. Samples are taken by walking into the field several steps, marking off 25 stems, and recording the number girdled. The operation is repeated three more times for a total of 100 stems. Percent girdled stems may be determined for ends, sides, and center of the field.

The number of girdled stems may not be an accurate estimate of *Dectes* populations because of the variation in the percent infested stems that are girdled. A better estimate of larval populations may be obtained by splitting 25 consecutive stems and counting the number of *Dectes* larvae (W. Campbell unpublished). Four sets of 25 stems for a 100 total stem count constitute one sample. Within field difference in infestation may be determined by splitting 100 stems on the ends, sides, and middle of the field. Three sets of samples taken on the same end of a field showed 36, 42, and 44% of the stems infested with larvae, but only 21, 27, and 27% were girdled (Table 17-3). Four 25-stem samples were as reliable as examining 60 m of row for determining percentage of girdled stems.

The proximity of the current crop of soybean to a field that was infested with

Table 17-3. Sampling for larvae of *Dectes texanus texanus* based on a 25-stem sample. Hyde County, North Carolina. 1976

Sample set	Number of girdled stems/25-stem sample					
	I	II	III	IV	Total	Mean
1	4	7	5	4	20	5.0
2	9	3	8	7	27	6.7
3	8	7	5	7	27	6.7

Sample set	Number of infested stems/25-stem sample					
	I	II	III	IV	Total	Mean
1	7	13	8	8	36	9.0
2	13	6	12	11	42	10.5
3	11	12	10	11	44	11.0

Dectes the previous year has an effect on *Dectes* population (W. Campbell unpublished). *Dectes* infestation decreased as the distance from the source of infestation increased. Infested stems/60 row-m averaged 87.9, 27.6 and 16.4 for soybean planted adjacent to emerging *Dectes,* two fields distant, and three fields distant from the source of infestation, respectively (Table 17-4). Each field was ca. 120 x 800 m (ca. 10 ha).

A soybean field that measured 52 x 800 m (ca. 4.2 ha) was sampled for distribution of *Dectes* infested stems. A heavy growth of weeds that included ragweed occupied the two ends of the field, and a thin stand of weeds occupied the sides of the field. The field was infested throughout; however, the center of the field averaged 25% infested stems while the ends averaged 29% and 39%. Girdled stems ranged from 3% to 36% per 25-stem sample with an average of 15% girdled stems. Approximately one-half of the infested stems were girdled.

A typical pattern of *Dectes* infestation is shown (Fig. 17-11) as it occurred across the end of a soybean field. The number of girdled stems in 60 row-m was recorded starting at a point 6 m from the end of the row. Samples were taken by recording the girdled stubble observed while walking slowly down rows 2, 4, 8, 16, 32 and 64 from opposite sides of the field.

Infestation of *Dectes* as represented by girdled stems was detected across the field. The south side of the field had a higher infestation than the north side. Girdled stems ranged from 22 to 165/60 row-m. Any row selected for sampling would have provided information on the presence of *Dectes* in this field (W. Campbell unpublished). Four 25-stem samples would provide an adequate sample and reduce scouting time for this pest.

Yield loss per hectare due to *Dectes* girdled stems may be estimated by multiplying the average number of girdled stems per row-m (1 m row width) by 10 000. Average yield of soybeans per plant is obtained and multiplied by the calculated total girdled plants per hectare. It is recommended that 10% be added for infested but not girdled stems.

Table 17-4. Effect of field location on *Dectes texanus texanus* infestation of soybean. Washington County, North Carolina. (1976) The size of each field was ca. 120 x 800 m

Site[a]	Number of *Dectes* girdled stems/60 row-m										Total	Mean
	I	II	III	IV	V	VI	VII	VIII	IX	X		
1	64	77	99	77	102	63	109	81	90	117	879	87.9
2	31	23	28	32	32	27	26	27	29	21	276	27.6
3	21	20	15	19	16	12	17	15	13	16	164	16.4

[a] 1—Adjacent to corn field (soybean 1975)
2—2nd field from corn (soybean 1975)
3—3rd field from corn (soybean 1975).

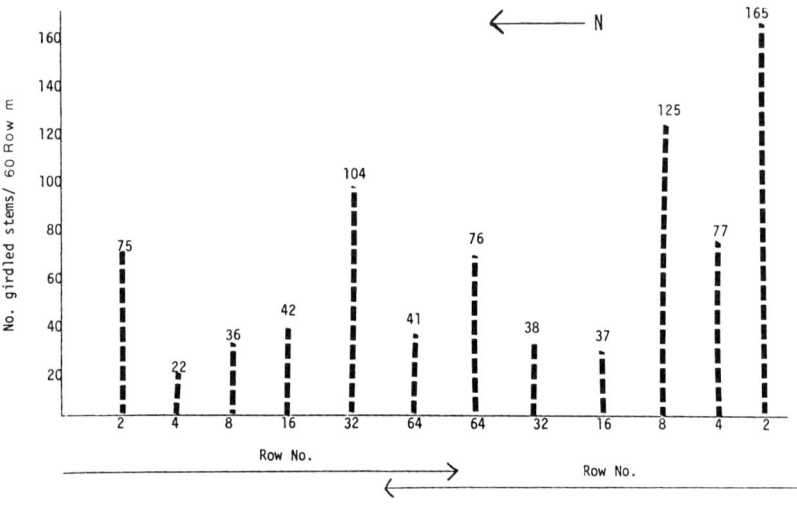

Figure 17-11. Dispersion of *Dectes texanus texanus* girdled stems within a soybean field. Washington County, North Carolina. 1976.

Soybean losses may also be estimated on a plant basis. The number of plants per acre or hectare may be computed using formulas 2-3 or 2-4 (see Chapter 2). If, for example, it is estimated that there are 210 000 plants per hectare and there are 10 girdled stems per 25-stem sample, then by using the following formula girdled stems per hectare may be calculated:

$$\frac{10 \text{ girdled stems}}{25\text{-stem sample}} = \frac{X}{210{,}000 \text{ stems (plants)}}$$

The yield loss or monetary loss is then estimated by multiplying the plant loss due to *Dectes* girdling by the bean weight or value per plant.

V. Concluding Remarks

Sampling for *Dectes* is most readily accomplished by examining soybean fields for girdled stems in the fall, winter, or spring and until the field is disked or plowed. The most reliable method of estimating *Dectes* infestation level is splitting stems lengthwise in September, October, or November prior to harvest. Soybeans are often harvested prior to *Dectes* larval girdling; therefore, larvae may be in the portion of the stem that is cut and passed through the combine.

Chemical control of this pest does not appear practical using current commercial pesticides (Campbell and Van Duyn 1977). Sampling methods described then may best be used to determine presence, abundance, patterns, and effect of cropping systems on population density and spatial patterns for use in pest management programs.

Delaying the harvest until well after soybean maturity increases the incidence of stem girdling by *Dectes*. Sampling for *Dectes* presence and abundance well in advance of soybean maturity will provide information for *Dectes* management by harvesting early to reduce girdling losses.

Although some of the techniques described in this chapter are probably peculiar to *Dectes*, it is expected that the general sampling principles are applicable to other species of stem borers.

Acknowledgments

This research was supported in part by the National Science Foundation and the Environmental Protection Agency, through a grant (NSF GB-34718) to the University of California. The findings, opinions, and recommendations expressed herein are those of the authors and not necessarily those of the University of California, the National Science Foundation, or the Environmental Protection Agency.

References

Bhattacharya, A. K., and Y. S. Rathore. 1980. Soybean insects—Problems in India. World Soybean Research, Westview Press. Boulder, Colorado. 897 p. Conf. II Proc., Raleigh, N. C. 291-302.

Campbell, W. V., and J. W. Van Duyn. 1977. Cultural and chemical control of *Dectes texanus texanus* on soybeans. J. Econ. Entomol. 70:256-258.

Corseuil, E., F. Z. da Cruz, and L. M. C. Meyer. 1974. Insetos nocivos a soja no Rio Grande do Sul. Univ. Fed. Rio Grande do Sul, Fac. Agron., Brazil. 36 p.

Daugherty, D. M., and R. D. Jackson. 1969. Economic damage to soybeans caused by a cerambycid beetle. Proc. N. Cent. Br. Entomol. Soc. Amer. 24:36.

Dillon, L. S. 1956. The neartic components of the tribes Acanthocinini (Coleoptera: Cerambycidae) Part III. Ann. Entomol. Soc. Amer. 49:332-355.

Falter, J. M. 1969. *Dectes* sp. (Coleoptera: Cerambycidae): a unique and potentially important pest of soybeans. J. Elisha Mitchell Sci. Soc. 85:123.

Grangrade, G. A. 1974. Insects of soybean. Jawaharlal Nehru Krishi Vishwa Vidyalaya, Tech. Bull. 24:88 p.

Gangrade, G. A. 1976. Assessment of effects on yield and quality of soybeans caused by major arthropod pests. Jawaharlal Nehru Krishi Vishwa Vidyalaya, Dep. Entomol., Tech. Rep. 145 p.

Genung, W. G., and V. E. Green, Jr. 1965. Some stem boring insects associated with soybeans in Florida. Fla. Entomol. 48:29-33.

Hatchett, J. H., D. M. Daugherty, J. C. Robbins, R. M. Barry, and E. C. Houser. 1975. Biology in Missouri of *Dectes texanus* a new pest of soybeans. Ann. Entomol. Soc. Amer. 68:209-213.

Jarvis, H., and J. H. Smith. 1946. Lucerne pests. Queensland Agr. J. 62:79-89.

Patrick, C. R. 1973. Observations on the biology of *Dectes texanus texanus* (Coleoptera: Cerambycidae) in Tennessee. J. Ga. Entomol. Soc. 8:277-279.

Chapter 18

Sampling *Epinotia aporema* on Soybean

Beatriz Spalding Correa Ferreira

I. Introduction

The borer *Epinotia aporema* (Walsingham) (Lepidoptera: Olethreutidae) has a wide distribution in the American Continents. It has been recorded from southern United States to Argentina including Mexico, Costa Rica, Guatemala, Peru, Uruguay (Clark 1954), Chile (Olalquiaga Fauré 1953), and Brazil (Biezanko 1961). In Brazil *E. aporema* causes damage to soybean from the southernmost state of Rio Grande do Sul to the state of Goias, located in the central part of the country.

E. aporema seems to be restricted to leguminous hosts. In Peru it is a pest of groundnuts (*Arachis hypogaea* L.) (Montero 1967). In Uruguay *E. aporema* was recorded by Morey (1972) and Biezanko et al. (1974) as occurring in many other crops such as alfalfa (*Medicago sativa* L.), broadbean (*Vicia faba* L.), soybean [*Glycine max* (L.)], common bean (*Phaseolus vulgaris* L.), pea (*Pisum sativum* L.) and clover (*Trifolium polymorphum* Poir). In Chile it is considered the second major pest for alfalfa and clover (Caballero 1972), and in Argentina it is also found on soybean (Rizzo 1972) and stored beans (Hayward 1958). Records from Brazil indicate that *E. aporema* is a pest of alfalfa, bean, soybean, and clover (Biezanko 1961).

A. Nature of Damage

In some areas of Brazil *E. aporema* is of major economic importance to soybean production. The larva feeds on vegetative and floral buds as well as pods. The larval habits as a stem borer and leaf roller make control procedures very difficult.

The injury caused by *E. aporema* is easily recognized. Newly hatched larvae move into the most tender buds where they feed on one of the youngest furled leaflets. As the larvae develop, the damaged apical leaflets remain attached to one another by a silken web forming a sort of a cartridge that characterizes the attack of these lepidopterous larvae. Injured leaves dry and finally die. The larvae move from the decaying parts to secondary buds, axils, and stems, opening galleries that obstruct sap movement. They also attack pods, damaging the seeds.

Little work has been done on *E. aporema*. Many aspects of its biology are not known, and the literature on this species is very limited. In Brazil, for a long time, the species was mistakenly identified as *Laspeyresia fabivora* Meyrick, due to similarities between the larvae of the two species as well as the type of injury to soybean. *E. aporema* was first reported in Brazil by Biezanko (1961) in the state of Rio Grande do Sul, but until 1973 it was considered to be a pest of only minor importance for soybean. Since then, however, heavy infestations of this pest were detected in the state of Paraná (Correa 1975), and in some areas it was considered the most frequent pest.

B. Life Cycle and Phenology

The life cycle of *E. aporema* consists of egg, five larval instars, pupa, and adult. The development from egg to adult ranges from 35 to 40 days (5 day egg stage, 14 to 20 day larval stage, and 15 day pupal stage) (Morey 1972).

The eggs are generally laid on tender leaflets of terminal buds. Eggs are oval shaped and very small (0.47 x 0.31 mm); soon after deposition they are light yellow. The quick-moving newly emerged larvae are yellowish-green with a shining black head and prothoracic shield. The first four larval instars are very similar in shape and color; the only noticeable difference is in size. They have a gelatinous aspect, and as they grow they become pale brown (Fig. 18-1). The fully developed larvae are reddish and ca. 11 mm wide. Pupation occurs in the soil. Adults of *E. aporema* are small, dark brown moths with silver spots on the wings (Fig. 18-2). Males are laterally dark and dorsally light in color, while the females are exactly the opposite.

Caballero (1972) reported the occurrence of four-to-five generations per year in Chile. The species overwinters as larvae in diapause or as adults. However, in Uruguay *E. aporema* was detected all year round on different host plants (Morey 1972). In Argentina the borer was reported to be found on soybean from late December until April (Rizzo 1971), while in Brazil it is active on the crop from November to April.

It is extremely important to correlate the occurrence and abundance of *E. aporema* with the various developmental stages of soybean. Generally, in the state of Paraná, high populations are detected at the end of the soybean vegetative stage through the flowering stage (R2), and these populations are considered to be economically important. After flowering the populations drop, tending to dis-

Figure 18-1. Larva of *Epinotia aporema*.

Figure 18-2. Adult of *Epinotia aporema*.

appear or remain at low levels until the end of the season. However, in other regions such as Passo Fundo (Rio Grande do Sul) and Brasilia (DF), high infestations are found at the end of the pod-set stage (R3-R5). Extensive damage to pods, buds, and axils results from these late infestations.

In southern Paraná high populations may be detected in January (Santos 1978) and February (Correa and Smith 1976; Calderon 1977). However, in the northern regions of the state highest infestations occur in November. Planting date is very important to the distribution of *E. aporema*. Gazzoni and Oliveira (1979) observed two generations on soybean planted in October, while only one generation was recorded on soybean planted in November.

The seasonal abundance of *E. aporema* is extremely variable from place to place and from year to year. In the state of Paraná, up to four generations per year may occur. The last generation is normally present on late plantings of soybean or on volunteer soybean plants germinating after harvest of the main crop. *E. aporema* overwinters as larvae in diapause on these plants, sheltered inside the leaves. During this period both larvae and adults are found on irrigated soybean grown in more northern Brazilian regions. The general seasonal life history of *E. aporema* is shown in Fig. 18-3.

II. Sampling for Adults

There are few studies on the efficiency of various methods for sampling *E. aporema* adults. This is mainly due to the fact that only recently has the importance of the species for the soybean crop been perceived. Adult sampling is also difficult because of the great variation in abundance of the pest from year to year in the different production areas.

Correa (1975) surveyed soybean fields using two different methods: collection and examination of plants (direct observations) and sweep net. Only with the second method were the adults collected. It was observed that to obtain greater efficiency of the method, the operation had to be done in the evening, since the activity of the pest increases at dusk. Caballero (1972) sampled adults of *Epinotia* sp. on alfalfa by counting the number of moths that flew each 25 steps within a field (flushing method, see Chapter 8). Since the adults of *E. aporema* are attracted by light, relative measurements of populations of this moth can be obtained by using light traps.

III. Sampling for Eggs

There is no reference in the literature concerning sampling for eggs of *E. aporema*. However, since the eggs of this pest are laid normally on the most tender leaflets of the buds and since they are very small, observation of the eggs still in the field is practically impossible. Absolute measurements of egg populations may be obtained by collecting entire plants along a given length of row or collecting

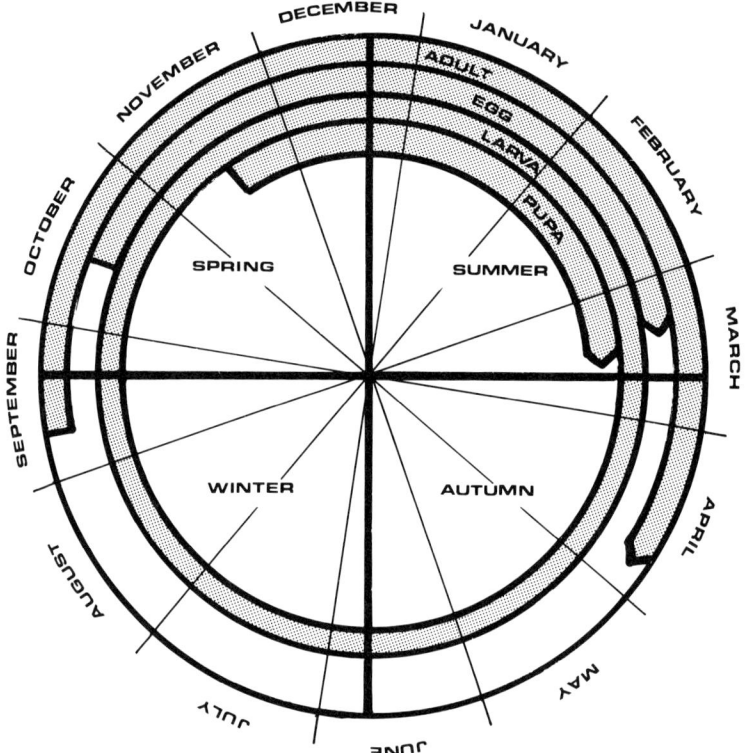

Figure 18-3. Seasonal occurrence of *Epinotia aporema* in Paraná, Brazil.

plants at random within the field and examining them under a stereomicroscope in the laboratory.

As the terminal buds are the preferred place for oviposition for this species, relative measurements of egg populations can be obtained by collecting at random a given number of buds and counting the eggs under a stereomicroscope in the laboratory. Normally these eggs are hidden and protected in the intense pilosity of the young buds.

IV. Sampling for Larvae

An absolute method for measuring larval populations of *E. aporema* is direct plant examination. Two variations of this method may be used: collecting plants along a measured length of row or examining single plants selected at random within a field. Both methods should also be monitored by examining whole plants in the laboratory (Correa and Smith 1976) or examining them directly in the field (Calderon 1977).

Table 18-1. Mean number of larvae of *E. aporema* collected by whole plant examination method in field and in laboratory (adapted from Calderon 1977)

Sampling date (in 1977)	Number of larvae/row meter	
	Field	Laboratory
Jan. 18	0.0	0.0
27	2.6	2.8
Feb. 2	6.4	6.8
8	8.0	8.2
14	7.8	8.0
22	10.0	10.6
March 1	4.2	4.4
8	4.2	4.0
15	3.0	3.4
22	1.8	2.0
29	0.4	0.6
April 5	0.2	0.2
Total	48.6	51.0

Relative sampling methods ordinarily used for the majority of insect pests on soybean such as sweep net, ground cloth, and vacuum net are not efficient for surveying larval populations of *E. aporema*. This inefficiency is mainly due to the location of the borer on the plant, with larvae well protected inside the buds.

Larval population data obtained by the method of examining plants in the field were compared with data obtained by examining plants in the laboratory (Calderon 1977). The number of larvae obtained with both methods was very similar (Table 18-1). This result indicates that although the first method is extremely dependent on the ability of the surveyor, it may be used to accurately estimate larval populations of *E. aporema* in soybean fields. However, methods of direct plant examination are difficult and time-consuming. Such methods are not recommended for use on sampling larvae in a pest management program. It is much simpler to estimate the population levels by counting the number of damaged terminal buds (indirect method). The injury caused by the borer on terminal buds is easily detectable, and in a short period of time relative data of larval populations may be obtained. Surveying soybean fields in the 1977 season, Correa Ferreira (unpublished) found a high correlation ($r=0.86$) between number of damaged apical buds and number of larvae per row meter (Table 18-2).

V. Sampling for Pupae

Sampling methods for pupae of *E. aporema* have not been investigated. However, as the pupae are normally located in the soil (1 or 2 cm below the soil surface) near the stems, measurements of pupal populations may be obtained by sampling a given area of soil.

Table 18-2. Mean number of larvae of *E. aporema* and terminal buds damaged sampled from soybeans using whole plant examination method and its respective correlation coefficient

	Mean number	
Sample	Larvae	Terminal buds damaged
1	23.6	10.3
2	15.0	7.8
3	14.7	6.1
4	13.1	7.2
5	13.0	9.2
6	10.5	7.2
7	7.5	5.9
8	6.0	3.4
9	4.6	2.3
10	3.8	4.0
correlation coefficient = 0.86		

VI. Concluding Remarks

The occurrence of *E. aporema* on soybean is fairly recent. However, the incidence of this species in soybean production areas of Brazil is increasing from year to year. The severity of damage to the crop by *E. aporema* seems also to be increasing.

The biology, population dynamics, seasonal abundance, and geographic distribution of the pest on soybean are not well known yet. There is need for additional studies to better understand the large inter-regional population fluctuations of this pest as well as the fluctuations that occur from year to year within the same region.

Due to the characteristic behavior and feeding habits of *E. aporema,* sampling methods specific for this species need to be developed. It is difficult to adopt methods that are suitable for other lepidopterous larvae. More research is needed on the cost of sampling (relative net precision), efficiency and reliability of the various sampling procedures.

References

Biezanko, C. M. de. 1961. XIII–Olethreutidae, Tortricidae, Phaloniidae, Aegeriidae, Gliphipterygidae, Yponomeutidae, Gelechiidae, Oecophoridae, Xylorictidae, Lithocolletidae, Cecidoseidae, Ridiaschinidae, Acrolophidae, Tineidae et Psychidae da Zona Sueste do Rio Grande do Sul. Arq. Entomol., Esc. Agron. "Eliseu Maciel," Pelotas (RS), series A: 1 a 16.

Biezanko, C. M. de, A. Ruffinelli, and D. Link. 1974. Plantas y otras substancias alimenticias de las orugas de los lepidopteros uruguayos. Rev. Cent. Ciencias Rur. 4:107-148.

Caballero, C. V. 1972. Reconocimento, biologia y control de las principales plagas que afectam los semilleros de alfalfa y trebol rosado, en Chile. Rev. Peru. Entomol. **15**:201-214.

Calderon, D. G. R. 1977. Ocorrência, danos e contrôle de *Epinotia aporema* (Walsingham, 1914) (Lepidoptera), em soja. M. S. thesis, Univ. Fed. Paraná, Curitiba, Brazil. 79 p.

Clarke, J. F. G. 1954. The correct name for a pest of legumes. Proc. Entomol. Soc. Wash. **56**:309-310.

Correa, B. S. 1975. Levantamento dos lepidópteros pragas e danos causados à soja. M. S. thesis, Univ. Fed. Paraná, Curitiba, Brazil. 120 p.

Correa, B. S., and J. G. Smith. 1976. Ocorrencia e danos de *Epinotia aporema* (Walsingham, 1914) (Lepidoptera: Tortricidae) em soja. An. Soc. Entomol. Brasil **5**:74-78.

Gazzoni, D. L., and E. B. de Oliveira. 1979. Distribuicão estacional de *Epinotia aporema* (Walsingham, 1914) e seu efeito sobre o rendimento e seus componentes, e caracteristicas agronomicas de soja cv. "UFV-1," semeada em diversas épocas. pp. 94-105 in Anais-I Seminário Nacional de Pesquisa da Soja. EMBRAPA/Cent. Nac. Pesq. Soja, Londrina, Brasil. Vol. 2. 389 p.

Hayward, K. 1958. Insectos tucumanos perjudiciales. Rev. Ind. Agr. Tucuman **42**:3-144.

Montero, M. P. 1967. Control químico de *Stegasta bosquella* Chamb. y *Epinotia aporema* Heinr. en maní. Rev. Peru. Entomol. **10**:56-61.

Morey, C. S. 1972. Biologia y morfologia larval de *Epinotia aporema* (Wals.) (Lepidoptera-Olethreutidae). Univ. Rep. Fac. Agron. Montevideo Bol. **123**:14 p.

Olalquiaga Fauré, G. 1953. Plagas de las leguminosas en Chile. Bol. Fitosanit. FAO **1**:174-176.

Rizzo, H. F. E. 1971. Catálogo de lepidopteros hallados en la Facultad de Agronomia y Veterinaria de Buenos Aires. Univ. Buenos Aires. Fac. Agron. Y Vet. Pub. Int. **2**:35 p.

Rizzo, H. F. 1972. Enemigos animales del cultivo de soja. Rev. Inst. Bolsa Cereales **2851**:1-6.

Santos, B. B. dos. 1978. Manejo dos insetos-pragas da soja no centro-sul do Paraná. M. S. thesis, Universidade Federal do Paraná, Curitiba, Brasil. 126 p.

Chapter 19

Sampling Threecornered Alfalfa Hopper on Soybean[1]

Arthur J. Mueller

I. Introduction

Early reports on the threecornered alfalfa hopper, *Spissistilus festinus* (Say) (Homoptera: Membracidae), (Figs. 19-1 and 19-2) indicated that the species was most abundant in the southern United States but that it was reported in all states except Maine, New Hampshire, Vermont, Washington, Oregon, Idaho, Nevada, and Utah (Wildermuth 1915). Van Duzee (1917) and Funkhouser (1927) disagreed with some of the state records given by Wildermuth because entomologists in the early part of the century often confused two or more species of *Spissistilus*. This confusion existed until Caldwell (1949) published a description for *Spissistilus festinus* (Say) based on genitalia. Caldwell gave the range as "across the southern United States and as far south as Costa Rica." North-south fluctuations probably are due to year-to-year weather variations. However, it is evident that the threecornered alfalfa hopper is present in most of the current soybean production areas of the United States although there are no records of the species in the midwestern states north of Missouri (M. Kogan, personal communication based on data of the Illinois Soybean Arthropods Reference Collection).

---

[1] Published with the approval of the Director, Arkansas Agricultural Experiment Station.

Figure 19-1. Threecornered alfalfa hopper adult.

A. Host Plants and Nature of Damage

As the common name implies, *S. festinus* is a pest of alfalfa. Although its favored host plants are members of Leguminosae, the earliest report of crop damage was to tomato (Oemler 1888). This insect species has a wide host range and has been reported to feed or reproduce on a variety of plants including Bermuda and

Figure 19-2. Threecornered alfalfa hopper nymph.

Johnsongrass, wheat, barley, oats, sweet, red and bur clovers, cowpeas, soybean, sunflower, cocklebur, and vetch.

Depending on the insect's developmental stage, the threecornered alfalfa hopper feeds by single random punctures of plant stems or by a continuous puncturing around the circumference of the stem forming a girdle (Moore and Mueller 1976). Laboratory feeding studies indicate that injury by the 1st and 2nd instars is in the form of single random punctures with no evidence of girdling. This type of feeding does not appear to seriously damage the plants. Stem girdling is first observed with the third stage nymph but fourth stage nymphs are the most injurious. The concentration of feeding at the plant base causes the characteristic girdling associated with lodging, breakage, and plant mortality (Fig. 19-3). A fourth stage nymph is capable of completely girdling the main stem in less than

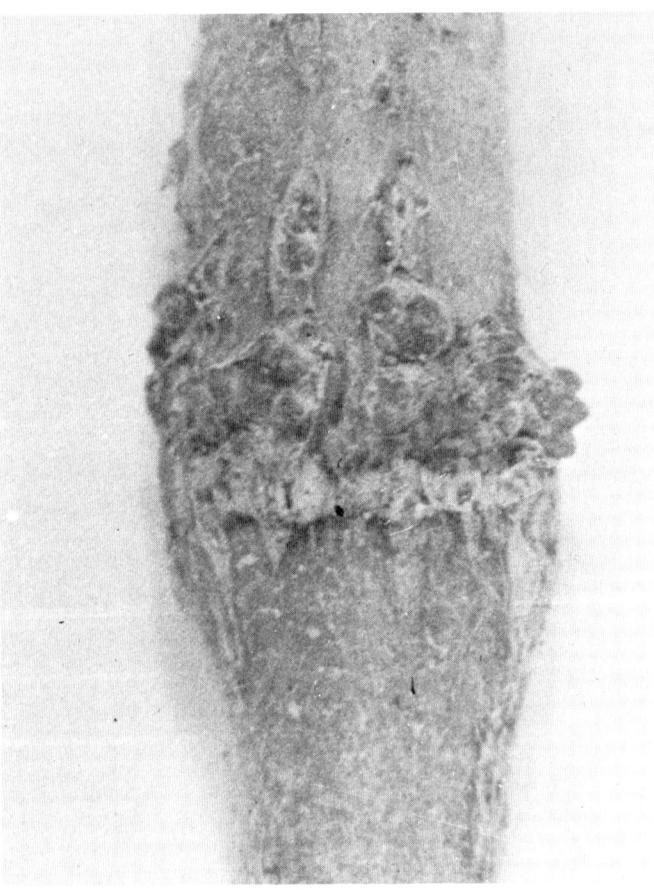

Figure 19-3. Characteristic basal main stem girdle. Note swelling above girdle and adventitious growth.

24 hours. The fifth stage nymphs girdle less than the 4th. Most 5th stage nymph and adult feeding is restricted to the upper petioles, though some is near the plant base.

Main stem girdling affects a single plant's yield by causing death, reduction in number and weight of beans, or harvesting difficulties due to lodging or breakage (Mueller and Dumas 1975). Girdling that causes most soybean plant lodging occurs before plants are 20-25 cm tall (Bailey et al. 1970, Davis and Laster 1968) (Fig. 19-4).

The young soybean plant may respond in three ways when girdled near the stem base (Moore and Mueller 1976). The most common response is plant death. Mortality normally is a slow process, requiring two-to-six weeks. Symptoms in sequence are a loss of plant vigor, cessation of growth, wilting from the top downward, a much reduced plant size, and finally death. Unless the majority of plants in a soybean field are girdled, yield reduction is not incurred because sur-

Figure 19-4. Types of feeding injury to young vegetative plant stages: upper, scars from random punctures and incomplete girdles; middle, complete girdle without scab; lower, complete girdle with swelling and adventitious growth above girdle.

viving plants are capable of compensating by lateral growth, filling in the vacant row space. This is true also if the girdled plants are broken over by cultivators or strong winds at these early vegetative plant growth stages. The second plant response is partial "scabbing over" of the girdle with wound tissue but without complete plant recovery (Fig. 19-5). Plants in this category continue to grow ver-

Figure 19-5. Typical plant responses if main stem girdle does not cause death: left, healthy ungirdled plant; center, plant with "scabbed over" girdle, plant has recovered and may produce normal complement of pods; right, spindly growth, plant will produce few if any pods.

tically but usually exhibit reduced height and vigor, become spindly, and produce few or no beans. Since these plants compete with adjacent plants for water, nutrients, and space, they may be considered as playing a role similar to a weed and may affect yields. The third plant response is a complete "scabbing over" of the girdle wound, after which the plant grows normally, with little if any reduction of bean pods (Fig. 19-5). However, the girdle is still a weakened area on the stem and the plant is susceptible to breakage later if pressure is applied at this area. This may cause a yield reduction since the beans may not develop fully or the plant may lodge and not be picked up by the combine at harvest. These plants then also fall into the category of weeds.

As plants increase in height and the main stem thickens, nymphal feeding is moved upward on the main stem and onto the leaf petioles (Fig. 19-6). No published studies have been conducted specifically to determine the effect of upper main stem and leaf petiole girdling on yield.

Caviness and Miner (1962) attempted to simulate threecornered alfalfa hopper injury by reducing the plant stand by 15, 30, and 45% at bloom and two weeks pre and post bloom. Although yields were reduced at all stages and stand reductions during this three year study, significant differences ($P < 0.01$) of the three year means occurred only at 30% stand reduction two weeks post bloom and 45% stand reduction at bloom and two weeks post bloom.

Tugwell et al. (1972) regulated threecornered alfalfa hopper populations by insecticide applications in field plots in Arkansas. Three schedules designed to produce three infestation levels plus an untreated check resulted in 17, 24, 34, and 42% main stem girdling with no significant differences in yield. However, mean yields of lodged plants not harvested by the combine were significantly different between the untreated check and the treated plots. The authors stated that, "If high winds or rain had caused injured plants to lodge late in the season, removing injured plants laden with beans too late for adjacent plants to compensate for their loss, the results may very well have been different."

Another source of possible damage due to girdling injury is increased susceptibility to fungal infection, e.g., *Sclerotium rolfsii* Saccardo (Herzog et al. 1975). Herzog et al. concluded that disease transmission by the insect was of little importance because simulated girdling damage also resulted in infection of seedling plants.

B. Life Cycle and Phenology

The life cycle of the threecornered alfalfa hopper consists of the egg, five nymphal stages, and the adult. The threecornered alfalfa hopper overwinters in both the adult and egg stages (Wildermuth 1915). In Arkansas adults have been observed during the winter months at the base of dock (*Rumex* spp.) and other weed plants in fields planted the previous season to soybean and under debris at field edges. On warm sunny days during the winter adults are active and have been found on above ground vegetation but were not observed to be feeding.

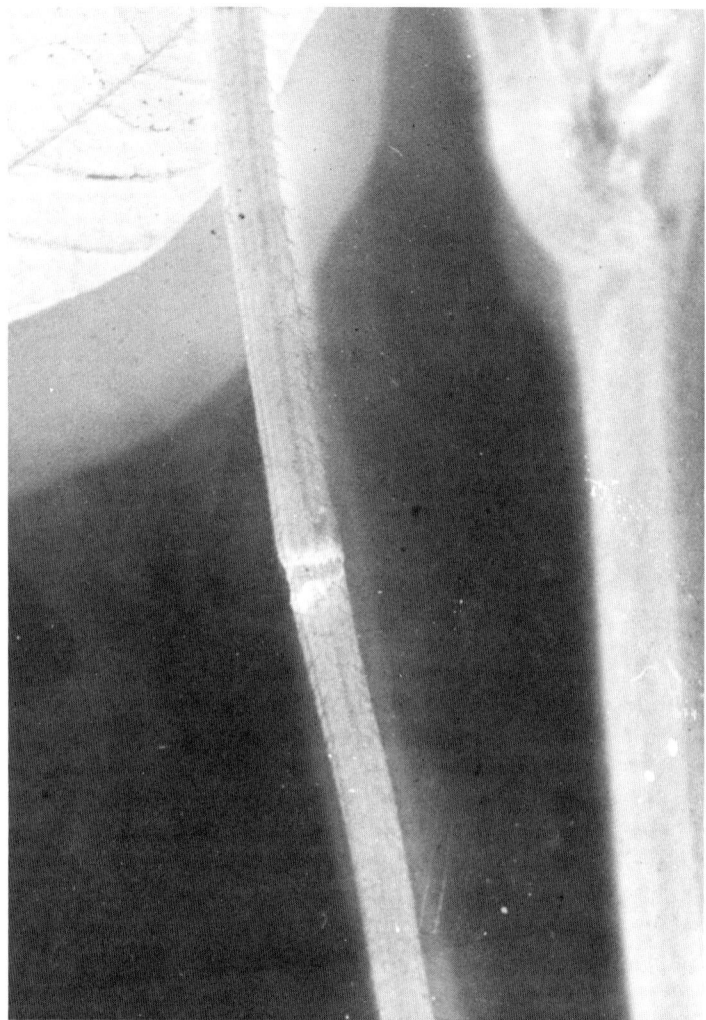

Figure 19-6. Leaf petiole girdle, occurs more frequently as main stem thickens.

A general survey for all insects was made each year from 1973 through 1976 in 15 soybean fields throughout the eastern soybean growing area in Arkansas. Samples were taken with a back pack suction sampler (D-Vac) and ground cloths. These data indicate that three overlapping generations occur per year in Arkansas (Fig. 19-7). Adults of the first generation normally peak from early to mid-July, the second generation in mid-August and the third in September, which provides adults for the overwintering population. An earlier generation(s) of this species may occur on hosts other than soybean prior to July. Wildermuth

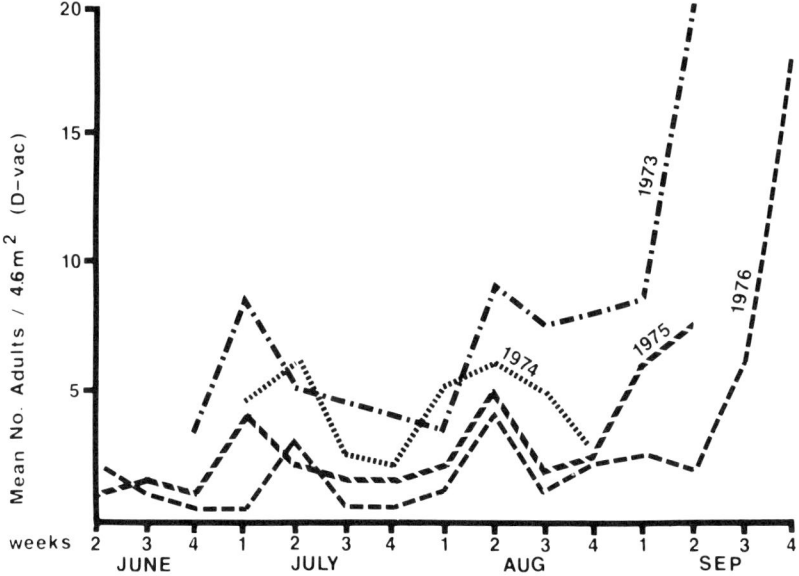

Figure 19-7. Mean number of threecornered alfalfa hopper adults collected from 15 soybean fields in 3 different areas in Arkansas.

(1915) reported up to four generations per year on alfalfa in southern Arizona. Kopp and Yonke (1973) reported only one generation per year in the northern states.

II. Sampling for Adults

Primarily two methods have been used for quick estimation of adult populations. The most frequently used and probably the best method is the sweep net. The adults are very active and fly readily when disturbed. As one walks through a soybean field, adults can be observed flying short distances, usually 3-4 meters. They fly both down and across rows to avoid the intruder. However, not all adults are in a position for quick flight. It is more difficult for those adults that are not at the plant canopy periphery to escape the sampler. Boyer (1967) compared three methods of sweeping with a standard 38 cm diameter sweep net.

Method 1. "Walk fast in row middle parallel with rows. Reach forward as far as possible and sweep the top of one row of beans pulling the net toward the surveyor in the manner of rowing a boat with a single oar. Approximately three feet of row are swept."

Method 2. "Walk fast in row middle parallel with rows. Sweep across the tops of the two rows between which the surveyor is walking."

Method 3. "Walk across rows. Reach over a row and sweep three feet of tops of plants of second row from surveyor."

Ten sweeps by each method were made 12 times near midday. Means, standard deviations, and coefficient of variation for the three methods were: Method 1- 9.58, 2.39 and 24.9%; Method 2–9.83, 3.54 and 36.0%; Method 3–8.0, 4.80 and 60.0%. Boyer concluded that Method 1 was best because it had the lowest coefficient of variation.

The comparative studies conducted by Hillhouse and Pitre (1974) did not count adults and nymphs separately. They concluded that sweeping across two rows was the least variable of the methods tested and had the highest net precision value. This method was similar to Boyer's Method 2 which had a slightly higher coefficient of variation than Boyer's Method 1.

Meisch and Randolph (1966) compared numbers of threecornered alfalfa hoppers collected at three times of day with a sweep net in alfalfa. Significantly greater numbers were collected between 1400-1500 h than between 0700-0800 or 1900-2000 h.

Preliminary studies indicate that higher numbers of adults are captured during early and late afternoon (1500 and 1900 h) with a sweep net at tops of plants when temperatures are higher, as opposed to morning samples (prior to 1000 h). However, factors other than temperature or time of day may be involved (A. Mueller unpublished).

Care must be taken when insects are counted in the field. When the sweep net is held open, threecornered alfalfa hopper adults are one of the first species to fly out. They hold to the inner sides of the sweep net, then very quickly exit with a jump and flight. This maneuver may be made from deep in the sweep net.

The other sampling method utilized by some researchers to estimate populations of adults is the D-Vac. However, no comparative studies of the D-Vac and other sampling methods have been reported with this species.

The ground cloth method is not adequate for sampling adults due to their flight behavior. Hillhouse and Pitre (1974) compared this method to several sweep methods. The relative variation was comparable to sweep net methods, 7.67, but mean numbers reported included adults and nymphs combined.

III. Sampling for Nymphs

The immature stages, especially instars three to five, are the most injurious to soybean. The most damaging feeding sites are low on the main stem. Due to relatively little movement and their location on the plant when they are causing damage, sampling methods for nymphs are limited. The two methods found to be best (A. Mueller unpublished) are direct visual observation and the ground cloth. Of these two, visual observation more closely estimates the absolute population densities and it is much more reliable in assessing damage but it is also more time-consuming per unit area.

Boyer (1967) recommended bending the plants over on 3 feet of row for locating nymphs. By this method stems near the ground level may be examined. When making visual observations, all sides of the stem should be carefully examined. The nymphs "play squirrel"; that is, when disturbed, they stop probing and move to the opposite side of the stem from the observer, thus increasing the changes of being overlooked. Data collected by this method can quickly be extrapolated to nymphs per plant, unit of row, or area.

The ground cloth is not an adequate method unless the nymphs are relatively high on the plant. Nymphs are difficult to dislodge by shaking the plants when they are near the base of the plant or when plants are small. Should they be dislodged, they usually fall on the ground and not on the ground cloth.

IV. Sampling for Eggs

No methods have been developed for assessing eggs on soybean. Moore and Mueller (1976) state that most oviposition occurs at the base of the lower petioles.

V. Indirect Sampling Procedures

Stem girdle counts is probably the most valuable sampling procedure for establishing economic thresholds for the threecornered alfalfa hopper. By visual observation, similar to counting nymphs, stem girdles can be easily and quickly recognized. The mean number of girdled plants (or ungirdled plants) per unit of row can quickly be determined. At the critical time for plant loss, plants ca. 25 cm tall or less, can be quickly scrutinized. Prior to making girdle counts, a quick method for determining if a number of plants are girdled, is to swing a sweep net across the rows near the ground level. If plants have been girdled, the stems will break and the plants will fall over. By this method, one can quickly decide if more intensive sampling to determine insect numbers is advisable.

VI. Concluding Remarks

The threecornered alfalfa hopper is present in most of the current soybean production areas of the United States. The most serious girdling caused by feeding occurs before plants are ca. 25 cm tall. Due to plant compensation and other unpredictable factors that may cause lodging, yield loss is very difficult to assess and consequently economic thresholds are vague.

Adult populations can be quickly estimated by the sweep net or D-Vac methods. The most often used and probably the best method is the sweep net. Various sweeping routines appear to be adequate for rapid field population estimates. The D-Vac is also used, principally by researchers, but no comparative studies have been reported. The ground cloth method is not as effective for sampling adults as either of the other two methods due to the threecornered alfalfa hopper adult flight behavior and preference for the upper part of the plant.

The most accurate method for sampling nymphs is visual observation but it is also the most time-consuming. The ground cloth may be used if the plants are not too short; it is less time-consuming than the visual method.

No methods have been developed for assessing eggs on soybean; however, it has been reported that the preferred ovipositional site is at the base of the leaf petioles.

By visual observation, similar to counting nymphs, stem girdles can be easily and quickly recognized. When plants are in early vegetative stages, the number of girdled plants can be estimated rapidly by swinging a sweep net across rows near the ground level. If plants are girdled, the stems will break and the plants will fall over. This will help the sampler to decide whether additional counts for determining insect populations are necessary.

There are some very obvious voids in sampling techniques for the threecornered alfalfa hopper. Some areas where research is lacking are: correlation of insect populations and incidence of girdling; determination of spatial patterns within the field relative to insect stage and numbers, plant growth stages, and seasonal variations; and the number of samples and sampling sites required to adequately estimate populations.

References

Bailey, J. C., L. B. Davis, and M. L. Laster. 1970. Stem girdling by the threecornered alfalfa hopper and height of soybean plants. J. Econ. Entomol. 63:647-648.

Boyer, W. P. 1967. Survey method for threecornered alfalfa hopper (*Spissistilus festinus*) in soybeans in Arkansas. USDA Coop. Econ. Insect Rep. 17:324-325.

Caldwell, J. S. 1949. A generic revision of the treehoppers of the tribe Ceresini in America north of Mexico, based on a study of the male genitalia. Proc. U. S. Nat. Mus. 98:491-521.

Caviness, C. E., and F. D. Miner. 1962. Effects of stand reduction in soybeans simulating threecornered alfalfa hopper injury. Agron. J. 54:300-302.

Davis, L. B., and M. L. Laster. 1968. Stage of growth when soybeans are damaged by the threecornered alfalfa hopper. Miss. Agr. Exp. Sta. Inform. Sheet 1016.

Funkhouser, W. D. 1927. Membracidae, general catalogue of the Hemiptera. fasc. 1, 581 p.

Herzog, D. C., J. W. Thomas, R. L. Jensen, and L. D. Newsom. 1975. Association of sclerotial blight with *Spissistilus festinus* girdling injury on soybean. Environ. Entomol. 4:986-988.

Hillhouse, T. L., and H. N. Pitre. 1974. Comparison of sampling techniques to obtain measurements of insect populations on soybeans. J. Econ. Entomol. 67:411-414.

Kopp, D. D., and T. R. Yonke. 1973. The treehoppers of Missouri: Part 2. Subfamily Similinae; Tribes Acutalini, Ceresini, and Polyglyptini. J. Kans. Entomol. Soc. 46:233-242.

Meisch, M. V., and N. M. Randolph. 1966. Preliminary studies of fluctuations of the threecornered alfalfa hopper and beet armyworm obtained in sweeps made in alfalfa at three periods of the day. J. Econ. Entomol. 59:1305-1306.

Moore, G. C., and A. J. Mueller. 1976. Biological observations of the threecornered alfalfa hopper on soybean and three weed species. J. Econ. Entomol. 69:14-16.

Mueller, A. J., and B. A. Dumas. 1975. Effects of stem girdling by the threecornered alfalfa hopper on soybean yields. J. Econ. Entomol. 68:511-512.

Oemler, A. A. 1888. Extracts from correspondence regarding a new tomato enemy in Georgia between Oemler and the Division of Entomology U.S.D.A. Insect Life 1:50.

Tugwell, P., F. D. Miner, and D. E. Davis. 1972. Threecornered alfalfa hopper infestations and soybean yield. J. Econ. Entomol. 65:1731-1733.

Van Duzee, E. P. 1917. Catalogue of the Hemiptera of America north of Mexico, excepting the Aphididae, Coccidae, and Aleurodidae. Univ. Calif. Tech. Bull. 2:828-842.

Wildermuth, V. L. 1915. Threecornered alfalfa hopper. J. Agr. Res. 3:343-362.

Chapter 20

Sampling Stem Flies in Soybean

Govind A. Gangrade and Marcos Kogan

I. Introduction

Stem flies or bean flies are small dipterans of the family Agromyzidae, mostly in the genera *Ophiomyia* and *Melanagromyza*. Several species of these and other closely related genera are well adapted to feeding on leguminous plants. The maggots of these flies are known either as leafminers or as petiole, stem, or root borers. The most comprehensive revision of the Agromyzidae of economic importance was published by Spencer (1973). The system of nomenclature and much of the basic biological information contained in this chapter were extracted from that treatise.

Eight species have been reported injuring *Glycine* spp.: *Ophyomyia phaseoli* (Tryon), *O. shibatsuji* (Kato), *O. centrosematis* (de Meijere), *Melanagromyza dolichostigma* de Meijere, *M. koizumii* Kato, *M. sojae* (Zehntner), *M. vignalis* Spencer, and *Japanagromyza tristela* (Thomson).

Identification of these species is difficult, but very detailed descriptions of adults and immature stages of several species are found in Kato (1961) and Spencer (1973).

A. Geographical Distribution

Authenticated reports of damage to soybean by stem flies indicate that they occur throughout the Orient, Africa, and Oceania. In Japan they seem to be most abundant in the more southern islands (Kato 1961). In China they have been reported as one of the most important pests of soybean in all cultivated areas of the Shantung Province (Res. Group 1978) and *O. phaseoli* has widespread occurrence in the soybean growing regions of Korea (Kogan 1977). An unidentified agromyzid and *M. sojae* attacked soybean in Indonesia (Hall 1924) and in Java *M. dolichostigma* and *M. phaseoli* were found inside the pith of soybean stems causing withering of tips (Goot 1930). Stem flies were also found in soybean in Sri Lanka (Ceylon) (Wikramasinge and Fernando 1962). In the Philippines, the bean fly, *O. phaseoli,* was reported mining leaves of plants at various stages of growth (Quesales 1918, Laan 1949), and despite the presence of high adult populations during the seedling stage, there were no records of damage to soybean (Rejesus 1976). Additional specific observations of damage to soybean were recorded from Thailand (Sepswasdi 1976), Egypt (Ali 1957), Tanganyika (Walker 1960), Tanzania and other parts of Africa (Waiyaki 1972), and Australia (Anonymous 1975). In India stem flies or bean flies are widespread throughout much of the soybean growing regions (Lall 1959, Gangrade 1974, Bhattacharya and Rathore 1977). A preliminary distribution map based on these records is shown in Fig 20-1.

B. Host Plants and Nature of Damage

The bean flies are predominantly legume feeders attacking both cultivated and wild species of *Glycine* and other herbaceous legumes. A summary of the host records, and distribution of the eight species of Agromyzidae listed above, is presented in Table 20-1 based on Spencer (1973).

Melanagromyza sojae seems to be the most common stem fly in India. Adults feed on the upper surface of leaves by making multiple punctures that appear as white spots mainly on leaves of soybean seedlings. These punctures are produced with the ovipositor, and after each puncture is made the fly turns around, apparently to feed on fluid that oozes from the injury. After repeated feeding punctures are produced, the fly moves down to the ventral surface of the leaf to lay an egg in a slit opened in the epidermis of the leaf by the ovipositor. The egg is not discernible from the leaf surface. Eggs are most frequently laid near the base of the leaf but can be found in other areas of the leaf blade, always near a vein.

Most injury to crops is caused by the maggots that mine leaves and tunnel petioles and stems. The newly emerged maggot enters the stem by making a gallery in the pith, eventually reaching down to the root zone. The fully grown maggot prepares, before pupating, an exit hole for the emergence of the adult fly. The exit hole is covered with a thin film of fluid secreted by the maggot. Only one (seldom two or three) maggot is found in an infested portion of the stem. A maggot of the fly damages 11.4-13.4 cm of stems in plants grown between June and October. The December-to-March crop larvae consume only 5.9 cm of stem length (G. Gangrade unpublished).

Infested stems turn red up to ca. 20-25 cm from the ground. Stem fly injury to

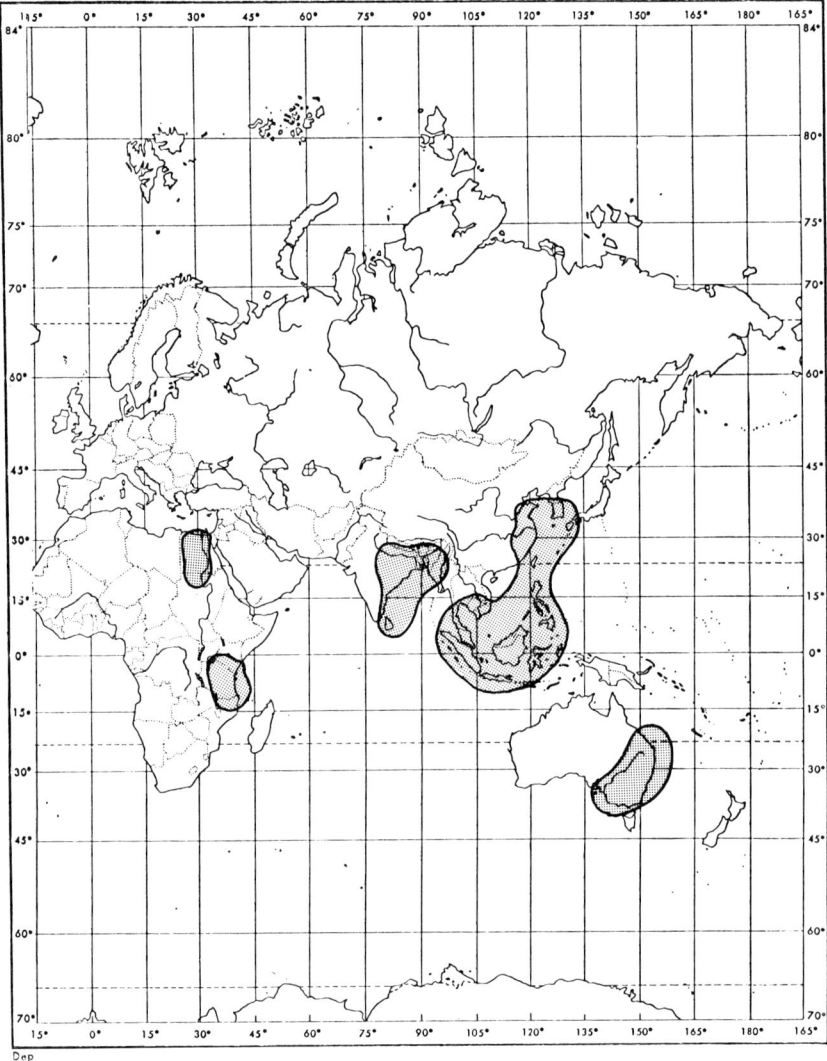

Figure 20-1. Distribution map of bean flies or stem flies (Diptera: Agromyzidae). Map shows areas in which the flies were recorded in association with soybean. The actual geographic range is wider than the one shown in the map.

young plants results in the death of plants up to about 30 days old. In these dying plants, leaves above the injured stems begin to dry and the plant has a drooping appearance as maggots proceed with the hollowing of the pith. Some plants may produce branches and recover from injury. Extensive infestations may sometimes require replanting of fields. Attacks later in the season produce no morphological changes or external symptoms. It is almost impossible to identify injured plants or distinguish them from healthy plants. In India *M. sojae* is a major pest of soybean planted on the rainy and spring season crop. In early growth stages infestations rarely exceed 30% but at this stage plants are most sensitive. As the crop grows, infestation levels of 70-100% may be recorded (Gangrade and Singh 1976, Bhattacharya and Rathore 1977, Lin et al. 1977).

C. Life Cycle and Phenology

The adult of *M. sojae* is a small black and metallic-green fly, ca. 2 mm long. Eggs, laid exclusively on the underside of leaves, hatch in 2-7 days. Larvae (Fig. 20-2A) start feeding on leaf tissue, but in 2-3 days move toward the stem passing through the petiole. There are three or four larval stages and the total larval development is completed in 10-15 days. Pupation occurs within the stem. The tunnel opened by the maggot is distinctly observed by cutting the lower stem longitudinally. The puparium is barrel-shaped (Fig. 20-2B), dark yellow or brown. Pupation lasts for 7-10 days. Emerging adults exit from the tunnel within the stem by a hole prepared by the mature larva (Bhattacharya and Rathore 1977).

The adult *O. phaseoli* is a small, shiny black fly. Oviposition takes place in young leaves either on the upper or lower surface of leaves. Females lay 100-300 eggs over a period of two weeks (Quesales 1918). Eggs hatch in 2-4 days. The larva forms a short leaf-mine but soon enters a vein and moves to the stem via the petiole. Duration of both the larval and the pupal stages is ca. 10 days each. The life cycle is therefore completed in 3-4 weeks. Pupation occurs within the tunnel in the lower portion of the stem. As many as eight exit holes have been observed per plant (Lin et al. 1977).

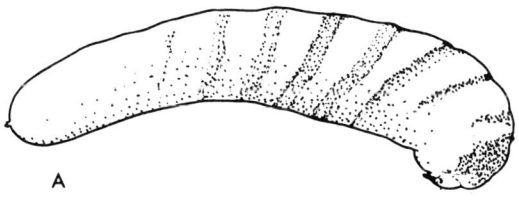

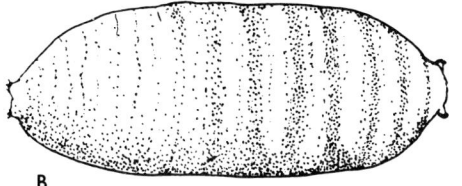

Figure 20-2. *Melanagromyza sojae*: (A) larva; (B) puparium. (After Kato 1961).

Table 20-1. Host records and general geographical distribution of eight species of Agromyzidae recorded in association with soybean (from Spencer 1973)

Stem fly species	Feeding habit [1]	Host plants	Geographic range
Japanagromyza tristella	LM	*Glycine max* (L.) Merr. *Pueraria hirsuta* Matsum. *P. phaseoloides* Benth. *P. thumbergiana* Benth.	Australia, Pacific, Tropical Asia, Africa, Israel
Melanagromyza dolichostigma	SB	*Calopogonium mucunoides* Desv. *Crotalaria juncea* L. *Glycine max* (L.) Merr. *Phaseolus calcaratus* Roxb. *P. vulgaris* L. *P. radiatus* L. *P. sublobatus* Roxb. *Pueraria javanica* Benth.	Java, Taiwan
M. koizumii	SB	*Glycine max* (L.) Merr. *Rhynchosia acuminatifolia* Makino	Japan
M. sojae	SB	*Aeschynomene indica* L. *Cajanus indicus* Spreng. *Flemingia* sp. *Glycine max* (L.) Merr. *Indigofera suffruticasa* Mill. *Indigofera sumatrana* Gaertn. *Medicago sativa* L. *Melilotus* sp. *Phaseolus calcaratus* Roxb. *P. radiatus* L. *P. sublobatus* Roxb. *Swainsonia galegifolia* (And.) R. Br.	Australia, Tropical Asia to Egypt and Africa
			Australia, Asia to Egypt and Africa

M. vignalis	Pod B	*Glycine* sp.	Africa
Ophiomyia centrosematis	SB	*Vigna catjang* Walp. *V. unguiculata* (L.) *Calopogonium mucunoides* Desv. *Centrosema pubescens* Benth. *Crotalaria mucronata* Desv. *Glycine max* (L.) Merr. *Phaseolus lunatus* L. *P. vulgaris* L. *Tephrosia candida* DC. *Vigna unguiculata* (L.)	Australia, Tropical Asia, East Africa
O. phaseoli	SB	*Cajanus indicus* Spreng. *Canavalia ensiformis* DC. *Crotalaria mucronata* Desv. *Crotalaria juncea* L. *Dolichos uniflorus* Lam. *Glycine max* (L.) Merr. *Lablab niger* L. *Phaseolus atropunpureus* DC. *P. calcaratus* Roxb. *P. lathyroides* L. *P. lunatus* L. *P. mungo* L. *P. panduratus* Mart. ex Benth. *P. radiatus* L. *P. semi-erectus* L. *P. vulgaris* L. *Vigna catjang* Walp. *V. unguiculata* (L.) Walp.	Australia, Pacific, Tropical Asia, Africa, Israel Australia, Pacific, Tropical Asia, Africa, Israel
O. shibatsuji	SB	*Glycine max* (L.) Merr. *Glycine ussuriensis* Reg. & Maack	Japan

[1] LM = leaf miner; Pod B = pod borer; SB = stem borer.

The phenology of most bean fly species on soybean has not been monitored in detail. A seasonal abundance curve of adult flies on soybean in Taiwan (Asian Vegetable Research and Development Center 1976) is shown in Fig. 20-3. There are two peaks of adult occurrence per crop season: one at germination and the other at the R1-R2 (bloom) stage. Population fluctuations seem to be influenced by soil moisture, soil pH, solar intensity, and relative humidity. The most important single factor in these fluctuations, however, seems to be the synchronization with the phenology of the soybean crop (Lin et al. 1977).

II. Sampling for Adults

Adult flies are very active fliers and are not easily collected by standard sampling methods. The sweep net is not efficient in sampling adult populations on soybean (Table 20-2). The sweep net method was compared to ground cloth catches and to direct observations of 100 plants. There was no correlation between adult population estimates among the various sampling methods and the level of infestation of plants recorded in the field. Black light and suction traps were used to monitor *Melanagromyza* spp. populations in Taiwan. Monthly total collections are shown in Fig. 20-3.

III. Sampling for Eggs, Larvae, and Puparia

The color of the egg and its placement embedded in the leaf tissue makes direct detection of eggs impossible. Testing of the lacto-phenol method of treatment of

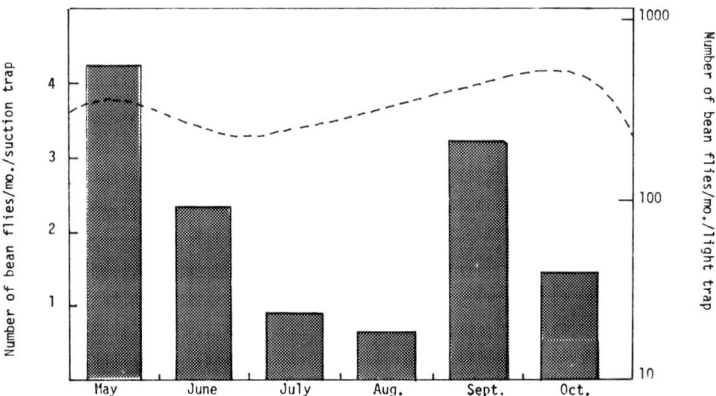

Figure 20-3. *Melanagromyza* spp. adult population fluctuation in Taiwan measured by black light (dashed line curve) and suction trap (bar diagram). (Redrawn from Rose et al. 1976 and Rose unpublished).

Table 20-2. Population of adult stem flies on soybean in India. Sampling by sweep net and ground cloth, compared to direct observation method

Sampling method	Number adult flies	Percent infested plants
Sweep net (10 sweeps)	4.25	
Ground cloth	1.7	85.8-90.6
Counts on 100 plants	101.8	

leaves is recommended; it produces adequate results for sampling leafhopper eggs.

Larval and pupal populations can be measured only by dissecting individual plants. Sample units of 100 plants were used to estimate larval populations in soybean. Soybeans were planted at two-week intervals throughout the year. One hundred plant samples were taken three and four weeks after planting. The plants were dissected and the number of bean fly larvae and pupae was counted. The annual fluctuation of larval populations is shown in Fig. 20-4 (Rose et al. 1976 and R. Rose, personal communication).

IV. Population Indices

At stages V1 through V3-V4, plants under normal moisture conditions show no external symptoms although some isolated plants may display withered leaves. Leaf color may gradually fade, but color change does not seem to be a good in-

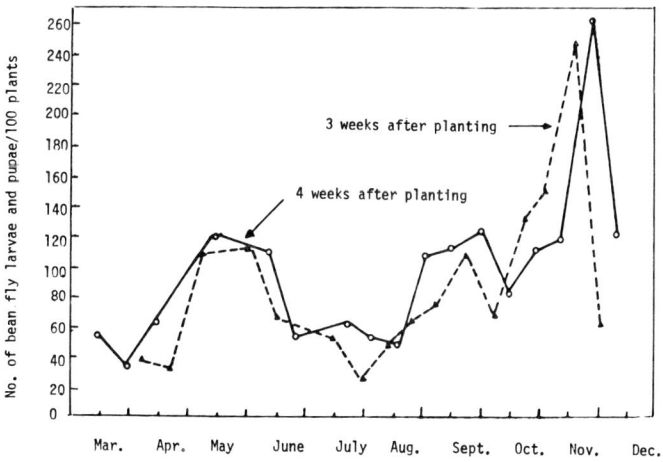

Figure 20-4. *Melanagromyza* spp. larval and pupal population fluctuation on soybean seedlings planted at two-week intervals throughout the year. (Rose et al. 1976 and Rose et al. unpublished).

dex. However, with the cessation of rains, which usually occur in September in India, injured plants rapidly dry up showing very clearly the extent of infestations.

The percent of infested plants determined by dissection was correlated with external symptoms in mung bean, but did not correlate with soybean. The symptoms used were leaf punctures by adult flies and thickness of the unifoliolate leaves (Rose et al. 1976 and Rose unpublished).

V. Concluding Remarks

Stem flies are an important component of soybean ecosystems in the Orient and Africa. The economic effect of early-season infestations seems to be unquestionable, but if infestations occur late in the season it seems that soybean can tolerate high populations with no apparent yield loss. Sampling procedures for all stages are still inadequate. Although dissection of seedlings gives an excellent absolute population estimate, it is not applicable for screening of resistant lines or varieties since the method is destructive. On the other hand, symptoms of bean fly infestation may not be very useful for population dynamics studies. These symptoms, however, would be perfectly adequate for resistance screening programs since plants surviving a heavy early infestation probably have some resistance qualities (if they are not merely escapees). For advancement of our knowledge on stem flies, however, new and more sensitive methods of detection should be developed. One possible avenue would be the use of infra-red photography for early detection of stress caused by larval feeding.

References

Ali, A. M. 1957. On the bionomics and control of bean fly *Agromyza phaseoli* Coq. (Diptera: Agromyzidae). Bull. Soc. Entomol. Egypte **41**:551-554.

Anonymous. 1975. Navy beans, French beans and soybeans. Summary of insect control recommendations. Queensland Agr. J. **101**:596-598.

Bhattacharya, A. K., and Y. S. Rathore. 1978. Survey and study of the bionomics of major soybean insects and their chemical control. G. B. Pant Univ. Dep. Entomol. Res. Bull. **107**:324 p.

Bhattacharya, A. K., and Y. S. Rathore. 1979. Soybean insects—Problems in India. in Proc. World Soybean Conference II, Raleigh, North Carolina (in press).

de Meijere, J. C. H. 1922. Contribution to the agromyzids of Java. Bijdr. Dierkd. **22**:17-24.

Gangrade, G. A. 1974. Insects of soybean. Jawaharlal Nehru Krishi Vishwa Vidyalaya Tech. Bull. **24**:88 p.

Gangrade, G. A., and O. P. Singh. 1976. Effect of stem fly, *Melanagromyza phaseoli* (Tryon), on yield of pods and grains of soybeans in India. Z. Angew. Entomol. **80**:438-441.

Goot, P. van der. 1930. The agromyzid flies of native leguminous plants in Java. Meded. Inst. Plziek. **78**:97 p.

Hall, C. J. J. van. 1924. Diseases and pests of cultivated plants in the Dutch East Indies in 1923. Meded. Inst. Plziek. **64**:47 p.

Kato, S. 1961. Taxonomic studies on soybean leaf and stem mining flies (Diptera: Agromyzidae) of economic importance in Japan, with descriptions of three new species. Bull. Nat. Inst. Agr. Sci. Ser. C **13**:171-206.

Kleinschmidt, R. P. 1960. New species of Agromyzidae from Queensland. Queensland J. Agr. Sci. **17**:321-337.

Kogan, M. 1977. Soybean entomology in Korea. Consultant report to the Crop Improvement Research Center and Institute of Agricultural Sciences, Office of Rural Development, Suweon, Korea. September 1977. College of Agr., Univ. of Ill., Urbana, 25 p.

Laan, P. A. van der. 1949. Chemical control of the beanfly, *Melanagromyza phaseoli* on soybeans. Meded. Alg. Proefst. Landbouwwet. **98**:28 p.

Lall, B. S. 1959. On the biology and control of the beanfly, *Agromyza phaseoli* Coq. Sci. Cult. **24**:531-532.

Lin, F. J., T. C. Wang, and C. Y. Hsieh. 1977. Population fluctuations of soybean miner, *Ophiomyia phaseoli* (Tryon). Bull. Inst. Zool. Acad. Sin. **17**:69-76.

Quesales, F. O. y. 1918. The bean fly. Philipp. Agr. **7**:2-31.

Rejesus, R. S. 1976. Insect pest diversity and succession in Asian soybeans. pp. 97-103 in R. M. Goodman, ed. Expanding the use of soybeans. Proc. of a Conference for Asia and Oceania. Univ. of Illinois, College of Agriculture, INTSOY Ser. 10. 261 p.

Research Group on the Integrated Control of Soybean Insects of Hui-Ming District, Shantung Province. 1978. Studies on the soybean agromyzid fly in Shantung Province (in Chinese with English summary). Acta Entomol. Sin. **21**:137-150.

Rose, R. J., C. S. Lin, S. P. Kung, H. W. Kao, C. Y. Su, and C. Y. Chang. 1976. Bean flies and their control. pp. 169-171 in R. M. Goodman, ed. Expanding the use of soybeans. Proc. of a Conference for Asia and Oceania. Univ. of Illinois, College of Agriculture. INTSOY Ser. 10. 261 p.

Sepswasdi, P. 1976. Control of soybean insect pests in Thailand. pp. 104-107 in R. M. Goodman, ed. Expanding the use of soybeans. Proc. of a Conference for Asia and Oceania. Univ. of Illinois, College of Agriculture. INTSOY Ser. 10. 261 p.

Spencer, K. A. 1973. Agromyzidae (Diptera) of economic importance. Dr. W. Junk B. V., Pub., The Hague. 418 p.

Waiyaki, J. N. 1962. The bean fly in Tanzania. Trop. Res. Misc. Rep. 786.

Walker, P. T. 1960. Insecticide studies on East African agricultural pests. III. Seed dressings for the control of the bean fly, *Melanagromyza phaseoli* (Coq.) in Tanganyika. Bull. Entomol. Res. **50**:781-793.

Wickramasinge, N., and H. E. Fernando. 1962. Investigations on insecticidal seed dressings, soil treatments and foliar sprays for the control of *Melanagromyza phaseoli* (Tryon) in Ceylon. Bull. Entomol. Res. **53**:223-240.

SECTION VII
Pod Feeders

Chapter 21

Sampling *Heliothis* spp. on Soybean

Ronald E. Stinner, J.R. Bradley, Jr., and John W. Van Duyn

I. Introduction

The species of the genus *Heliothis* may represent the most important insect pest complex world-wide with respect to potential crop loss. These noctuid species exhibit wide host ranges, high fecundity, and great vagility. In the United States, two species, the corn earworm, *H. zea* (Boddie), and the tobacco budworm, *H. virescens* (Fabricius), are considered serious pests of many agriculturally important crops. A comprehensive bibliography of these two species was recently prepared by Kogan et al. (1978), and studies on economic thresholds and sampling *Heliothis* spp. on cotton, corn, soybean, and other host plants were compiled in a publication by the Southern Regional Cooperative Project S-59 (Sterling 1979).

Other species of *Heliothis* may be damaging to soybean in Europe, Asia and Oceania. *H. armigera* (Hübner) is the most important Old World species. *H. punctigera* Wllgr. is an Australian species, and *H. viriplaca adaucta* Btlr. occurs in Japan. These two species apparently coexist with *H. armigera* within certain zones of their geographic range.

A. Geographical Distribution

H. zea and *H. virescens* are limited in their distribution to the western hemisphere (Kogan et al. 1978), but both range rather widely over North, Central, and South America (Fig. 21-1). *Heliothis* spp. are major pests of soybean in the southern United States and are most abundant in a geographic band that in-

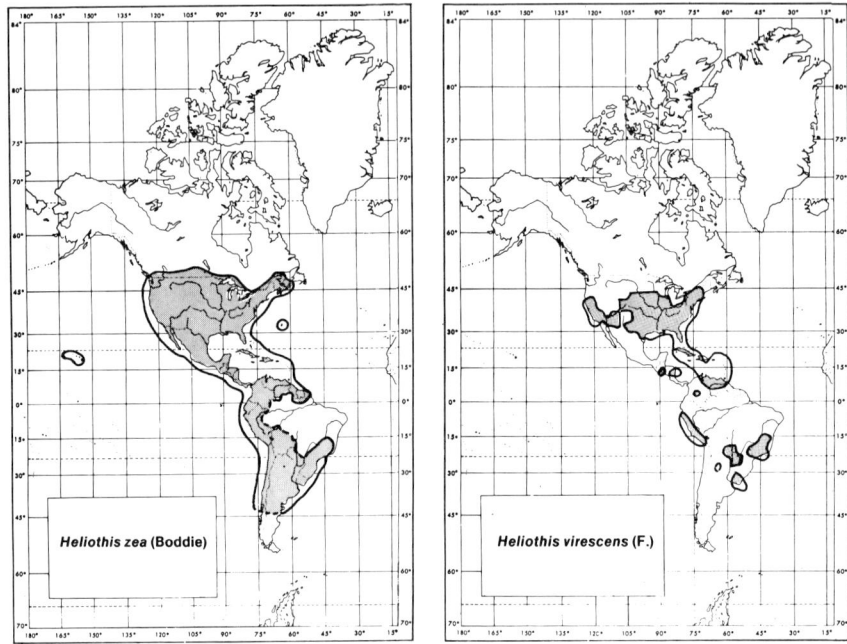

Figure 21-1. Distribution of *Heliothis zea* and *H. virescens* in the western hemisphere (reproduced by permission from Kogan et al. 1978).

cludes the coastal plain areas of Alabama, Florida, Georgia, South Carolina, North Carolina, and southern Virginia. In much of this area *Heliothis* (mainly *zea*) is the most important soybean insect pest. The importance of *Heliothis* decreases north and west of this area, presumably because of low overwintering survival (north), unsuitable soil types (north and west), and low availability of corn (west). Although states of the mid-south, north to Missouri and west to Oklahoma, do not experience *Heliothis* in high numbers annually, this insect may be a serious pest in some years and in localized areas. In soybean growing regions outside the coastal plain and mid-south areas, *Heliothis* is usually an infrequent and minor pest.

The ratio of *H. zea* to *H. virescens* on soybean is usually heavily skewed in favor of *H. zea* although the ratio greatly fluctuates with the year and location. In those locales where corn is a major field crop (e.g., in North Carolina) *H. zea* appears to typically compose at least 90% of the population. In five years (1971-1975) of sampling in soybean, the percentage of *H. virescens* larvae ranged from 0% to 20%, averaging less than 5%. However, in areas where cotton is predominant (e.g., in west Mississippi) *H. virescens* may compose the majority of the population. *H. virescens* is typically more abundant on soybean in the mid-south. The *H. zea/H. virescens* ratio may also fluctuate greatly with crop phenology. When soybean is vegetative and neighboring corn is immature, the ratio of *Heliothis* on soybean may favor *H. virescens*.

B. Host Plants and Nature of Damage

The list of host plants of the two *Heliothis* spp. is very broad. The feeding habits indicate that these are polyphagous pests. One or both species can cause economic damage to soybean, cotton, tobacco, field and sweet corn, tomatoes, grain sorghum, and peanuts as well as many other agricultural, fruit, vegetable, and ornamental crops. Included among the known hosts are also a large number of uncultivated plants (Quaintance and Brues 1905; Barber 1937; Neunzig 1963, 1969; Graham and Robertson 1970; Roach 1975). With the deletion of corn and the addition of tobacco as primary hosts, the same host range would probably apply equally well to *H. virescens*.

Both species feed on leaves and fruit of host plants, with the latter preferred. On soybean, most eggs are laid on or around blooms and terminal growth. Early instar larvae feed first on blossoms, foliage, and sometimes very small pods. *H. zea* is primarily a foliage feeder until seed enlargement begins within the pods. At this time the caterpillars feed predominantly on developing seeds (Fig. 21-2). If pods are not available due to advanced maturity or decimation, larvae will complete their development on foliage. Severely infested fields may have the plants completely stripped of foliage and pods.

C. Life Cycle and Phenology

Eggs, larvae, and pupae of *H. zea* and *H. virescens* are similar, and the following descriptions from Deitz et al. (1976) hold for both species.

> Egg: height 0.50-0.55 mm, diameter 0.56-0.57 mm; subspherical with flattened base and prominent ribs; whitish to cream-colored, later with reddish-brown band becoming gray before hatching. Larvae: length 1.5 mm (first instar) to 42.3 mm (sixth instar); newly emerged larva translucent yellowish-white with brown to dark brown head capsule; second instar yellowish-brown frequently with orange and brown longitudinal lines with head capsule reddish-brown to brown; color third to last (fifth or sixth) instars highly variable; body with ground color light greenish-yellow to reddish or brown, usually with distinct longitudinal white or cream-colored bands, chalazae and parts of meso- and metathoracic legs usually shiny black, head from brown to orange: *abdomen with five pairs of prolegs*. Pupa: length 18.0-22.0 mm, width 5.0-6.0 mm; shiny reddish-brown to dark brown.

Although the eggs of the two species are virtually indistinguishable (Neunzig 1964), the larvae and pupae can be separated. The larvae of both species curl up when disturbed. Later instars of *H. virescens* possess spinules on the chalazae of setae D1 of abdominal sgements 1, 2, and 8, which are absent in *H. zea* (Fig. 21-3A,B). A mandibular retinaculum or retinacular scar is characteristic of *H. virescens* but absent in *H. zea* (Neunzig 1969, Fig. 21-3C,D). Early instar larvae

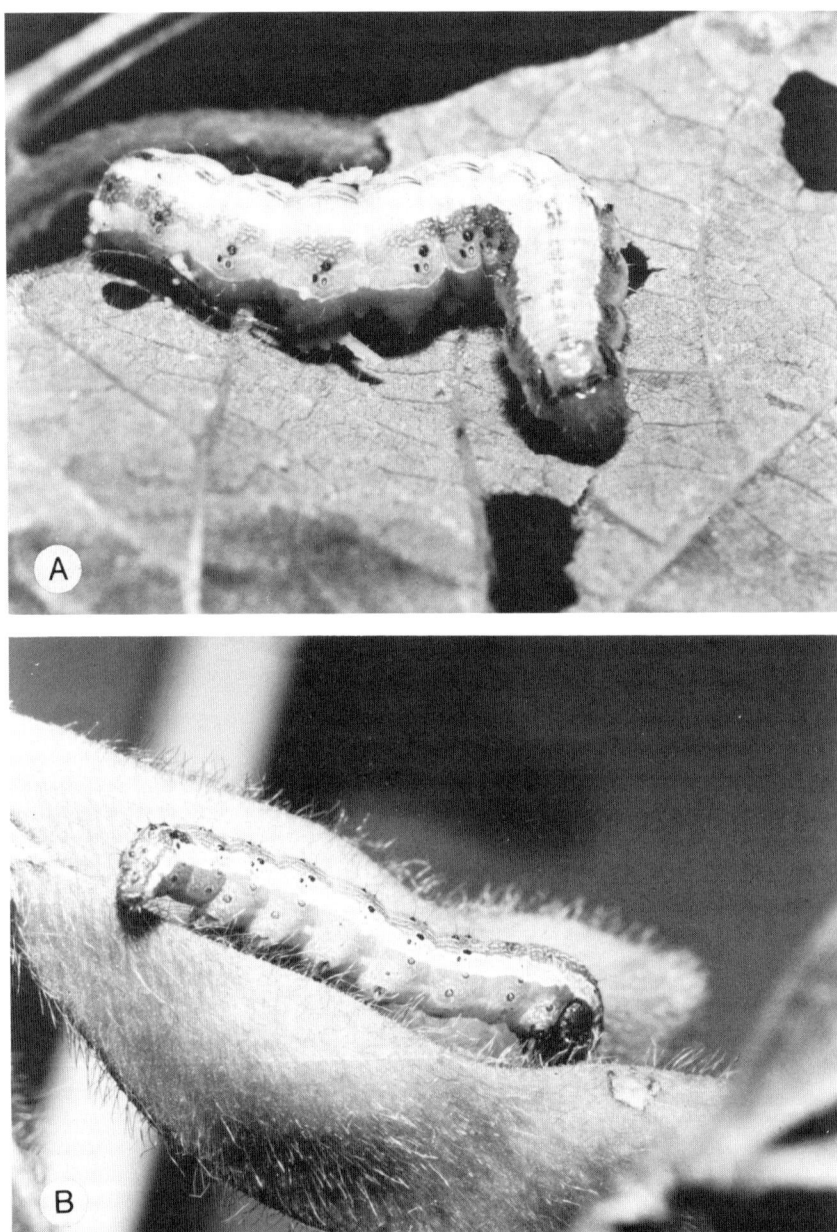

Figure 21-2. Injury to soybean. (A) Foliage feeding; (B) pod feeding.

Sampling *Heliothis* spp. on Soybean

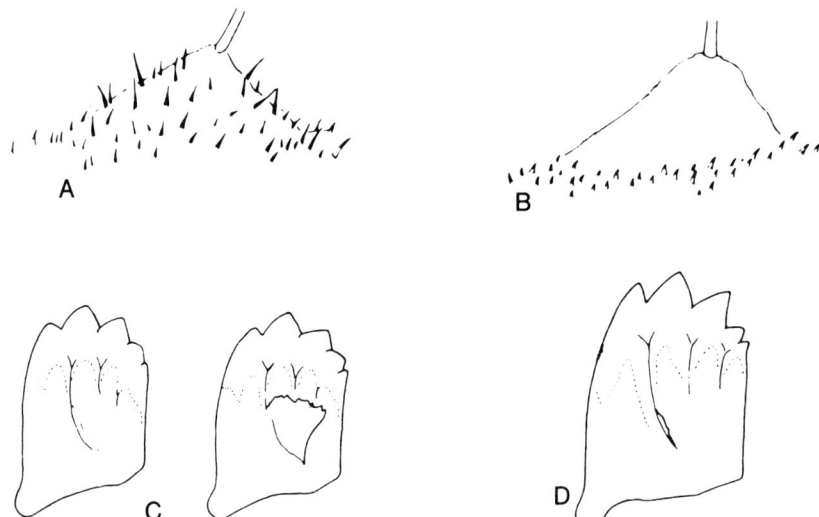

Figure 21-3. Larval characteristics: setae D1 of abdominal segment 8 of (A) *H. virescens;* (B) *H. zea* from Neunzig (1969); (C) mandible with retinaculum or retinacular scar - *H. virescens;* (D) retinacular scar absent - *H. zea.*

are more difficult to separate, but keys for their separation have been published (Neunzig 1964).

H. virescens pupae tend to be smaller than *H. zea.* The spiracles of *H. zea* pupae are relatively large, lengthwise about 1/7 the length of the segment (Neunzig 1960), and a maxillary palp sclerite is visible; in *H. virescens,* the spiracles are smaller (approx. 1/15 length of segment, Fig. 21-4) and the sclerite is not visible.

H. zea and *H. virescens* adults have a wing span of 36.0-41.0 mm; forewings of males are generally light yellowish-olive and those of females are yellowish-brown to pinkish-brown. Hindwings of both sexes are white with broad dark brown outer marginal bands and, commonly, with narrow brown intermarginal bands. Vesture of the head and thorax is the same color as the forewing, and the abdomen is paler. Adults of *H. zea* lack the oblique bands on the forewings seen on *H. virescens* (Fig. 21-5); *H. virescens* adults also have a green cast (rather than brown) to the forewing and body color.

Although both species may have continuous generations in southern Florida, in most areas *Heliothis* overwinter as diapausing pupae in earthen cells at depths to 15 mm (Barber 1941, Slosser et al. 1975). Adults emerging in the spring oviposit on wild hosts, corn, and tobacco. The actual date of spring emergence varies from year to year and from north (early May) to south (early March). Although successful overwintering of diapausing pupae of both species seems limited to regions where the frost line is about 5-10 mm in the soil, the high vagility of adult moths enables invasion of sweet corn in southern Canada by mid-summer.

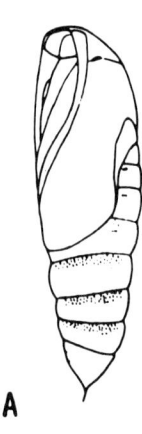
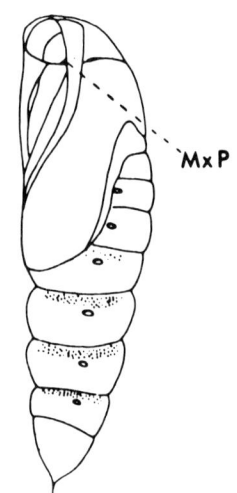

Figure 21-4. Pupae of (A) *H. virescens* and (B) *H. zea* (from Neunzig 1969).

Up to 3000 eggs/female are laid singly, hatching in 3-10 days, depending on temperature. There are five or six larval instars (seven have been infrequently reported). Developmental periods of corn earworms vary considerably according to larval food (Isely 1935). The prepupa drops from the plant and burrows into the soil, forming a J-shaped tunnel and pupal cell. About 8-12 days after pupation the adult emerges from the tunnel. Generation time averages 3-4 weeks during the summer, with a preovipositional period of 2-5 days.

In the southeast (North Carolina), the first generation of corn earworm is produced largely on seedling corn and to a lesser extent on tobacco and wild hosts such as toadflax. The second generation develops almost exclusively within corn ears while the third and fourth generations develop on cotton, soybean, peanut, tobacco, and sorghum (Neunzig 1969, Lincoln 1972).

Ovipositing moths are particularly attracted to hosts in the peak of flowering (Johnson et al. 1975). Although low levels of both *H. zea* and *H. virescens* may be found feeding on soybean foliage throughout the vegetative period (usually from May into July) moths do not appear to be strongly attracted to soybean until it blooms. It has been established that the occurrence of large-scale *Heliothis* outbreaks on soybean is associated with the following simultaneous events: (1) peak bloom, (2) open crop canopy, and (3) a large peak moth flight. The importance of specific crops to any generation, particularly that which damages soybean, within a local area depends then on the availability of preferred hosts for oviposition within the local crop complex. The combination of high mobility and fecundity (with host selection closely tied to crop phenology and plant growth characteristics) can lead to the "overnight" explosions of the pest often encountered in individual fields.

In addition to the above characteristics, the ability of *H. virescens*, in particular, to develop resistance to a broad range of pesticides has further enhanced its pest status. In parts of northern Mexico the total loss of cotton production has been

Sampling *Heliothis* spp. on Soybean

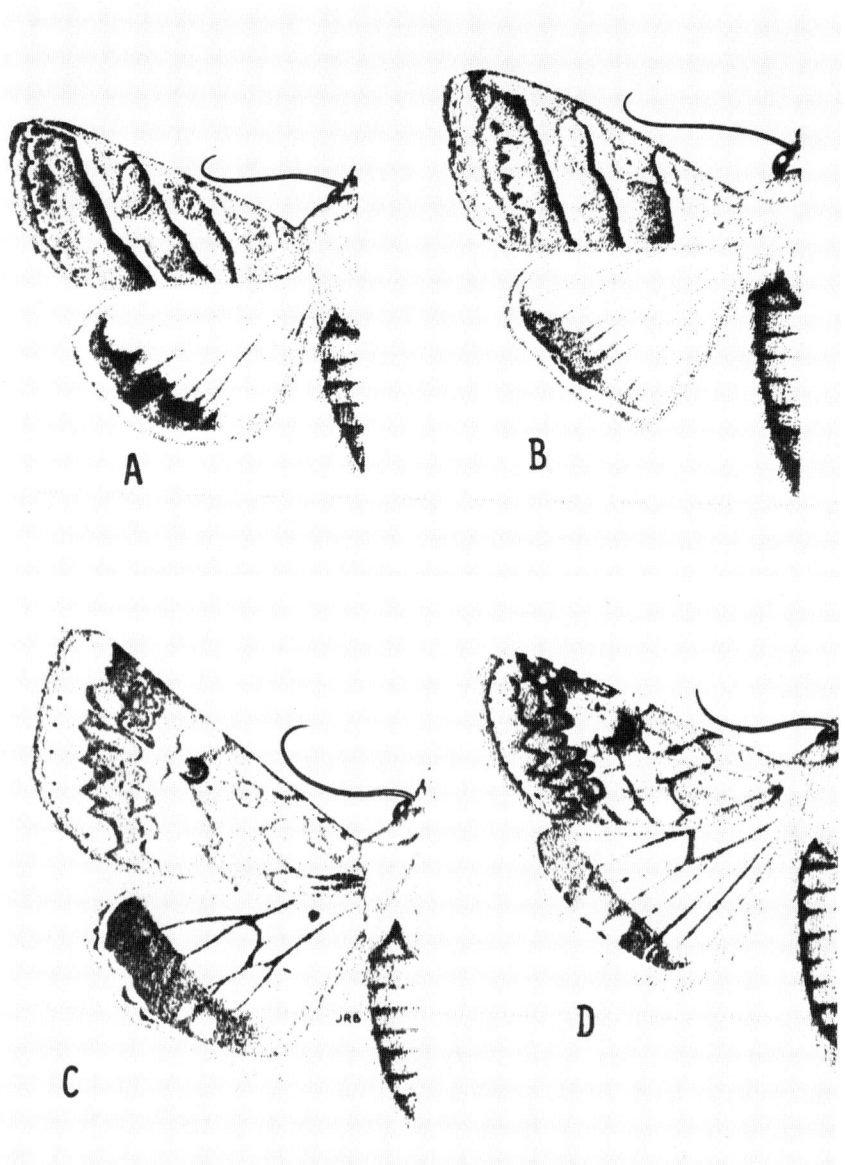

Figure 21-5. *Heliothis* adults: (A) *H. virescens* ♂; (B) *H. virescens* ♀; (C) *H. zea* ♂; (D) *H. zea* ♀ (from Neunzig 1969).

attributed to destruction caused by this species (Adkisson 1972). In other areas, drastic changes in cotton production systems had to be made to reduce the destructive potential of the tobacco budworm as pesticides became ineffective (Bottrell and Adkisson 1977).

All of these characteristics lead one to conclude that these two species are close to the perfect "r-strategists." However, both species are also cannibalistic to some extent, with *H. zea* much more so, possibly having evolved on the limited food supply provided by the small grain "ears" of maize ancestors in Central or South America (Stinner et al. 1977).

II. Sampling for Adults

Adult *Heliothis* may be sampled by flushing or by using "sugar lines." However, the most common survey technique, particularly for *H. zea*, has been the blacklight trap. Trap location has been shown to greatly affect numbers caught. The oviposition potential at the trap location also appears to be related to the male/female ratio of trapped moths. Recently a synthetic pheromone trap was developed for *H. virescens*; it will probably replace the blacklight trap as the major adult sampling tool. Due to the mobility of these adults, all these techniques provide relative density information only. The development of radar techniques (including Doppler radar) offers new promise for absolute density measures. Such techniques have already been used with limited success for *H. armigera* (Hubner) in Africa (Rainey and Joyce 1972, Schaefer 1976, Rainey 1978).

The problem of determining whether the adults are of local origin or are migrants provides a further complication that has not yet been satisfactorily resolved. However, the use of isoenzyme techniques (Sell et al. 1974) and X-ray bombardment (D'Auria and Bennett 1975) provide at least initial tools for this determination.

III. Sampling for Eggs

The sampling of *Heliothis* eggs in any row crop or on native hosts is accomplished by direct visual examination; however, visual counts for absolute densities are difficult on blooming soybean due to the great amount of foliage, pubescence characteristics, and blossom color (Boyer et al. 1963, Johnson et al. 1975). Relative estimates of egg abundance may therefore be made by examination of the highly attractive foliage and bloom terminals (Morrison et al. 1979). This may be achieved by excising terminals, placing them in refrigerated conditions, and examining them in the laboratory. Numbers of eggs in the field rapidly change due to oviposition, predation, and hatching. When sampling for comparative purposes, these factors should be closely considered.

IV. Sampling for Larvae

Sampling of *H. zea* larvae in soybean has been achieved by visual examination (single-plant and linear row), by sweep net, suction devices, and ground cloth (Boyer and Dumas 1963, Falter and Van Duyn 1973, Turnipseed 1974, Barnes et al. 1974, Boyer 1974, Deitz et al. 1976). The most commonly used technique has been the ground cloth. Numbers of large larvae collected are usually near-estimates of absolute numbers/linear row unit. Although larger larvae are most easily detected, first instars can be sampled with reasonable accuracy for comparative purposes. When sampling small larvae, the worker must observe the drop cloth very closely for an adequate period of time to allow for larval movement. Partitioning of the ground cloth into six or eight equal compartments with lines is very useful when counting small larvae. Small larvae may be destroyed on the cloth while counting.

In much of the southern United States, infestation levels vary greatly with crop phenology and canopy development and with higher populations found in open-canopied and blooming fields (Falter and Van Duyn 1973, Barnes et al. 1974, Stinner et al. 1977). Generalizations regarding sampling efficiency of various techniques in soybean in totum are extremely difficult due to the problems of making comparisons among fields and regions that differ in planting dates, cultivars, row width, growth forms, and timing of phenological events. With row plantings of "standard" width, a comparison of three sampling methods (D-Vac, sweep net, ground cloth) (Shepard et al. 1974) showed that samples of low mean numbers yielded higher coefficients of variability (CV). Also, linear regression of CV on the sampling means revealed high correlation, and analysis of covariance of weighted means revealed no significant difference among the three techniques. However, another study comparing sampling techniques showed the ground cloth, beat method, consistently gave significantly larger mean numbers of large, small, and large + small *H. zea* larvae with the lowest CV for large and large + small larvae; the CV for small larvae was intermediate (Turnipseed 1974). Of two variations of vacuum net sampling (vertical and horizontal), sweep net sampling (across and along the row), and ground cloth (shake and beat), the plant shaking variation of the ground cloth gave the lowest CV for small larvae.

When sampling for *Heliothis* larvae in a pest management context, biased and sequential sampling can be used. With knowledge of the planting date, cultivar, row-spacing, and growth characteristics of soybean fields and of the occurrence and abundance of the F_3 moth flight by light trap, predictions of high, moderate, or low probability of economic infestations can be made for each field. Early-planted closed-canopied fields, blooming before the moth flight, would be of low probability for infestation. Late-planted, open-canopied fields that bloom in concert with the moth flight have a high probability of being infested. Fields of intermediate conditions may suffer economic infestations only during years of high moth abundance and are therefore of moderate infestation probability.

Upon assessing the probability of infestation, the sampling effort is allocated to high risk fields. The sampling plan should avoid field margins and areas of abnormal growth.

Sequential sampling plans have not been developed for *Heliothis* spp. on soybean although several are available for *Heliothis* spp. on cotton (Allen et al. 1972, Sterling 1973, Sterling and Pieters 1975, Pieters and Sterling 1975). For soybean, however, intuitive, nonstatistical forms of sequential sampling may be used in circumstances in which very high or low larval numbers are encountered. These intuitive plans are further facilitated when a population estimate relative to the threshold is readily apparent when considering the attractiveness of the field and other factors that favor *Heliothis* infestations. There are no reported studies on the number of samples required for reliable estimates but experience in pilot pest management projects has given satisfactory procedures for making management decisions. A minimum of three samples (2 row-m/sample) per field with one sample for every 0.6 ha over the minimum number is suggested. However, the number may be limited to three where very high or low populations are present even in fields designated to receive high numbers of samples.

In cases of populations closely above or below threshold levels, a greater number of samples must be taken. A 50% increase in samples is suggested with a minimum of five samples/field.

V. Sampling for Pupae

Pupal sampling has proven quite difficult but has been accomplished by several means: soil sifting, soil scraping, and estimation of emergence of prepupae from fruiting forms. Soil sifting provides the most precise information, but is quite expensive (both in time and sample number limitations). Soil scraping (removing top 1.5 cm of soil) is much more rapid and allows the observer to count entrance tunnels to pupal chambers. This technique, however, can be used only in nonrocky soils and where large numbers of similar noctuids (i.e., fall armyworm) do not occur. Estimation of pupal numbers based on prepupal emergence from the host has been used successfully for *H. zea* on corn, where ears are collected and held over small areas of soil that is subsequently sifted (Caron et al. 1978).

For overwintering studies, a common technique for pupal sampling is to place full-grown, media-reared larvae under inverted cups on the soil surface. Frequent inspection, possibly with replacement of lost of dead individuals, gives a known population entering the winter. Subsampling by soil-sifting can be done to obtain mortality/survival over time (Slosser et al. 1975), or cages may be used to capture either in-season or overwintered emerging moths.

A similar technique to determine the emergence of adult moths from various crops is the use of soil covers or cages (Roach and Ray 1976). While yielding a good assessment of the source of moth generations, this technique requires considerable labor, and is, at best, an indirect measure of pupal abundance.

VI. Sampling Considerations and Concluding Remarks

In much of the southeastern United States, it is third generation *Heliothis* populations that cause economic damage to soybean. Second generations moths emerging from corn fields during late July and early August and oviposit in soybean fields, particularly open-canopied and flowering fields. Larval population surveys and phenological records illustrate the relationship of soybean canopy establishment and flowering to *Heliothis* spp. populations during "normal" seasons.

In North Carolina the pattern of *H. zea* in soybean is clumped and affected by many factors. Blooming date and canopy formation are probably the major influencing factors in this pattern. Soybean fields that bloom while moths are abundant are much more attractive for oviposition than fields that are without flowers, especially fields that are past flowering as opposed to pre-bloom fields. Canopy closing appears to affect larval abundance by reducing attractiveness for oviposition and possibly also egg-larval survival. Because of these factors *Heliothis* numbers vary tremendously among fields of different cultivars, planting dates, and extent of canopy development (as affected by fertility, other pests, planting date, etc.).

Soybean cultivars of maturity group V ('Dare', 'York', 'Forrest', 'Essex', 'Coker 136', etc.) that are planted before mid-May in North Carolina usually flower before a major *Heliothis* moth flight and thus usually escape heavy infestation. Maturity group VI soybeans ('Lee 68' & 74, 'Pickett 71', 'Davis', 'FFR 666', 'McNair 600', etc.) are also seldom damaged if they are planted before mid-May and grow vigorously, since they tend to bloom and establish a thick canopy early. Cultivars of maturity groups VII and VIII ('Bragg', 'Ransom', 'Hutton', 'FFR 777', 'Hampton', etc.) most frequently receive high *Heliothis* populations especially if they are planted after the first of June and if they exhibit poor growth. However, even these late maturing cultivars are much less likely to develop *Heliothis* populations when planted by mid-May.

During the growing seasons of 1971-1975, extensive sampling of soybean in three areas of North Carolina demonstrated this consistent relationship among full bloom, canopy development, and oviposition by *Heliothis*. Larval populations consistently peaked at the medium pod stage (R4-R5) (oviposition two weeks prior, at full bloom), and populations in open-canopied fields were generally consistently higher than in closed-canopy fields (Fig. 21-6). Soybean with an intermediate canopy development generally supported intermediate populations. These observations led to the recommendation that, where other insects such as the Mexican bean beetle (which prefer early-planted fields) are not a problem, soybean should be planted as early as possible or, if planted later, that it be planted with narrow rows, heavier seeding rates, broadcast, or drilled to develop a full canopy as rapidly as possible.

Although high densities of *Heliothis* on early-planted soybean are very uncommon, computer simulations (using HELSIM, Stinner et al. 1977) had predicted

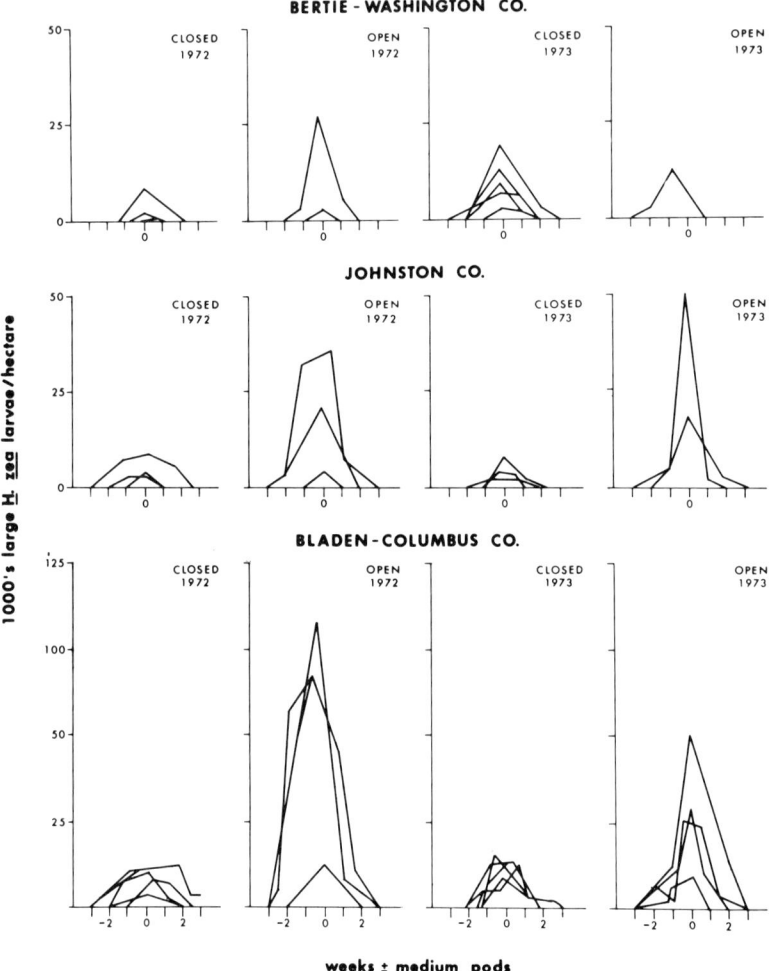

Figure 21-6. *Heliothis* spp. larval populations on open and closed canopy soybean fields in 3 regions of North Carolina (1972 and 1973).

that under hot and dry conditions, *Heliothis* populations on early-planted soybean would exceed those normally encountered on late-planted soybean. These simulations included direct effects of temperature on *Heliothis* development as well as indirect effects on crop growth rate and canopy development. During the summer of 1977, much of North Carolina was subjected to prolonged hot, dry weather. In those areas, there were numerous observations of high infestation levels of *Heliothis* in early-planted soybean, but not in late-planted soybean. The explanation for this occurrence is threefold: (1) Corn matured much more rapidly. In addition, in many areas over 40% of the total corn acreage (the later planted fields) dried up without producing ears. (2) Since blooming in soybean is largely photoperiodically controlled, early soybean bloomed only 2-3 days earlier than normal as opposed to the corn that matured 2-3 weeks earlier. In addition, the prolonged drought prevented normal canopy development and many early planted fields had greatly reduced foliage at time of bloom, the most attractive state for *Heliothis*. (3) Overwintered *Heliothis* adults emerged about two weeks earlier than normal and development was about 20% faster than at usual temperatures, resulting in the peak F_2 moths coincident with early blooming soybean, rather than two weeks later than peak-bloom of early soybean (the usual situation).

The lack of attractive late corn, the availability of highly attractive, open-canopied, blooming soybeans, and the earlier F_2 moth flight coincident with peak bloom in these early-planted fields resulted in major infestations of early soybean by *Heliothis* in the drought areas. In those regions where there was adequate rainfall, *Heliothis* was found at high densities only in late-planted fields, as had been observed and simulated for normal rainfall (under normal or higher than normal temperatures).

Acknowledgments

The research reported herein was supported in part by NSF grants GB-28855 and GB-38271X, and by NSF/EPA grant GB-34718. Thanks are expressed to Dr. H. H. Neunzig and the North Carolina Agricultural Experiment Station for permission to reproduce Figs. 21-3, 21-4, 21-5. Appreciation is also due Ms. Sue Cohill for manuscript preparation.

References

Adkisson, P. L. 1972. The integrated control of the insect pests of cotton. Proc. Tall Timbers Conf. Ecol. Anim. Contr. Habitat Manage. 4:175-188.
Allen, J., D. Gonzalez, and D. V. Gokhale. 1972. Sequential sampling plans for the bollworm, *Heliothis zea*. Environ. Entomol. 1:771-780.
Barber, G. W. 1937. Seasonal availability of food plants of two species of *Heliothis* in eastern Georgia. J. Econ. Entomol. 30:150-158.
Barber, G. W. 1941. Hibernation of the corn earworm in southeastern Georgia. USDA Tech. Bull. 791:17 p.
Barnes, G., B. F. Jones, and W. P. Boyer. 1974. Control insects on soybeans. Ark. Agr. Ext. Serv. Leafl. 193(rev.):6 p.
Bottrell, D. G., and P. L. Adkisson. 1977. Cotton insect pest management. Annu. Rev. Entomol. 22:451-481.
Boyer, W. P. 1974. Soybean insect scouting in Arkansas. Ark. Coop. Ext. Serv. Bull. pp. 1-3.
Boyer, W. P., and W. A. Dumas. 1963. Soybean insect survey as used in Arkansas. USDA Coop. Econ. Insect Rep. 13:91-92.

Boyer, W. P., W. H. Whitcomb, G. C. Dowell, and R. Bell. 1963. Notes on *Heliothis* in Arkansas. USDA Coop. Econ. Insect Rep. **13**:109-111.

Caron, R. E., J. R. Bradley, Jr., R. H. Pleasants, R. L. Rabb, and R. E. Stinner. 1978. Overwinter survival of *Heliothis zea* produced on late-planted field corn in North Carolina. Environ. Entomol. **7**:193-196.

D'Auria, J. M., and R. Bennett. 1975. X-rays and trace elements. Chemistry **48**:17-19.

Deitz, L. L., J. W. Van Duyn, J. R. Bradley, Jr., R. L. Rabb, W. M. Brooks, and R. E. Stinner. 1976. A guide to the identification and biology of soybean arthropods in North Carolina. N.C. Agr. Exp. Sta. Tech. Bull. **238**:264 p.

Falter, J., and J. Van Duyn. 1973. Soybean insect control. N.C. Agr. Ext. Serv. Insect Notes 1(rev):5 p.

Graham, H. M., and O. T. Robertson. 1970. Host plants of *Heliothis virescens* and *H. zea* (Lepidoptera: Noctuidae) in the Lower Rio Grande Valley, Texas. Ann. Entomol. Soc. Amer. **63**:1261-1265.

Graham, H. M., N. S. Hernandez, Jr., and J. R. Llanes. 1972. The role of host plants in the dynamics of populations of *Heliothis* spp. Environ. Entomol. **1**:424-431.

Isely, D. 1935. Relation of hosts to abundance of cotton bollworm. Ark. Agr. Exp. Sta. Bull. **320**:30 p.

Johnson, M. W., R. E. Stinner, and R. L. Rabb. 1975. Ovipositional response of *Heliothis zea* (Boddie) to its major hosts in North Carolina. Environ. Entomol. **4**: 291-297.

Kogan, J., D. K. Sell, R. E. Stinner, J. R. Bradley, Jr., and M. Kogan. 1978. The literature of arthropods associated with soybean. V. A bibliography of *Heliothis zea* (Boddie) and *H. virescens* (F.) (Lepidoptera: Noctuidae). Univ. of Illinois, College of Agriculture, INTSOY Ser. 17. 242 p.

Lincoln, C. 1972. Seasonal abundance. pp. 2-7 in Distribution, abundance, and control of *Heliothis* species in cotton and other host plants. S. Coop. Ser. Bull. **169**:92 p.

Morrison, D. E., J. R. Bradley, Jr., and J. W. Van Duyn. 1979. Populations of corn earworm and associated predators after applications of certain soil-applied pesticides to soybeans. J. Econ. Entomol. **72**:97-100.

Neunzig, H. H. 1960. The pupae of *Heliothis zea* and *Heliothis vierscens* (Lepidoptera: Noctuidae). Ann. Entomol. Soc. Amer. **53**:551-552.

Neunzig, H. H. 1963. Wild host plants of the corn earworm and the tobacco budworm in eastern North Carolina. J. Econ. Entomol. **56**:135-139.

Neunzig, H. H. 1964. The eggs and early-instar larvae of *Heliothis zea* and *Heliothis virescens* (Lepidoptera: Noctuidae). Ann. Entomol. Soc. Amer. **57**:98-102.

Neunzig, H. H. 1969. The biology of the tobacco budworm and the corn earworm in North Carolina with particular reference to tobacco as a host. N. C. Agr. Exp. Sta. Tech. Bull. **196**:76 p.

Pieters, E. P., and W. L. Sterling. 1975. Sequential sampling cotton squares damaged by boll weevils or *Heliothis* spp. in the Coastal Bend of Texas. J. Econ. Entomol. **68**:543-545.

Quaintance, A. L., and C. T. Brues. 1905. The cotton bollworm. USDA Div. Entomol. Bull. **50**:155 p.

Rainey, R. C. 1978. The evolution and ecology of flight: the "oceanographic" approach. pp. 33-48 in H. Dingle, ed. Evolution of insect migration and diapause. Springer-Verlag, N. Y. 284 p.
Rainey, R. C., and R. J. V. Joyce. 1972. The use of airborne Doppler equipment in monitoring wind-fields for airborne insects. 7th Int. Aerospace Instrum. Symp. Cranfield. 8.1-8.4.
Roach, S. H. 1975. *Heliothis* spp.: Larvae and associated parasites and diseases on wild host plants in the Pee Dee area of South Carolina. Environ. Entomol. 4:725-728.
Roach, S. H., and L. Ray. 1976. Pattern of emergence of adult *Heliothis* from fields planted to cotton, corn, tobacco, and soybeans. Environ. Entomol. 5:628-630.
Schaefer, G. W. S. 1976. Radar observations of insect flight. Symp. Royal Entomol. Soc. London. 7:157-197.
Sell, D. K., G. S. Whitt, and L. K. Lee. 1974. Inheritance of est-II phenotypes in the corn earworm. J. Hered. 65:243-244.
Shepard, M., G. R. Carner, and S. G. Turnipseed. 1974. A comparison of three sampling methods for arthropods in soybeans. Environ. Entomol. 3:227-232.
Slosser, J. E., J. R. Phillips, G. A. Herzog, and C. R. Reynolds. 1975. Overwinter survival and spring emergence of the bollworm in Arkansas. Environ. Entomol. 4:1015-1024.
Sterling, W. L. 1973. Sequential sampling for cotton insects. Folia Entomol. Mex. 25-26:55-56.
Sterling, W. L., ed. 1979. Economic thresholds and sampling of *Heliothis* species on cotton, corn, soybeans and other host plants. S. Coop. Ser. Bull. 231: 159 p.
Sterling, W. L., and E. P. Pieters. 1975. Sequential sampling for key arthropods of cotton. Tex. Agr. Exp. Sta. Dep. Entomol. Tech. Rep. 24:21 p.
Stinner, R. E., R. L. Rabb, and J. R. Bradley, Jr. 1977. Natural factors operating in the populations dynamics of *Heliothis zea* in North Carolina. Proc. Int. Congr. Entomol. 15:622-642.
Turnipseed, S. G. 1974. Sampling soybean insects by various D-Vac, sweep, and ground cloth methods. Fla. Entomol. 57:217-223.

Chapter 22

Sampling Lepidopterous Pod Borers on Soybean

Takashi Kobayashi and Toshio Oku

I. Introduction

Lepidopterous pod borers known as major soybean pests are: (1) Tortricidae (Eucosminae), i.e., the soybean pod borer (*Leguminivora glycinivorella* Matsumura) and the bean podworm complex (*Matsumuraeses phaseoli* Matsumura and *M. falcana* Walsingham); and (2) Pyralidae (Phycitinae), i.e., the limabean pod borer (*Etiella zinckenella* Treitschke). Diagnostic descriptions for identification of these species are presented in the Appendix to this chapter and are based mainly on Ishikura et al. (1952), Naito and Masaki (1962) and Issiki et al. (1969).

A. Geographical Distribution and Host Plants

The soybean pod borer, *Leguminivora glycinivorella* (Fig. 22-1A), is recorded from the temperate Far East in the north to ca. 30°N. The crop injury is heavier in the colder zones where populations are univoltine (Naito and Masaki 1962, Kobayshi and Oku 1976). Obraztsov (1967) regarded *L. glycinivorella* to be conspecific to *L. anticipans* Meyrick (=*parastrepta* Meyrick) of tropical India. The identity of these species, however, seems doubtful, because they have a disjunct distribution and marked differences in larval feeding habits. *L. glycinivorella* is a typical pod borer, that does not feed on other plant organs and feeds only on soybean and a few wild legumes of the genera *Glycine*, *Pueraria*, or *Sophora*; while *L. anticipans* is a flower-webber of the mango (*Mangifera*). Matsumoto and Kurosawa (1958b) suggested that the soybean pod borer may move from wild

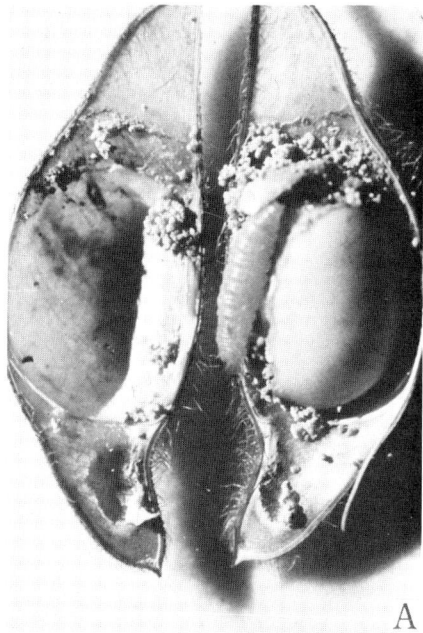
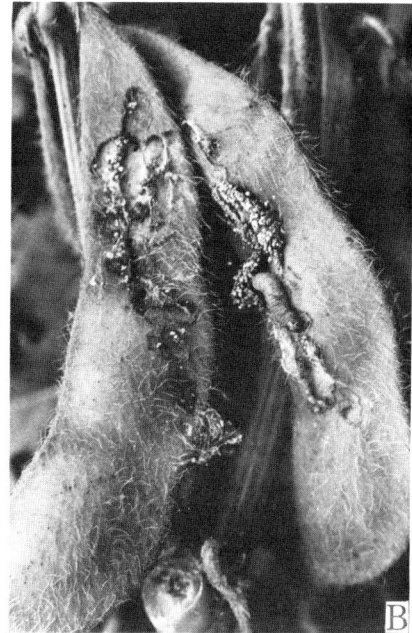
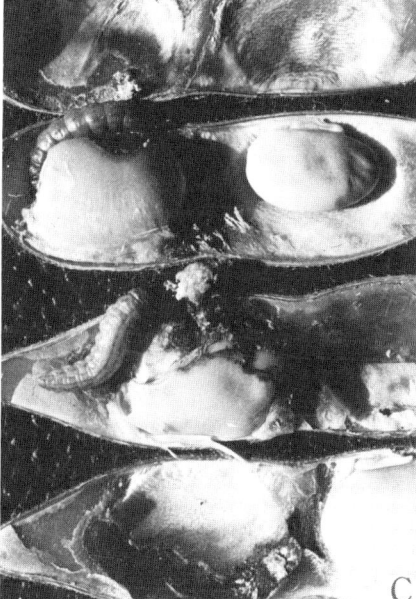

Figure 22-1. The mature larvae of pod borers on soybean pods: (A) the soybean pod borer, *Leguminivora glycinivorella*; (B) a bean podworm, *Matsumuraeses phaseoli*; and (C) the limabean pod borer, *Etiella zinckenella*.

hosts into soybean fields. It seems, however, that the pod borer usually multiplies within soybean fields.

The bean podworms of the *Matsumuraeses* complex (Fig. 22-1B) are all endemic to eastern Asia. *M. phaseoli* and *M. falcana* are well-known species found in association with soybean (Razowski and Yasuda 1975), but precise identification of their immature stages is still impossible. As there are several other species of *Matsumuraeses* feeding on wild legumes, it is likely that some additional species may be found to attack soybean. *M. phaseoli* and *M. falcana* appear to have a wide host range within the Leguminosae. The larva injures various plant organs and sometimes bores into pods. *M. phaseoli* is known from the temperate Far East with no known records from tropical Asia. Distribution of *M. falcana* extends more south to Formosa and Nepal.

The limabean pod borer, *Etiella zinckenella* (Fig. 22-1C), is widely known as a pod borer of many leguminous plants in the genera *Glycine, Phaseolus, Lupinus, Crotalaria, Pisum, Vigna, Robinia,* etc., from warmer regions of the world. Soybean injury has been reported from eastern Asia, southern Europe, and Central and South America. Geographic differences of the host specificity have been suggested by Naito and Masaki (1962).

Pod borer problems in soybean are, therefore, most serious in Eastern Asia, where all major pod borers so far mentioned occur. In this region, attacks to soybean by *Matsumuraeses* spp. are sporadic and, usually, negligible in the main soybean producing areas. In eastern Asia, therefore, *L. glycinivorella* and *E. zinckenella* are the predominant pod boring species. In southern Europe and in Central and South America only *E. zinckenella* is a damaging pod borer species in soybean. Since *L. glycinivorella* prefers the cooler soybean producing areas and *E. zinckenella* the warmer ones, the two species coexist and have an economic impact only in a limited intermediate region.

B. Life Cycle and Phenology

The adult soybean pod borer lays eggs mainly on the pod of hairy types and on the stipule and calyx of glabrous ones. The recorded peak period of oviposition is usually four weeks after full-bloom (R2) in northern Japan (Endo 1967), but it may vary with locality and possibly with soybean variety. Duration of the egg stage is five to ten days in the field. The larva injures only pods and undergoes five stages during the feeding period, which lasts slightly more than one month. When fully grown the larva emerges from the pod and spins a cocoon underground. Winter is passed by the mature larva in a diapause state in the cocoon. Pupation of the overwintering larva occurs in the summer after a post-hibernation dormant period. The species is univoltine in the cooler areas and bivoltine, at least partially, in the warmer areas of its distribution. In the univoltine zones (areas of China and Japan between 37°N and 44°N), the peak of adult occurrence is retarded from north to south as shown in Fig. 22-2A-C (Okada 1948, Nishijima and Kurosawa 1953, Nagi 1959, Hsu et al. 1965, Endo

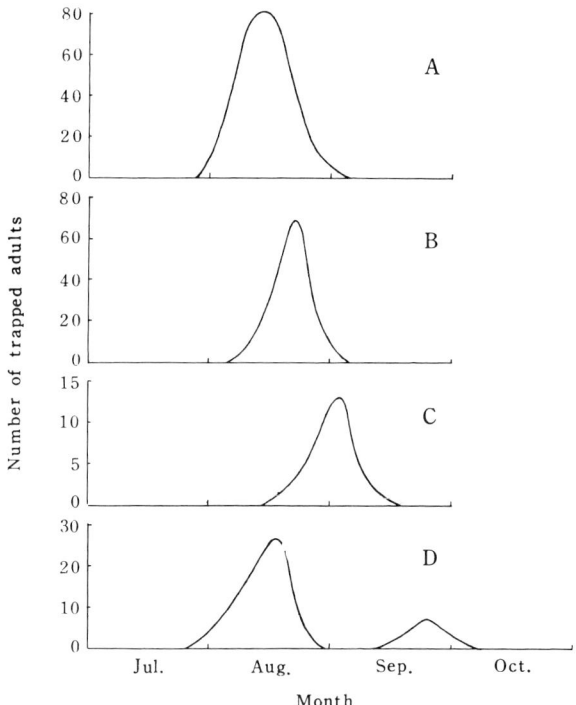

Figure 22-2. Schematized seasonal occurrence of the soybean pod borer adult in several localities: (A) northeast China, about 44°N (Hsu et al. 1965); (B) Iwate Prefecture, Japan, about 40°N (Nagai 1959); (C) Fukushima Prefecture, Japan, about 37°N (Endo 1967); and (D) Saitama Prefecture, Japan, about 36°N (Naito and Masaki 1962).

1967). In Japan's Kanto Plain, part of the partially bivoltine zone, there are two separate peaks of adult emergence (Fig. 22-2D), and the rate of adult emergence in the second period increases when summer temperature is high (Ninomiya et al. 1957, Naito and Masaki 1962).

The general life cycle of the limabean pod borer is similar to that of the soybean pod borer with slight differences in ovipositional and feeding habits. The peak of limabean pod borer oviposition is reached between five to ten days after pod-setting (R3-R4) in southwestern Japan (Ishikura et al. 1953). *E. zinckenella* is multivoltine over the regions of normal abundance. The Far East can be divided into several zones according to the number of generations per year estimated from thermal constants for development based on studies of the Kanto population in central Japan (Fig. 22-3) (Naito and Masaki 1962). In the Kanto district, peas and the earliest soybean varieties are attacked by the first generation, and later-podding soybean varieties are damaged by the succeeding generations.

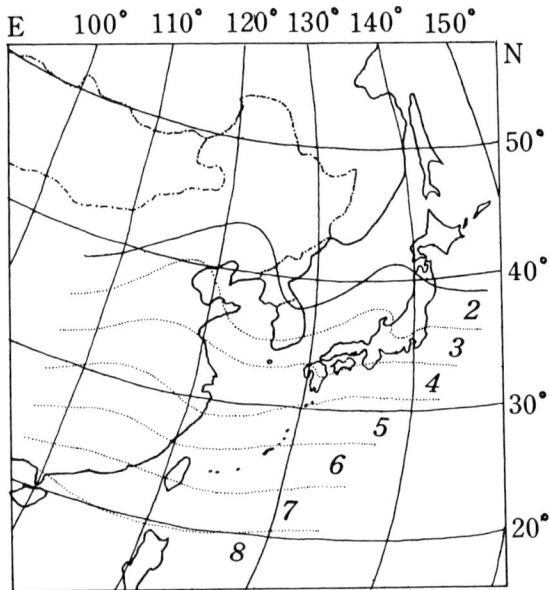

Figure 22-3. Zonation of the Far East in accordance to the annual number of generations of the lima-bean pod borer estimated from heat-unit accumulation. The northern limit of distribution and the annual number of complete generations are represented by a thick line and arabic numerals, respectively. (Adapted from Naito and Masaki 1962 and supplemented with Kobayashi and Oku 1976).

An example of the seasonal abundance of adults in this province is shown in Fig. 22-4 (Takano et al. 1956). Larval abundance in a soybean field is greatly influenced by whether or not adequate food plants have continuously been present in the vicinity.

Bean podworm, *Matsumuraeses* spp., adults oviposit on the foliage of host plants. Larval injury is first observed as webbed shoots. As plants mature the larvae web together pods, leaves and branches. Most older instars exhibit the habit of boring into pods, petioles, branches, or stems, but they do not lose the webbing habit. In southwestern Japan a bean podworm, presumably *M. falcana*, has three or four annual generations on various beans successively cultured during the warmer season. This podworm passes winter as a dormant pupa or young

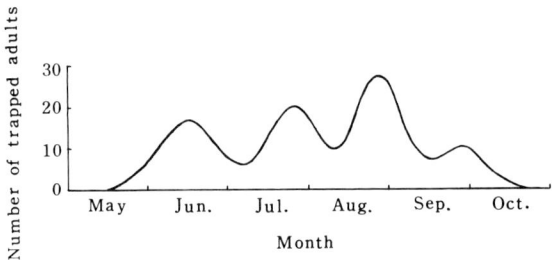

Figure 22-4. Schematized seasonal occurrence of the lima-bean pod borer adult in Ibaragi Prefecture, Japan, about $36°N$. (Adapted from Takano et al. 1956).

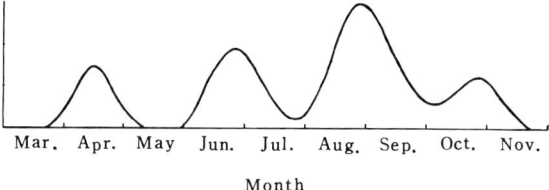

Figure 22-5. Schematized seasonal occurrence of a bean podworm adult, presumably *Matsumuraeses falcana*, in Kagawa Prefecture, Japan. (Adapted from Asuyama et al. 1955).

larva feeding on broad bean (Asuyama et al. 1955; Issiki et al. 1969; Kobayashi 1972,1976). The seasonal moth abundance is schematically shown in Fig. 22-5. In northern Japan the podworms may, at least in part, move from wild legumes into crop fields because the overwintering generation emerges as adults before beans are planted in the fields.

II. Sampling for Adults

Uchida and Okada (1940) used a field cage covering 1 m^2 to measure absolute populations of the soybean pod borer moth emerging from the overwintering underground cocoon. No absolute methods for adults in flight have been devised for any of the pod borer species.

Relative methods most commonly used are light traps of various types. Tungsten lamps are moderately attractive to pod borer adults, but they are insufficiently attractive to the limabean pod borer (Naito and Masaki 1962). Ultraviolet lamps, mercury-vapour lamps, and fluorescent lamps are highly attractive to all pod borer moths, though the range of attraction has not been established.

Molasses bait traps often recommended for collection of tortricid moths are unstable in their attractiveness to the soybean pod borer adult. High moth catches were obtained in only a few instances (Okada 1948, Kuwabara 1950, Ninomiya et al. 1957).

As the soybean pod borer adult female usually starts ovipositional activity during late afternoon hours before sunset (Uchida and Okada 1937), sweep net sampling in late afternoon is effective for collecting the moths. A sample unit of fifty sweeps is employed as a standard method for estimation of the seasonal sequence of moth activity in Hokkaido, Japan (Nishijima and Kurosawa 1953, Matsumoto and Kurosawa 1958a). It is, however, difficult to eliminate the influence of such personal factors as speed and method of sweeping on the moth catches (see Chapter 2). The effect of weather on the efficiency of these relative methods of sampling pod borer adult moths has not been ascertained.

III. Sampling for Eggs

The soybean pod borer moth lays eggs singly on various plant organs. The preferred oviposition site differs between hairy and glabrous soybean varieties, and preference is consistent within each pubescent type. In general, much more larval injury is caused by the pod borer to hairy soybean varieties than to glabrous ones (Uchida and Okada 1940, Okada 1948, Kuwabara 1950, Nishijima and Kurosawa 1953, Matsumoto 1962, Hsu et al. 1965). Varietal differences are mainly due to selection of oviposition sites by the adult female, which has a marked preference for the pods of hairy varieties (Nishijima 1954b, 1960a). In hairy varieties, about 80% or more of the eggs are found on the pods and the remaining 20% on other organs. In glabrous varieties, on the other hand, most eggs are deposited on the stipule and on the calyx, with a very small proportion of the eggs oviposited on pods. In both cases, approximate estimates of the egg population can be obtained by examination of the main oviposition site alone. The number of eggs laid on the single pod is usually less than three. Rarely is this level exceeded; however, a record of 24 per pod has been noted (Okada 1948). On hairy soybean deposition of eggs is found mainly on pods longer than 1 cm or 2 cm, and oviposition is usually higher on larger pods. Pods may escape oviposition if they reach the yellowing stage (R6-R7) without oviposition-inducing stimuli (Okada 1948, Nishijima 1954b, Matsumoto 1962, Endo 1967). Consequently, patterns of ovipositional sequence and of egg dispersion over the soybean canopy may vary not only with moth occurrence but also with pod growth (Sato and Inoue 1962, Endo 1967). Hence, the start of the sampling period should be determined according to the stage of pod development rather than to moth occurrence. Dispersion of eggs on a plant is not uniform, as expected from the ovipositional behavior. A simple stratification based on the height of the plant or the order of nodes for egg sampling may be inadequate because of the dynamic nature of the dispersion. Whole plant examination seems the more adequate method to use to avoid the unreliability of stratification of samples in estimating egg populations.

The limabean pod borer lays eggs singly preferentially on immature pods. Ovipositional preference and the dispersion of eggs in the soybean canopy are, therefore, primarily determined by pod growth in relation to moth occurrence, as is also the case of the soybean pod borer (Ishikura et al. 1952, Naito and Masaki 1962, Tsutsui 1969). A major difference, however, in the oviposition habits of both species, is the location of *E. zinckenella* eggs on host plants. Eggs of the limabean pod borer are usually deposited on the basal part of pods or the calyx of the soybean.

Eggs of *Matsumuraeses* spp. are laid on the foliage, but the spatial patterns on the soybean plant have not been investigated. Spatial patterns of eggs among plants or within fields have not been studied for any of the podborer species.

IV. Sampling for Larvae and Pupae

Absolute numbers of *M. glycinivorella* larvae boring into pods may be estimated by the number of larval entrance holes, because larvae do not move to other pods during the feeding period. The entrance hole (Fig. 22-6A) is very characteristic. It is marked by a brownish discoloration before the pod dries, and boring occurs on a small swell of pod tissue along the sutural lines of the soybean pod (Uchida and Okada 1937, Nishijima and Kurosawa 1953, Nishijima 1954a).

Processes related to larval population decrease are monitored by dissection of pods sampled every ten days or so (Nishijima 1954a,c). Mortality factors, except for natural enemies, are classified as follows:

(a) Death during penetration through the pod tissue (pod without larval feeding marks on the inner surface)—a dead first instar observed in the interrupted mine in fresh pod tissue around the entrance hole.
(b) Death shortly after penetration (pod with a brownish mark on the inner surface along a feeding scar extending from an entrance hole)—a dead first or second instar present in the seed chamber is visible after the pod fully matured; this type of mortality is higher in "Nagaha" (long-leafed) than in other soybean strains.
(c) Death from cannibalism by older larvae (pod with conspicuous feeding scars on seeds, with a dead fourth or fifth instar in the seed chamber—the dead larvae show signs of wounds; this type of mortality increases with larval density.
(d) Death due to shedding of immature pods containing larvae.

Mortality factors (a) and (b) are not related to the morphological characteristics of the pod (Nishijima 1954c), although some mortality may result from

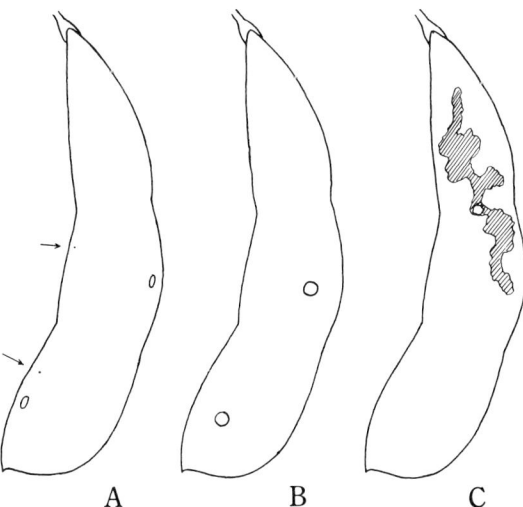

Figure 22-6. Surface feeding scars on soybean pods: (A) the soybean pod borer, *Leguminivora glycinivorella*; (B) the limabean pod borer; *Etiella zinckenella*; and (C) a bean podworm *Matsumuraeses* sp.

insufficient food when the larva feeds on very small seeds such as sterile seeds. It has been noted that sterile seeds are frequently produced by larval feeding on the funicle (Nishijima 1959). The "embryoless" seed due to this cause is found most abundantly in the second basal seed chamber, while the "physiological" sterile is found in the first basal seed chamber. In addition to these mortality factors, older instars may die presumably due to excessive heat during the hottest periods in the bivoltine zone (Naito and Masaki 1962).

Absolute population of mature larvae is indirectly measured by counting the elliptical exit holes (Fig. 22-6A) appearing mainly along the sutural lines of pods (Nishijima 1954a). Covering individual plants with a fine mesh net is also useful to collect larvae that emerge from infested pods (Uchida and Okada 1937). Inspection of collections from covered plants should be made in the afternoon because the larval emergence occurs almost exclusively during the forenoon (Okada 1948). Earthen cocoons of the mature larvae may be detected by sieving upper soil-layer to a 10 cm depth (Okada 1948). The same method may be applicable for sampling pupae in the cocoon.

Larval and pupal habits of the limabean pod borer are similar to those of the soybean pod borer, except for the fact that the former can move from pod to pod during the larval stage (Ishikura et al. 1953). Studies on the behavior of the younger instars are not available. Indirect sampling methods based on detection of larval feeding scars are applicable only for older instars. Characteristics of injured pods and the larval exit hole are illustrated in comparison to those of the soybean pod borer (Fig. 22-6B).

Sampling for *Matsumuraeses* spp. larvae, which bore into pods only occasionally, may be adapted from those used for other foliage feeders.

The number of injured seeds on harvested plants is a useful index of the relative larval abundance (Matsumoto and Kurosawa 1958a, Naito and Masaki 1962, Kobayashi and Oku 1976). For interpretation of data obtained by sampling seeds, it is worth noting that a pod borer does not necessarily injure a single seed. Older larvae often move from seed to seed within a pod. Furthermore, the limabean pod borer frequently moves from pod to pod. A random sample of 10 to 20 plants per plot is customarily taken in Japan. This is an adequate sample unit from the standpoint of time and cost.

The spatial pattern and the sampling error of this method applied for the soybean pod borer have been investigated. Morishita's (1959) I_δ index is plotted against the number of injured seeds per plant (Fig. 22-7), based on the adopted 10-plant samples from unsprayed soybean fields in Iwate Prefecture, Japan. From Fig. 22-7 it is seen that the pattern of injured seeds is more or less clustered and approaches the Poisson distribution towards higher injury levels, particularly above 10 injured seeds per plant. From this same set of samples, a standard error below 10% of the mean was obtained from samples with the mean exceeding 15.

Field tests on soybean varietal differences to pod borer attack were conducted using small (4 to 8 m^2) plots. Larval counts on glabrous soybean were greater than expected, on the basis of numbers of eggs laid. Eggs laid on hairy varieties

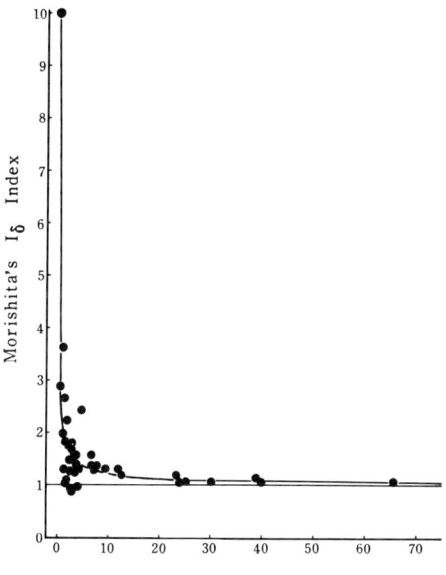

Figure 22-7. Relation between the number of seeds injured by the soybean pod borer per plant and its spatial pattern defined by Morishita's I_δ index. Based on samples of 10 plants from unsprayed soybean fields in Iwate Prefecture, Japan: $I_\delta = 1$:Poisson, $I_\delta > 1$:toward negative binomial, and $I_\delta < 1$:toward normal. (Morishita 1959).

far exceeded the number laid on the glabrous varieties (Nishijima and Kurosawa 1953, Matsumoto 1962). The discrepancy may in part be attributed to an interplot dispersion of the larvae soon after emerging from eggs (Nishijima 1960b). In such experiments, therefore, it seems necessary to exclude samples from peripheral plants at least within ca. 2 m from the border of each plot. Using this precaution, one can then use the number of injured seeds as a criterion for the level of the soybean resistance based on the oviposition preference by the pod borer.

V. Concluding Remarks

There are adequate data on sampling methods for the various developmental stages of the soybean pod borer and the limabean pod borer. However, reliable conversion factors for relative data on adult populations by trapping or by sweep net sampling are still unavailable. There have been fewer studies on *Matsumuraeses* spp. than on other pod borer species. The correct identification of immature stages as well as development of more efficient sampling methods for *Matsumuraeses* spp. also need additional research. These shortcomings have not been serious up to the present because podworm injury is not widespread. Potential changes in the pest situation of these polyphagous species, however, may result from some modification of the agricultural environment.

The main emphasis of existing pod borer research has been placed in biological aspects, which are indispensable for development of chemical control methods, development of varieties resistant to pod borers, or analysis of resistance factors.

A major difficulty in the development of sequential sampling techniques arise from the lack of data on the spatial patterns of various developmental stages in the soybean field. Furthermore, there is need for developing sampling methods for populations on other host plants. Such studies are particularly needed if we are to elucidate the seasonal and annual population dynamics of the limabean pod borer and *Matsumuraeses* spp., which cannot complete the annual cycle on a single soybean crop and often move among various plant species.

APPENDIX 22-1: Diagnoses for Lepidopterous Soybean Pod Borers and Larval Injury[1]

A. Descriptions of Eggs, Larvae, and Cocoons

The soybean pod borer (*L. glycinivorella*): eggs semi-ovate; surface slightly wrinkled; about 0.46 x 0.32 mm; mainly laid on pods of hairy soybean varieties and on stipule and calyx of glabrous varieties. Mature larvae 8-12 mm long; head greyish-brown; body pale yellowish, turning bright reddish-orange, not dotted; streaked dorsally; pinaculum somewhat grey-brownish (Fig. 22-8A). Cocoon about 10x4 mm formed with webbed particles in upper soil-layer mainly to a depth of 10 cm from ground surface.

The limabean pod borer (*E. zinckenella*): eggs semi-ovate; surface slightly punctate; about 0.53x0.41 mm; mainly laid on pods, particularly on basal parts and on calyx. Mature larvae 15-18 mm long; head yellowish-brown, darker behind; body pale green, turning deep green or purple with reddish tinge, dotted with black on thoracic shield; streaked with darker color dorsally; pinaculum obscure (Fig. 22-8B). Cocoon about 13x5 mm, its nature and location similar to those of the soybean pod borer.

The bean podworms (*Matsumuraeses* spp.): eggs semi-ovate (no detailed observations and measurements); laid on foliage. Mature larvae 14-17 mm long; head orange-brown; body pale yellowish, turning more or less tinged with brown, not dotted or streaked dorsally; pinaculum conspicuous, grey-brownish (Fig. 22-8C). Special cocoon not formed, pupation taking place in larval feeding site; overwintering cocoon in dead plant parts.

B. Key to Larvae Based on Setal Plan and Crochets of Prolegs

(For terminology used in the key see Hinton 1946.)
1. First thoracic segment with two L setae and D1 seta apparently more dorsad than XD1 seta and more cephalad than D2 seta; 8th abdominal segment with SD seta above spiracle . . . .
 . . . . The limabean pod borer (*Etiella zinckenella*) (Fig. 22-9A).

[1] Adapted mainly from Ishikura et al. 1952, Naito and Masaki 1962, Issiki et al. 1969.

Sampling Lepidopterous Pod Borers on Soybean 433

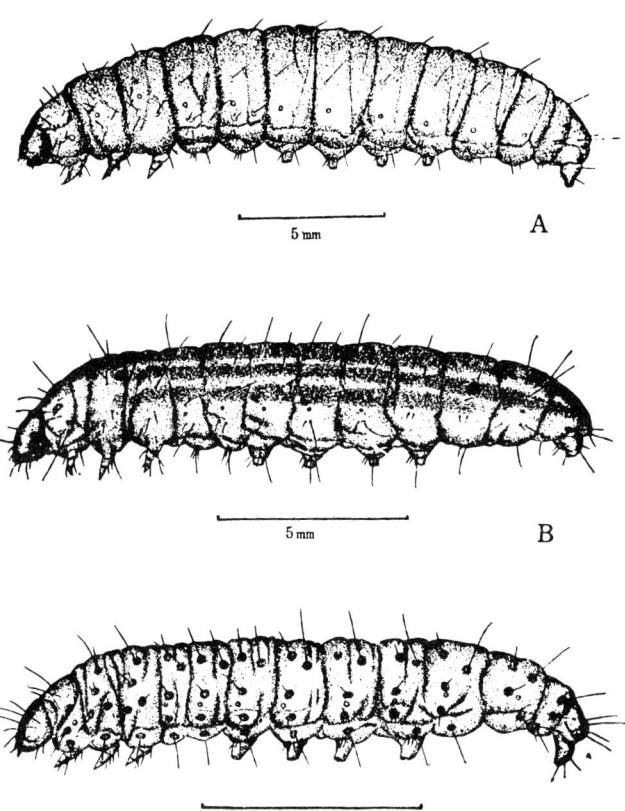

Figure 22-8. General appearance of mature pod borer larvae: (A) the soybean pod borer, *Leguminivora glycinivorella*, (B) the limabean pod borer, *Etiella zinckenella;* and (C) a bean podworm, presumably *Matsumuraeses falcana*.

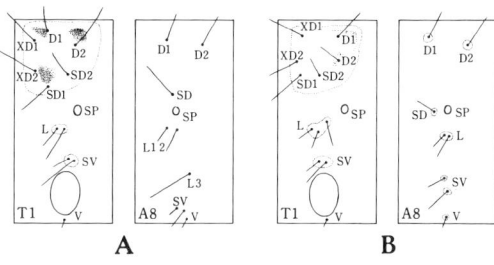

Figure 22-9. Larval chetotaxy of pod borers: (A) the limabean pod borer, *Etiella zinckenella*; and (B) the soybean pod borer, *Leguminivora glycinivorella*. The general plan of B adaptable to the bean podworms, *Matsumuraeses* spp. (T1: 1st thoracic segment, A8: 8th abdominal segment, SP: spiracle; setal nomenclature after Hinton 1946).

First thoracic segment with three L setae and D1 seta hardly dorsad and cephalad than XD1 and D2, respectively; 8th abdominal segment with SD seta cephalad to spiracle .... 2 (Fig. 22-9B)

2. Crochets of proleg partly biordinal ....
.... The soybean pod borer (*Leguminivora glycinivorella*)

   Crochets of proleg entirely uniordinal ....
   .... The bean podworms (*Matsumuraeses* spp.)

C. Key to Seed Injury by Some Lepidoptera and Diptera Pests

1. Large circular holes cut off pods. ....
   .... Noctuid worms (*Heliothis viriplaca, Pyrrhia umbra*, etc.)

   No large circular holes cut off pods......................... 2

2. Pod malformed with conspicuous thickening around injured seed chamber filled with symbiotic fungi ....
   .... The soybean pod gall-midge (*Asphondylia* sp.)

   Pod not malformed 3

3. Pods usually webbed together or with other parts; irregular feeding scar spread over pod surface (see Fig. 22-6C), extending into seed chamber of which inner-side is conspicuously stained dark brown-greyish; fecal pellets rather irregular in size and form, dirty greyish-ochreous in color.........
   .... The bean podworms (*Matsumuraeses* spp.)

   Pods not webbed together; without surface feeding scar except for entrance and exit holes; inner side of seed chamber lightly darkened along feeding scar of young larva, not stained, unless extremely wet; fecal pellets of mature larva nearly uniform in shape and size like sawdust, orange-ochreous or brownish in color.. 4

4. Exit hole of mature larva elliptical, about 1.8 mm long, located along sutural line of pod; entrance hole of hatching larva opened on a minute swell along sutural line (see Fig. 22-6A); seed injured along sutural part; fecal pellets of mature larva ca. 0.4 mm diameter, somewhat elongate, orange-ochreous in color ...... The soybean pod borer (*Leguminivora glycinivorella*);

   Exit hole of larva (not limited to mature larva) circular, opened at various positions on the pod wall (see Fig. 22-6B), about 2 mm diameter when larva fully developed (no detailed observations of entrance hole); seed injured irregularly, often entirely consumed; fecal pellets of mature larva about 0.8 mm diameter, not elongate, variable in color but usually dark reddish-brown..
   .... The limabean pod borer (*Etiella zinckenella*).

D. Identification of Adults

If necessary to examine genitalia for precise identification of trapped adults, refer to the following publications: for the soybean pod borer–Okada (1948), Obraztsov (1960), Hsu et al. (1965); for the limabean pod borer–Heinrich (1956); and for the bean podworms (*Matsumuraeses* spp.)–Razowski and Yasuda (1975) (descriptions made before this publication are inadequate, since there is a confusion among species).

References

Asuyama, H., K. Fukunaga, M. Hori, H. Ishikura, A. Kawada, H. Muko, H. Sugawara, and H. Yuasa. 1955. Sakumotsu-byōgaichū Handobukku (Agricultural pest handbook). (In Japanese) Yōkendō, Tokyo.
Endo, T. 1967. Studies in biology and insecticidal control of the soybean pod borer. (In Japanese) Bull. Fukushima Agr. Exp. Sta. 3:85-96.
Heinrich, C. 1956. American moths of the subfamily Phycitinae. U. S. Nat. Mus. Bull. 207:581 p.
Hinton, H. E. 1946. On the homology and nomenclature of the setae of lepidopterous larvae. Trans. Royal Entomol. Soc. London 97:1-37.
Hsu, C. F., S. K. Kou, Y. M. Hang, C. Feng, Y. Chang, and Y. C. Lee. 1965. A study on the soybean pod borer [*Leguminivora glycinivorella* (Mats.) Obraztsov]. (In Chinese with English summary) Acta Entomol. Sin. 14:461-475.
Ishikura, H., N. Nagaoka, I. Honda, and M. Fujita. 1953. Studies on insect pests of soybean. II. Ecological studies on limabean pod borer, *Etiella zinckenella* Treitschke, in Shikoku district with experiment on its control. (In Japanese with English summary) Bull. Shikoku Agr. Exp. Sta. 1:186-216.
Ishikura, H., N. Nagaoka, T. Kobayashi, and M. Fujita. 1952. Studies on insect pests of soybean. I. Insect pest fauna of soybean in Shikoku district. (In Japanese with English summary) Bull. Chugoku-Shikoku Agr. Exp. Sta. 1:134-150.
Issiki, S., A. Mutuura, Y. Yamamoto, I. Hattori, H. Kuroko, T. Kodama, T. Yasuda, S. Moriuchi, and T. Saito. 1969. Early stages of Japanese moths in colour. vol. II. (In Japanese) Hoikusha, Osaka.
Kobayashi, T. 1972. Biology of insect pests of soybean and their control. Jap. Agr. Res. Quart. 6:212-218.
Kobayashi, T. 1976. Insect pests of soybean in Japan and their control. PANS 22:336-349.
Kobayashi, T., and T. Oku. 1976. Studies on the distribution and abundance of the invertebrate soybean pests in Tohoku district, with special reference to insect pests infesting seeds. (In Japanese with English summary) Bull. Tohoku Nat. Agr. Exp. Sta. 52:49-106.
Kuwabara, T. 1950. Varietal difference of the podborer injury in soybean. (In Japanese) Daizu-Zosan Shiryo, Hokkaido 2:1-21.

Matsumoto, S. 1962. Studies on varietal differences of soybean with respect to injury caused by the soybean podborer, *Leguminivora glycinivorella*. (In Japanese with English summary) Hokkaido Agr. Exp. Sta. Rep. **58**:58 p.

Matsumoto, S., and T. Kurosawa. 1958a. On the yearly deviations of injuries to the soybean caused by the soybean pod borer, *Grapholitha glycinivorella* Matsumura. (In Japanese with English summary) Jap. J. Appl. Entomol. Zool. **2**:93-99.

Matsumoto, S., and T. Kurosawa. 1958b. Notes on a new host-plant of the soybean pod borer, *Grapholitha glycinivorella* Matsumura. (In Japanese with English summary) Jap. J. Appl. Entomol. Zool. **2**:189-191.

Morishita, M. 1959. Measuring of the dispersion of individuals and analysis of the distributional patterns. Mem. Fac. Sci. Kyushu Univ. (E) **2**:215-235.

Nagai, R. 1959. Control effect of simultaneous insecticide spray in an extensive area against the soybean podborer injury. (In Japanese) Soc. Plant Prot. North Jap. Annu. Rep. **10**:143-145.

Naito, A., and J. Masaki. 1962. Studies on the distribution of lima bean pod borer, *Etiella zinckenella* Treitschke, and soybean pod borer, *Grapholitha glycinivorella* Matsumura. (In Japanese with English summary) J. Central Agr. Exp. Sta. **2**:145-228.

Ninomiya, T., H. Takezawa, and T. Akiyama. 1957. Biological notes on the soybean podborer in Kanagawa Prefecture. (In Japanese) Kanto-Tosan Plant Prot. Soc. Annu. Rep. **4**:31-32.

Nishijima, Y. 1954a. Studies on the injuries of soybean varieties and the decreasing process of the larval population of the soybean pod borer, *Grapholitha glycinivorella* Matsumura. (In Japanese with English summary) Mem. Fac. Agr. Hokkaido Univ. **2**:112-124.

Nishijima, Y. 1954b. On the boring of hatching larva and the location of egg deposition of the soybean pod borer, *Grapholitha glycinivorella* Matsumura. (In Japanese with English summary) Mem. Fac. Agr. Hokkaido Univ. **2**:127-132.

Nishijima, Y. 1954c. Studies on the larval mortality of the soybean pod borer, *Grapholitha glycinivorella* Matsumura, with special reference to the growth and thickness of the soybean pod and seed. (In Japanese with English summary) Mem. Fac. Agr. Hokkaido Univ. **2**:133-140.

Nishijima, Y. 1959. The occurrence of embryoless seeds in soybean varieties in relation to the feeding of the soybean pod borer, *Grapholitha glycinivorella* Matsumura. (In Japanese with English summary) Jap. J. Appl. Entomol. Zool. **3**:183-189.

Nishijima, Y. 1960a. Host plant preference of the soybean pod borer, *Grapholitha glycinivorella* Matsumura (Lep., Eucosmidae). Entomol. Exp. Appl. **3**:38-47.

Nishijima, Y. 1960b. Studies on the soybean pod borer, *Grapholitha glycinivorella* Matsumura, with special reference to the egg-laying number and the larval entrance to soybean varieties. (In Japanese with English summary) Mem. Fac. Agr. Hokkaido Univ. **3**:26-31.

Nishijima, Y., and T. Kurosawa. 1953. Some factors affecting varietal differences of soybean to attack by the soybean pod borer, *Grapholitha glycinivorella* Matsumura. (In Japanese with English summary) Bull. Hokkaido Agr. Exp. Sta. **65**:39-51.

Obraztsov, N. S. 1960. Die Gattungen der palaearktischen Tortricidae. II. Unterfamilie Olethreutinae. Tijdschr. Entomol. **103**:111-143.

Obraztsov, N. S. 1967. Die Gattungen der palaerktischen Tortricidae. III. Addenda und Corrigenda. Tijdschr. Entomol. **110**:13-36.

Okada, I. 1948. Studies on the soybean pod borer (*Grapholitha glycinivorella* Matsumura). (In Japanese) Kanchi-Nogaku **2**:193-239.

Razowski, J., and T. Yasuda. 1975. On the Laspeyresiini genus *Matsumuraeses* Issiki (Lepidoptera, Tortricidae). Acta Zool. Cracov. **20**:89-106.

Sato, K., and H. Inoue. 1962. Studies on the soybean podborer injury. I. Relation between the position of pod in the soybean canopy and the rate of injured seeds. (In Japanese) Hokunō **29**:1-3.

Takano, S., T. Takano, and K. Kimizaki. 1956. Seasonal prevalence of the limabean podborer in Ibaragi Prefecture. (In Japanese) Kanto-Tosan Plant Prot. Soc. Annu. Rep. **3**:43.

Tsutsui, K. 1969. Genshoku Sakumotsugaichu-bojo. (Agricultural pests and their control in colour.) (In Japanese) Ienohikari-kyōkai, Tokyo.

Uchida, T., and I. Okada. 1937. Die Lebensweise von *Grapholitha glycinivorella* Matsumura in Mandschurei (Vorhersagung). (In Japanese) Kontyu **11**:331-343.

Uchida, T., and I. Okada. 1940. Die Wichtigkeit der Bekämpfung des Sojabohne-Schädlings, *Grapholitha glycinivorella* Matsumura, in Mandschukuo und einige dafür dienliche Untersuchungen. (In Japanese) Koshurei-Nojishikenjō Kenkyu-Jihō **32**:107-134.

Chapter 23

Sampling Phytophagous Pentatomidae on Soybean[1]

James W. Todd and Donald C. Herzog

I. Introduction

At least 40 species of stink bugs have been found to occur worldwide on soybean; however, damage has actually been documented for only a few of these. There is at least one important stink bug species in every major soybean producing region of the world. Because of the breadth of the subject and diversity of species involved, this chapter must be restricted somewhat in its scope. Many species have similar life cycles, damage potential, habits, and seasonal occurrence; therefore, it was necessary to select a "case-in-point" species. The southern green stink bug, *Nezara viridula* (L.), was chosen because of its extremely broad host range, geographical distribution, and abundance on soybean. If notable exceptions occur for other species, these will also be discussed.

[1] Original research reported here was supported in part by the Georgia Soybean Commodity Commission and by the National Science Foundation and Environmental Protection Agency through a grant (NSF GB-34718) to the University of California. The findings, opinions and recommendations expressed herein are those of the authors and not necessarily those of the Georgia Soybean Commodity Commission, the University of California, the National Science Foundation, or the Environmental Protection Agency.

A. Geographical Distribution

The southern green stink bug is by far the most cosmopolitan of the pentatomids attacking soybean. It is found throughout the tropical and subtropical regions of Europe, Asia, Africa, and America (Lethierry and Severin 1893). Numerous records have been published for the West Indies, northern portion of South America, Central America, Mexico, and the southern portion of the United States. Southeast Asia probably is the original home of the southern green stink bug based on the polymorphism of the species in that region (Yukawa and Kiritani 1965, Kiritani 1970). Undoubtedly the insect has spread to various parts of the world via some unknown means and is still spreading to new areas (Fig. 23-1). *N. viridula* invaded Australia in 1916 (Wilson 1960), New Zealand in 1944 (Cumber 1949), and Hawaii in 1961 (Mitchell 1965). Distant (1880) stated that the wind has probably greatly assisted the insect in its wide dissemination; while sailing from India he found several specimens that landed aboard ship when it was more than a hundred miles southwest of Madagascar. The *N. viridula* was also collected aboard a weather ship located 500 km south of the Japanese mainland (Asahina and Turuoka 1970).

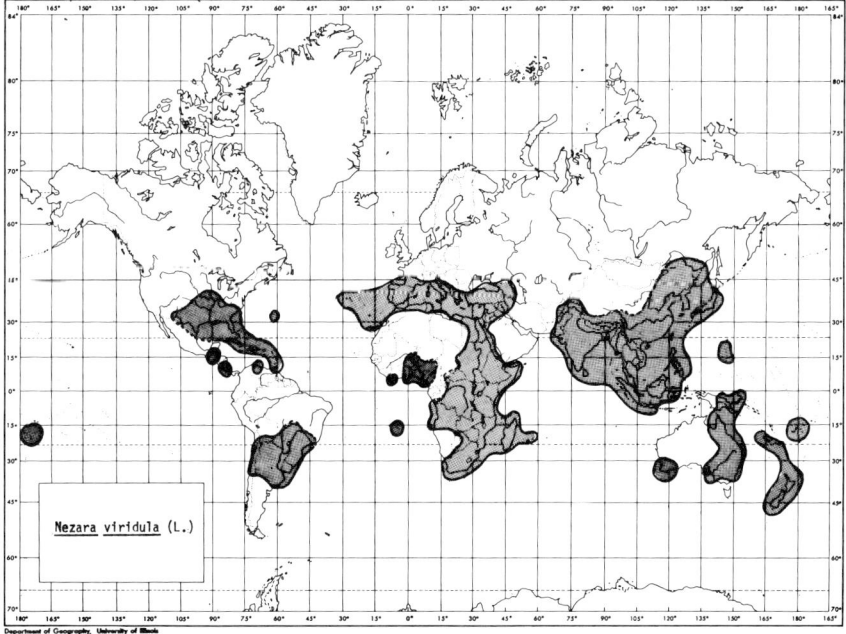

Figure 23-1. Worldwide distribution map of *Nezara viridula*.

B. Host Plants and Nature of Damage

The southern green stink bug is highly polyphagous, attacking both monocots and dicots. The comprehensive listing of host plants for this insect by Hoffman (1935) is the first reference these authors have seen that states that *N. viridula* feeds on soybean. Hoffman listed three families of monocotyledons with the Gramineae (eight species) being the most important. Of the 29 families of dicotyledons listed, by far the most important is the Leguminosae with 27 species, followed by the Cruciferae with 8 species, and Malvaceae and Solanaceae with 6 species each. Additionally, Jones (1918) and Drake (1920) have cited the insect as feeding on various plant species of economic importance (Table 23-1). The long list of food plants shows that the insect is a very general feeder and that it can subsist on a great variety of plants. However, it probably does not breed, at least in large numbers, on all these plants and only occasionally feeds on a number of them. Observations recorded in the literature and some that these authors have made indicate that the southern green stink bug has a decided preference for a few plants, particularly some of the legumes. This preference varies with the development of the preferred food plants in different seasons of the year. The greatest expression of preference by far is during the period of fruit formation. After the fruit have matured the plants become unattractive, and the bugs move to more succulent plants.

The mouthparts of the southern green stink bug (as in all Hemiptera) are modified to form a structure for piercing and sucking. Both nymphs and adults obtain their food by puncturing the tissues of plants with their beaks and extracting the plant juices. They may attack all parts of a plant—stems, foliage (particularly leaf veins), flowers, and fruit—but greatly prefer the young tender growth and fruiting structures. The feeding punctures caused by the insertion of the rostrum form minute, hard, brownish or blackish spots on the plant. The loss of plant juices, the injection of powerful digestive enzymes, and the provision of entry sites for pathogenic and decay organisms at the feeding puncture are the principal means by which the feeding process of the bug is detrimental to the plant. Feeding by stink bugs during the early stage of seed formation can result in shriveled, deformed, and undersized seeds, whereas, feeding at a later stage on the larger, though still green, seed produces only a noticeable black mark in a depression (Kilpatrick and Hartwig 1955, Turner 1967). Additionally, the feeding damage on mature fruit may mar its appearance and lower its market value. Immature fruit is much retarded in growth and greatly distorted and discolored by hard callouses that form around the punctures (Drake 1920). The above description of southern green stink bug damage is typical of that described in many of the early records of injury to garden and vegetable crops in the United States (Jones 1918, Drake 1920). Other workers have described similar results from studies with the green stink bug, *Acrosternum hilare* (Say), and the brown stink bug, *Euschistus servus* (Say) (Blickenstaff and Huggans 1962, Daugherty et al. 1964, Miner 1966).

Table 23-1. Common cultivated and wild host plants of *N. viridula* [after Jones (1918), Drake (1920), and Hoffman (1935)]

Family	Common name	Scientific name
Amaranthaceae	pigweed	*Amaranthus* spp.
Chenopodaceae	lambsquarter	*Chenopodium* spp.
Compositae	sunflower	*Helianthus* spp.
Cruciferae	cabbage	*Brassica oleracea*
	cauliflower	*Brassica oleracea*
	collard	*Brassica oleracea*
	mustard	*Brassica* spp.
	radish	*Raphanus sativus*
	turnip	*Brassica rapa*
Cucurbitaceae	cucumber	*Cucumis sativus*
	squash	*Cucurbita* spp.
Cyperaceae	nutsedge	*Cyperus esculentus*
Euphorbiaceae	castor bean	*Ricinus communis*
Gramineae	corn	*Zea mays*
	rice	*Oryza sativa*
	sugar cane	*Saccharum officinarum*
	wheat	*Triticum* spp.
Juglandaceae	pecan	*Carya illinoensis*
Leguminosae	bean	*Phaseolus* spp.
	beggarweed	*Desmodium* spp.
	clover	*Trifolium* spp.
	cowpea	*Vigna sinensis*
	crotalaria	*Crotalaria* spp.
	peanut	*Arachis hypogea*
	peas	*Pisum sativum*
	soybean	*Glycine max*
Malvaceae	cotton	*Gossypium hirsutum*
	okra	*Hibiscus esculentus*
Passifloraceae	passion flower	*Passiflora incarinata*
Phytolaccaceae	pokeweed	*Phytolacca decandra*
Polygonaceae	sorrel	*Rumex* spp.
Rosaceae	peach	*Prunus persicus*
	wild blackberry	*Rubus* spp.
	wild plum	*Prunus* spp.
Rutaceae	grapefruit	*Citrus paradisi*
	lemon	*Citrus limon*
	lime	*Citrus aurantifolia*
	orange	*Citrus* spp.
Solanaceae	eggplant	*Solanum melongena*
	pepper	*Capsicum* spp.
	potato	*Solanum tuberosum*
	tobacco	*Nicotiana tabacum*
	tomato	*Lycopersicon esculentum*
Vitaceae	wild grape	*Vitis* spp.

Increased foliar retention was observed where high levels of stink bug damage was inflicted on developing soybean seeds (Daugherty et al. 1964, Gomes 1966, Rizzo 1972). Daugherty et al. (1964) referred to this as a "duddy" condition in the field, which resulted in patches of plants on which the foliage remained green until killed by frost. This condition is believed to be due in part to the failure of seeds to develop properly.

Soybean seeds damaged by southern green stink bug have a slightly higher protein content and a slightly lower oil content than nondamaged beans (Miner 1961, 1966). Qualitative changes in soybean oil are also observed as the severity of stink bug damage increases; linoleic, palmitic, stearic, and oleic acids increase while linolenic decreases (Todd et al. 1973).

Germination is reduced following heavy infestations of stink bugs. It seems that the location of a stink bug puncture is probably more important than the number of punctures (Jensen and Newsom 1972). One puncture in the radicle-hypocotyl axis of the seed can prevent germination, whereas several punctures in the cotyledons may affect the vigor of the plant but not prevent germination.

Stink bug damage not only affects soybean directly through reduction in yield and quality, but also through providing an entry point for the introduction of disease organisms such as the yeast spot disease, *Nematospora coryli* Peglion. Kilpatrick and Hartwig (1955) reported a high percentage of "stink bug-injured" seeds along with isolations from punctured and nonpunctured seeds that yielded a number of different fungi. Their data indicated, however, that stink bug injury was not necessary for fungus infection. Six species of pentatomids were capable of transmitting *N. coryli* into developing soybean seeds (Daugherty 1967). Ragsdale (1977) isolated several bacterial genera from organs of the southern green stick bug and subsequently showed insect transmissibility of five bacteria pathogenic to soybean.

The infestation level of stink bugs on soybean that will justify control measures has been investigated for the southern green stink bug (Duncan and Walker 1968, Todd and Turnipseed 1972, Cherry 1974), the green stink bug (Blickenstaff and Huggans 1962, Daugherty et al. 1964, Miner 1966, and Yeargan 1977), and the brown stink bug (Blickenstaff and Huggans 1962, Daugherty et al. 1964). The brown stink bug in Missouri caused characteristic seed damage and reduced yields at an initial infestation rate of 1 adult/plant (Blickenstaff and Huggans 1962). Miner (1966) compared the damage to soybean caused by the green, southern green, and the brown stink bugs in field cages. He found that the type and severity of damage by the green stink bug was identical to damage by the southern green stink bug. He also reported that brown stink bugs similarly caged caused much less damage than the southern green stink bug. Based on this and other related work, Miner (1966) concluded that an appreciable infestation when pods are small, possibly 1 bug/2 row-m, was a threat to yield. Additionally, he stated that 1 bug/row-m can cause sufficient damage to justify treatment if buyers discount for stink bug damage. Duncan and Walker (1968) maintained an infestation of southern green stink bug for a seven-week period at densities of 2.5, 1.1, and 0.7 bugs/row-m and reported 62.1, 45.6, and 33.1% damaged seeds, respectively. Todd and Turnipseed (1974) reported significant reductions in seed yield and significant increases in seed damage from population densities of 3.3, 9.8, and 16.4/row-m. They further reported that seeds harvested from cages containing 40 pods infested on four dates with a single bug (fourth instar through adult) exhibited 63.9-78.5% damage. Germination, emergence, and seedling survival were reduced significantly by all degrees of *N. viridula* damage. Yeargan (1977) reported similar results with caged populations of *A. hilare* in Kentucky. These results indicate the potential of these insects to inflict serious damage on soybean in the field.

The nature and extent of stink bug damage to soybean depends on the stage of seed development at which feeding occurs. Pods punctured during early endosperm formation are often largely drained of their contents, which may result in empty pods with the remaining seeds being flat or severely atrophied. Feeding that occurs during later stages of endosperm development is accompanied by subsequent formation of a sunken deformed area with puncture marks where the stylets pierced the seed coat. The flesh of the cotyledons is often sunken and deformed beneath the seed coat, with a whitish chalky area where the cell contents have been removed. Frequently, a dark discoloration may be present around the punctured area, and the inner membrane of the seed coat may be fused to the cotyledons rather than being free as is normal (Miner 1966).

II. Life Cycle and Phenology

The life history of the southern green stink bug has been studied by a number of workers in different parts of the world (Jones 1918, Drake 1920, Cumber 1949, Everett 1950, Mitchell and Mau 1969). The work of Drake (1920) and Jones (1918) at Gainesville, Florida and Baton Rouge, Louisiana, respectively, is the most applicable to the southern United States. The southern green stink bug, like most other pentatomids, overwinters in the adult stage mainly under litter, bark, and other objects that offer protection. Rosenfeld (1911) listed it among the insects taken from Spanish moss in Louisiana during December and January. In Florida and the other states bordering the Gulf of Mexico, hibernation of the southern green stink bug is only partial, with a few of the adults remaining on succulent plants throughout the entire winter (Drake 1920). Watson (1918) stated: "They are abundant in October, plentiful in November, common in December, but rather scarce in January and February." Although they may become active during mild periods throughout the winter, there are no winter broods and only adults are found during this season of the year. In the early spring, as the days begin to warm, the bugs emerge from their hibernating quarters in search of food. Mating begins almost immediately upon emergence from hibernation (Drake 1920). The male and female usually remain in copulation for a considerable period of time, firmly joined to one another at the tips of their abdomens and with their heads facing in opposite directions. Under natural conditions copulation may be repeated several times until the eggs have been deposited. Newly emerged adults, after feeding a few days, reach sexual maturity. The eggs are generally laid in regularly shaped compact clusters in which the individual eggs are arranged in very regular rows and firmly glued together. At the time of deposition the eggs are light yellowish white or cream colored. Early in the incubation period the eggs begin to turn pinkish, and a red crescent-shaped spot appears on the operculum. These colors gradually grow deeper and more conspicuous until eclosion. The average incubation time is about six days in summer, but during early spring and late fall the period is often extended to two or three weeks. Figure 23-2 shows a schematic representation of the life cycle.

All stink bug species have five nymphal instars. During the first stage the nymphs cluster on or near the egg shells. Southern green stink bug nymphs have not been observed to feed while clustered; but just before or subsequent to molting, the nymphs become active, scatter more or less, and begin to feed (Drake 1920). Aggregation of early instars of southern green stink bug seems to have an effect on the rate of development and mortality (Kiritani 1964, Kiritani et al. 1965, Kiritani and Kimura 1965,1966). The gregarious habit of nymphal bugs disappears by the fourth instar. The nymphs, like the adults, are usually found on those portions of the plant on which they prefer to feed—the tender growing shoot and especially the developing fruit. The antennae of the nymphs are each composed of four segments; the adults have five. No wings are present on the nymphal instars, although the fifth instar has wing pads. All nymphs from any egg cluster usually spend about the same amount of time in the first and second stage, but the length of time for the other stages is quite variable. During the summer the period from egg to adult is about 35 days with temperature conditions having an important bearing on this (Jones 1918, Drake 1920). Drake (1920) stated that the duration of the different stages is lengthened during the cool weather of the spring and fall. Jones (1918) reported an average preoviposition period of four weeks for the bug in Louisiana. Drake (1920) and Jones (1918), at Gainesville, Florida and Baton Rouge, Louisiana, respectively, both observed four generations per year. Drake (1920) further reported that probably five generations occur in the southern portion of Florida.

Various aspects of the life history and behavior of the southern green stink bug in Japan have been studied in depth by Kobayashi (1959), Kariya (1961), Kiritani and Hokyo (1962, 1965), Kiritani (1963, 1964, 1965, 1970), Kiritani et al. (1963, 1965, 1966a, 1966b), and Kiritani and Kimura (1965). Generally, the reports on life cycle and generation time are much the same, with variations in development time and number of eggs and egg clusters due to seasonal temperatures at a location.

Kobayashi (1959) described the diagnostic characteristics of the various instars of the genus *Nezara* and prepared a key to immature stages of the species, *N. antennata* and *N. viridula*. Mitchell and Mau (1969) published a comprehensive study on the importance of the antennae in the mating process of *N. viridula*. Also in their studies, adults reached sexual maturity in 5-17 days and viable eggs were deposited 7-8 days following mating. The average life span for virgin and mated females was 60 and 30 days, respectively.

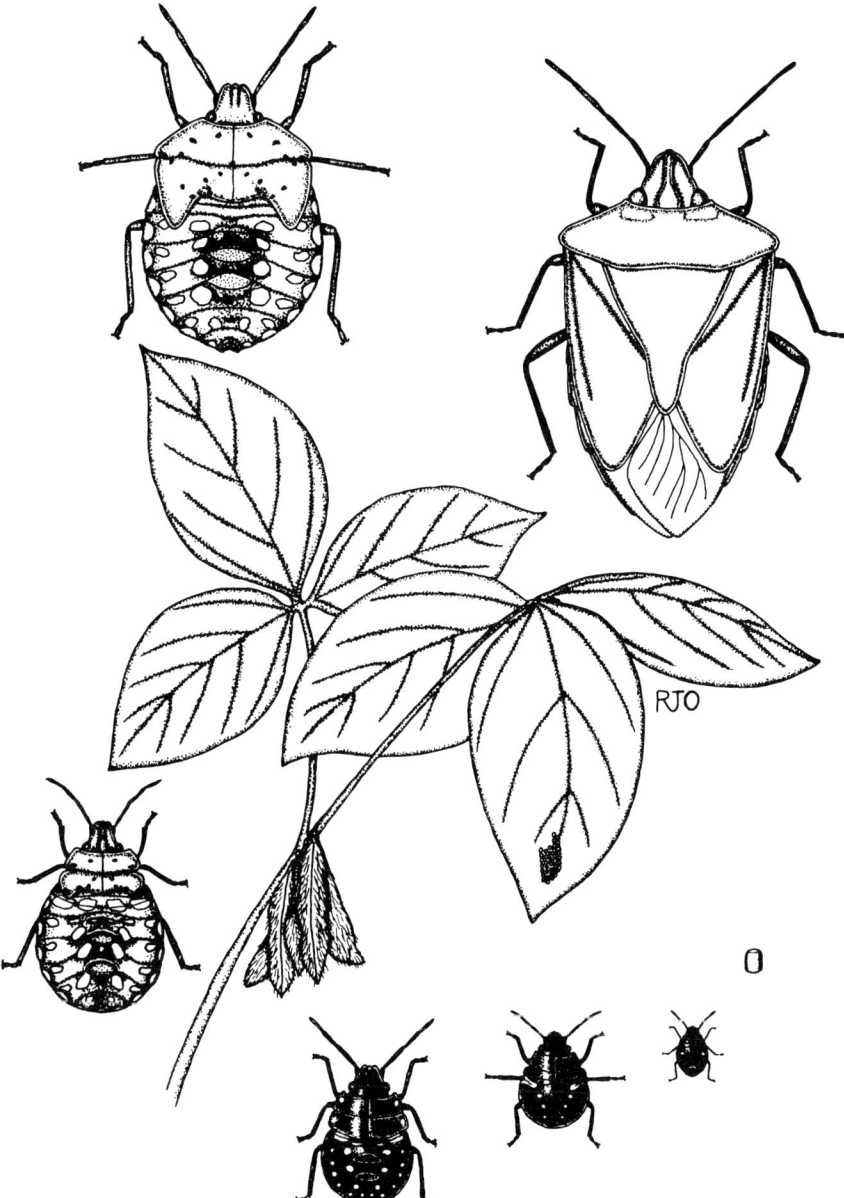

Figure 23-2. Typical life history schema of *Nezara viridula*.

Synchronization of host plant and pest phenologies is an extremely important consideration in the characterization of pest populations. This is particularly important with respect to members of the stink bug complex on soybean since most species are primarily pod or seed feeders, and population peaks must therefore coincide with seed developmental stages of the primary host species. During the critical periods in crop phenology, considerable sampling effort must be expended to adequately monitor pest population trends.

Brief discussions of the life history and important host plants were given, but it is necessary to reiterate some of these considerations relative to pest population dynamics and crop phenology. A generalized schema for a typical seasonal host sequence of *N. viridula* in the southeastern coastal plain of the United States is given in Fig. 23-3. Overwintering in *N. viridula* is accomplished by the adult in diapause (Pitts 1977, Harris and Todd unpublished). When active during warm periods in winter they can be found feeding primarily on crucifers and small grains. As the weather begins to warm in spring, the adults move into clover, early vegetables, corn, and tobacco where they feed or oviposit. The resultant nymphs and adults constitute the first generation. Tomatoes, leguminous and cruciferous vegetables, and okra become attractive in April, May, and June, and these provide the major food sources and oviposition sites during early and midsummer. By the time soybean becomes attractive in late July and August, third generation adults are present and migrate into soybean, which provides the only major source of food in late summer and early fall (late July through November). Third and fourth generation adults in soybean produce fourth and fifth generation nymphs. Soybean is attractive to stink bugs primarily after flowering and pod set although a few individuals can be found in the crop throughout the growing season. The phenological relationship between growth stages of soybean plants and stink bug populations is illustrated in Fig. 23-4.

Kiritani et al. (1965) studied dispersal of *N. viridula* in relation to feeding and oviposition on a variety of crops in Japan. They reported that marked and released adults were recaptured within a few days in fields ca. 1000 m away. Census data taken in Japan indicate that rape, radish, wheat, and barley are primary sites for feeding and mating and that gravid females disperse to other kinds of host plants, e.g., potato, beans, etc., for oviposition. This behavior illustrates the importance of synchronization between the life cycle of the insect and seasonal succession of the host plant.

Two different field situations concerning stink bug population trends were encountered in Louisiana soybean in 1976 (Fig. 23-5). The early-maturing variety, 'Forrest', supported a stink bug population composed of equal numbers of *N. viridula* and other species, while the variety 'Bragg', which matures about a month later, supported a population composed almost completely of *N. viridula*. Figure 23-5 also shows that as an early planting senesces and becomes unattractive for feeding and oviposition, adults move to a more reproductively and nutritionally acceptable host. It is this behavioral response that Newsom and Herzog (1977) have utilized in controlling *N. viridula* in an early-planted, early-maturing trap crop, thereby preventing dispersal to the surrounding, later maturing, larger planting.

Sampling Phytophagous Pentatomidae on Soybean 447

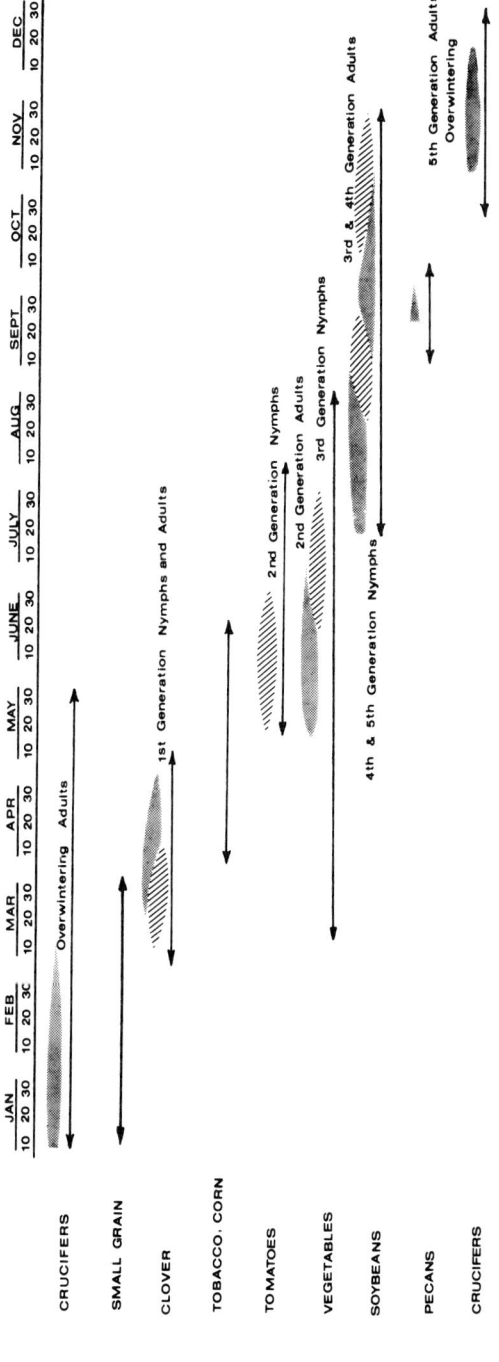

Figure 23-3. Seasonal host sequence of various phytophagous stink bug species in the southeastern United States. Time lines indicate periods when host plants are available and suitable for stink bug feeding. Movement of adults among various hosts for feeding and oviposition is common.

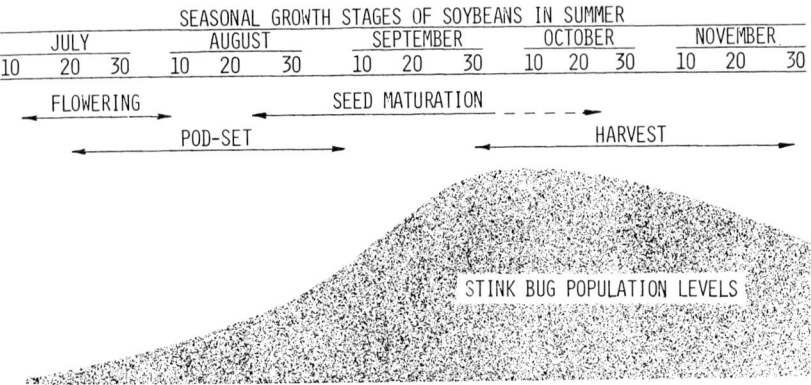

Figure 23-4. Phenological relationship between growth stages of soybean plants and typical population trends of stink bug species in the southeastern United States.

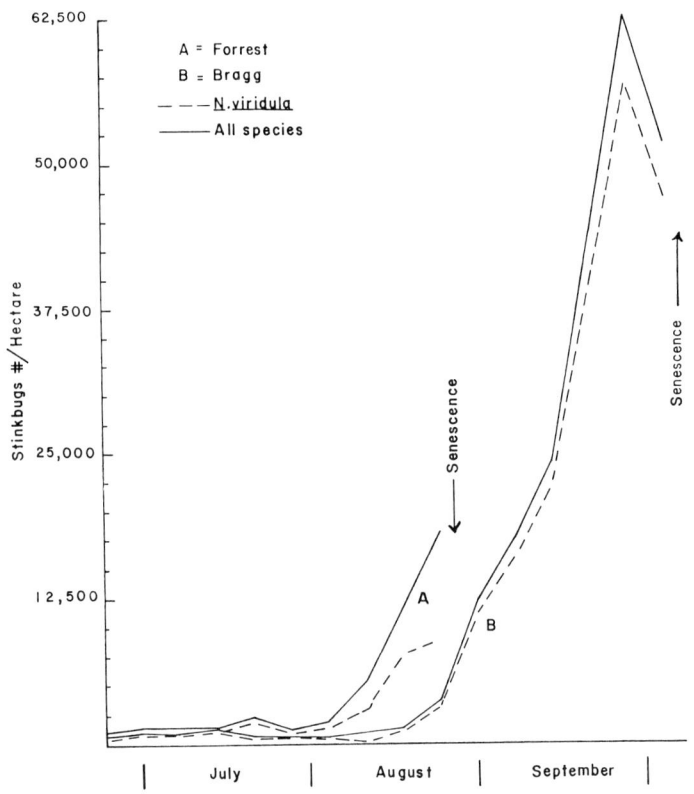

Figure 23-5. Seasonal fluctuations of stink bug populations on (A) 'Forrest' and (B) 'Bragg' soybean in Louisiana in 1976.

III. Sampling for Adults and Nymphs

Since stink bug adults and nymphs are similar in habit and occurrence, sampling procedures for these stages will be discussed simultaneously. As in the previous sections, emphasis will be placed on the southern green stink bug. All sampling procedures are generally applicable to other phytophagous pentatomids, and only the dispersion characteristics are not necessarily comparable among species.

Absolute quantitative measurements of stink bug populations are best made utilizing the fumigation cage method described in Chapter 2. It is the opinion of the authors that the ground cloth method of sampling so nearly approximates the absolute measurement of stink bug populations that it may be considered as absolute and can be used with a minimal loss in accuracy because of the relative ease of sample acquisition as compared to the traditional absolute methods. This hypothesis was supported by Marston et al. (1976). However, problems involved with population dispersion severely limit the applicability of either the absolute or ground cloth methods to sampling stink bugs on soybean. The problem of spatial patterns will be discussed below.

In their study Marston et al. (1976) first utilized a fumigation cage that enclosed a 1 m section of row. This method was used for comparison with relative sampling methods. Because of dissatisfaction with numbers of insects collected, they modified their fumigation cage to encompass 2.4 row-m. Comparison of the two yielded mean numbers of stink bugs/4.6 row-m of 0.9 and 2.0 for the basic and modified cages, respectively. Although analysis of their data yielded no statistically significant difference between the two "absolute" methods, it would seem that the magnitude of the difference (2.25X) is too large to be overlooked. This leads one to believe that neither of the "absolute" methods can be relied on as being truly absolute.

The blacklight trap shows promise as a survey tool for monitoring stink bug adult population fluctuations. Preliminary results of Herzog (unpublished) indicate that *N. viridula, A. hilare*, and *E. servus* are attracted to blacklight in sufficient numbers to allow qualitative and quantitatively relative studies on their populations. Other phytophagous species found in smaller numbers include *E. tristigmus* (Say), *Banasa dimidiata* Say, *Thyanta custator* (Fabricius), *Proxys punctulatus* (Palisot de Beauvois), and *Piezodorus guildinii* Westwood. Cantelo et al. (1973,1974) reported excellent results of blacklight survey on St. Croix for *N. viridula, Acrosternum marginatum* (Palisot de Beauvois), and *Loxa flavicollis* (Drury), all of which have been recorded in the literature as occurring in soybean, and for *Loxa pilipes* Horvath, which has not been recorded from soybean.

Another possibly useful survey tool is the male sex pheromone reported by Mitchell and Mau (1971) to occur in *N. viridula*. Although at least one unpublished attempt has failed to verify its occurrence (J. W. Thomas, personal communication) more recent work (Harris and Todd 1980) indicates that attraction between sexes is by way of an aggregation phermone produced by males. It is probable that one or more male or female pheromones do occur in *N. viridula* and that others remain to be discovered in other stink bug species. Such pheromones could be utilized in conjunction with blacklight trapping, electric grids, or sticky traps for population monitoring.

A. Comparison of Relative Sampling Methods for Adults and Nymphs

The earliest reference seen comparing sampling methods for stink bugs concerned research conducted on cotton. Race (1960) compared 100 sweep samples with catch achieved using a tractor-mounted mechanical collection device (the Nisbett "Bug Catcher") that was run for the same distance required to make 100 sweeps. Catch of *Chlorochroa sayi* Stal, *C. ligata* (Say), and *T. custator* taken by the two methods indicated that the sweep net collections failed to provide a dependable stink bug population index on cotton plants during the latter half of the growing season. Race (1960) recommended that the sweep net sampling method be used only in conjunction with visual inspections of plants in the field.

Relative measurements of stink bug populations on soybean have been made by visual means or with the ground cloth, sweep net, or vacuum net methods. Very little literature is available that compares the various sampling methods.

Duncan (1968) compared ground cloth and sweep net methods for sampling adults and all nymphal instars of *N. viridula*. He concluded that fourth and fifth instar nymphs and adults were sampled with equal efficiency by both techniques, while second and third instar nymphs were sampled more efficiently by the ground cloth technique. First instar nymphs were not sampled adequately with either of the methods utilized.

Marston et al. (1976) compared three relative sampling methods with an "absolute" fumigation cage method. They logically grouped all species and developmental stages as would be the case with sampling in pest management. However, they apparently failed to separate predatory from phytophagous species. The fumigation cage and ground cloth methods collected significantly larger numbers of stink bugs than did the vacuum net and sweep net methods: 3.9, 4.9, 1.9, and 1.3/sample, respectively.

Rudd and Jensen (1977) compared sweep net and ground cloth sampling methods for *N. viridula*. Results of regression analyses showed that numbers of stink bugs captured by the two methods were highly correlated, ranging from $r = 0.835$ for *N. viridula* females to $r = 0.982$ for fourth instar *N. viridula*, with $r = 0.998$ for third, fourth, and fifth instars and adults combined. All correlations were statistically significant ($P \leq 0.01$).

The times required for acquisition of samples were reported by Rudd and Jensen (1977) as 21.2 and 48.3 seconds/sample for the sweep net and ground cloth methods, respectively. They concluded that the sweep net was superior to the ground cloth for most sampling objectives based on the time required per observation.

The ratio of the standard deviation to the mean ($s/\bar{x}$) is an unbiased estimator

Table 23-2. Comparison of means of $s/\bar{x}$ by sweep net and ground cloth methods for *N. viridula* (adapted from Rudd and Jensen 1977)

Life stage	$s/\bar{x}$ Sweep net	$s/\bar{x}$ Ground cloth	Q^2
2nd instar	2.45	2.94	1.42
3rd instar	1.69	2.50	2.19
4th instar	1.59	2.30	2.07
5th instar	1.25	1.95	2.40
Adult	1.53	1.85	1.46
Combined [a]	1.39	1.77	1.61

[a] 3rd, 4th, 5th instars and adults.

of data set variability. This ratio was reported by Rudd and Jensen as shown in Table 23-2. The sample (method) with the smaller $s/\bar{x}$ ratio provides the most information and thereby requires significantly smaller samples (numbers of observations) to obtain the same amount of information. In Table 23-2, column Q^2 gives the squared ratio of $s/\bar{x}$ by ground cloth to $s/\bar{x}$ by sweep net derived by Rudd and Jensen (1977). This quantity is the ratio of the number of observations required by the two methods to yield the same amount of information. The data show that in all cases, a larger number of observations are required to measure a population mean using the ground cloth than the sweep net sampling method. This factor ranges from 42% for second instar to 140% for fifth instar nymphs. For combined populations (third, fourth, fifth instars and adults), 61% more samples are required using the ground cloth method.

B. Calibration of Sampling Procedures

Relative sampling methods should be calibrated to some absolute method that gives accurate population estimates. The ground cloth method is often considered to approximate an absolute measurement and may be used as a standard for calibration in conducting a sampling program for pest management purposes.

Since there are so many variations of the sweep net method, and since even with a standardized sweeping method individual variation exists as to force of the sweep, distance between sweeps, etc., it is advisable that every sampler adopting the sweep net method of sampling calibrate his individual variation with the ground cloth method. The same would hold true for anyone adopting the D-Vac sampling method.

To accomplish this calibration it is necessary to take a number of paired comparison samples. Rudd and Jensen (1977) found it difficult, if not impossible, to achieve a high degree of correlation when comparing methods on an observation/observation basis. They were able to overcome this difficulty by comparing sampling methods on a per field basis. A regression analysis was conducted using means calculated for each method in each field on each sampling date. In this way they achieved a high degree of correlation. Because of the strength of the correlations, they were able to calculate factors for the conversion of numbers of *N. viridula* caught in sweep samples to number per row-m, based on regression analysis. Based on the economic threshold level of 3.3 bugs/row-m currently recommended by most states (Kogan 1976), Table 23-3 gives, for various developmental stages, numbers of bugs per-sweep and for samples of varying numbers of sweeps to equal the threshold level.

It should perhaps be stressed at this point that this research was conducted with *N. viridula*. It would be unwise to relate these data to other species of phytophagous pentatomids. It is probable that behavior differs sufficiently among species so that the regression and, therefore, the conversion factor, would be significantly different for other species. But, for pest management purposes in the southeastern United States where *N. viridula* is the predominant species, it is probable that this conversion could be used without incurring significant error.

The regression equation $Y = a + bX$, then, has the following components:

Y = number caught/2 m shake sample
a = the Y intercept (approximately and, in most cases, statistically 0)
X = number caught/some number (25, 50, 100) of sweeps
b = the calibration or conversion factor

This method of calibration should be used. Ten or more observations should be taken by each sampling method in each field sampled on each sampling date. Thirty or more such sets of observations should be taken, covering a wide range of populations, different varieties, different locations, and chronological times. Regression analysis of these data should yield an accurate, highly correlated factor for calibration of sweep net or D-Vac samples with ground cloth samples.

Table 23-3. Conversion of economic threshold levels to numbers of *N. viridula* obtained by the sweep net method (adapted from Rudd and Jensen 1977)

Life stage	3.3 Stink bugs/row-m			
	No./Sweep	No./25 Sweeps	No./50 Sweeps	No./100 Sweeps
2nd instar	0.26	6.4	12.8	25.5
3rd instar	0.43	10.7	21.4	42.8
4th instar	0.38	9.5	19.0	38.1
5th instar	0.47	11.8	23.5	47.0
Adult ♂	0.37	9.2	18.5	36.9
Adult ♀	0.44	11.1	22.2	44.4
Adults combined	0.39	9.7	19.4	38.8
All stages combined	0.37	9.2	18.3	36.6

C. Spatial Patterns

The spatial patterns of the species under study has a very significant impact on the type of sampling method applicable to that species and the number of sampling units required to measure a population mean with a given degree of precision (see Chapter 3). Until this publication, relatively little attention had been given to this aspect of sampling insects on soybean. This has been especially true concerning members of the stink bug complex.

The only research concerning spatial patterns of stink bug species was conducted on rice in Japan. Of the species considered, only *N. viridula* occurs on soybean. Nakasuji et al. (1965) reported that under most conditions invading females of *N. viridula* disperse at random into rice fields. A similar situation is seen in soybean (Herzog unpublished). Invading females apparently disperse at random into vegetative soybean or trap crops of early-planted, early maturing soybean (Newsom and Herzog 1977). However, one to several males are often found in close proximity to the females resulting an overdispersed or clumped dispersion of males and of the adult population as a whole. In other cases invading adults have been seen to appear first in a corner or edge of a field, exhibiting an "edge effect" similar to that reported by Nakasuji et al. (1965). Exceptions to this may be due to conditions of surrounding crops or native vegetation. Under certain conditions shortfall of invading adults or overflow into adjacent areas may occur. Nonrandom patterns can also result from physiological processes relating to mating behavior causing males in most cases to exhibit a clumped pattern (Nakasuji et al. 1965). Their data indicate that more males than expected were located near mating pairs, seemingly refuting the presence of a male sex pheromone reported by Mitchell and Mau (1971). Subsequent research in Georgia (Harris and Todd 1980) has shown that both male and female *N. viridula* were significantly more attracted to cages containing males than to cages containing females or to control cages. These data suggest that an aggregation pheromone produced by males results in the concentration of both sexes into the same area.

The dispersion of egg masses in a field is primarily determined by the dispersion of the ovipositing females and by their degree of activity and pattern of movement. *N. viridula* lays eggs in relatively large masses with a relatively long period elapsing between successive ovipositions (Kiritani and Hokyo 1965, Kiritani and Kimura 1966), and the females leave the plant on which it has oviposited soon after an egg mass is laid (Kiritani et al. 1965). Because the dispersion of females and their redistribution is random, egg masses are also deposited at random, fitting the Poisson distribution (Hokyo and Kiritani 1962). However, individual eggs exhibit a highly clumped pattern by virtue of being deposited in large masses.

First instar nymphs have not been observed feeding on plant tissue and seldom leave the egg mass unless disturbed (Drake 1920, Kiritani 1964). Second instars begin feeding on plant tissue but remain congregated on or near the egg mass. Third instars disperse as a group from the egg mass, and usually retain the clustered identity of the cohort. The clumped pattern of the nymphs is retained throughout the early instars and the degree of aggregation is diminished only through mortality or disturbance. The degree of aggregation has been shown to decrease abruptly in the fourth instar on rice due to increased ability and tendency to disperse. Further dispersal occurs in the fifth instar, populations approaching the random or Poisson distribution (Hokyo and Kiritani 1962). Early season populations on soybean, however, are rarely of sufficient magnitude for late instar nymphs to achieve the truly random pattern (Herzog unpublished).

N. viridula population patterns on rice have been reported as being clumped according to the negative binomial distribution (Nakasuji et al. 1965) and as being random according to the Poisson distribution (Hokyo and Kiritani 1962). However, the literature contains no reports on spatial patterns of stink bugs on soybean. Further analysis of the data presented by Rudd and Jensen (1977) revealed that *N. viridula* exhibits a negative binomial distribution on soybean (W. G. Rudd, personal communication). The dispersion characteristic ($\underline{k}$) derived for 100 sweep samples was 1.25. For 2-m ground cloth samples this parameter was calculated as 0.997 based on a single set of observations. Since these are the only data available, the dispersion of *N. viridula* is considered to be according to the negative binomial for calculation of sequential sampling plans.

D. Sequential Sampling Plans

When sampling stink bug populations, several factors influence or dictate the number of sampling units that must be taken to answer questions about a population: (1) the reason for sampling, (2) the sampling distribution, (3) the true population mean, (4) the level of precision desired, and, in some cases, (5) the sampling method used. Ruesink and Kogan (1975) have listed the major sampling objectives as data collection for research purposes and for making management decisions. Research requires precise parameter quantification, while the making of management decisions requires only the classification of field situations into control decision categories.

When stink bug populations are sampled with the objective of making management decisions concerning the probability of economic injury being incurred, it is only necessary to classify the population as exceeding or not exceeding economically damaging levels. Sequential sampling plans may be designed that will make it possible to make this decision subject to a prespecified risk level. Since the spatial pattern or dispersion of the population in the field significantly influences parameters necessary to the construction of sequential plans, this relationship must be defined and understood. See Chapter 3 for a thorough discussion of dispersion and sampling distributions. Since the data of Rudd and Jensen (personal communication, see Spatial Patterns section above) are the only available, the dispersion of *N. viridula* is considered to be according to the negative binomial for calculation of sequential sampling plans, with $\underline{k}$ = 1.25 and 0.997 for sweep net and ground cloth, respectively.

The economic threshold of infestation presently recommended for control of stink bugs in most states is 3.3 bugs/row-m throughout reproductive crop development (Kogan 1976). However, the recommendation in Alabama, Georgia, and Florida is to control 1.1 bugs/row-m from bloom through midpod-fill and 3.3 bugs/row-m from midpod-fill through maturity. The population level recommended for control of stink bugs on soybean grown for seed in Louisiana and Georgia is 1.1 bugs/2 row-m at any time during the reproductive development of the crop. As has been noted, 3.3 bugs/row-m = 36/100 sweeps, 18/50 sweeps; 1.1 bugs/row-m = 12/100 sweeps, 6/50 sweeps; 1.1 bugs/2 row-m = 6/100 sweeps, 3/50 sweeps.

The final factor that must be considered before sequential sampling plans are developed is the selection and standardization of the sampling method. Data on the sampling efficiency of the various methods notwithstanding, choice of the sampling method will probably continue to be made according to the personal bias of the sampler. While standardization is not as important with the ground cloth method, it is essential with the sweep net. Jensen (1976) describes a standardized sweeping method.

Using methods and formulas supplied by Waters (1955) and Onsager (1976) (see Chapter 4) and based on Louisiana field data (Herzog unpublished), several sequential sampling plans for management have been constructed for several threshold levels that can be used with ground cloth or sweep net sampling methods.

The first set of sampling plans was constructed for economic threshold levels of 3.3 bugs/row-m. The population level, below which economic damage is not expected to occur during the sampling interval and below which controls can be deferred, is set at 1/2 of the economic threshold or 1.6 bugs/row-m. The fact that stink bugs are dispersed in a clumped manner (negative binomial) requires that the risk factor be set at relatively high levels. Figures 23-6 and 23-7 show sequential sampling plans for use with ground cloth and sweep net methods, respectively. In each figure the inner and outer paired decision lines denote risk levels of $\alpha = \beta = 0.2$ and $\alpha = \beta = 0.1$ (20 and 10%), respectively. Tabular forms of the same plans are shown in Tables 23-4 and 23-5.

Sequential sampling plans for threshold levels of 1.1 bugs/row-m and 1.1 bugs/ 2 row-m are presented in Figs. 23-8 to 23-11. For calculation of these plans, risk levels were set at $\alpha = \beta = 0.2$ and the population levels below which treatments would not be required were set at 1/2 of the economic threshold, 1.1 bugs/ 2 row-m and 1.1 bugs/4 row-m. The corresponding tabular forms for these plans are given in Tables 23-6 and 23-7.

The average sample number curves shown in Figs. 23-12 and 23-13 give the theoretical mean number of observations required to reach a management decision at any particular population level. Examination of these curves reveals that in all cases, larger numbers of observations are required to reach a management decision when using the ground cloth sampling method as compared to the sweep net. This relationship is quantified in Table 23-8. This table shows a 1.5- 2.4-fold increase in numbers of samples required when comparing the ground

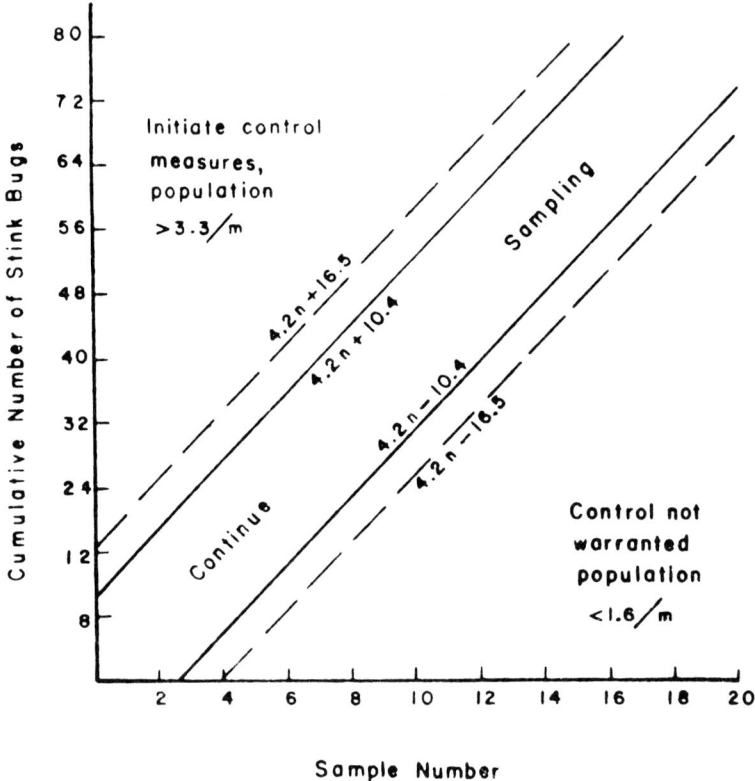

Figure 23-6. Sequential sampling plans for *N. viridula* sampled by the ground cloth method, representing a threshold of 3.3 bugs/row-m and at 10% (outer decision lines) and 20% (inner decision lines) risk levels.

cloth with the sweep net method. This difference in numbers of observations required is due to two major factors: the magnitude of the dispersion parameter (k) characterizing the sampling distributions of the two methods and the number of individuals captured per observation by the sampling method being used to characterize the population. The greater the degree of clumping, the larger will be the number of observations required to reach a decision. The sweep net method is clearly superior in this respect for, although stink bugs exhibit a highly clumped pattern, a single sweep net sample averages population estimates of several 1-m sections of row, thereby increasing the dispersion parameter. The ground cloth sample, on the other hand, is a single 1, or 1.3, or 2.0, or 2.3 m section of row. A 100 sweep sample estimates the population present in approximately 91 m of row.

Often, intensive sampling must be conducted in order to precisely define population parameters such as the mean and variance at a wide range of population densities. Such would be the case when studying population dynamics for the

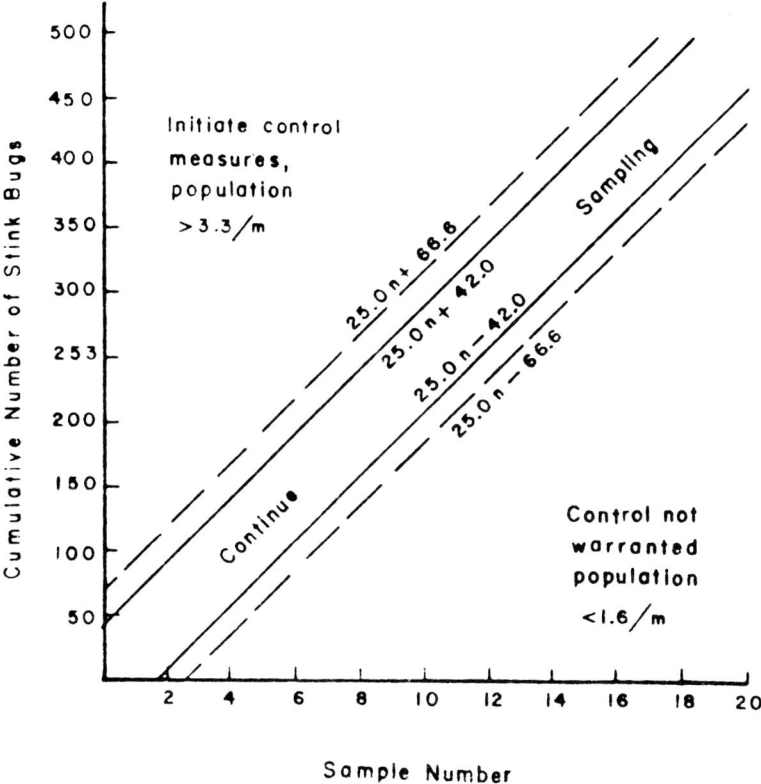

Figure 23-7. Sequential sampling plans for *N. viridula* sampled with the sweep net method, representing a threshold of 3.3 bugs/row-m and at 10% (outer decision lines) and 20% (inner decision lines) risk levels.

purpose of construction or validation of simulation models. Here it is also desirable to know at what point the parameters have been estimated with the desired level of precision. The number of observations required cannot be estimated intuitively. Equations are often seen that allow the estimation of sample number, such as the following from Ruesink and Kogan (1975):

$$N = 25_t^2 \, a\bar{x}^{b-2}$$

where 25 = (100/accuracy as a percent of the mean)^2, t is the two tailed probability level at P = 0.05 with n- 2 degrees of freedom (n = total number of observations used in computing a and b: modified from Ruesink and Kogan (1975) by Ruesink, personal communication), $\bar{x}$ is the mean, and a and b are the intercept and regression coefficient derived by linear regression of log s^2 (variance) on log $\bar{x}$ (mean). This method requires a prior knowledge of the mean and vari-

Table 23-4. Sequential table for sampling stink bugs on soybean by the ground cloth method (Threshold level 3.3 bugs/row-m)

Number of 2-m shakes	Cumulative number of stink bugs			
	20% Risk		10% Risk	
	≤	≥	≥	≤
1	—	15	21	—
2	—	19	25	—
3	2	23	30	—
4	6	28	34	0
5	10	32	38	4
6	14	36	42	8
7	18	40	46	12
8	23	44	50	17
9	27	49	55	21
10	31	53	59	25
11	35	57	63	29
12	39	61	67	33
13	44	65	71	38
14	48	70	75	42
15	52	74	80	46
16	56	78	84	50
17	60	82	88	54
18	65	86	92	58
19	69	91	97	63
20	73	95	101	67

Decision labels (printed vertically in the table): ≤ columns — "CONTROL NOT WARRANTED"; between-columns — "CONTINUE SAMPLING"; 20% Risk ≥ column — "INITIATE CONTROLS".

ance of the population being sampled. Since these two parameters are not often known before sampling is begun, it would be advantageous to develop a plan that does not rely on a prior knowledge about the population.

Chapter 5 of this book provides methods and formulas for developing sequential estimation plans that allow measurement of population means at predetermined levels of precision. When sampling from a randomly dispersed population (Poisson), the collection of a prespecified number of individuals fulfills the requirements of the plan without regard for the sampling method utilized, the size of the observational unit, or the number of sampling units taken. However, the situation becomes much more complicated when sampling from populations of species, such as stink bugs, that are dispersed in a clumped manner (negative binomial). Because of the clumped nature of the population, differing numbers of individuals must be collected at the different density levels to adequately characterize that population with the desired degree of precision. Such plans are applicable to any sampling method and to observational units of any size, provided that the sampling distribution remains the same for all situations.

Table 23-5. Sequential table for sampling stink bugs on soybean by the sweep net method (Threshold level 3.3 bugs/row-m)

Number of 100-sweep samples	Cumulative number of stink bugs							
	20% Risk				10% Risk			
	≥		≤		≤		≥	
1	67		—	C	—		92	
2	92		8	O	—		117	
3	117	I	33	N	8	C	142	I
4	142	N	58	T	33	O	167	N
5	167	I	83	R	58	N	192	I
6	192	T	108	O	83	T	217	T
7	217	I	133	L	108	I	242	I
8	242	A	158		133	N	267	A
9	267	T	183	N	158	U	292	T
10	292	E	208	O	183	E	317	E
11	317		233	T	208		342	
12	342	C	258		233	S	367	C
13	367	O	283	W	258	A	392	O
14	392	N	308	A	283	M	417	N
15	417	T	333	R	308	P	442	T
16	442	R	358	R	333	L	467	R
17	467	O	383	A	358	I	492	O
18	492	L	408	N	383	N	517	L
19	517	S	433	T	408	G	542	S
20	542		458	E	433		567	
				D				

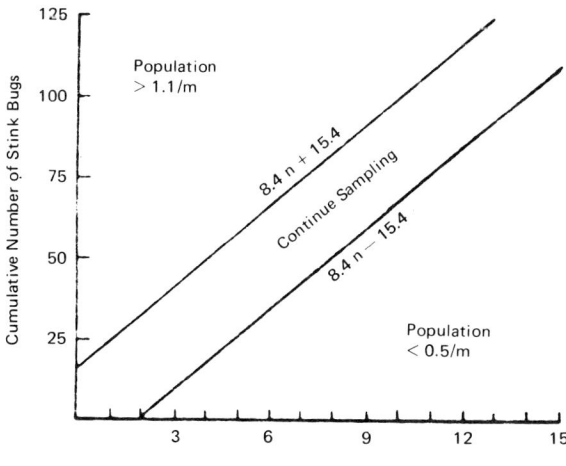

Figure 23-8. Sequential sampling plan for *N. viridula* sampled by the sweep net method and representing a threshold of 1.1 bugs/row-m.

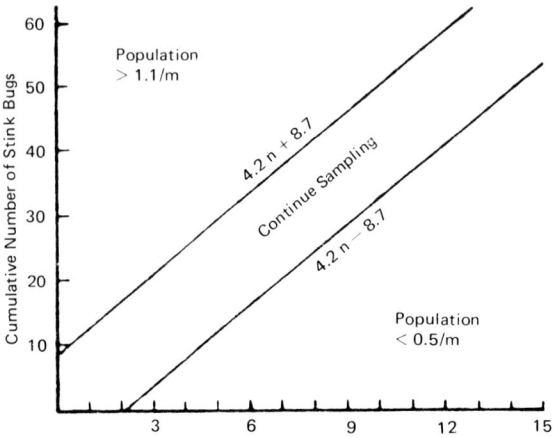

Figure 23-9. Sequential sampling plan for *N. viridula* sampled by the ground cloth method and representing a threshold of 1.1 bugs/row-m.

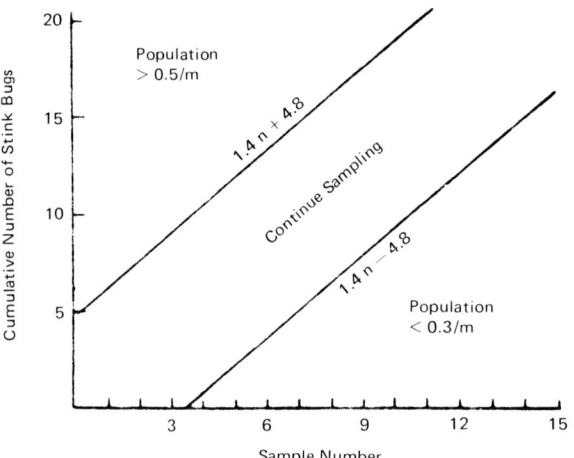

Figure 23-10. Sequential sampling plan for *N. viridula* sampled by the sweep net method and representing a threshold of 1.1 bugs/2 row-m.

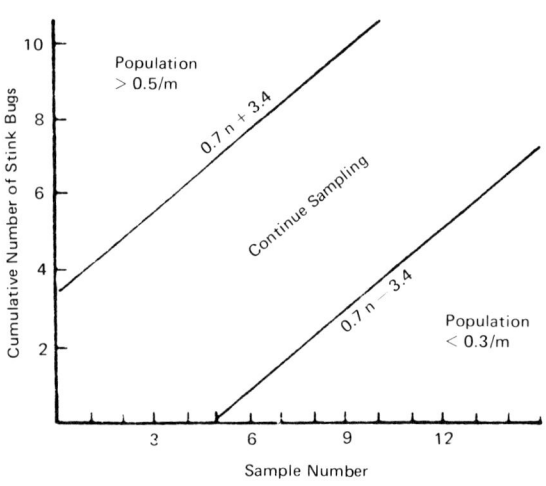

Figure 23-11. Sequential sampling plan for *N. viridula* sampled by the ground cloth method and representing a threshold of 1.1 bugs/2 row-m.

Table 23-6. Sequential table for sampling stink bugs on soybean [Threshold level 1.1 bugs/2 row-m (Risk level = 20%)]

Number of samples	Cumulative number of stink bugs			
	100 sweeps		2-m shake	
	≤	≥	≥	≤
1	C —	24	7	— C
2	O 1	33	8	— O
3	N 9 C	41 I	10 C	— N
4	T 17 O	49 N	11 O	0 T
5	R 26 N	58 I	12 N	2 R
6	O 34 T	66 T	14 T	3 O
7	L 43 I	74 I	15 I	5 L
8	51 N	83 A	17 N	6
9	N 59 U	91 T	18 U	7 N
10	O 68 E	100 E	19 E	9 O
11	T 76	108	21	10 T
12	84 S	116 C	22 S	12
13	W 93 A	125 O	24 A	13 W
14	A 101 M	133 N	25 M	14 A
15	R 109 P	141 T	26 P	16 R
16	R 118 L	150 R	28 L	17 R
17	A 126 I	158 O	29 I	19 A
18	N 135 N	166 L	31 N	20 N
19	T 143 G	175 S	32 G	21 T
20	E 151	183	34	23 E
	D			D

Sequential estimation plans are presented in Figs. 23-14 to 23-16. These plans are designed for *N. viridula* and will yield confidence intervals equal to the population mean ($\bar{x} \pm .5\bar{x}$) at 0.8, 0.9, and 0.95 levels of probability. Plans calculated to allow mean estimation within confidence interval lengths less than the mean prove unworkable because of the extremely large number of samples required at low population levels. To use these plans with either the sweep net or ground cloth sampling methods, one must consider the effect of sampling method on the dispersion characteristic (k). As stated above, this parameter takes the value of 1.25 for 100-sweep samples and 1.00 (0.997) for 2-m ground cloth samples. This necessitates the calculation of separate decision lines for the two sampling methods (in reality for the differing dispersion parameters) as shown in Figs. 23-14 to 23-16. This method provides that at a given level of probability, a population mean can be estimated by collecting K individuals from the population as determined by the current sample size N. In other words, continue sampling and plotting cumulative number of individuals collected until a point above the decision line has been reached. At this point the population mean has been estimated with a confidence interval equal to the mean at the chosen level of precision.

Table 23-7. Sequential table for sampling stink bugs on soybean [Threshold level 1.1 bugs/2 row-m (Risk level = 20%)]

Number of samples	Cumulative number of stink bugs							
	100 sweeps				2-m shake			
	≥			≤	≤		≥	
1		13		–	C	–	5	
2		18		–	O	–	5	
3	I	22	C	3	N	–	C	6	I
4	N	26	O	8	T	–	O	7	N
5	I	30	N	12	R	0	N	7	I
6	T	34	T	16	O	0	T	8	T
7	I	39	I	20	L	1	I	9	I
8	A	43	N	24		2	N	10	A
9	T	47	U	29	N	2	U	10	T
10	E	51	E	33	O	3	E	11	E
11		55		37	T	4		12	
12	C	60	S	41		5	S	12	C
13	O	64	A	45	W	5	A	13	O
14	N	68	M	50	A	6	M	14	N
15	T	72	P	54	R	7	P	15	T
16	R	76	L	58	R	7	L	15	R
17	O	81	I	62	A	8	I	16	O
18	L	85	N	67	N	9	N	17	L
19	S	89	G	71	T	10	G	17	S
20		93		75	E	10		18	
					D				

The availability of these plans provides further opportunity for comparison of sweep net with ground cloth sampling methods. Table 23-9 lists number of observations that must be taken by sweep net and ground cloth methods to estimate a population mean within one confidence interval equal to the mean ($\bar{x} \pm .5\bar{x}$). Sampling units currently being used in sampling soybean insects are included for comparison. Since the dispersion parameter (k) is not known for 50-sweep or 2.3 m, 1.3 m, or 1 m ground cloth samples, these units were assigned the k known for their counterparts. It is possible that in some cases the reality may not be significantly different. Both the k = 1.00 and k = 1.25 are tabulated for the ground cloth variations to illustrate the consequence of greater clumping on number of samples required to precisely define the mean. Calculations are based on numbers of individuals per sample that yield means of 0.01, 0.05, 0.1, 0.5, and 1 bug/0.3 row-m for example, 0.1 bug/0.3 row-m = 1.9/50 sweeps, 3.7/100 sweeps, 0.8/2.3 m shake, 0.6/2 m shake, 0.4/1.3 m shake.

The sweep net method in almost all cases has a distinct and decided advantage over the ground cloth method. Table 23-9 illustrates that 95% confidence in a population mean of 0.01 ± 0.005 can be derived by taking 55 samples of 100

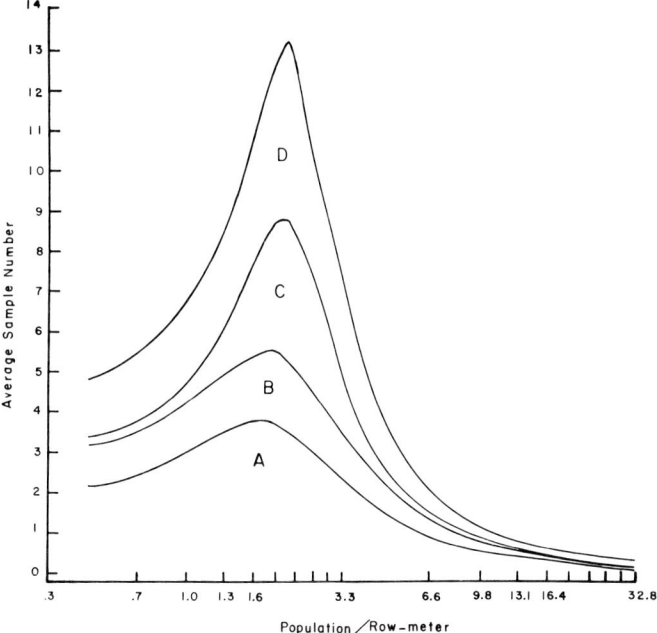

Figure 23-12. Average sample number curves for *N. viridula* sampled with sweep net (A and C) and ground cloth (B and D) sampling methods at 20% (A and B) and 10% risk levels (C and D).

sweeps, while an excess of 120 shake samples of any size would be required to yield the same degree of precision. A confidence level of 80% for the same mean can be achieved following 25 and 97 samples of 100 sweeps and 2.3 m shakes, respectively.

The development of such sequential estimation plans will be of great benefit to researchers, especially when working with populations that are distributed in a clumped manner. These plans allow estimation of populations at predetermined levels of precision and risk without a prior knowledge of the population mean and variance.

IV. Sampling for Eggs

The phytophagous stink bug species most frequently oviposit on the underside of soybean leaves. However, it is not uncommon to find egg masses deposited on the upper leaf surface, on the stems, or on pods. While many foliage inhabiting pests deposit eggs singly and at random on the crop, stink bugs, like the Mexican bean beetle, deposit eggs in clusters or masses. The egg masses are most often found containing eggs in multiples or near-multiples of 14 because of ovariole number present in females. Egg mass sizes among the pentatomids range from 7 or less to more than 150.

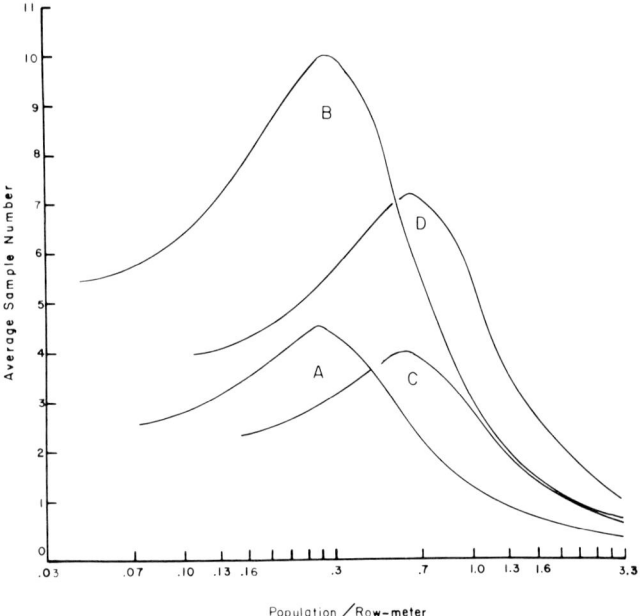

Figure 23-13. Average sample number curves for *N. viridula* sampled with sweep net (A and C) and ground cloth (B and D) for 1.1 bugs/2 row-m (A and B) and 1.1 bugs/row-m (C and D).

Table 23-8. Average number of samples required to categorize a population at point of maximum sampling requirement and at threshold levels of 3.3, 1.1, and 0.5 stink bugs/row-m

Threshold level	Population level	Average sample number					
		Sweep		Shake		Difference	
		10%	20%	10%	20%	10%	20%
3.3/m	maximum	8.7	3.7	13.2	5.4	1.5X	1.5X
	threshold	4.9	2.3	7.2	3.4	1.5X	1.5X
1.1/m	maximum		3.9		7.2		1.8X
	threshold		2.5		4.9		1.9X
0.5/m	maximum		4.5		10.0		2.2X
	threshold		2.9		7.1		2.4X

Sampling Phytophagous Pentatomidae on Soybean

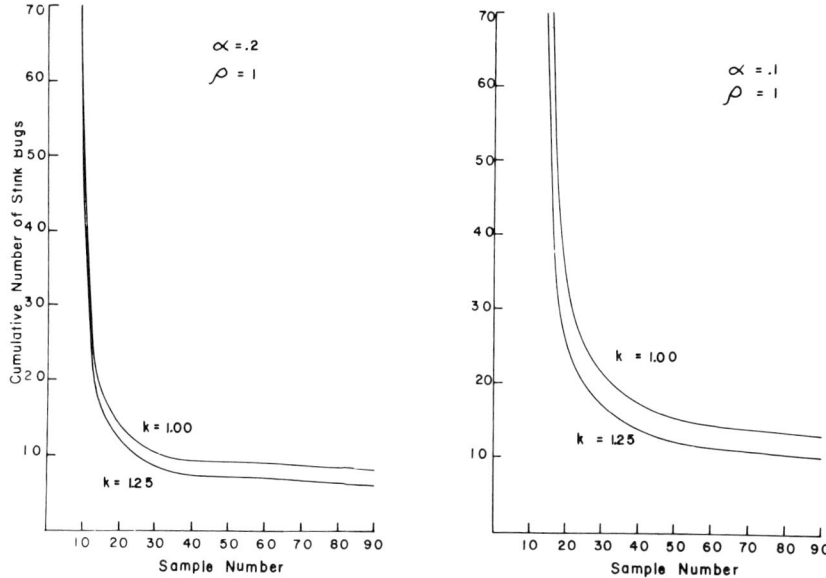

Figure 23-14. Sequential plan for estimation of *N. viridula* populations at $\rho = 1.0$ and $\alpha = 0.2$.

Figure 23-15. Sequential plan for estimation of *N. viridula* populations at $\rho = 1.0$ and $\alpha = 0.1$.

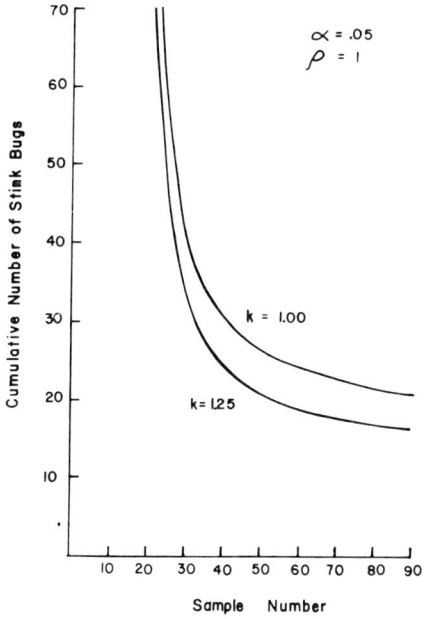

Figure 23-16. Sequential plan for estimation of *N. viridula* populations at $\rho = 1.0$ and $\alpha = 0.05$.

Table 23-9. Number of samples required to describe populations at $\bar{x} \pm 0.5 \bar{x}$ by sweep net and ground cloth sampling methods

Population level	Risk level	50 sweeps k=1.25	100 sweeps k=1.25	2.3m shake		2.0m shake		1.3m shake		1.0m shake	
				k=1.25	k=1.00	k=1.25	k=1.00	k=1.25	k=1.00	k=1.25	k=1.00
.01/0.3m	0.2	40	25	83	97	105	>120	>120	>120	>120	>120
	0.1	62	40	>120	>120	>120	>120	>120	>120	>120	>120
	0.05	90	55	>120	>120	>120	>120	>120	>120	>120	>120
.05/0.3m	0.2	17	12	24	28	28	33	38	46	49	60
	0.1	21	18	38	43	45	52	58	70	72	88
	0.05	33	25	52	61	63	73	83	101	104	123
.1/0.3m	0.2	13	11	17	18	19	21	23	28	28	33
	0.1	19	16	25	28	29	33	38	43	44	51
	0.05	26	20	35	39	40	46	52	60	63	73
.5/0.3m	0.2	10	9	10	11	14	15	16	17	17	18
	0.1	15	<15	15	17	17	17	18	20	20	21
	0.05	21	<20	20	22	22	23	25	27	28	30
1/0.3m	0.2	10	8	9	9	10	10	11	11	11	12
	0.1	<12	<12	*	*	*	*	15	17	17	18
	0.05	<17	<11	*	*	*	*	*	*	22	26

*Cannot be determined from the sampling plans, but <15 at 0.1 and <22 at 0.05.

The sampling of stink bug eggs on soybean and most other crops is labor intensive. Such sampling must be accomplished by means of direct visual examination of individual plants. Assume a mean egg mass size of 50 and 725 masses produced per hectare (36,250 eggs/hectare = 3.3/row-m). That means that on the average the foliage of 15 row-m must be examined to discover each egg mass. Now consider a similar population level of eggs of the soybean looper, velvetbean caterpillar, green cloverworm, or corn earworm. The egg population here would also average 3.3/row-m, but because the eggs are deposited singly and essentially at random, probably no more than 0.7 row-m would have to be searched to find each egg.

The size of the sampling unit chosen for use is greatly influenced by the fact that the eggs are deposited in clusters. With the latter case described above, observational units of 0.3, 1.0, or 1.7 row-m could conveniently be used. In the stink bug example described formerly, the use of these sample units would result in the majority of the observations containing 0 eggs. In fact, it is not impossible that a precisely defined lepidopteran egg population estimate could be derived with the same amount of labor required to find the first stink bug egg mass.

The specific identification of egg masses encountered in sampling is also likely to present difficulty. According to Whitemarsh (1917), the eggs of *A. hilare* have about 65 chorial processes (spines) in a halo around the top of the egg, while Jones (1918) and Drake (1920) report that the chorial processes on the eggs of *N. viridula* range from 28 to 32. Esselbaugh (1946) reports that chorial processes on eggs of the genus *Euschistus* range from 25 to 38.

V. Indirect Sampling Procedures

Surveys of crop damage due to insect feeding are important in the initial phases of both experimental work and assessment programs in relation to insect pest management. Since stink bugs feed primarily on the seeds of soybean, damage is measured as the number of punctured seeds (per unit area) and the relative degree of injury along with the percentage of seeds in each injury category (according to severity). This information along with estimates of population density can be tested for correlation to provide valuable inputs for the validation and refinement of economic injury levels. Often, experience of this type is most valuable for farm managers or pest management consultants due to the confidence with which they can subsequently view the damage potential of given population densities. This is the main value of indirect sampling procedures such as surveying for evidence of damage in the field. It provides an additional insight into the workings of the soybean agroecosystem that will be invaluable in making pest management decisions.

Assessment of severity of stink bug damage can be made during bean development or after maturity. In either case, pods must be opened, seeds removed, and visual assessment of damage made utilizing predetermined criteria. Injury categories range from no injury, through light and medium, to heavy or severe injury.

These categories can be described as follows:

Light injury—seed with puncture marks with no deformity of the seed coat or endosperm

Medium injury—seed with puncture marks and mild deformity but no reduction in size

Heavy injury—seed with puncture marks, grossly deformed and somewhat reduced in size and weight

Severe injury—seed with puncture marks, grossly deformed and drastically reduced in size and weight

The latter two categories are sometimes combined into one and referred to as heavy or severe injury. Beans in this category are of no practical value for oil, meal, or seed.

Additionally, seeds injured to such an extent are usually blown out with the crop residue in the harvesting process. Thus, assessments of stink bug damage made from the harvested sample might be misleading due to the absence of seed with severe injury.

Examination of developing or mature intact soybean pods on the plant is an impractical and inaccurate method of indirect sampling since puncture marks are not always visible or may be obscured by disease lesions or other masking characteristics (see Chapter 1 for additional discussion of damage assessment).

VI. Concluding Remarks

Stink bug sampling in soybean is among the most difficult tasks faced by pest managers and researchers. This is due mainly to the clumped patterns early in the life cycle and a more nearly random pattern in later stages. Coupled with this is the fact that damage is not readily visible in the growing crop. This is in contrast to most other important foliage inhabiting soybean pest species. These considerations necessitate greater sampling effort to adequately characterize or measure stink bug population levels.

To date no research has been published that defines the dispersion characteristics of any stink bug species on soybean. These parameters must be defined for all stink bug species being studied so that the accuracy and efficiency of sampling methods can be determined. They are necessary for the construction of sequential sampling plans for management and sequential plans for population estimation.

Blacklight traps provide an important tool in monitoring populations and movements of many insect species including stink bugs. These parameters can serve as important indicators of impending population buildups in the field, especially with regard to movements into and out of an area for feeding and oviposition. Much research is needed to validate the utility of this sampling method as a survey tool for soybean insects.

APPENDIX 23-1. Generic Key to the Adult Stink Bugs Known to Occur in Soybean

A necessary prerequisite to an efficient and effective sampling program is the accurate identification of species encountered. It is essential to distinguish among predatory, phytophagous, and innocuous species groups, but it is also necessary to distinguish among phytophagous species. We present here a key to separate genera of stink bugs known to occur in soybean fields. It was necessary to combine and modify generic keys presently available in the literature (Blatchley 1926, Hoffman 1971, Rolston 1974) to derive a key suitable for field use. This key is intended only as a guide for field identification of pentatomids collected from soybean and is by no means to be considered a definitive taxonomic work. A more rigorous taxonomic treatment is the principal taxonomic literature pertaining to the pentatomine genera consisting of the classic work of Blatchley (1926) along with the works of Van Duzee (1917), Schouteden (1907), Hart (1919), Froeschner (1941), Hoffman (1971), Rolston (1974), and Slater and Baranowski (1978). The damage potential on soybean has been established for only a few of the species included. It should be noted that although all have been found to occur in the soybean field, several species may have been feeding on weed hosts in the field or nearby. Finally, it is likely that others occur in soybean but were not included in the key. Anatomical structures used as diagnostic characters in the key are shown in Fig. 23-17.

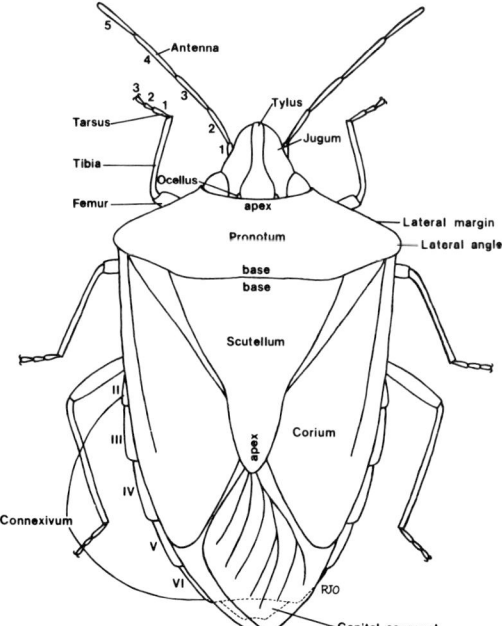

Figure 23-17. *Nezara viridula* dorsal view illustrating parts used in description of a pentatomid.

1. First segment of rostrum short and thick, free, only its base being between the bucculae, which converge and unite behind or beneath the rostrum (Fig. 23-18A); base of rostrum close to end of tylus; osteolar opening usually without a distinct auricle, but extended as a wide curved plateau raised above the metapleural surface (Fig. 23-20H,I,J); predaceous species (Subfamily Asopinae)..................................*Stiretrus*
Alcaeorrhynchus
Perillus
Euthyrhynchus
Podisus
First segment of rostrum slender, embedded between the bucculae, which are wide and parallel (Fig. 23-18B); base of rostrum distinctly separated from end of tylus; osteolar opening and peritreme extend as a narrow groove flush with metapleural surface; phytophagous species..........2

2. Metasternum with a broad median smooth area, its anterior end forked, prolonged forward between the middle coxae, and strongly notched posteriorly to receive the ventral spine (Fig. 23-19C)...............*Edessa bifida*
E. meditabunda
E. rufomarginata
Metasternum without a median smooth area as described above (Fig. 23-19A,B) 3

3. Second ventral abdominal segment produced forward in a stout spine or tubercle toward or between the hind coxae (Fig. 23-19).............4
Second ventral segment at middle not produced forward in a stout spine or tubercle.......................................9

4. Jugae surpassing tylus (Fig. 23-18C) 5
Jugae not surpassing tylus, or, if slightly so, jugae not contiguous in front of tylus (Fig. 23-18D) 6

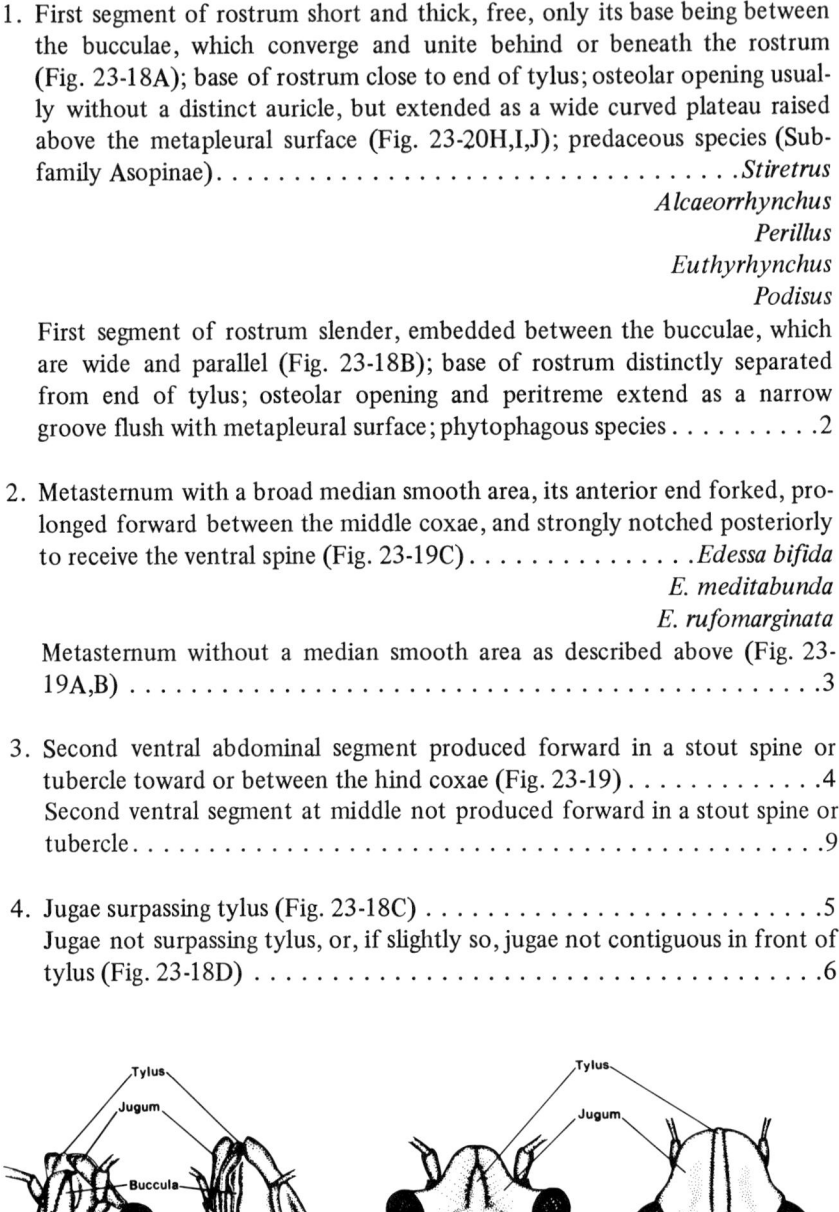

Figure 23-18. Dorsal and ventral views of the heads of certain pentatomids. (a) *Podisus maculiventris*, (b) *Rhytidolomia senilis*, (c) *Edessa bifida*, (d) *Piezodorus guildinii*. (b after Parshley 1915).

5. Tips of jugae spine-like or pointed, not approaching or contiguous in front of tylus; humeri ending in a strong upward and outward projecting spine; mesosternum with a prominent spine-like median carina; color dull greenish-yellow.............................. *Arvelius albopunctatus*
Tips of jugae rounded, approaching or contiguous in front of tylus; humeri not spined, mesosternum not carinate; color reddish brown to dull clay-yellow............................. *Dendrocoris fruticicola*
D. melacanthus

6. Ventral spine reaching or slightly surpassing middle coxae; spiracles large, black; color straw yellow to light green or green ..... *Piezodorus guildinii*
P. hubneri
P. pallescens
Ventral spine not reaching middle coxae 7

7. Larger, length 12 mm or more; second antennal segment more than half the length of fifth..8
Smaller, length not over 11 mm; second antennal segment less than half the length of fifth; ventral spine short, tuberculate .... *Banasa dimidiata*
B. euchlora

8. Spine of second ventral abdominal segment very short and rounded or blunt (Fig. 23-19A); osteolar canal short, subtruncate, not reaching middle of its supporting plate (Fig. 23-20B)................... *Nezara antennata*
N. viridula

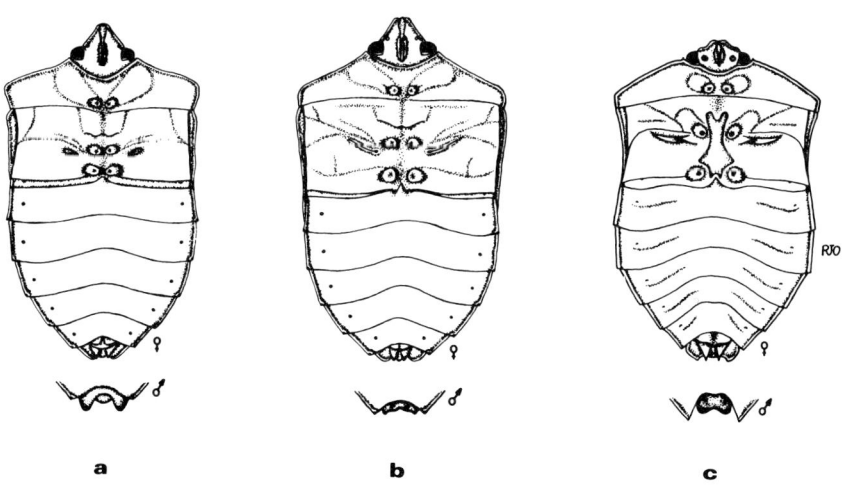

Figure 23-19. Ventral views of certain female and male pentatomids. (a) *Nezara viridula*, (b) *Acrosternum hilare*, (c) *Edessa bifida*.

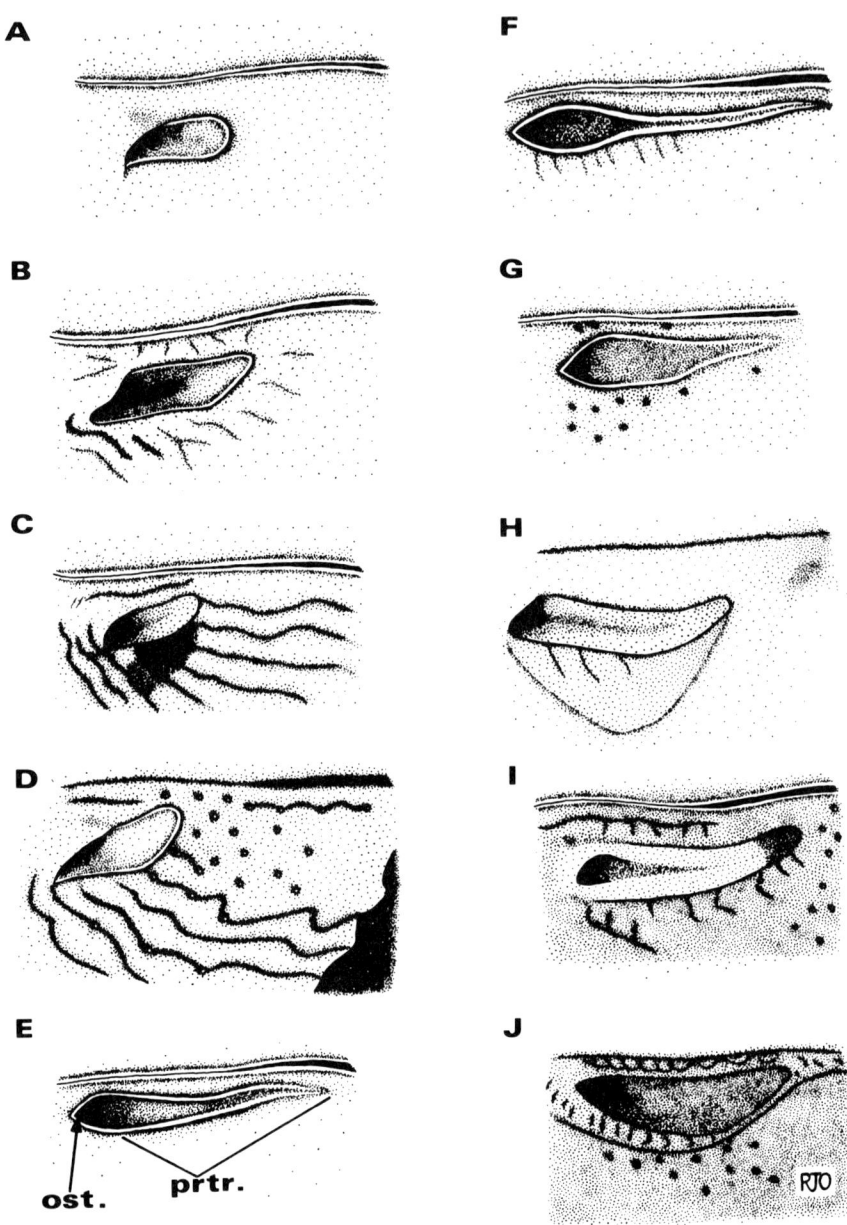

Figure 23-20. Osteolar openings of certain pentatomids. (A) *Euschistus servus*, (B) *Nezara viridula*, (C) *Oebalus pugnax*, (D) *Proxys punctulatus*, (E) *Thyanta calceata*, (F) *Acrosternum hilare*, (G) *Holcostethus limbolarius*, (H) *Euthyrhynchus floridanus*, (I) *Podisus maculiventris*, (J) *Stiretrus anchorago* (ost. "osteolar opening"; prtr. "peritreme").

Spine of second ventral segment distinct (Fig. 23-19B); osteolar canal long, tapering, much surpassing middle of its supporting plate (Fig. 23-20F) . . . .
.... *Acrosternum acutum*
A. armigera
A. hilare
A. marginatum
A. pennsylvanicum

9. Osteolar opening without a distinct auricle, peritreme extended as a narrow tapering groove lying flush on the metapleural surface (Fig. 23-20E,G) . . 10
Osteolar opening usually with a short curved auricle having a rounded tip, peritreme not extended as a groove, or, if so, the groove very short and ending abruptly (Fig. 23-20A,C,D) 11

10. Jugae distinctly longer than tylus and broadly contiguous in front of it; upper body brown, pronotum with narrow, smooth, ivory-white margins . . . .
.... *Holcostethus [=Peribalus] limbolarius*
Jugae equalling or shorter than tylus; body dull green or brownish, pronotum not margined, or at most with reddish or black margins ....
.... *Thyanta antiguensis*
T. calceata
T. custator
T. pallidovirens accerra
T. perditor

11. Tibia of posterior legs without evident groove along the dorsal (posterior) surface 12
Tibia of posterior legs with prominent groove on dorsal surface 16

12. Tylus much longer than jugae; humeri prominently spined; color black with apex of scutellum white to yellow .... *Proxys punctulatus*
Tylus equalling or shorter than jugae; color not as above............ 13

13. Jugae parallel or divergent in front of tylus, pointed apically, color light to dark brown .... *Dichelops furcatus*
Jugae convergent or contiguous in front of tylus................. 14

14. Length not over 15 mm; jugae with tips rounded, rarely longer than tylus and then converging and contiguous in front of it; lateral margins of pronotum not crenulate 15
Larger, length 20 or more mm; jugae with tips subacute, surpassing tylus but not converging in front of it; lateral margins of pronotum crenulate
.... *Loxa flavicollis*

15. Humeral spines projecting forward; uniformly light brown or straw yellow
 .... *Oebalus [=Solubea] ornatus*
 O. pugnax
 O. poecilus
 O. ypsilon-griseus
 Humeri rounded, or with spines projecting laterally; color grayish-brown with margins of scutellum and anterior edge of pronotum yellow ....
 .... *Mormidea lugens*

16. Length of first antennal segment subequal to distance from its base to apex of head 17
 First segment of antennae much shorter. .... *Tibraca limbativentris*

17. Lateral margins of pronotum smooth; humeral angles broadly rounded
 .... *Hymenarcys nervosa*
 Lateral margins of pronotum crenulate; humeral angles subacute to spiniform .... *Euschistus atrox*
 E. crassus
 E. crenator
 E. heros
 E. ictericus
 E. impictiventris
 E. obscurus
 E. quadrator
 E. servus
 E. tristigmus
 E. variolarius

Acknowledgments

The authors wish to thank the many persons who assisted in one or more phases of the preparation of this manuscript.

Special thanks go to Drs. L. D. Newsom and R. L. Jensen who reviewed the manuscript and made many helpful suggestions. Also, to Mr. Cecil Smith who made many helpful suggestions relative to the key to the genera.

Appreciation is extended to Mr. Russ Ottens who prepared all of the original illustrations, and to Mrs. Jan Smith who prepared many of the charts and graphs.

To our secretaries—Miss Laura Atkins, Mrs. Elaine Belk, Ms. Terri Hilton, Mrs. Beverly Lastinger, and Mrs. Margaret Norman—our appreciation for typing and proofing the manuscript.

The authors are particularly indebted to Miss Sheran Thompson for her invaluable assistance in completing many phases of the manuscript.

References

Asahina, S., and Y. Turuoka. 1970. Records of the insects visited a weather-ship located at the ocean weather station "Tango" on the Pacific, V. Insects captured during 1968. Kontyu 38:318-330.

Blatchley, W. S. 1926. Heteroptera or true bugs of eastern North America. Nature Publ. Co., Indianapolis. 1116 p.

Blickenstaff, C. C., and J. L. Huggans. 1962. Soybean insects and related arthropods in Missouri. Mo. Agr. Exp. Sta. Res. Bull. 803:51 p.

Cantelo, W. W., J. L. Goodenough, A. H. Baumhover, J. S. Smith, Jr., J. M. Stanley, and T. J. Henneberry. 1974. Mass trapping with blacklight effect on isolated populations of insects. Environ. Entomol. 3:389-395.

Cantelo, W. W., J. S. Smith, Jr., A. H. Baumhover, J. M. Stanley, T. J. Henneberry, and M. B. Peace. 1973. Changes in the population levels of 17 insect species during a 3 1/2-year blacklight trapping program. Environ. Entomol. 2:1033-1038.

Cherry, E. T. 1974. Economic threshold studies of stink bugs on soybeans. Tenn. Farm Home Sci. Progr. Rep. 87.

Cumber, R. A. 1949. The green vegetable bug *Nezara viridula*. N. Z. J. Agr. 79:563-564.

Daugherty, D. M. 1967. Pentatomidae as vectors of yeast spot disease of soybeans. J. Econ. Entomol. 60:147-152.

Daugherty, D. M., M. H. Neustadt, C. W. Gehrke, L. E. Cavanah, L. F. Williams, and D. E. Green. 1964. An evaluation of damage to soybeans by brown and green stink bugs. J. Econ. Entomol. 57:719-722.

Distant, W. L. 1880. Insecta. Rhynchota. Hemiptera-Heteroptera. Biologia Centralia-Americana, London 1(14):78-79.

Drake, C. J. 1920. The southern green stink bug in Florida. Fla. State Plant Bd. Quart. Bull. 4:41-94.

Duncan, R. G. 1968. Feeding relationships of *Nezara viridula* (L.) (Hemiptera: Pentatomidae) on soybeans. M. S. thesis, Louisiana State University, Baton Rouge. 52 p.

Duncan, R. G., and J. R. Walker. 1968. Some effects of the southern green stink bug on soybeans. La. Agr. 12:10-11.

Esselbaugh, C. O. 1946. A study of the eggs of the Pentatomidae (Hemiptera). Ann. Entomol. Soc. Amer. 39:667-691.

Everett, P. 1950. Spread of green vegetable bug. N. Z. J. Agr. 80:145-146.

Froeschner, R. C. 1941. Contributions to a synopsis of the Hemiptera of Missouri, Pt. 1. Amer. Midland Natur. 26:122-146.

Gomes, J. E. 1966. Retencão foliar em soja. Gov. do Estado do RGS, Sec. Agr., Servico de Informacão e Divulgacão Agricola, Brazil. 15 p.

Harris, V. E., and J. W. Todd. 1980. Male-mediated aggregation of male, female, and 5th instar southern green stink bugs and concomitant attraction of a tachinid parasitoid, *Trichopoda pennipes*. Entomol. Exp. Appl. (in press).

Hart, C. A. 1919. The Pentatomidae of Illinois with keys to the Nearctic genera. Ill. Natur. Hist. Surv. Bull. 13:157-223.

Hoffman, R. L. 1971. The insects of Virginia: No. 4. Shield bugs (Hemiptera: Scutelleroidae: Scutelleridae, Corimelaenidae, Cydnidae, Pentatomidae). Va. Polytech. Inst. Res. Div. Bull. 67:1-61.

Hoffman, W. E. 1935. The food plants of *Nezara viridula* (L.) (Hemiptera: Pentatomidae). Proc. Int. Congr. Entomol. (Madrid). **6**:811-816.

Hokyo, N., and K. Kiritani. 1962. Sampling design for estimating the population of the southern green stink bug, *Nezara viridula* (Pentatomidae, Hemiptera) in the paddy field. Jap. J. Ecol. **12**:228-235.

Jensen, R. L. 1976. Are you doing a complete job of sweeping the field? Agri-Fieldman. **32**:31-32, 34.

Jensen, R. L., and L. D. Newsom. 1972. Effect of stink bug-damaged soybean seeds on germination, emergence, and yield. J. Econ. Entomol. **65**:261-264.

Jones, T. H. 1918. The southern green plant bug. USDA Bull. **689**:27 p.

Kariya, H. 1961. Effect of temperature on the development and mortality of the southern green stink bug, *Nezara viridula* and the oriental green stink bug, *N. antennata*. Jap. J. Appl. Entomol. Zool. **5**:191-196.

Kilpatrick, R. A., and E. E. Hartwig. 1955. Fungus infection of soybean seed as influenced by stink bug injury. Plant Dis. Rep. **39**:177-180.

Kiritani, K. 1963. The change in reproductive system of the southern green stink bug, *Nezara viridula*, and its application to forecasting of the seasonal history. Jap. J. Appl. Entomol. Zool. **7**:327-337.

Kiritani, K. 1964. The effect of colony size upon the survival of larvae of the southern green stink bug, *Nezara viridula*. Jap. J. Appl. Entomol. Zool. **8**:45-54.

Kiritani, K. 1965. The natural regulation of the population of the southern green stink bug, *Nezara viridula* L. Proc. Int. Congr. Entomol. (London). **12**:375.

Kiritani, K. 1970. Studies on the adult polymorphism in the southern green stink bug, *Nezara viridula* (Hemiptera: Pentatomidae). Res. Pop. Ecol. **12**:19-34.

Kiritani, K., and N. Hokyo. 1962. Studies on the life table of the southern green stink bug, *Nezara viridula*. Jap. J. Appl. Entomol. Zool. **6**:124-140.

Kiritani, K., and N. Hokyo. 1965. Variation of egg mass size in relation to the oviposition pattern in Pentatomidae. Kontyu. **33**:427-432.

Kiritani, K., and K. Kimura. 1965. The effect of population density during nymphal and adult stages on the fecundity and other reproductive performances. Jap. J. Ecol. **15**:233-236.

Kiritani, K., and K. Kimura. 1966. A study of the nymphal aggregation of the cabbage stink bug, *Eurydema rugosum* Motschulsky (Heteroptera: Pentatomidae). Appl. Entomol. Zool. **1**:21-28.

Kiritani, K., N. Hokyo, and K. Kimura. 1963. Survival rate and reproductivity of the adult southern green stink bug, *Nezara viridula*, in the field cage. Jap. J. Appl. Entomol. Zool. **7**:113-124.

Kiritani, K., N. Hokyo, K. Kimura, and F. Nakasuji. 1965. Imaginal dispersal of the southern green stink bug, *Nezara viridula* L., in relation to feeding and oviposition. Jap. J. Appl. Entomol. Zool. **9**:291-297.

Kiritani, K., N. Hokyo, and S. Iwao. 1966a. Population behavior of the southern green stink bug, *Nezara viridula*, with special reference to the developmental stages of early-planted paddy. Res. Pop. Ecol. **8**:133-146.

Kiritani, K., N. Hokyo, and K. Kimura. 1966b. Factors affecting the winter mortality in the southern green stink bug, *Nezara viridula* L. Ann. Soc. Entomol. Fr. **2**:199-207.

Kobayashi, T. 1959. The developmental stages of some species of the Japanese Pentatomidae (Hemiptera). VII. Developmental stages of *Nezara* and its allied genera. Jap. J. Appl. Entomol. Zool. 3:221-231.

Kogan, M. 1976. Soybean disease and insect pest management. pp. 114-121 in R. M. Goodman, ed. Expanding the use of soybeans. Proc. of a Conf. for Asia and Oceania. Univ. of Illinois, College of Agriculture, INTSOY. Ser. 10. 261 p.

Lethierry, L., and G. Severin. 1893. Catalogue général des Hèmiptères. Tomo 1. Hétéroptères Pentatomidae. F. Hayez, Académie Royale de Belgique, Bruxelles. 286 p.

Marston, N. L., C. E. Morgan, G. D. Thomas, and C. M. Ignoffo. 1976. Evaluation of four techniques for sampling soybean insects. J. Kans. Entomol. Soc. 49:389-400.

Miner, F. D. 1961. Stink bug damage to soybeans. Ark. Farm Res. 10:12.

Miner, F. D. 1966. Biology and control of stink bugs on soybeans. Ark. Agr. Exp. Sta. Bull. 708:40 p.

Mitchell, W. C. 1965. An example of integrated control of insects: Status of the southern green stink bug in Hawaii. Agr. Sci. Rev. 3:32-35.

Mitchell, W. C., and R. F. L. Mau. 1969. Sexual activity and longevity of the southern green stink bug, *Nezara viridula*. Ann. Entomol. Soc. Amer. 62:1246-1247.

Mitchell, W. C., and R. F. L. Mau. 1971. Response of the female southern green stink bug and its parasite, *Trichopoda pennipes*, to male stink bug pheromones. J. Econ. Entomol. 64:856-859.

Nakasuji, F., N. Hokyo, and K. Kiritani. 1965. Spatial distribution of three plant bugs in relation to their behavior. Res. Pop. Ecol. 7:99-108.

Newsom, L. D., and D. C. Herzog. 1977. Trap crops for control of soybean pests. La. Agr. 20:14-15.

Onsager, J. A. 1976. A sequential sampling plan for classifying infestations of southern potato wireworm. Amer. Potato J. 51:313-317.

Parshley, H. M. 1915. Systematic papers on New England Hemiptera. II. Synopsis of the Pentatomidae. Psyche. 22:170-177.

Pitts, J. R. 1977. Effect of temperature and photoperiod on *Nezara viridula* L. M. S. thesis, Louisiana State University, Baton Rouge. 89 p.

Race, S. R. 1960. A comparison of two sampling techniques for *Lygus* bugs and stink bugs on cotton. J. Econ. Entomol. 53:689-690.

Ragsdale, D. W. 1977. Isolation and identification of bacteria and fungi transmitted during feeding and from various organs of *Nezara viridula* (L.). M. S. thesis, Louisiana State University, Baton Rouge. 48 p.

Rizzo, H. F. 1972. Enemigos animales del cultivo de la soja. Rev. Inst. Bolsa Cereales. 2851:1-6.

Rolston, L. H. 1974. Revision of the Genus *Euschistus* in middle America (Hemiptera: Pentatomidae, Pentatomini). Entomol. Amer. 48:1-102.

Rosenfeld, J. H. 1911. Insects and spiders in Spanish moss. J. Econ. Entomol. 4:398-409.

Rudd, W. G., and R. L. Jensen. 1977. Sweep net and ground cloth sampling for insects in soybeans. J. Econ. Entomol. 7:301-304.

Ruesink, W. G., and M. Kogan. 1975. The quantitative basis of pest management: sampling and measurement. pp. 309-351 in R. L. Metcalf, and W. H. Luckmann, eds. Introduction to insect pest management. John Wiley and Sons, N. Y. 587 p.

Schouteden, H. 1907. Fam. Pentatomidae, Sub fam. Graphosomatinae. Genera Insectorum, Fasc. **30**:1-93.

Slater, J. A., and R. M. Baranowski. 1978. How to know the true bugs (Hemiptera–Heteroptera). Wm. C. Brown Co., Dubuque, Iowa. 256 p.

Todd, J. W., and S. G. Turnipseed. 1972. Effects of southern green stink bug damage on yield and quality of soybeans. J. Econ. Entomol. **67**:421-426.

Todd, J. W., M. D. Jellum, and D. B. Leuck. 1973. Effects of southern green stink bug damage on fatty acid composition of soybean oil. Environ. Entomol. **2**:685-689.

Turner, J. W. 1967. The nature of damage by *Nezara viridula* (L.) to soybean seed. Queensland J. Agr. Anim. Sci. **24**:105-107.

Van Duzee, E. P. 1917. Catalogue of the Hemiptera of America north of Mexico. Univ. Calif. Publ. Tech. Bull. **2**:1-902.

Waters, E. E. 1955. Sequential sampling in forest insect surveys. For. Sci. **1**:68-79.

Watson, J. R. 1918. Insects of a citrus grove. Fla. Agr. Exp. Sta. Bull. **148**:165-167.

Whitemarsh, R. D. 1917. The green soldier bug *Nezara hilaris*. Ohio Agr. Exp. Sta. Bull. **310**:517-552.

Wilson, F. 1960. Review of the biological control of insects and weeds in Australia and Australian New Guinea. Commonwealth Institute of Biological Control, Ottawa. pp. 29-30.

Yeargan, K. V. 1977. Effects of green stink bug damage on yield and quality of soybeans. J. Econ. Entomol. **70**:619-622.

Yukawa, J., and K. Kiritani. 1965. Polymorphism in the southern green stink bug. Pac. Insects. **7**:639-642.

SECTION VIII
Natural Control Agents

Chapter 24

Sampling Parasitoids of Soybean Insect Pests

Norman L. Marston

I. Introduction

Programs for sampling parasitoids of soybean insects usually are designed to determine the effectiveness of the parasitoids in controlling pests rather than to estimate populations of the parasitoids per se. Researchers also are often concerned with the relative effectiveness of several parasitoids attacking a host, especially when they are considering the parasitoids for introduction to suppress a pest species. The measure usually used in such studies is the percentage parasitism of a stage or group of stages of the host. Samples of hosts are taken in the field and either dissected or reared in the laboratory to determine the species of parasitoids present and the percentage of hosts parasitized by each species.

There are certain inherent difficulties that may greatly distort such assessments of the value of parasitoids in an ecosystem or may bias comparisons between parasitoid species attacking a particular host. These problems will be discussed first in this chapter to provide a background for interpreting the information on sampling for parasitism of specific pests, which follows. The limited information regarding sampling procedures for adult parasitoids is presented last.

A. Potential Problems in Estimating Parasitism from Field-Collected Hosts

Simmonds (1948) and A. J. Mueller (personal communication) have described several of the sources of bias in estimating percentage parasitism. These and other potential problems may be placed into six basic categories.

1. Interruption of Exposure to Parasitoids by Sampling

Perhaps most obvious is the bias that results when hosts are removed from the field for rearing or dissection so that they are no longer exposed to parasitoids. For example, a newly hatched lepidopteran larva would be exposed to parasitoids for only a fraction of its life cycle, and the percentage parasitism of a sample composed of a preponderance of early instars, as at the beginning of a brood, would greatly underestimate the true rate of parasitism had the hosts been left in the field. The bias would be magnified where a high rate of predation occurs in the field, since few larvae would reach the late stages and samples would contain mostly young individuals that had had little exposure to parasitoids.

2. Temporal Variation of Parasitism

Samples taken at only one or a few points in time may not accurately measure the overall effect of a parasitoid through a time period, especially if the parasitoid population is fluctuating greatly. For example, percentage parasitism of a sample taken just as a generation of parasitoids is beginning to emerge would not accurately measure the impact of that species on the host during succeeding weeks when the parasitoid becomes more abundant. Most workers studying parasitism of soybean insects have avoided this problem by sampling at intervals throughout the season and noting the trend in parasitism.

3. Spatial Variation of Parasitism

Percentage parasitism may vary among sampling sites within an ecosystem. This source of error may be avoided by sampling at several sites and evaluating the variability of the data. However, parasitism may also vary among microhabitats within the ecosystem, and if the sampling procedure is not equally effective in all microhabitats, a consistent bias in estimation of percentage parasitism may result. The bias would be especially great if a portion of the host population were inaccessible to parasitoids (e.g., within soybean pods) and this portion of the population were not effectively sampled.

4. Variation of Parasitism with Host Age

This source of error is most often encountered when estimates of percentage parasitism are based on a broad range of host stages (as has often been done with lepidopteran larvae), but the parasitoids are restricted to particular stages or vary in their ability to parasitize each host stage. For example, if a parasitoid attacked only late instars of a pest, an estimate of percentage parasitism based on total larvae would not estimate the population reduction inflicted by the parasitoid as well as an estimate based only on the susceptible stages. The error would be magnified if a high rate of predation or disease occurred in the field since this would skew the distribution of ages in the samples toward the nonsusceptible stages.

This source of bias may be complicated by the effectiveness of the sampling procedure if some stages are more efficiently sampled than others. For example, the age distribution in the sample would not accurately reflect the age distribution of the host in the field if early instars were difficult to discern in the debris of a sample. An adjunct to this source of bias is differential survival of host stages under the artificial rearing conditions of the laboratory. For example, early instars of lepidopteran larvae may have a high rate of mortality when placed on artificial medium, so estimates of percentage parasitism based on surviving larvae would be biased toward the rate of parasitism of older larvae.

5. Change in Host Longevity, Survival, or
   Behavior with Parasitism

The likelihood that parasitized vs. unparasitized larvae would be included in a sample may be affected by a change in the duration of the parasitized stages of a host. A lengthening of the susceptible host stage, as often occurs, would result in an overestimate of percentage parasitism since parasitized hosts would be more likely to be collected than unparasitized hosts. An opposite and perhaps even greater bias may occur if the susceptible stage of the host is long (e.g., diapausing larvae or pupae) so that more than one parasitoid generation could develop on the stage.

Greater mortality of parasitized than unparasitized hosts during rearing would lead to an underestimate of percentage parasitism. This source of bias may be readily avoided by dissecting all larvae that die of unknown causes during rearing to determine if they were parasitized. Also, undetected mortality of hosts in the field due to mutilation by the parasitoid would increase the underestimate of parasitoid effectiveness.

Parasitized hosts may behave differently than unparasitized hosts, and these variations may change the likelihood that parasitized hosts would be included in samples, thereby biasing the estimates of parasitism.

6. Competition with Other Parasitoids or Diseases

This factor would lead to an underestimate of percentage parasitism in all cases. For example, if a parasitoid attacked third instars and emerged from fifth instars, the percentage parasitism of fourth instars would accurately estimate the population reduction due to the parasitoid. But, if a number of the fourth instars in a sample were killed by other species of parasitoids or disease before they reached the fifth stage, they could not produce the parasitoid in question. Thus, these larvae should be deleted from consideration to accurately estimate percentage parasitism.

B. Methods for Compensating Sample Bias

Avoiding all these sources of error in a research program would be virtually impossible. It would require a detailed knowledge of the ecology and behavior of the parasitoids and their hosts, and the effects of parasitism on the host's development and behavior. In addition, a thorough knowledge of the effectiveness of the sampling procedure for collecting various stages of parasitized and unparasitized hosts would be required. However, some meaningful improvements in technique may reduce several of the sources of bias, resulting in estimates of parasitism that are closer to reality than those usually obtained. Also, the data may be adjusted to partially remove at least some of the sources of bias, providing estimates of parasitism that are closer to reality than a simple expression of percentage parasitism of sampled hosts. To perform such adjustments the unique host-parasitoid relationships and experimental design should be examined to determine which sources of bias seem most important and to decide what modifications of procedure and data adjustments would be appropriate to reduce the error.

The consequences of some of the sources of error and some possible data adjustments may be illustrated by a set of data obtained from weekly samples of a population of green cloverworms, *Plathypena scabra* (F.), collected on soybean near Columbia, Missouri during August 1975 (Table 24-1). The larvae were collected on polyethylene ground cloths. The collections probably were biased in favor of large larvae since these were easiest to detect in the debris and may have been more readily dislodged onto the ground cloth. The four weeks' data were grouped together for ease of illustration; it was recognized, however, that parasitism may have varied considerably from week to week.

Of the total larvae collected, 23.4% (103/440) were parasitized and 0.5% (2/440) were diseased. The effect of the fourth source of bias (in part) is immediately apparent in that many of the larvae, particularly early instars, did not survive due to injury during collection or inability to develop on the diet. These larvae could not have produced parasitoids, and when they are deleted from consideration, the estimate of the total parasitism is increased to 31.3% (103/329).

The data for the tachinid fly, *Winthemia sinuata* Reinhard, illustrate several of the sources of bias. First, this species lays its eggs on sixth instars, for the most part (Lentz and Pedigo 1974), so that an estimate of percentage parasitism based on total surviving larvae, 1.8% (6/329), would be less realistic than an estimate based on six instars, only 13.6% (6/44). However, the apparent parasitism (AP) of sixth instars still underestimates the population reduction due to the parasitoid, since exposure of the larvae to the parasitoid was interrrupted when the larvae were brought into the laboratory. The actual population reduction due to

Table 24-1. Fate of larvae of the green cloverworm, *Plathypena scabra* (F.), collected on soybean. Columbia, MO. 1975

Fate of larvae	Number of larvae of indicated instar						Total number of larvae
	1	2	3	4	5	6	
Deaths due to:							
Injury or lack of feeding	46	22	17	9	12	5	111
Apanteles marginiventris (Cresson)	0	3	15	13	5	0	36
Rogas nolophanae Ashmead	0	4	11	26	6	0	47
Protomicroplitis facetosa (Weed)	0	0	1	1	11	0	13
Winthemia sinuata Reinhard	0	0	0	0	0	6	6
Voria ruralis (Fallen)	0	0	0	0	1	0	1
Bacillus thuringiensis (Berliner)	0	0	0	0	1	0	1
Nomuraea rileyi (Farlow) Sampson	0	0	0	0	1	0	1
Survivors	11	33	55	45	42	38	224
Total number of larvae	57	62	99	94	79	49	440

the parasitoid is here called effective parasitism (EP). A rough correction for this underestimate may be made by assuming that the larvae were, on the average, halfway through the instar and that parasitism would have continued at the same rate through the remainder of the stage had the larvae been left in the field. If AP is expressed as a fraction, then estimated effective parasitism (EEP) would be

$$EEP = AP + AP(1 - AP) \qquad (24\text{-}1)$$

Formula 24-1 assumes that an additional fraction of the larvae that had not been parasitized during the first half of the stage would have been attacked during the remainder of the stage if the larvae had not been removed to the laboratory. Substituting $1 - AS$ (AS = apparent survival) for AP, the formula reduces to

$$EEP = 1 - AS^2 \qquad (24\text{-}2)$$

For *W. sinuata*, $AS = 1 - 0.136 = 0.864$ and $EEP = 1 - 0.864^2 = 0.254$ or 25.4%.

The many factors that may affect such an EEP become apparent when the assumptions made in computing it are examined:

1. All ages of larvae within the stage were equally susceptible to parasitism.
2. Predation and disease were negligible.
3. Parasitized and unparasitized larvae developed at the same rate and were equally likely to be collected.
4. No larvae were parasitized before molting to the sixth stage.
5. Rate of entry into the stage and rate of pupation were constant.
6. The parasitoid population and search rate remained constant.

These assumptions are obviously unrealistic, but the effects of the variables may be off-setting to some extent. According to Lentz and Pedigo (1974), fifth instars may be attacked in the laboratory, and the flies are reluctant to oviposit on sluggish larvae. *W. sinuata* eggs are laid externally and would be cast with the skin of the host before they hatched if laid during the later portion of the stage. Both of these factors, and possibly a greater likelihood that parasitized larvae would be included in samples due to their slower rate of development, would cause EEP to be an overestimate. On the other hand, losses due to predation or disease in the stage would cause the mean proportional age of the larvae in that stage to be less than 0.5, so that EEP would be an underestimate, especially if parasitized larvae were sluggish and more likely to be preyed upon than unparasitized larvae. Generally, the EEP would more closely estimate EP than would AP unless many larvae entering the stage were already parasitized or the parasitoid had a strong preference for young larvae within the stage.

Effective parasitism by the braconid wasps—*Rogas nolophanae* Ashmead, *Apanteles marginiventris* (Cresson) and *Protomicroplitis facetosa* (Weed)—may be estimated somewhat differently than was the parasitism by *W. sinuata*. The host presumably develops through one or more larval stages between oviposition and emergence of the adult parasitoids. Thus, the percentage parasitism of an intermediate stage would provide an accurate measure of EP. For example, *R. nolophanae* prefers to oviposit in third instars (Lentz and Pedigo 1974) and emerges from fifth instars for the most part, so that the percentage parasitism of fourth instars may provide an accurate EEP. An additional problem exists, however, in that many fourth instars produce *A. marginiventris* and die shortly afterward, thereby precluding parasitism by *R. nolophanae*. Therefore, these larvae should be excluded from the computation of percentage parasitism along with the larvae that died when initially placed on the diet. Ninety-four fourth instars were collected. Excluding the 9 larvae that died without initiating feeding and the 13 larvae from which *A. marginiventris* emerged, EEP would be 26/72 = 36.1%, as contrasted to the AP based on total larvae of 10.7% (47/440).

Apanteles marginiventris parasitizes first and second instars, primarily, and emerges mostly from fourth instars (A. J. Mueller, personal communication), so the percentage parasitism of those third instars that developed on the diet should provide a relatively unbiased estimate of population reduction due to this species, 15/82 = 18.3%. The fact that some fifth instars produced *A. marginiventris* may indicate that it also attacks third instars to some extent. In this case, the EEP based on third instars would underestimate EP somewhat since the larvae would have been removed from exposure to the parasitoids by sampling.

Protomicroplitis facetosa emerges mostly from fifth instars, whereas the instars attacked are not known. It seems apparent that the percentage parasitism of third and fourth instars would underestimate the percentage parasitism by this species, so the fifth instar may be the best to use for estimating EP. Here, parasitism by both *A. marginiventris* and *R. nolophanae* probably would kill larvae before *P. facetosa* could emerge, so larvae from which these parasitoids emerged should be deleted from consideration as well as those that died without feeding. EEP would be 11/56 = 19.6%. Those larvae killed by the tachinid fly, *Voria ruralis* (Fallen) and the two diseases are included in the denominator since they may have killed the larvae in the sixth stage so that parasitism by *P. facetosa* would not have been precluded.

All green cloverworm instars probably are attacked by *V. ruralis* since Elsey and Rabb (1970) found that it will attack all instars of the cabbage looper (though its effectiveness in developing in each instar varies). It kills late in the host's larval cycle, however, so parasitism by the three hymenopterans probably would preclude mortality due to *V. ruralis*. Excluding larvae parasitized by the three hymenopterans and those dying due to injury or lack of feeding, parasitism by *V. ruralis* would be 1/227 = 0.4%. A further rough correction for the fact that sampling interrupted exposure to parasitism, as done for *W. sinuata*, would increase EEP to 0.9%.

The EEP's are not additive since they are computed from a restricted group of the larvae rather than from the entire sample. The combined effect of two parasitoids, cumulative effective parasitism (CEP), expressed as a fraction, would be

$$CEP = EEP_1 + EEP_2 (1 - EEP_1), \qquad (24\text{-}3)$$

where EEP_1 and EEP_2 are the estimated effective parasitism for parasitoids 1 and 2, respectively. Substituting 1 - ES (estimated survival) for EEP, the formula reduces to

$$CEP = 1 - (ES_1 \times ES_2) \qquad (24\text{-}4)$$

where ES_1 and ES_2 are the estimated proportion of each group of larvae not parasitized by parasitoid 1 and parasitoid 2, respectively.

For more than two species, the formula becomes

$$CEP = 1 - (ES_1 \times ES_2 \times \ldots ES_n). \qquad (24\text{-}5)$$

AP's (excluding larvae that died without feeding), EEP's and CEP's for the five parasitoids discussed above are shown in Table 24-2. Even allowing for error in the estimates, it is apparent that the AP's grossly underestimate the true population reductions due to the parasitoids. The CEP for all species, 0.690, is more

Table 24-2. Computed parameters for the parasitism of *Plathypena scabra* larvae parasitized by four species of Hymenoptera and one species of Diptera

Parasite species	Apparent parasitism (AP)	Estimated effective parasitism (EEP)	Cumulative effective parasitism (CEP)
A. marginiventris	0.046	0.183	0.183
R. nolophanae	0.079	0.361	0.478
P. facetosa	0.033	0.196	0.580
V. ruralis	0.003	0.009	0.584
W. sinuata	0.018	0.254	0.690

than double the AP of the total sample, 0.313 (103/329). It is also important to note the change in relative ranking of the parasitoid species as mortality factors, with the most striking change being that of *W. sinuata*.

II. Measuring Percentage of Parasitism of Soybean Pests

A. Parasitoids of Lepidoptera

Most of the studies of parasitism of soybean insects relate to lepidopteran larvae. The early works (e.g., Sherman 1920, Hill 1925, Hinds and Osterberger 1931, Kretzschmar 1948) simply reported the species of parasitoids, and sometimes the incidence of apparent parasitism, without describing the methods used to collect and hold the larvae. Greater attention has been given to detail as the effort devoted to research in parasitoid-host relationships has increased in the United States in the late 1960s and 1970s.

The sweep net has been used most often as a device for collecting larvae for estimating parasitism. The sweep net provides only relative population estimates. For ecological studies, however, it is important to have reliable estimates of host populations and, despite recent attempts to calibrate sweep net procedures (Hillhouse and Pitre 1974, Marston et al. 1976, Hammond and Pedigo 1976, Rudd and Jensen 1977), it still is difficult to accurately estimate field populations by this sampling method (see Chapter 2). Its efficiency also may vary for various sizes of larvae, which would bias estimates of apparent parasitism toward the rates observed for certain instars.

Whole-plant examination has been used as an alternative to the sweep net procedure for collecting host larvae (Burleigh 1971, Smith et al. 1976). This method is advantageous insofar as it provides the sampler with an estimate of population per unit area, but it is difficult and time consuming to collect enough larvae to accurately measure percent parasitism when populations are low.

Modifications of the ground cloth technique (Boyer and Dumas 1963) also provides estimates of populations per unit area and are less time consuming than

Sampling Parasitoids of Soybean Insect Pests   489

Figure 24-1. Sampling for soybean lepidopteran larvae with a canvas drop-cloth.

whole-plant sampling. The technique (see Chapter 2) was used by G. J. Sanders (personal communication) in a study of the parasitism of lepidopteran larvae in South Carolina soybean and Whiteside et al. (1967) in Delaware (in part) (Fig. 24-1). The larvae collected on the ground cloth are transferred to containers for transport to the laboratory. The basic method is tedious under field conditions and it is difficult to find all of the small larvae on the cloth. A modified procedure uses a 3-m polyethylene sheet attached to boards on either side, instead of a canvas cloth (Marston unpublished). After the sheet is spread between two rows, a 3-m length of 1.3-cm diameter conduit is used to bend the plants on one side over the drop-sheet and vibrate them to dislodge the host larvae. The sides of the sheet are then joined, forming a tube, and one end is raised to allow the larvae and debris to slide into a paper sack or an organdy bag. The larvae are then taken to the laboratory where they are removed from the debris and transferred to artificial medium or leaves.

Where it is important to retrieve all larvae on a group of plants, as in studies on estimation of parasitism and predation of introduced larvae during a time interval, a procedure was used in which the plants are brought into the laboratory before the larvae are removed. Several layers of white paper toweling are unrolled on the ground adjacent to the plants with the larvae. The plants are then bent over the toweling, clipped at ground level, rolled up in the toweling (Fig. 24-2), and placed in plastic bags. In the laboratory, the toweling is unrolled onto a table and the plants are vibrated in a polyethylene "shaker" (Fig. 24-3) to dis-

Figure 24-2. Rolling plants infested with lepidopteran larvae into paper toweling for transport to the laboratory.

lodge the larvae into an enamel pan for counting and transfer to rearing containers. Larvae adhering to the toweling are easily detectable on the white background.

Several workers have noted that it is important to place larvae in coolers for transportation to the laboratory to prevent death due to overheating. This is also important when larvae are not isolated from predators in the field. Sanders (personal communication) noted that *Heliothis* spp. should be isolated from other larvae since they are cannibalistic. These problems may be avoided by placing larvae on artificial diet in the field as described by Calkins and Sutter (1972).

If the objective of the survey is to assess species composition of the parasitoid fauna attacking a pest, it may be most efficient to group larvae in large containers such as the 3.8-liter ice cream cartons (Whiteside et al. 1967, Barry 1970). Otherwise, most workers have isolated larvae in plastic petri dishes or plastic jelly cups so that they may be individually observed. Compartmented plastic trays (Ignoffo and Boening 1970) are particularly convenient where large numbers of larvae are being monitored daily (Fig. 24-4).

It is important to discriminate the stage of each larva at the time of collection to accurately estimate percent parasitism (K. V. Yeargan, personal communi-

Figure 24-3. Vibrating soybean plants in a polyethylene "shaker" to dislodge lepidopteran larvae.

cation). It is equally important to note the instar from which the parasitoid emerges when studying all aspects of the parasitoid-host relationship. The compartmented trays are particularly convenient in this regard since the stage of each larva can be rapidly determined by measuring the width of its head capsule microscopically with an ocular micrometer through the Mylar covering of the tray. The species and stage can be written directly on the Mylar with a felt-tipped marker.

The problem of diet for larvae held for emergence of parasitoids in the laboratory has not been completely resolved. Whiteside et al. (1967), Barry (1970), and Lentz and Pedigo (1975) used soybean leaflets for rearing the green cloverworm, *Plathypena scabra* (F.). Sanders (personal communication) also used soybean leaflets for her study of the parasitoids of soybean Lepidoptera in South Carolina, but took the precaution of washing them in 1% sodium hypochlorite (followed by two rinses with water) to prevent infection from naturally-occurring pathogens on the leaves. Yeargan (personal communication) found that alfalfa trifoliolates were more convenient than soybean in his studies of the parasitoids of lepidopteran larvae in Kentucky.

Artificial media have been used by several workers with varying degrees of success. Burleigh (1971,1972) used the media of Berger (1963) or Burton

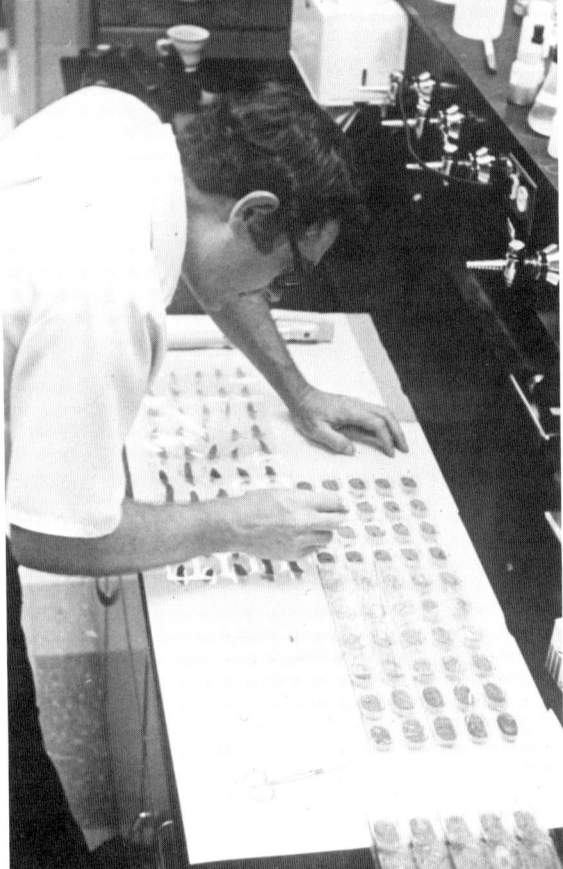

Figure 24-4. Compartmented plastic diet tray for rearing lepidopteran larvae to measure percentage parasitism.

(1969) for his study of the parasitoids of the soybean looper, *Pseudoplusia includens* (Walker), in Louisiana. D. C. Herzog (personal communication) also used Berger's diet for the soybean looper, but used the diet of Greene et al. (1976) for rearing the velvetbean caterpillar, *Anticarsia gemmatalis* (Hübner). W. M. Brooks (personal communication) states that many of the species of lepidopteran larvae collected by Deitz et al. (1976) in North Carolina soybean would not develop well on Berger's medium, so they added soybean leaf meal (1.5% by weight) obtained by drying field-collected leaves in an oven and pulverizing them in a blender. E. J. Armbrust (personal communication) reports that four media were used for a study of the parasitoids of soybean Lepidoptera in Illinois (Roberts et al. 1977). A pinto bean medium (Burton 1969) was used for the black cutworm, *Agrotis ipsilon* (Hufnagel); the fall armyworm, *Spodoptera frugiperda* (J. E. Smith); the armyworm, *Pseudaletia unipuncta* (Haworth); and the yellow-striped armyworm, *Spodoptera ornithogalli* (Gueneé). Alfalfa meal (1.5%) was added to Berger's medium for rearing the soybean looper, the cabbage looper, *Trichoplusia ni* (Hübner); the celery looper, *Anagrapha falcifera* (Kirby); *Autographa precationis* (Gueneé); and *Colias* spp. Soy protein (5.2%) was added to Berger's medium for rearing the green cloverworm. Sodium ascorbate and acetic acid (2.1% ea.) were added to Berger's medium for rearing the corn earworm, *Heliothis zea* (Boddie) and the tobacco budworm, *Heliothis virescens* (F.). A modification of Vanderzandt's (1962) medium has been used to rear lepidopteran larvae from Missouri soybean, mostly green cloverworms. The larvae develop slowly on the diet, however, with high mortality of early instars. Also, due to the slow rate of development, there is a problem of mold developing in the cells.

Little work has been done on the parasitoids of lepidopteran larvae attacking soybean outside the United States. Gangrade (1974) reported that the brown-striped semilooper, *Mocis undata* F., a pest of soybean in India, is attacked by three dipteran parasitoids and three hymenopteran parasitoids, but he did not describe his sampling procedures.

Gujrati et al. (1973a, 1973b) studied the biology of the soybean leaf-miner, *Stomopteryx subsecivella* Zell., in India, by collecting leaf-mines in the field, removing the larvae from the mines, and placing them in petri dishes with a pile of tender, green leaves. The larvae constructed mines in the central leaves. Two ectoparasitoids were recorded.

Little specific information is available on means of estimating the rates of parasitism of lepidopteran pupae, though many larval-pupal parasitoids have been reared from pupae collected as larvae in the studies described above. Pupae of the green cloverworm occasionally are collected along with larvae in soybean, and their parasitoids may be reared in the same way as those of the larvae. Khan et al. (1976) was able to obtain estimates of parasitism of pupae of the corn earworm, *H. zea*, by the nematode *Chromonema heliothidis* Khan (Steinernematidae), by allowing larvae to pupate in field soil, recovering the pupae, and holding them for emergence of nematodes. Alternatively, they isolated nematodes from field soil samples in a Baermann funnel and exposed these unidentified nematodes to larvae of *H. zea* reared on artificial diet to determine species that could potentially infect *H. zea*.

Estimating rates of parasitism of naturally-occurring eggs of the Lepidoptera is difficult due to the impracticability of locating them on the plants. Nevertheless, Barber (1936) noted high rates of parasitism of the corn earworm and the velvetbean caterpillar by *Trichogramma minutum* Riley in soybean, though he did not describe his method for collecting and holding the eggs. Lentz and Pedigo (1975) were able to circumvent the problem by placing plants from the field in pots, exposing them for 10 hr to ovipositing green cloverworm in the laboratory, and returning them to the field for 72 hr. The plants were then returned to the laboratory where the surviving eggs were removed to 30-ml jelly cups and incubated for parasitoid emergence. Lewis et al. (1975) used artificially applied eggs in their evaluations of the effects of kairomones on the activity of the egg parasitoid, *Trichogramma achaeae* Nagaraja and Nagarkatti. In one set of experiments they applied corn earworm eggs to soybean plants with a camel's hair brush, using a solution of Plantgard as an adhesive (Nordlund et al. 1974). In another case, they were able to collect enough native eggs of the corn earworm and the vetvetbean caterpillar to measure the effect of their treatments. In both experiments parasitism was determined by examining dissected eggs under a microscope for presence of immature stages of the parasitoid as described by Hoffman et al. (1970).

Artificially applied eggs were used to measure parasitism by *Trichogramma pretiosum* Riley in Missouri. Free eggs of the cabbage looper were obtained by brushing them from Teflon-coated oviposition substrates (Hoffman et al. 1968). The eggs were then individually applied to the undersides of marked leaflets with a camel's-hair brush coated with a 1% solution of methyl cellulose. After 24 hr, the leaflets were returned to the laboratory and sections of leaflets with surviving eggs were removed with a number 2 cork borer and placed individually in cylindrical cells drilled in acrylic plastic blocks. The cells were sealed with Mylar film and incubated until the fate of the eggs could be determined. This method is advantageous insofar as it is possible to expose eggs for a known period of time and, given a record of temperature and its effect on rate of development, to compute expected total parasitism during the entire stage.

B. Parasitoids of Coleoptera

Among the Coleoptera, greatest attention has been given to the parasitoids of the Mexican bean beetle, *Epilachna varivestis* Mulsant (Nichols and Kogan 1972, see Chapter 9). Landis and Howard (1940) summarize the many years of effort devoted to the unsuccessful attempt to effect biological control of the beetle by introducing and establishing the tachinid fly, *Aplomyiopsis (=Paradexodes) epilachnae* (Aldrich). Little information on sampling procedures was reported in these studies. More recently, the parasitoid, *Pediobius foveolatus* (Crawford) (Hymenoptera: Eulophidae), has received considerable attention (Rust 1977). The parasitoid was originally introduced from India where its biology was studied by Lall (1961). To estimate percent parasitism in the field he collected third and fourth instar hosts and fed them in the laboratory on fresh brinjal leaves in cages. Mummified larvae were transferred to 2.5-cm diameter tubes for emergence of parasitoids. Parasitized larvae were found on the under surface of the leaves during the winter and on stems near ground level during the summer. They were recognizable by their greyish-brown color. In the United States, Stevens et al. (1975) studied the effect of inoculative releases of the parasitoid on suppression of the beetle. They estimated rate of parasitism in the field by collecting third and fourth instar larvae, prepupae, and mummies from infested areas. The mummies were segregated to avoid parasitism of live larvae by emerging parasitoids in transit to the laboratory. Live larvae were reared in 0.5-1 cartons on soybean leaves until they pupated or mummified. Mummies were placed in 30-ml media cups for emergence of adult parasitoids. Those mummies from which no parasitoids emerged were dissected to verify parasitism.

Several studies of parasitism of the bean leaf beetle, *Cerotoma trifurcata* (Forster) (Chrysomelidae) (see Chapter 10) have been conducted. *Celatoria diabroticae* Shimer (Diptera: Tachinidae) was recorded as a parasitoid of adults by Eddy and Nettles (1930) and Isely (1930), but they did not describe sampling procedures. Tugwell et al. (1973) estimated parasitism of field-collected beetles by dissecting them in the laboratory and noting internal parasitoids. They found 19.6% of those collected were parasitized, probably by *C. diabroticae*. Herzog (1977) measured rate of parasitism by *C. diabroticae* and *Hyalomyodes triangulifer* (Loew) (Diptera: Tachinidae) by collecting adults in the field and rearing them on soybean leaflets or soybean seedlings in 26 x 55-mm shell vials capped with ventilated plastic stoppers. Dead beetles were dissected to determine if parasitoid larvae were present in the abdomen. Callow adults were excluded in the field since these would have had little exposure to parasitism.

Dectes texanus LeConte (Coleoptera: Cerambycidae), a stem borer, has recently become a pest of soybean in certain areas of the United States (see Chapter 17). Its biology was studied by Hatchett et al. (1975) from larvae collected in the stems of giant ragweed. Larvae were reared to maturity, or until the emergence of parasitoids, on artificial media (Hatchett et al. 1973) in 25-ml plastic cups. They recorded seven species of hymenopteran parasitoids.

The girdle beetle, *Oberea brevis* Swederus (Coleoptera: Lamiidae), inflicts similar damage to soybean in India. Kapoor et al. (1972) noted that 16% of mature larvae (presumably obtained by dissecting girdled stems) were parasitized by four species of external hymenopteran parasitoids.

C. Parasitoids of Diptera

The bean flies of the family Agromyzidae are serious pests of soybean and other types of beans throughout Africa, Asia, and Australia (Spencer 1973). Quesales (1918) reared two species of parasitoids from *Ophiomyia phaseoli* (Tryon) (=*Agromyza destructor* Malloch) infesting *Phaseolus* spp. in the Philippines. To rear the parasitoids, puparia of the fly were dissected from the stems and put in vials with cotton plugs with a small amount of moist soil in the bottom. Greathead (1969) conducted the most comprehensive study of the parasitoids attacking bean flies. Working with *Phaseolus vulgaris* in East Africa he recorded three species of flies, *Ophiomyia (=Melanagromyza) phaseoli* (Tryon), *O. spencerella* (Greathead) and *O. centrosematis* (deMeij.) and nine species of hymenopteran parasitoids. Parasitism was measured by randomly collecting bean plants from experimental plots, cutting off the lower 13 cm of the stem and placing this portion in 15.2 x 2.5-cm "specimen tubes" with gauze tops. Samples were taken 25-30 days after emergence of plants since most oviposition by the flies occurred on seedlings and the flies began to emerge ca. 30 days after appearance of the plants. Whenever a parasitoid emerged, the plant was examined for the puparium to determine the species of host from which it had issued; and after emergence was complete, each plant was dissected and all unemerged puparia were examined for evidence of parasitism. In soybean, Gangrade (1974) reported three species of parasitoids from *O. phaseoli* in India. Rose et al. (1976) monitored parasitism of *Melanagromyza sojae* (Zehntner) and *O. phaseoli* in Taiwan during the spring and early summer of 1975. Nearly 50% of puparia of both species of flies were parasitized by the most prevalent of the four species of hymenopteran parasitoids during April and August. Neither of these authors described their sampling procedure.

D. Parasitoids of Hemiptera

Numerous studies have been conducted on the biological control of stink bugs, particularly the southern green stink bug, *Nezara viridula* (L.), a cosmopolitan species attacking many important crops (DeWitt and Godfrey 1972, see Chapter 23), but few of these studies have dealt with parasitoids of stink bugs in soybean. In the United States, Deitz et al. (1976) reared the tachinid, *Trichopoda pennipes* (F.) from the green stink bug, *Acrosternum hilare* (Say); and he reared *Cylindromyia binotata* (Bigot), *C. euchenor* (Walk.), and *Euthera tentatrix* Loew, from the brown stink bug, *Euschistus servus* (Say), collected in soybean in North Carolina, but they did not describe the methods used to collect and hold the bugs for emergence of parasitoids. In Kentucky, K. V. Yeargan (personal communication) has studied the parasitoid complex associated with the green stink bug in soybean by collecting them with a sweep net and holding them for emergence of parasitoids in paper cartons with snap beans as a food source. D. C. Herzog (personal communication) states that R. M. McPherson has recorded nine species of parasitoids from eight stink bug species present on soybean in Louisiana. The bugs were collected in the field, brought to the laboratory, and held on snap beans in clear plastic containers (30 x 41 x 13 cm) for emergence of parasitoids.

Panizzi and Smith (1976) reared two parasitoids from stink bugs on soybean in Brazil. *Telenomus mormidea* Costa Lima (Hymenoptera: Scelionidae) was reared from egg masses of *Piezodorus guildinii* (Westwood) collected in the field and isolated in 7 x 4-cm plastic vials, and the tachinid, *Eutrichopodopsis nitens* Blanchard, was reared from a nymph by unspecified means.

E. Parasitoids of Homoptera

Dysart (1966) investigated the parasitoid complex attacking the bandedwing whitefly, *Trialeurodes abutiloneus* (Haldeman) (Aleyrodidae). He isolated sections of leaves bearing puparia of the whitefly by pinning them to the underside of a cork that was then inserted into the neck of a glass vial. When large numbers of parasitized puparia were used, they were placed in 3.8-1 ice cream cartons. Four species of hymenopteran parasitoids were recorded. Watve and Clower (1976) evaluated the effects of parasitoids of the bandedwing whitefly in several agroecosystems in Louisiana. Hosts were collected by bringing infested leaves into the laboratory where they were placed in 3.8-1 ice cream cartons. They were held at 22-23°C for 10-12 days after which adult parasitoids that had emerged were separated from the debris. Extent of parasitism was determined by examining the dorsal aspect of pupal exuviae. Exits made by the parasitoid were elliptical or circular, whereas emergence exits of whitefly imagos formed an "inverted T" in the dorsum of the thoracic region. They also recorded four species of hymenopteran parasitoids.

The potato leafhopper, *Empoasca fabae* (Harris) (Cicadellidae), is a potential pest of soybean (Ogunlana and Pedigo 1973, see Chapter 12). There have been no studies of its parasitoids in soybean, but Subba Rao et al. (1965) studied the parasitoids of a related species, *E. devastans* Distant, in India. Four species of hymenopteran parasitoids were reared from eggs dissected from the parenchymatous tissue of the leaf veins of field-collected leaves of bhindi. DeLong (1971) summarized work with parasitoids of other species of leafhoppers.

III. Estimating Populations of Adult Parasitoids

Little effort has been devoted to study of populations of adult parasitoids of soybean pests. The general surveys of soybean arthropods (e.g., Kretzschmar 1948, Deitz et al. 1976) have recorded the species of parasitoids collected in soybean, but no published studies have attempted to determine the number of parasitoids in fields or to relate populations of native adults to percentage parasitism of hosts.

Studies of populations of parasitoids and their relationship to percentage parasitism of the green cloverworm are currently being undertaken in Missouri. A modified cage-aerosol technique was first used in which a section of row was enclosed in a polyethylene tent (Fig. 24-5) and fumigated with pyrethrins from an aerosol, but this procedure proved too laborious for estimating low populations. Subsequently, a D-Vac procedure (Fig. 24-6) was used in which the cone is maneuvered slowly from the bases of the plants upward over both sides of a section of row to estimate populations per unit area. Marston et al. (1976) showed that this D-Vac procedure collected more parasitic Hymenoptera than the cage-aerosol procedure, probably because such minute, light-bodied forms become entangled in the hairs on the plants and are difficult to dislodge after being killed by the fumigant.

Individual species of parasitic Hymenoptera generally have occurred in low numbers in these studies ($<$1/row-m), even though the green cloverworm population may be heavily parasitized. Thus, a sample of at least 15 row-m would be

Figure 24-5. Polyethylene fumigation cage for collecting soybean arthropods.

advisable as a starting point until sufficient information is accumulated to determine standard errors of population estimates for individual species. Even larger samples are advisable when populations are low, as prior to flowering. Such large samples create a problem insofar as the D-Vac loses effectiveness as the collecting bag fills with debris. This effect may be reduced by periodically removing debris from the collecting bag while the machine is in operation, or by splitting the sample into several small subsamples. Or one may compensate for the reduction in efficiency by adjusting the data with a calibration ratio computed from the ratio of numbers in small and large subsamples (e.g., 3 row-m and 12 row-m).

One of the greatest difficulties in measuring populations of adults of minute forms such as the egg parasitoids, *Trichogramma* spp., is separating them from the debris in collections. To resolve this problem, first collections are spread in 24 x 40-cm enamel pans and the adhering arthropods are brushed from the shredded leaves and other large debris. The samples are then stored in 238-ml cardboard cartons. When the arthropods in the samples are to be counted, the sample is segregated with sieves (usually 12-, 30- and 200-mesh; U. S. Standard Sieve Series). If a quantity of dirt and fine debris is present in the portion of the sample on the fine sieve (and sometimes the middle sieve, as well) the arthropods

Figure 24-6. D-Vac machine for sampling soybean arthropods.

are removed from the debris by extracting them in a layer of xylene while leaving the debris in a layer of ice (Marston and Hennessey 1978). Small parasitoids, such as *Trichogramma,* are easily recognizable on the filter papers to which they are transferred from the xylene.

Traps may be useful in measuring relative abundance of adult parasitoids in soybean. G. T. Schmidt (personal communication) has tested the relative efficiency of rotary (Fig. 24-7), Malaise (Fig. 24-8), barrier, and black-light traps for collecting entomophages in soybean. In general, Malaise traps collected the greatest numbers of tachinids (Diptera) while the rotary traps most effectively sampled populations of hymenopteran parasites. M. E. Irwin (personal communication) has used sticky traps to obtain data on penetration of parasitoids into soybean in Illinois.

Figure 24-7. Rotary trap for collecting insects flying above soybean.

Figure 24-8. Malaise trap for collecting insects in soybean.

Acknowledgments

I wish to thank the many individuals who generously contributed their unpublished techniques for inclusion in this chapter. I am also grateful to G. T. Schmidt, D. L. Hostetter, and D. J. Isenhour for providing photographs.

References

Barber, G. W. 1936. Efficiency of *Trichogramma minutum* Riley in relation to population density of its host. J. Econ. Entomol. **29**:631.

Barry, R. M. 1970. Insect parasites of the green cloverworm in Missouri. J. Econ. Entomol. **63**:1963-1965.

Berger, R. S. 1963. Laboratory techniques for rearing *Heliothis* species on artificial media. USDA Agr. Res. Serv. Ser. **33-84**:13 p.

Boyer, W. P., and B. A. Dumas. 1963. Soybean insect survey as used in Arkansas. USDA Coop. Econ. Insect Rep. **13**:91-92.

Burleigh, J. G. 1971. Parasites reared from the soybean looper in Louisiana 1968-1969. J. Econ. Entomol. **64**:1550-1551.

Burleigh, J. G. 1972. Population dynamics and biotic controls of the soybean looper in Louisiana. Environ. Entomol. **1**:290-294.

Burton, R. L. 1969. Mass rearing the corn earworm in the laboratory. USDA Agr. Res. Serv. Ser. **33-134**:8 p.

Calkins, C. O., and G. R. Sutter. 1972. A method of collecting and maintaining insect larvae to establish incidence of parasitism and disease in field populations. Environ. Entomol. **1**:264-265.

DeLong, D. M. 1971. The bionomics of leafhoppers. Annu. Rev. Entomology **16**:179-210.

DeWitt, N. B., and G. L. Godfrey. 1972. The literature of arthropods associated with soybeans. II. A bibliography of the southern green stinkbug *Nezara viridula* (Linnaeus) (Hemiptera: Pentatomidae). Ill. Natur. Hist. Surv. Biol. Notes **78**:23 p.

Deitz, L. L., J. W. Van Duyn, J. R. Bradley, Jr., R. L. Rabb, W. M. Brooks, and R. E. Stinner. 1976. A guide to the identification and biology of soybean arthropods in North Carolina. N. C. Agr. Exp. Sta. Tech. Bull. **238**:264 p.

Dysart, R. J. 1966. Natural enemies of the bandedwing whitefly, *Trialeurodes abutilonea* (Hemiptera: Aleyrodidae). Ann. Entomol. Soc. Amer. **59**:28-33.

Eddy, C. O., and W. C. Nettles. 1930. The bean leaf beetle. S. C. Agr. Exp. Sta. Bull. **265**:25 p.

Elsey, K. D., and R. L. Rabb. 1970. Biology of *Voria ruralis* (Diptera: Tachinidae). Ann. Entomol. Soc. Amer. **63**:216-222.

Gangrade, G. A. 1974. Insects of soybeans. Jawaharlal Nehru Krishi Vishwa Vidyalaya Tech. Bull. **24**:88 p.

Greathead, D. J. 1969. A study in East Africa of the bean flies (Dipt., Agromyzidae) affecting *Phaseolus vulgaris* and of their natural enemies, with the description of a new species of *Melanagromyza* Hend. Bull. Entomol. Res. **59**:541-561.

Greene, G. L., N. C. Leppla, and W. A. Dickerson. 1976. Velvetbean caterpillar: a rearing procedure and artificial medium. J. Econ. Entomol. **69**:487-488.

Gujrati, J. P., G. A. Gangrade, and K. N. Kapoor. 1973a. Notes on studies on the newly recorded parasites of the leaf-miner *Stomopteryx subsecivella* Zell. (Gelechiidae: Lepidoptera). Sci. and Cult. **39**:470-471.

Gujrati, J. P., K. N. Kapoor, and G. A. Gangrade. 1973b. Biology of soybean leaf-miner, *Stomopteryx subsecivella* (Lepidoptera: Gelechiidae) Entomologist **106**:187-191.

Hammond, R. B., and L. P. Pedigo. 1976. Sequential sampling plans for the green cloverworm in Iowa soybeans. J. Econ. Entomol. **69**:181-185.

Hatchett, J. H., D. M. Daugherty, J. C. Robbins, R. M. Barry, and E. C. Hauser. 1975. Biology in Missouri of *Dectes texanus*, a new pest of soybeans. Ann. Entomol. Soc. Amer. **68**:209-213.

Hatchett, J. H., R. D. Jackson, and R. M. Barry. 1973. Rearing a weed cerambycid, *Dectes texanus*, on an artificial medium, with notes on biology. Ann. Entomol. Soc. Amer. **66**:519-522.

Herzog, D. C. 1977. Bean leaf beetle: parasitism by *Celatoria diabroticae* (Shimer) and *Hyalomyodes triangulifer* (Loew). J. Ga. Entomol. Soc. **12**:64-68.

Hill, C. C. 1925. Biological studies on the green cloverworm. USDA Bull. **1336**: 17 p.

Hillhouse, T. L., and H. N. Pitre. 1974. Comparison of sampling techniques to obtain measurements of insect populations on soybeans. J. Econ. Entomol. **67**:411-414.

Hinds, W. E., and B. A. Osterberger. 1931. The soybean caterpillar in Louisiana. J. Econ. Entomol. **24**:1168-1173.

Hoffman, J. D., L. R. Ertle, W. B. Brown, and F. R. Lawson. 1970. Techniques for collecting, holding, and determining parasitism of lepidopterous eggs. J. Econ. Entomol. **63**:1367-1369.

Hoffman, J. D., S. H. Long, and W. A. Dickerson. 1968. Teflon and cornstarch-coated paper as substrates for oviposition by cabbage looper moths. J. Econ. Entomol. **61**:1760.

Ignoffo, C. M., and O. P. Boening. 1970. Compartmented disposable plastic trays for rearing insects. J. Econ. Entomol. **63**:1696-1697.

Isely, D. 1930. The biology of the bean leaf beetle. Ark. Agr. Exp. Sta. Bull. **248**:20 p.

Kapoor, K. N., J. P. Gujrati, and G. A. Gangrade. 1972. Parasites of *Oberea brevis* (Coleoptera: Lamiidae). Ann. Entomol. Soc. Amer. **65**:755.

Khan, A., W. M. Brooks, and H. Hirschman. 1976. *Chromonema heliothidis* n. gen, n. sp. (Steinernematidae, Nematoda), a parasite of *Heliothis zea* (Noctuidae, Lepidoptera), and other insects. J. Nematol. **8**:159-168.

Kretzschmar, G. P. 1948. Soybean insects in Minnesota with special reference to sampling techniques. J. Econ. Entomol. **41**:586-591.

Lall, B. S. 1961. On the biology of *Pediobius foveolatus* (Crawford) (Eulophidae: Hymenoptera). Indian J. Entomol. **23**:268-273.

Landis, B. J., and N. F. Howard. 1940. *Paradexodes epilachnae*, tachinid parasite of the Mexican bean beetle. USDA Tech. Bull. **721**:31 p.

Lentz, G. L., and L. P. Pedigo. 1974. Life history phenomena of *Rogas nolophanae* and *Winthemia sinuata*, parasites of the green cloverworm. Ann. Entomol. Soc. Amer. **67**:678-680.

Lentz, G. L., and L. P. Pedigo. 1975. Population ecology of parasites of the green cloverworm in Iowa. J. Econ. Entomol. **68**:301-304.

Lewis, W. J., R. L. Jones, P. A. Nordlund, and A. N. Sparks. 1975. Kairomones and their use for management of entomophagous species. I: Evaluation for increasing rates of parasitization by *Trichogramma* spp. in the field. J. Chem. Ecol. **1**:343-347.

Marston, N. L., and M. K. Hennessey. 1978. Extracting arthropods from plant debris with xylene. J. Kans. Entomol. Soc. **51**:239-244.

Marston, N. L., C. E. Morgan, G. D. Thomas, and C. M. Ignoffo. 1976. Evaluation of four techniques for sampling soybean insects. J. Kans. Entomol. Soc. **49**:389-400.

Nichols, M. P., and M. Kogan. 1972. The literature of arthropods associated with soybeans. I. A bibliography of the Mexican bean beetle, *Epilachna varivestis* Mulsant (Coleoptera: Coccinellidae). Ill. Natur. Hist. Surv. Biol. Notes **77**: 20 p.

Nordlund, D. A., W. J. Lewis, H. A. Gross, Jr., and E. A. Harrell. 1974. Description and evaluation of a method for field application of *Heliothis zea* eggs and kairomones for *Trichogramma*. Environ. Entomol. **3**:981-984.

Ogunlana, M. O., and L. P. Pedigo. 1973. Pest status of the potato leafhopper on soybeans in central Iowa. J. Econ. Entomol. **67**:210-202.

Panizzi, A. R., and J. S. Smith. 1976. Observacoes sobre inimigos naturais de *Piezodorus guildinii* (Westwood, 1837) (Hemiptera, Pentatomidae) em soja. Am. Soc. Entomol. Brasil **5**:11-17.

Quesales, F. O. Y. 1918. The bean fly. Philipp. Agr. **7**:2-31.

Roberts, S. J., W. K. Mellors, and E. J. Armbrust. 1977. Parasites of lepidopterous larvae in alfalfa and soybean ecosystems in central Illinois. Great Lakes Entomol. **10**:87-93.

Rose, R. I., C. S. Lin, S. P. Kung, H. W. Kao, C. Y. Su, and C. Y. Chang. 1976. Bean flies and their control. pp. 169-171 in R. M. Goodman, ed. Expanding the use of soybeans. Proc. of a Conference for Asia and Oceania. Univ. of Illinois, College of Agriculture, INTSOY Ser. 10. 261 p.

Rudd, W. G., and R. L. Jensen. 1977. Sweep net and ground cloth sampling for insects in soybeans. J. Econ. Entomol. **70**:301-304.

Rust, R. W. 1977. Evaluation of trap crop procedures for control of Mexican bean beetle in soybeans and lima beans. J. Econ. Entomol. **70**:630-632.

Sherman, F. 1920. The greenclover worm (*Plathypena scabra* Fabr.) [sic.] as a pest on soybeans. J. Econ. Entomol. **13**:295-303.

Simmonds, F. J. 1948. Some difficulties in determining by means of field samples the true value of parasitic control. Bull. Entomol. Res. **39**:435-440.

Smith, J. W., E. G. King, and J. V. Bell. 1976. Parasites and pathogens among *Heliothis* species in the central Mississippi delta. Environ. Entomol. **5**:224-226.

Spencer, K. A. 1973. Agromyzidae (Diptera) of economic importance. Dr. W. Junk B. V., Pub., The Hague. 418 p.

Stevens, L. M., A. L. Steinhauer, and J. R. Coulson. 1975. Suppression of Mexican bean beetle on soybeans with annual inoculative releases of *Pediobius foveolatus*. Environ. Entomol. **4**:947-952.

Subba Rao, B. R., B. Parshad, A. Ram, R. P. Singh, and M. L. Srivastava. 1965. Studies on the parasites and predators of *Empoasca devastans* Distant (Homoptera: Jassidae). Indian J. Entomol. **27**:104-106.

Tugwell, P., E. P. Rouse, and R. G. Thompson. 1973. Insects in soybeans and a weed host (*Desmodium* sp.). Ark. Agr. Exp. Sta. Rep. Ser. **214**:18 p.

Vanderzant, E. S., C. D. Richardson, and S. W. Fort, Jr. 1962. Rearing of the bollworm on artificial diet. J. Econ. Entomol. **55**:140.

Watve, C. M., and D. F. Clower. 1976. Natural enemies of the bandedwing whitefly in Louisiana. Environ. Entomol. **5**:1075-1078.

Whiteside, R. C., P. P. Burbutis, and L. P. Kelsey. 1967. Insect parasites of the green cloverworm in Delaware. J. Econ. Entomol. **60**:326-328.

Chapter 25

Sampling Predaceous Hemiptera on Soybean

Michael E. Irwin and Merle Shepard

I. Introduction

Predaceous Hemiptera are usually more abundant in soybean fields than all other insect predators combined (Kretzschmar 1948, Blickenstaff and Huggans 1962, Tugwell et al. 1973, Deitz et al. 1976). Complexes of predaceous Hemiptera change in composition between geographical areas but the predominant groups in most areas where soybean is grown include Nabidae (*Nabis* spp.), Lygaeidae (*Geocoris* spp.), Anthocoridae (*Orius* spp.), Pentatomidae [*Podisus* spp., *Stiretrus anchorago* (F.)], and Reduviidae (*Zelus* spp., *Sinea* spp., and others).

Within soybean fields, the relative mix of genera of predaceous Hemiptera varies with locality and time of season. The genera *Nabis*, *Geocoris* and *Orius* are almost always more abundant than the others (Blickenstaff and Huggans 1962, Tugwell et al. 1973, Deitz et al. 1976, Pacheco 1976). Table 25-1 illustrates the percentage catch per genus of predaceous Hemiptera for the season in soybean fields in different localities, along with sampling methods and a comparison of collections of predaceous Hemiptera with those of all other predators (where data were available). In some surveys, *Nabis* spp. were more abundant if catches were summed through the season (Balduf 1923, Pacheco 1976, Correa et al. 1977, Pitre et al. 1978). In some, *Geocoris* spp. were more abundant (Tugwell et al. 1973, Deitz et al. 1976); in others, *Orius insidiosus* (Say) was most abundant (Blickenstaff and Huggans 1962, Barry 1973, Pitre et al. 1978). Alfalfa fields in Oklahoma had 6.7 times more *Orius* spp. than *Nabis* spp. and 5.5 times more *Nabis* spp. than either *Geocoris* spp. or *Sinea diadema* (F.); total predaceous Hemiptera were 2.8 times more abundant than all other species of predators encountered for the season by sweep net samples (Fenton and Howell 1957). Hay

Table 25-1. Composition of predaceous Hemiptera as compared to other arthropod predators at various geographical locations

Geographical[a] location	Sampling method	Percent of total Hemiptera							Total predaceous Hemiptera	Percentage of total predators	References
		Orius spp.	*Nabis* spp.	*Geocoris* spp.	*Podisus* spp.	*Sinea* spp.	*Zelus* spp.	*Stiretrus* spp.			
Mexico, Sonora	sweep net	8	59	29	—	1	3	—	1819	40	Pacheco 1976
U.S.A., Arkansas	sweep net	28	13	58	1	1	1	—	2313	74	Tugwell et al. 1973
U.S.A., Missouri	sweep net	78	16	6	1	1	—	—	1540	73	Blickenstaff and Huggans 1962
U.S.A., Missouri	D-Vac	82	15	3	—	—	—	—	157271	—	Barry 1973
U.S.A., North Carolina	D-Vac + ground cloth	68	21	11	—	—	—	—	31684	59	Deitz et al. 1976
U.S.A., Ohio	sweep net	25	37	1	6	31	—	—	167	47	Balduf 1923
U.S.A., Minnesota	sampling cylinder	91	9	—	—	—	—	—	216	89	Kretzschmar 1948
U.S.A., South Carolina	ground cloth	1	35	34	—	—	—	—	70	—	Shepard et al. 1974a
U.S.A., Kentucky	ground cloth	—	63	37	—	—	—	—	19	—	Raney and Yeargan 1977

[a] In all areas of Brazil where soybean sampling has been reported, *Nabis* spp. was the most abundant hemipteran predator, followed by *Geocoris* spp. (Correa et al. 1977).

alfalfa in northern California had the following mixture of predaceous Hemiptera: *Orius tristicolor* (White) (38.3%), *Nabis* spp. (29.2%), *Geocoris atricolor* (Montandon) (22.8%), *G. pallens* (Stål) (8.9%), *G. punctipes* (Say) (0.4%), *Zelus renardii* (Kolenati) (0.1%), *Sinea diadema* (0.1%), all others (0.1%) (Benedict and Cothran 1975).

Climatic factors and available host, to a large extent, influence the geographic range and abundance of these various groups of predators. Within North America, *O. insidiosus* is apparently most abundant in the central and northern Midwest (Missouri, Illinois, Kentucky, Minnesota) and adjacent to cotton in the Delta Region of Mississippi (Pitre et al. 1978). *Geocoris* spp. are more numerous in the south and southeast (and southwest: no records from soybean, but many from cotton, e.g., Gonzalez et al. 1977). *Nabis* spp. were most abundant in Brazil (Correa et al. 1977), and in Sonora, Mexico (Pacheco 1976); and in the United States, in Ohio (Balduf 1923) the Delta Region of Mississippi away from cotton fields (Pitre et al. 1978).

All predaceous Hemiptera feed on a wide range of hosts and may extend this polyphagy to plant feeding to some extent (Stoner 1970, Stoner et al. 1975). Such plant feeding causes no damage to row crops but almost certainly has survival value for the predators by maintaining populations where prey are scarce or absent (Salas-Aquilar and Ehler 1977). This ability has implications in natural control: if insect prey are scarce, these species can survive for a considerable amount of time by sucking plant parts and seeds (Sweet 1960).

The phenology of the complex of predaceous genera in a soybean field is sometimes asynchronous. In Kentucky the population density of *O. insidiosus* peaked earliest (early to mid-July), the density of *Nabis* spp. peaked near midseason (early August), and the density of *Geocoris* spp. peaked late (late August, early September) (Raney and Yeargan 1977). But in North Carolina *Orius* and *Geocoris* populations peaked in early August while *Nabis* and a second *Geocoris* population peaked in mid-September (Deitz et al. 1976). In South Carolina, population levels of *Geocoris* and *Nabis* peaked in late September (Shepard et al. 1974b). In Brazil *Nabis* was by far the most abundant hemipteran predator with population peaks in mid- to late January (midseason) in Paraná and Santa Catarina (Correa et al. 1977). In Rio Grande do Sul, Paraná, and Goias, the number of *Nabis* was highest in early to mid-February with an additional peak in early March in some locations in the three states. *Geocoris* spp. was second in abundance to *Nabis* in all Brazilian areas sampled. In Mississippi, population densities of *Nabis* were greatest between early and late July; of *Geocoris*, depending on year and location, between late June and early August (usually late July); and of *Orius*, between mid- and late June, with a second peak between mid-July and early August (Pitre et al. 1978).

Population increases of predaceous Hemiptera often follow the buildup of their prey. *Orius insidiosus* was commonly found in association with large populations of *Sericothrips variabilis* (Beach) (Raney and Yeargan 1977), and in some cases, densities of *O. insidiosus* tracked those of *S. variabilis* (Fig. 25-1) (Irwin and Kuhlman 1979). Population densities of *Nabis* spp. peaked when populations of

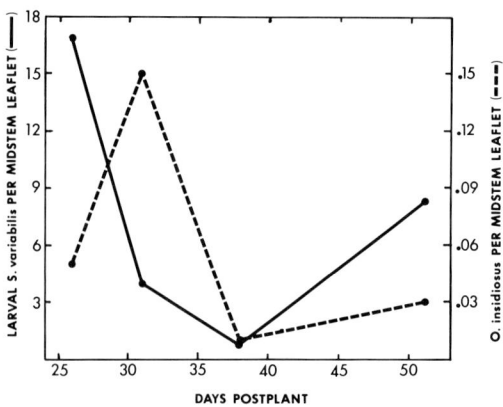

Figure 25-1. Mean densities of larval *Sericothrips variabilis* (Beach) and all stadia of *Orius insidiosus* (Say) per center leaflet of midstem "Williams" soybean trifoliolates in the control plots, Dixon Springs Agr. Exp. Sta., Simpson, Ill., 1975 (after Irwin and Kuhlman 1979).

green cloverworm [*Plathypena scabra* (F.)] were highest in Kentucky (Raney and Yeargan 1977).

II. Species Composition, Geographical Distribution, and Life Cycle

A. *Nabis* spp. (Damsel Bugs)—Hemiptera: Nabidae

There are over 20 species in the family Nabidae in America north of Mexico (Blatchley 1926), 13 of which belong to the genus *Nabis* (Slater and Baranowski 1978). Species within the genus *Nabis* are often the most abundant group of hemipterous predators in soybean. Adults of this genus are 5.6 to 9.0 mm in length, slender, and gray, tan, or brown (Fig. 25-2). Their front legs are slightly raptorial, the beak is four-segmented with the first joint very short, and antennae are also four-segmented (Blatchley 1926).

Blatchley (1926) provided a detailed description of this family and keys for separating subfamilies and species occurring in the eastern United States. Within the family, 17 species occur in the eastern United States but only about 5 are important enough as predators in soybean to be considered here. Benedict and Cothran (1975) devised a key to separate *N. americoferus* Carayon and *N. alternatus* Parshley. Deitz et al. (1976) modified a key after Harris (1928) to separate males of the common species of *Nabis* (see Section V, Appendix).

Adults and nymphs of *Nabis* spp. feed on almost any soft-bodied arthropod and the eggs of many species. To some extent, *Nabis* also feed on plants but cause no damage (Ridgway and Jones 1968).

In soybean in the southeastern United States, *N. roseipennis* Reuter comprises over 90% of all *Nabis* species (Turnipseed 1974, Deitz et al. 1976). Other species of *Nabis* that often occur in soybean are *N. alternatus, N. americoferus, N. capsiformis* Germar, and *N. deceptivus* Harris. In addition to these species, *N. kalmii* Reuter occurs in soybean in Missouri (Blickenstaff and Huggans 1962),

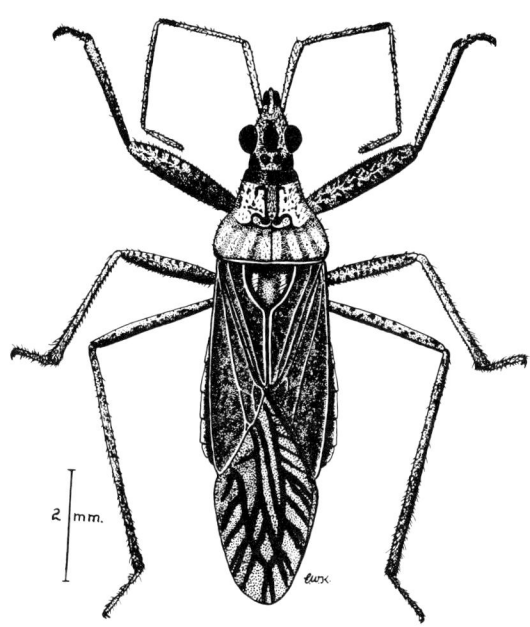

Figure 25-2. Habitus drawing of *Nabis roseipennis* Reuter.

and *N. paranensis* Harris is found in soybean in Brazil (International Reference Collection of Soybean Associated Arthropods, Illinois Natural History Survey).

Nabis spp. overwinter as adults in winter grain, alfalfa and other ground cover (Deitz et al. 1976). *N. roseipennis* prefers shady habitats while *N. americorferus* favors sunny, drier habitats (Harris 1928). Between 15 and 65 eggs are deposited in plant tissue by each female *N. roseipennis*. After hatching, the predator passes through five nymphal stages, reaching adulthood in ca. 50 days (Deitz et al. 1976).

B. *Geocoris* spp. (Big-Eyed Bugs) — Hemiptera: Lygaeidae

Geocoris is a large genus containing ca. 125 species from all parts of the world. The most important species are: *G. punctipes, G. sobrinus* (Blanchard), and *G. thoracicus* (Fieber) from the Neotropical Region; *G. atricolor, G. bullatus* (Say), *G. decoratus* Uhler, *G. discopterus* Stål, *G. pallens, G. punctipes,* and *G. uliginosus* (Say) from the Nearctic Region; *G. ater* (F.), *G. grylloides* (L.), *G. lineolus* (Rambur), and *G. megacephalus* (Rossi) from the Palearctic Region; *G. flaviceps* (Burmeister), *G. ochropterus* (Fieber), and *G. varius* (Uhler) from the Oriental Region; *G. amabilis* Stål, *G. pallidipennis* (Costa), and *G. ruficeps* (Gemar) from the Ethiopian Region (Slater 1964).

About 19 species of *Geocoris* occur in America north of Mexico (Slater and Baranowski 1978). Adults of this genus are 2.8 to 4.2 mm long, oval, and robust and have a broad head with large, curved, backward projecting eyes (Fig. 25-3).

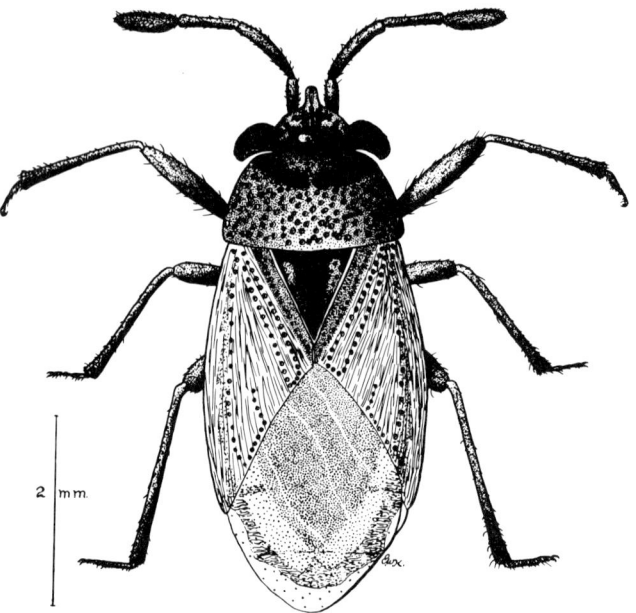

Figure 25-3. Habitus drawing of *Geocoris punctipes* (Say).

The femora are only slightly swollen, unarmed, and not truly raptorial. Within the United States and Canada, the distribution of various species of *Geocoris* seems discrete. For instance, *G. pallens* occurs along the Pacific slope from Washington through California, *G. punctipes* occurs throughout the south from California to North Carolina, and *G. bullatus* occurs from the Pacific Northwest eastward to New England and Quebec. A key to separate three of the most abundant species of *Geocoris* occurring in soybean fields in the eastern, southern, and midwestern United States is given in Section V, Appendix.

Most early literature refers to *Geocoris* as destructive to crops (Usinger 1936), but by the late 1930s, *Geocoris* was regarded as beneficial. Sweet (1960) recorded the polyphagous habit of *Geocoris* and suggested that the ability to subsist on plant juices in part could account for their high densities in certain areas.

Geocoris spp. are known to prey on several groups of arthropods: Aphididae, Diaspididae, Miridae, Curculionidae, Chrysomelidae, Sphingidae, Gelechiidae, Noctuidae, and Tetranychidae, to name a few (Tamaki and Weeks 1972). Species of *Geocoris* can be found in many crop communities: legumes, cotton, sugar beet, vegetables, citrus, ornamentals, tobacco, and many others, and they are often the most abundant hemipterous predator.

Prey consumption by *G. punctipes* under laboratory conditions differed depending on developmental stage and sex. Late instar nymphs ate more prey than earlier instars; fifth instar females ate more than fifth instar males (Crocker et al. 1975). *Geocoris punctipes* laid varying numbers of eggs depending on species of prey: tubermoth eggs plus green beans (264), *Lygus* eggs plus green bean (134), tubermoth larvae plus green beans (190), *Lygus* nymphs plus green beans (35). The highest percentage of egg hatch was with diets containing prey eggs: tubermoth eggs plus green beans (71%), *Lygus* eggs plus green bean (61%). A diet of tubermoth eggs alone was not completely satisfactory. Only a 37% egg hatch was recorded and an average reduction of 47% in egg laying was found when compared to a similar test with green beans (Dunbar and Bacon 1972a). Furthermore, both *G. pallens* and *G. bullatus* survived longer if offered an insect (aphid) plus plant (sunflower seed) diet than either alone (Tamaki and Weeks 1972).

Like almost all predaceous hemipterans, *Geocoris* spp. undergo five nymphal stadia. *Geocoris punctipes* females, under laboratory conditions of 22.5°C and 50% RH and in a 14L:10 hr D photoperiod laid 178 eggs or 5.5 eggs per oviposition day. Development occurred in the following number of days: eggs (9.9), nymphs (26.8, male; 27.7, female), adults (41.5, male; 67.7, female). Females averaged 5.2 days preoviposition period and 30.8 days oviposition period (Champlain and Sholdt 1967). *Geocoris punctipes* oviposited and had a better rate of increase at lower temperatures than did either *G. atricolor* or *G. pallens* (Dunbar and Bacon 1972b). A comparative study of developmental biology of *G. bullatus* and *G. pallens* showed that egg hatching was optimal at between 27°C and 41°C for both species, and these data agree quite closely with similar data for *G. punctipes* (Tamaki and Weeks 1972). Although insecticide applications in soybean fields differentially affected population density of *Geocoris* spp. in South Carolina (Turnipseed 1972), development of surviving nymphs within different insecticide treatments was apparently unaffected (Walker et al. 1974).

Overwintering records of individuals of the genus *Geocoris* are known from Indiana, where adults of *G. discopterus* Stål were discovered in mullein leaves, and adults of *G. uliginosus* were commonly encountered beneath logs along roadsides (Blatchley 1895). Although *G. pallens* overwintered as adults in soil ca. 2.5 cm below the surface in cultivated fields, *G. bullatus* was found to overwinter in the egg stage in peach tree litter in Washington (Tamaki and Weeks 1972). Dispersal patterns of *G. punctipes* in soybean fields were found to be along rows for nymphs but were nondirectional for adults. Within a release field, dispersal was found to be slow, and very few ^{32}P labeled specimens were recovered more than 10 m from the release site (Shepard et al. 1974c).

C. *Orius* spp. (Minute Pirate Bugs and Flower Bugs)
 —Hemiptera: Anthocoridae

Orius spp. are extremely small, adults ranging from 1.5 to 2.2 mm in length (Fig. 25-4). Females insert eggs into the tissue of soybean petioles (M. Irwin unpublished) and possibly in young pods after pod set.

A detailed description of the family is given by Blatchley (1926) with keys to subfamilies, genera, and species occurring in eastern North America. Slater and

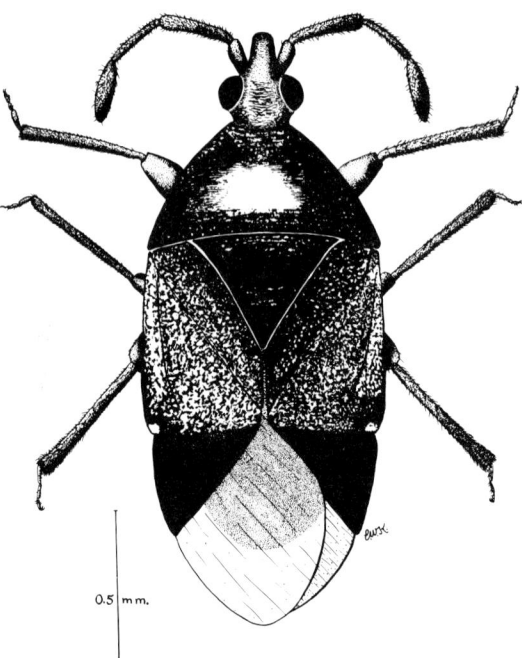

Figure 25-4. Habitus drawing of *Orius insidiosus* (Say).

Baranowski (1978) give updated keys to genera. Within the family Anthocoridae, members of the genus *Orius* occur as predators in soybean fields around the world. In some areas, species of *Anthocoris* are probably also important predators in soybean. In North America, however, *Orius* is the dominant genus of Anthocoridae in soybean fields.

Detailed taxonomic treatments of *Orius* exist for the Nearctic Region (Kelton 1963) and for the western hemisphere (Herring 1966). Two species occur commonly in soybean fields in the western hemisphere. In the midwestern, eastern, and southeastern portions of the United States, *O. insidiosus* predominates, while *O. tristicolor* is more common in the west and southwest. Both species occur in Central and South America and in the West Indies. Separation of these two species by descriptive characters is provided in Section V, Appendix.

The preferred prey of *Orius* spp. has been shown to be various species of thrips (Thysanoptera: Thripidae) (Dicke and Jarvis 1962, Rajasekhara and Chatterji 1970, Salas-Aquilar and Ehler 1977, Irwin and Kuhlman 1979). But species of *Orius* are known to attack and consume a wide variety of prey, including each other (Dicke and Jarvis 1962). Among the more common prey are spider mites of the genus *Tetranychus* (Askari and Stern 1972a,b), aphids in soybean fields including *Aphis glycines* Matsumura and *Aulacorthum solani* (Kaltenbach) in Japan (Oku and Kobayashi 1966), and eggs of Lepidoptera including *Heliothis zea* (Boddie) (Barber 1936), *Anticarsia gemmatlis* (Hübner) (Buschman et al. 1977), *Heliothis virescens* (F.) (Lingren and Wolfenbarger 1976), *Ostrinia nubil-*

alis (Hübner) (Dicke and Jarvis 1962), young lepidopterous larvae of the genera *Sciropophaga* and *Proceras* (Miller 1956), and Cicadelllidae including nymphs of *Nephotettix* sp. (Manley 1974). They are also known to attack and consume plant bugs (Miridae), whiteflies (Aleyrodidae), lacewing eggs (Chrysopidae) (Deitz et al. 1976), Collembola, and Cecidomyiidae (Dicke and Jarvis 1962).

Orius insidiosus overwinters as adults in protected areas where a covering of leaves and rubbish exist (Marshall 1930, Barber 1936). They are active in early spring and can occasionally be found on dandelion flowers where they apparently search for prey. They move into soybean fields as soon as a population of thrips becomes established (Mayse et al. 1978b, Irwin and Kuhlman 1979) and are the earliest species of insect predator to colonize soybean fields in Illinois (Mayse et al. 1978a; M. Irwin, field observations). Nymphs and adults of *O. insidiosus* prefer to occupy terminal buds of soybean plants during the daytime (Mayse et al. 1978b, Irwin and Kuhlman 1979). Terminal buds are the preferred colonization sites of *Frankliniella tritici* (Thysanoptera: Thripidae) (Irwin et al. 1979) and it is suspected that *O. insidiosus* may prey preferentially on *F. tritici* (M. Irwin, field observation).

D. Reduviidae (Assassin Bugs)—Hemiptera: Reduviidae

Reduviidae is a large family of predaceous hemipterans. Most species are active predators on various groups of insects, and some are prominent in soybean fields. The two genera most often encountered in soybean fields in North America are *Zelus* (Figs. 25-5, 25-6) and *Sinea*.

1. *Zelus* spp.

The genus *Zelus* contains ca. 13 species in the United States, most of which occur in the southern and southwestern portions (Slater and Baranowski 1978). The most common and widespread species of *Zelus* is *Z. exsanguis* (Stål), which occurs from New England to the Pacific and south to Florida and Texas. The International Reference Collection of Soybean-Associated Arthropods, Illinois Natural History Survey, houses specimens of *Z. socius* Uhler from soybean in the United States, *Z. longipes* (L.) from soybean in Colombia, and *Zelus* spp. from Brazil and Colombia.

Zelus socius is univoltine. An average of 26 eggs are deposited in a somewhat circular mass, and a female can lay up to eight egg masses with a total of 230 eggs. It takes about two months for an individual to develop from egg to adult (Swadener and Yonke 1973).

Zelus exsanguis overwinters in the northern United States as a striking green nymph with golden abdominal markings (Slater and Baranowski 1978). *Zelus socius* was found to overwinter in Missouri as fourth and fifth instar nymphs in leaf litter or clumps of grass (Swadener and Yonke 1973).

Figure 25-5. *Zelus* sp. preying on *Cerotoma facialis* (Erichson) (Coleoptera: Chrysomelidae) in a soybean field, Portoviejo, Ecuador.

2. *Sinea* spp.

The genus *Sinea* contains ca. 10 species in the United States, all but two occurring in the southern and southwestern states. *Sinea diadema* occurs throughout the United States and is the species most often encountered in soybean fields (Slater and Baranowski 1978). The International Reference Collection of Soybean-Associated Arthropods, Illinois Natural History Survey, has several specimens of *Sinea* from soybean fields: *S. spinipes* (Herrich-Schaeffer) and *S. diadema* from the United States, *S. complexa* Caudell from Brazil, and *S.* sp. nr. *coronata* Stål from Mexico. A key for separating *Zelus* and *Sinea* is presented in Section V, Appendix.

Sinea diadema eggs are laid in a double row on stems and leaves in masses averaging 8-12 eggs. A single female can lay 412 eggs under laboratory conditions. The average length of time for an individual to reach the adult stage from the time the egg was laid is about 54 days. *Sinea diadema* was reported to overwinter in rosettes of common mullein, *Verbascum thapsus* (L.), in Kansas (Readio 1924).

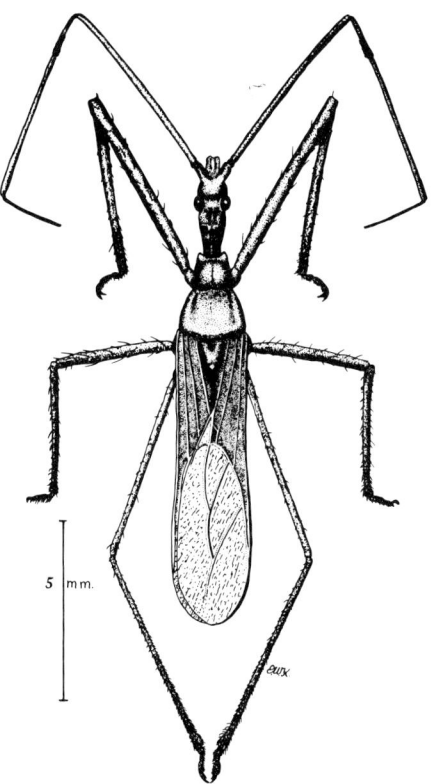

Figure 25-6. Habitus drawing of *Zelus cervicalis* Stål.

3. Other species

Within the International Reference Collection of Soybean-Associated Arthropods, Illinois Natural History Survey, are the following reduviid species from soybean fields: *Arilus cristatus* (L.) from Missouri, United States; *Atrachelus cinereus* (F.) from North Carolina, United States; *A. cinereus crassicornis* (Burmeister) from Sinaloa, Mexico and Rio Grande do Sul, Brazil; *Doldina bicarinala* Stål from Colombia; *Emesaya brevipennis* (Say) from Colombia; *Emesaya modica* McAtee and Malloch from Sinaloa, Mexico; *Empicoris orthoneuron* McAtee and Malloch from Mississippi, United States; *Hiranetis branconiformis* (Burmeister) from Colombia; *Pselliopus ornatipennis* from Santa Catarina, Brazil; *Rasahua hamatus* (F.) from Rio Grande do Sul, Brazil; *Repipta flavicans* (Amyot and Serville) from Santa Catarina, Brazil; and *R. taurus* (F.) from Colombia and Belize.

E. Pentatomidae (Stink Bugs) — Hemiptera: Pentatomidae

Members of this family are mostly plant feeders but a few are predaceous on other insects. There is no easy way to distinguish predaceous from phytophagous

Figure 25-7. *Podisus maculiventris* (Say) preying on larval *Spodoptera exigua* (Hübner) (Lepidoptera: Noctuiidae) in a soybean field, South Carolina, United States.

Pentatomidae. Slater and Baranowski (1978) have separated the predaceous genera from several phytophagous genera (see Section V, Appendix).

Among the predaceous species, two are more often encountered in soybean fields in the United States than all other predaceous Pentatomidae: *Podisus maculiventris* (Say) (Fig. 25-7) and *Stiretrus anchorago* (Fig. 25-8). The genus *Podisus* contains ca. 10 species in the United States (Slater and Baranowski 1978). The International Reference Collection of Soybean-Associated Arthropods, Illinois Natural History Survey, contains two species of *Podisus* associated with soybean: *P. maculiventris* (Fig. 25-9) from the United States and *P. nigrispinus* (Dallas) from Colombia and Brazil. *S. anchorago* (Fig. 25-10) is less commonly encountered in soybean fields than is *P. maculiventris* and is confined to the eastern and southeastern United States (Slater and Baranowski 1978).

Amyotea malabarica (F.) is a predator of soybean pests in India. The dorsal surface of most of the head, thorax and forewings are bright orange-red. It proved to be an effective predator of *Nezara viridula* (L.) (Hemiptera: Penta-

Figure 25-8. *Stiretrus anchorago* (F.) preying on *Epilachna varivestis* Mulsant (Coleoptera: Coccinellidae) in a soybean field, South Carolina, United States.

tomidae) in the laboratory, consuming up to 54 individuals during a life cycle (Singh et al. 1973). *A. malabarica* was introduced into the United States, but was not released into soybean fields because of the potential phytophagous habit of young nymphs.

The developmental biology and bionomics of *P. maculiventris* have been investigated (Landis 1937, Warren and Wallis 1971, Shiyomi 1974, Hokyo and Kawauchi 1975, Waddill and Shepard 1975a,b), as has the developmental biology of a related species, *P. placidus* Uhler (Oetting and Yonke 1971). The rate of development and subsequent percentage mortality of *P. maculiventris* depended on the type of prey taken. Insect eggs were most nourishing, followed by larvae of several prey species. The better prey were those with higher metabolic rates. Furthermore, nymphs died more rapidly when their developmental rate was slow (Landis 1937). The developmental time of *P. maculiventris* was found to correlate with temperature; at 10°C development ceased (Warren and Wallis 1971). *P. maculiventris* females lay an average of about 313 eggs each and have a preoviposition period of 10.2 days and an oviposition period of 42.6; eggs are laid in masses that average 18 eggs each (Warren and Wallis 1971). *P. maculiventris* nymphs tend to disperse along, rather than across, rows (Waddill and Shepard 1975a), and adults overwinter in ground duff, under bark scales, or other protected places (Warren and Wallis 1971).

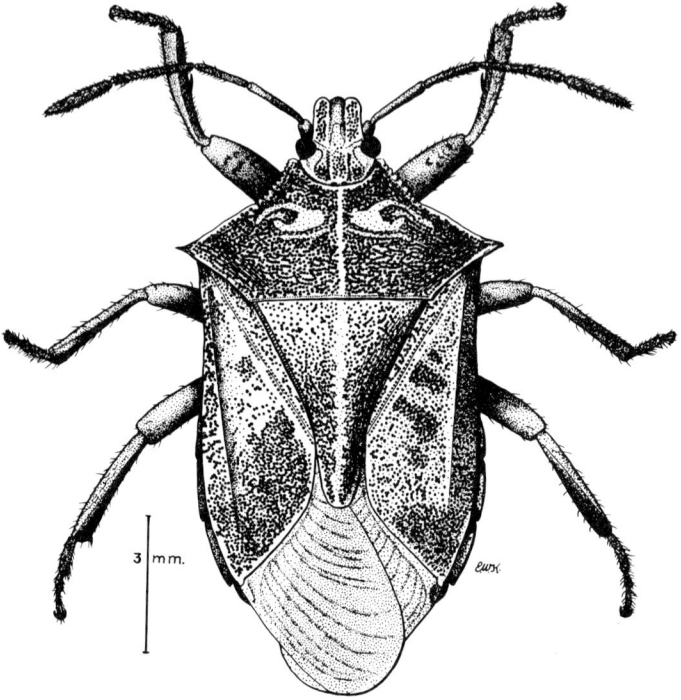

Figure 25-9. Habitus drawing of *Podisus maculiventris* (Say).

III. Sampling Predator Populations

A. Reasons for Sampling Predators

Pest management has become more sophisticated over the past several years. Economic injury levels have become more realistic; application rates and kinds of insecticides have become more selective. Yet, in soybean insect pest management, a lack of knowledge of predator impact and population estimation has hampered the integration of natural mortality factors into economic injury level assessments. Impact of predators on pest populations is currently being studied in several localities. When more is known about the relationship between predator density and pest density, predator density may become an integral part of the management system. This has been accomplished in the cotton ecosystem in Australia, where management decisions relative to pests that attack cotton take into account population levels of the pests, damage to the squares, and population levels of key predators (Sterling 1977). Population estimation of predators must be based on accurate sampling techniques that are repeatable in field situations.

Sampling Predaceous Hemiptera on Soybean

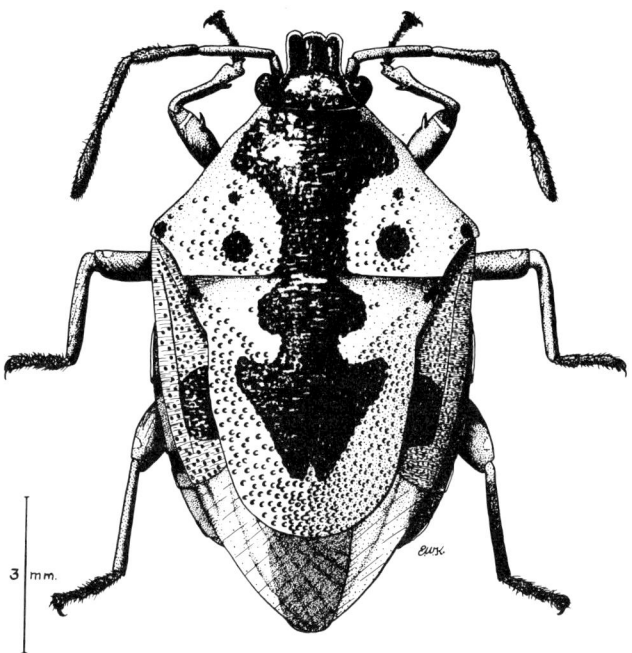

Figure 25-10. Habitus drawing of *Stiretrus anchorago* (F.).

One of the difficulties in estimating populations of some of the groups of predators such as *Podisus, Stiretrus, Zelus,* and *Sinea* is that their populations are usually low and variable and their occurrence patchy. Numbers of these predators may increase only late in the growing season when adequate prey populations have become established. However, this complex of predators, individuals of which are much larger in size than specimens of *Nabis, Geocoris,* or *Orius,* are capable of successfully attacking and consuming larger pests. Because numbers of *Orius, Nabis,* and *Geocoris* are usually higher than those of pentatomids and reduviids, developing sampling methods to estimate their population densities is feasible. Although quantitative assessment of the specific role of these predators has not been determined, enough evidence is available on pest population impact to make population density approximations useful.

Good sampling techniques should show the effects of insecticides, applied for control of specific pests, on predators. The "systems" approach to pest management requires reliable estimates of predator densities as well as other natural mortality agents to initialize and validate simulation models of pest-predator interactions.

B. Factors Influencing Sampling

Although general sampling methods have been considered, there are some factors that apply more specifically to developing sampling plans for predaceous Hemiptera. For example, *Orius* spp. have been found to aggregate in the terminals of soybean plants (Mayse et al. 1978a), and in cotton, *G. pallens* and *G. punctipes* adults preferred pink bollworm eggs attached to upper and terminal portions of cotton plants (Irwin et al. 1974). *Geocoris* spp. were present in the upper portion of cylindrical screened cages during mid-morning when cages were placed adjacent to soybean plants; fewer were there during mid-afternoon. Ground cloth samples yielded significantly more nymphs of *Geocoris* spp. during mid-morning (Shepard et al. 1974c). This is in general agreement with other studies that show differences in numbers of predators collected at different times of the day (Dumas et al. 1962, 1964). Although numbers of *Nabis* spp. sampled at 0 700 hr, 1 300 hr, or 1 800 hr were not significantly different, numbers of *O. insidiosus* and *Geocoris* spp. were larger when sampled at 0 700 hr. More *Nabis*, *Geocoris* and *Orius* were collected from cotton plants during the morning (Gonzalez et al. 1977). The predatory activity of *N. roseipennis* was greatest in the mornings and evenings with maximum predation occurring at 0 700 hr (Donahoe and Pitre 1977). In English grassland habitats, 7.6 times more nabids were collected per 100 sweeps at night than by day (Fewkes 1961). Those predators actively searching for prey in the plant canopy are more likely to be collected by conventional sampling techniques.

Adjacent crops can greatly influence the abundance and time of occurrence of predaceous Hemiptera in soybean fields. In the Black Belt of Mississippi, population densities of *O. insidiosus* increased later in soybean fields if they were distant (>0.8 km) from cotton fields, whereas population densities of *Geocoris* spp. built up later in soybean fields adjacent to cotton fields; both *Geocoris* spp. and *Nabis* spp. were less abundant in soybean adjacent to cotton (Pitre et al. 1978).

Soybean plant height also must be considered when designing a sampling program for predaceous Hemiptera (Dumas et al. 1964), although this factor is difficult to separate from other phenological events such as the buildup of prey species. Populations of *Geocoris* and *Orius* were greatest in cotton at greater plant height (Leigh et al. 1974), but this relationship has not been reported for soybean.

C. Comparison of Sampling Methods for Predaceous Hemiptera

Ordinarily, sampling hemipterous predators is conducted at the same time as sampling for phytophagous soybean arthropods. The most economical technique for sampling predators, from a practical standpoint, is usually the sampling method chosen for target pests.

In general, only species of *Geocoris, Nabis,* and *Orius* reach high enough densities in soybean fields to allow comparisons of sampling techniques. The ground cloth and sweep net are commonly used techniques for sampling predaceous hemipterans. A comparison of the sweep net with a cylinder sampler, a near absolute technique, revealed that *Nabis* spp. were not sampled as reliably with the sweep net as with the cylinder sampler (Kretzschmar 1948). However, use of the latter technique would not be practical in most cases where estimates of predators are required for pest management decisions.

Correlation between sweep net and ground cloth methods were not as strong for *Geocoris* spp. in soybean as for pest species because *Geocoris* spp. were more mobile, but sampling by sweep net required less time (Rudd and Jensen 1977). The ground cloth method produced higher numbers of *Geocoris* and *Nabis* nymphs and with smaller coefficients of variation than did the sweep net and D-Vac. There were, however, no significant differences between the three techniques (ground cloth, D-Vac, and sweep net) when numbers of adults of these predators were compared (Shepard et al. 1974a). When numbers of adults and nymphs of *Geocoris* and *Nabis* were combined, the ground cloth technique yielded higher numbers of individuals with lower coefficients of variation. In fact, the highest numbers of *Geocoris* spp. and *Nabis* spp. with the lowest coefficients of variation were found with the ground cloth when foliage was beaten as opposed to shaken (Turnipseed 1974).

A preliminary comparison of absolute with sweep net, D-Vac, and ground cloth samples for *O. insidiosus* showed that all sampling methods except the D-Vac correlated reasonably well with one another (M. Irwin unpublished). During the sampling of populations of *Sericothrips variabilis*, nymphs and adults of *O. insidiosus* were also captured. The sampling technique entailed hand picking center leaflets of midstem leaves and terminals and processing them in various ways. *Orius insidiosus* was most often encountered in the terminal buds of the plants, at least during the daytime (Irwin and Kuhlman 1979). This suggests that methods that concentrate on the upper portion of the plants are better suited to obtaining larger numbers of specimens. Although it has not been tested, one of the easiest and perhaps most practical methods for sampling *O. insidiosus* would be to randomly gather a number of soybean terminals, strike them over a cloth or enamel pan, and count the number of nymphs and adults that are dislodged. Direct observation has been used successfully to estimate population density of *O. insidiosus* in soybean fields in Illinois (Mayse et al. 1978a,b). For special studies, especially of colonization rates, this technique may be most appropriate. It has the advantage that it can be accomplished throughout the life of the crop, something other relative techniques are unable to do.

This discussion points to a general conclusion: the sampling techniques that work well for the pest species seem to work well for hemipterous predators. The simplest and usually most reliable technique is the ground cloth. It is fast and efficient for *Geocoris, Nabis,* and *Orius,* and it can be accomplished while sampling for pest species such as defoliating lepidopterous larvae, adult bean leaf beetles, stink bug pests, and many others. Sweep net sampling is also reasonably effective and could substitute for the ground cloth. The D-Vac sampling technique correlates least favorably with absolute sampling methods for most predaceous Hemiptera and should be avoided.

D. Sequential Sampling Plans for Predaceous Hemiptera

The mathematical distribution of predaceous Hemiptera should be considered when developing a sampling plan for this group. For sequential sampling plans, knowledge of spatial patterns is essential. Although data regarding spatial distributions for predaceous Hemiptera are scarce, populations of *Nabis* and *Geocoris* in soybean conform to the Poisson (Waddill et al. 1974).

Sequential sampling plans were formulated for *Geocoris* spp. and *Nabis* spp. in soybean fields in South Carolina. The plans were based on predator populations considered, from a three-year study, to be high, medium, or low. The resulting sequential models are diagrammed in Fig. 25-11 (*Geocoris* spp.) and Fig. 25-12 (*Nabis* spp.) (Waddill et al. 1974). Decision making between low and medium populations generally required fewer samples than between medium and high population levels for both genera in South Carolina. Furthermore, the probability of a correct decision (i.e., into which population category does the real population belong) increases with a higher population for all three categories and for both genera; probability of detecting low vs. medium population densities naturally requires lower densities of predators than does a similar decision between medium and high densities (Waddill et al. 1974).

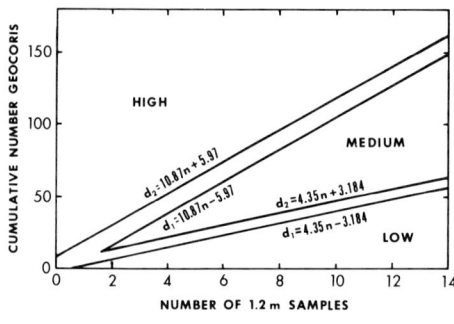

Figure 25-11. Sequential sampling plan for *Geocoris* spp. d = cumulative number of *Geocoris* spp., n = number of samples taken (after Waddill et al. 1975b).

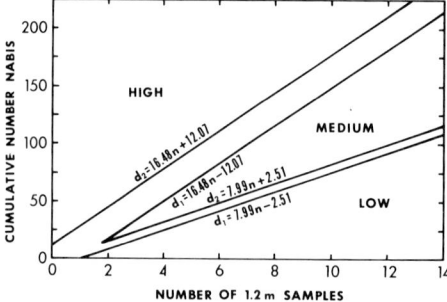

Figure 25-12. Sequential sampling plan for *Nabis* spp. d = cumulative number of *Nabis* spp., n = number of samples taken (after Waddill et al. 1975b).

IV. Concluding Remarks: Impact of Predaceous Hemiptera on Prey Populations

Because hemipterous predators make up an important component of the total predatory complex in soybean, several field and laboratory studies have been conducted to assess their impact on soybean insect pests. Both *Nabis* spp. and *Geocoris* spp. fed on eggs plus first, second, and third instars of the Mexican bean beetle, *Epilachna varivestis* Mulsant. In field cages, first instar Mexican bean beetles were reduced 62 and 72% by *Geocoris punctipes* and *Nabis* spp., respectively (Waddill and Shepard 1974). *Nabis alternatus* was reported to be a more effective predator of cabbage loopers, *Trichoplusia ni* (Hübner), and corn earworms, *H. zea*, than was *Chrysopa carnea* Stephens (Barry et al. 1974). Further, pest population regulation by *N. alternatus* was reflected by increased yields from soybean in cages containing this predator.

Some of the most effective predators of eggs of the velvetbean caterpillar, *A. gemmatalis*, in Florida soybean were *Nabis* spp. as determined by radiotracer techniques in the field (Buschman et al. 1977). In South Carolina, *Nabis* spp. were by far the most important predators of eggs and first instar *Heliothis* spp. in soybean (McCarty et al. 1980). First instar *Heliothis* on cotton plants were consumed at the rate of 1.9, 7.9, and 38.2 per day by individual first, third, and fifth instars of *N. roseipennis*, respectively (Donahoe and Pitre 1977).

A comparison of predation by the pentatomids *P. maculiventris* and *S. anchorago*, on Mexican bean beetle larvae revealed that *P. maculiventris* accepted a wider range of prey stages than did *S. anchorago*. Although predation by both species increased with increasing temperature, *S. anchorago* displayed a distinct preference for larvae of Mexican been beetles over those of the velvetbean caterpillar (Waddill and Shepard 1975b). In laboratory tests, it was found that the more even the predator's ability to attack, the more prey were killed by the predator (*P. maculiventris*) (Shiyomi 1974).

Zelus renardii preyed on a wide range of insect species in cotton fields in Texas. Of these, 32% were pest species and 21% were potentially beneficial. It is believed that the dense population of *Z. renardii* that occurred in cotton during the 1976-77 season was at least partially responsible for preventing outbreaks of lepidopterous larvae (Ables 1978).

The importance of *Orius* spp. as predators has been recognized for many years (Marshall 1930, Barber 1936, Whitcomb and Bell 1964). The impact of *Orius* spp. on pest populations varies from region to region. In Florida, where *Orius* is sparse, only two specimens of *O. insidiosus* were recovered from the field during a study of predators of velvetbean caterpillar eggs (Buschman et al. 1977). On the other hand, populations of *Orius* spp. exerted intense predation pressure on aphids, *Aulacorthum solani*, in soybean fields in Morioka, Japan (Oku and Kobayashi 1966).

The relationship between *Orius* and its prey, *Sericothrips variabilis*, in small field plots of soybean in southern Illinois provides some evidence that *Orius* responds in a density-dependent manner to prey populations and, therefore, may suppress thrips populations (see Fig. 25-1) (Irwin and Kuhlman 1979).

A comparison of predator efficiency in cages over soybean in Missouri revealed that *N. alternatus* was more efficient than *G. punctipes* or *Chrysopa carnea* (Barry et al. 1974). A combination of *Nabis* and *Orius* caused significant mortality in eggs and early instars of the green cloverworm on soybean (Pedigo et al. 1972). Predatory ability of *Nabis* has also been reported for suppression of green cloverworm larvae (Pedigo et al. 1972), and Mexican bean beetle larvae (Waddill and Shepard 1974). In South Carolina large nabid nymphs devoured an average of 20 *Heliothis zea* eggs per day (Turnipseed 1972).

APPENDIX 25-1. Diagnostic keys are presented for the most common groups of predaceous Hemiptera found in soybean fields

A. Key for the Identification of the Most Common *Nabis* spp.
(after Harris 1928 and Deitz et al. 1976)

1. Pro- and mesothoracic femora armed with short, blunt teeth ventrally; male clasper with broad blade and projection on ventral margin (Fig. 25-13A); dark robust species . . . . *N. deceptivus* Harris
   Pro- and mesothoracic femora unarmed; shape of male clasper variable; body color and size variable. 2

2. Cells (not veins) of forewings without dark spots; male clasper with semicircular blade (Fig. 25-13B). . . . . *N. capsiformis* Germar
   Cells of forewing with dark spots; male clasper variable. 3

3. Male clasper elongate and sinuate (Fig. 25-13C); color brownish. . . . .
   . . . . *N. roseipennis* Reuter
   Male clasper not sinuate; color grayish. 4

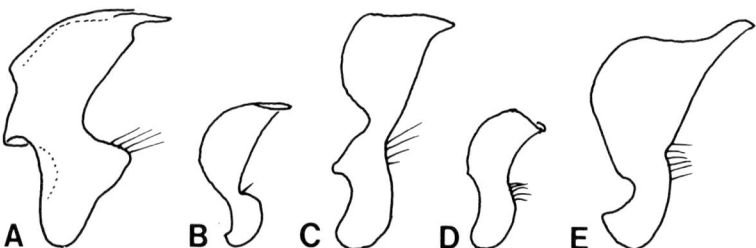

Figure 25-13. Claspers of male terminalia of *Nabis* spp. (A) *N. deceptivus* Harris, (B) *N. capsiformis* Germar, (C) *N. roseipennis* Reuter, (D) *N. alternatus* Parshley, (E) *N. americoferus* Carayon (after Deitz et al. 1976).

4. Male clasper small with narrow blade (less than width of dorsal aspect of eye) (Fig. 25-13D)..........................*N. alternatus* Parshley
Male clasper larger with moderately broad blade (Fig. 25-13E)...........
....*N. americoferus* Carayon

B. Key for the Identification of Three Major Species of *Geocoris*
(after Slater and Baranowski 1978)

1. Vertex of head smooth and polished...............*G. punctipes* (Say)
Vertex of head, at least in part, rugulose or granulose................2

2. Scutellum usually bicolored, usually longer than it is wide..............
....*G. bullatus* (Say)
Scutellum usually unicolorus, nearly equilateral ....*G. uliginosus* (Say)

C. Descriptive Characters for the Separation of Two *Orius* species
(after Herring 1966)

1. Ground color of hemelytra yellow; fuscous markings, if any, limited to cuneus and base of clavus; head and pronotum sometimes yellow; calluses poorly developed, both calluses and pronotum rugulose or coarsely rugulose-punctate, shape of female pronotum as in Fig. 25-14A; male clasper as in Fig. 25-14B ....*O. insidiosus* (Say)
Ground color of hemelytra not yellow, usually darker; fuscous markings prominent; clavus usually entirely brown to black; calluses distinctly elevated, smooth and shining; sides of pronotum of female straight, at least at middle (Fig. 25-14C); male clasper as in Fig. 25-14D ...*O. tristicolor* (White)

D. Key for the Separation of the Two Major Genera of Reduviidae
(after Slater and Baranowski 1978)

Sides of mesopleuron with a small tubercle or fold projecting in front of posterior margin of pleuron (Fig. 25-15) ....*Sinea*
Sides of mesopleuron without a tubercle or fold ....*Zelus*

E. Key to the Identification of the Important Predaceous Pentatomidae
in Soybean Fields (after Slater and Baranowski 1978)

1. Scutellum greatly enlarged, nearly reaching posterior end of abdomen.....2
Scutellum much smaller, usually triangular, remote from posterior end of abdomen and with forewing broadly exposed laterally 3

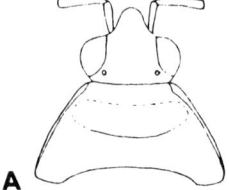
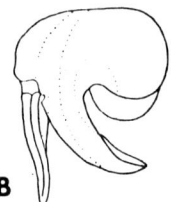
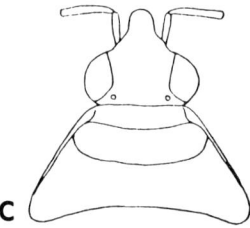
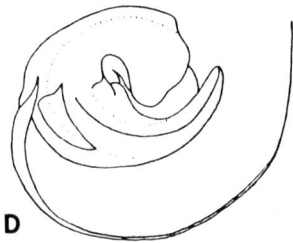

Figure 25-14. Female pronota (A and C) and claspers of male terminalia (B and D) of *Orius insidiosus* (Say) (A and B) and *Orius tristicolor* (White) (C and D) (after Herring 1966).

2. Color variegated with dark blue and either white, orange-yellow or red .... *Stiretrus*
   Color variable, not as above. .... phytophagous genera

3. First segment of labium lying in a groove between bucculae so that it is often obscured and the beak often has the appearance of not beginning at front of head; bucculae parallel, not coming together posteriorly on underside of head (Fig. 25-16A). .... phytophagous genera

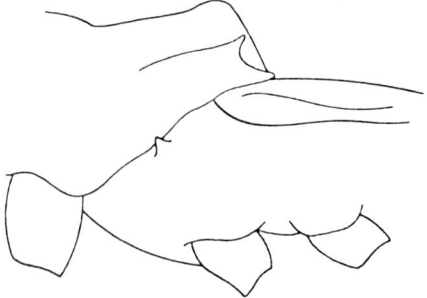

Figure 25-15. Lateral view of mesopleuron of *Sinea* sp. (after Slater and Baranowski 1978).

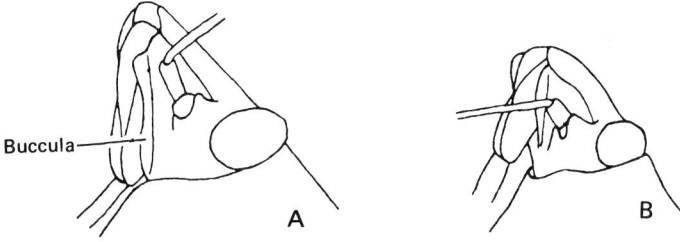

Figure 25-16. Bucculae on Pentatomidae. (A) on nonpredaceous genera, (B) on predaceous genera (after Slater and Baranowski 1978).

First segment of labium freely exposed, not lying in a groove formed by bucculae; bucculae coming together posteriorly on underside of head (Fig. 25-16B) (predaceous genera)..................................4

4. Front femur with a small tubercle or spine below on distal third (sometimes deteriorated; if so, colors bright red and black); usually marked with bright colors: red, white, yellow, or orange......................*Perillus*
Front femur unarmed below, usually brown or gray-colored species 5

5. Abdomen below with one or two series of small black spots running midway between midline and lateral margins; usually less than 14 mm in length ....*Podisus*
Abdomen below lacking row(s) of black spots; larger species, usually more than 14 mm in length.............................*Apateticus*

References

Ables, J. R. 1978. Feeding behavior of an assassin bug, *Zelus renardii*. Ann. Entomol. Soc. Amer. **71**:476-478.

Askari, A., and V. M. Stern. 1972a. Biology and feeding habits of *Orius tristicolor* (Hemiptera: Anthocoridae). Ann. Entomol. Soc. Amer. **65**:96-100.

Askari, A., and V. M. Stern. 1972b. Effect of temperature and photoperiod on *Orius tristicolor* feeding on *Tetranychus pacificus*. J. Econ. Entomol. **65**: 132-135.

Balduf, W. V. 1923. The insects of the soybean in Ohio. Ohio Agr. Exp. Sta. Bull. **366**:145-181.

Barber, G. W. 1936. *Orius insidiosus* (Say), an important natural enemy of the corn earworm. USDA Tech. Bull. **504**:24 p.

Barry, R. M. 1973. A note on the species composition of predators in Missouri soybeans. J. Ga. Entomol. Soc. **8**:284-286.

Barry, R. M., J. H. Hatchett, and R. D. Jackson. 1974. Cage studies with predators of the cabbage looper, *Trichoplusia ni*, and corn earworm, *Heliothis zea*, in soybeans. J. Ga. Entomol. Soc. **9**:71-78.

Benedict, J. H., and W. R. Cothran. 1975. A faunistic survey of Hemiptera-Heteroptera found in northern California hay alfalfa. Ann. Entomol. Soc. Amer. **68**:897-900.

Blatchley, W. S. 1895. Notes on the winter insect fauna of Virgo County, Indiana. II. Psyche **7**:267-270.

Blatchley, W. S. 1926. Heteroptera or true bugs of eastern North America. The Nature Pub. Co., Indianapolis. 1116 p.

Blickenstaff, C. C., and J. L. Huggans. 1962. Soybean insects and related arthropods in Missouri. Mo. Agr. Exp. Sta. Bull. **803**:51 p.

Buschman, L. L., W. H. Whitcomb, R. C. Hemenway, D. L. Mays, Nguyen Ru, N. C. Leppla, and B. J. Smittle. 1977. Predators of velvetbean caterpillar eggs in Florida soybeans. Environ. Entomol. **6**:403-407.

Butler, G. D., Jr. 1966. Development of several predaceous Hemiptera in relation to temperature. J. Econ. Entomol. **59**:1306-1307.

Champlain, R. A., and L. L. Sholdt. 1967. History of *Geocoris punctipes* (Hemiptera: Lygaeidae) in the laboratory. Ann. Entomol. Soc. Amer. **60**:881-883.

Correa, B. S., A. R. Panizzi, G. G. Newman, and S. G. Turnipseed. 1977. Distribuição geografica e abundancia estacional dos principais insectos-pragas da soja e seus predadores. An. Soc. Entomol. Brasil **6**:40-50.

Crocker, R. L., W. H. Whitcomb, and R. M. Ray. 1975. Effects of sex, developmental stage, and temperature on predation by *Geocoris punctipes*. Environ. Entomol. **4**:531-534.

Deitz, L. L., J. W. Van Duyn, J. R. Bradley, Jr., R. L. Rabb, W. M. Brooks, and R. E. Stinner. 1976. A guide to the identification and biology of soybean arthropods in North Carolina. N. C. Agr. Exp. Sta. Bull. **238**:264 p.

Dicke, F. F., and J. L. Jarvis. 1962. The habits and seasonal abundance of *Orius insidiosus* (Say) (Hemiptera-Heteroptera: Anthocoridae) on corn. J. Kans. Entomol. Soc. **35**:339-344.

Donahoe, M. C., and H. N. Pitre. 1977. *Reduviolus roseipennis* behavior and effectiveness in reducing numbers of *Heliothis zea* on cotton. Environ. Entomol. **6**:872-876.

Dumas, B. A., W. P. Boyer, and W. H. Whitcomb. 1962. Effect of time of day on surveys of predaceous insects in field crops. Fla. Entomol. **45**:121-128.

Dumas, B. A., W. P. Boyer, and W. H. Whitcomb. 1964. Effect of various factors on surveys of predaceous insects in soybeans. J. Kans. Entomol. Soc. **37**:192-201.

Dunbar, D. M., and O. G. Bacon. 1972a. Feeding, development, and reproduction of *Geocoris punctipes* (Heteroptera: Lygaeidae) on eight diets. Ann. Entomol. Soc. Amer. **65**:892-895.

Dunbar, D. M., and O. G. Bacon. 1972b. Influence of temperature on development and reproduction of *Geocoris atricolor, G. pallens,* and *G. punctipes* (Heteroptera: Lygaeidae) from California. Environ. Entomol. **1**:596-599.

Fenton, F. A., and D. E. Howell. 1957. A comparison of five methods of sampling alfalfa fields for arthropod populations. Ann. Entomol. Soc. Amer. **50**:606-611.

Fewkes, D. W. 1961. Diel vertical movements in some grassland Nabidae (Heteroptera). Entomol. Mon. Mag. **97**:128-130.

Gonzalez, D., D. A. Ramsey, T. F. Leigh, B. S. Ekbom, and R. van den Bosch. 1977. A comparison of vacuum and whole-plant methods for sampling predaceous arthropods on cotton. Environ. Entomol. **6**:750-760.

Harris, H. M. 1928. A monographic study of the hemipterous family Nabidae as it occurs in North America. Entomol. Amer. **9**(n.s.):1-97.

Herring, J. L. 1966. The genus *Orius* of the Western Hemisphere (Hemiptera: Anthocoridae). Ann. Entomol. Soc. Amer. **59**:1093-1109.

Hokyo, N., and S. Kawauchi. 1975. The effect of prey size and prey density on the functional response, survival, growth and development of a predatory pentatomid bug *Podisus maculiventris* Say. Res. Pop. Ecol. **16**:207-218.

Irwin, M. E., R. W. Gill, and D. Gonzalez. 1974. Field-cage studies of native egg predators of the pink bollworm in southern California cotton. J. Econ. Entomol. **67**:193-196.

Irwin, M. E., and D. E. Kuhlman. 1979. Relationships among *Sericothrips variabilis*, systemic insecticides, and soybean yield. J. Ga. Entomol. Soc. **14**:148-154.

Irwin, M. E., K. V. Yeargan, and N. L. Marston. 1979. Spatial and seasonal patterns of phytophagous thrips in soybean fields with comments on sampling techniques. Environ. Entomol. **8**:131-140.

Kelton, L. A. 1963. Synopsis of the genus *Orius* in America North of Mexico. Can. Entomol. **95**:631-636.

Kretzschmar, G. P. 1948. Soybean insects in Minnesota with special reference to sampling techniques. J. Econ. Entomol. **41**:586-591.

Landis, B. J. 1937. Insect hosts and nymphal development of *Podisus maculiventris* Say and *Perillus bioculatus* F. (Hemiptera: Pentatomidae). Ohio J. Sci. **37**:252-259.

Leigh, T. F., D. W. Grimes, W. L. Dickens, and C. E. Jackson. 1974. Planting pattern, plant population, irrigation, and insect interactions in cotton. Environ. Entomol. **3**:492-496.

Lingren, P. D., and D. A. Wolfenbarger. 1976. Competition between *Trichogramma pretiosum* and *Orius insidiosus* for caged tobacco budworms on cotton treated with chlordimeform sprays. Environ. Entomol. **5**:1049-1052.

Manley, G. V. 1974. Immature stages and biology of *Orius tantillus* (Motschulsky), (Hemiptera: Anthocoridae), inhabiting rice fields in west Malaysia. Entomol. News **87**:103-109.

Marshall, G. E. 1930. Some observations on *Orius (Triphleps) insidiosus* (Say). J. Kans. Entomol. Soc. **3**:29-32.

Mayse, M. A., M. Kogan, and P. W. Price. 1978a. Sampling abundances of soybean arthropods: Comparison of methods. J. Econ. Entomol. **71**:135-141.

Mayse, M. A., P. W. Price, and M. Kogan. 1978b. Sampling methods for arthropod colonization studies in soybean. Can. Entomol. **110**:265-274.

McCarty, M. T., M. Shepard, and S. G. Turnipseed. 1980. Identification of predaceous arthropods in soybeans using autoradiography. Environ. Entomol. (in press).

Miller, N. C. E. 1956. The biology of the Hemiptera. Leonard Hill, Ltd., London. 162 p.

Oetting, R. D., and T. R. Yonke. 1971. Immature stages and biology of *Podisus placidus* and *Stiretrus fimbiatus* (Hemiptera: Pentatomidae). Can. Entomol. **103**:1505-1516.

Oku, T., and T. Kobayshi. 1966. Influence of the predation of *Orius* spp. (Hemiptera: Anthocoridae) on the aphid population in a soybean field. An example of interrelation between a polyphagous predator and its principal preys. Jap. J. Appl. Entomol. Zool. **10**:89-94.

Pacheco, F. 1976. Seasonal and daily fluctuation of soybean insect populations in the Yaqui Valley, Sonora, Mexico. pp. 584-593 in L. D. Hill, ed., World Soybean Research. Proc. World Soybean Res. Conf., Interstate Print., Danville, Ill. 1073 p.

Pedigo, L. P., J. D. Stone, and G. L. Lentz. 1972. Survivorship of experimental cohorts of the green cloverworm on screenhouse and open-field soybean. Environ. Entomol. 1:180-186.

Pitre, H. N., T. L. Hillhouse, M. C. Donahoe, and H. C. Kinard. 1978. Beneficial arthropods on soybeans and cotton in different ecosystems in Mississippi. Miss. Agr. For. Exp. Sta. Tech. Bull. **90**:9 p.

Rajasekhara, K., and S. Chatterji. 1970. Biology of *Orius indicus* (Hemiptera: Anthocoridae), a predator of *Taeniothrips nigricornis* (Thysanoptera). Ann. Entomol. Soc. Amer. **63**:364-367.

Raney, H. G., and K. V. Yeargan. 1977. Seasonal abundance of common phytophagous and predaceous insects in Kentucky soybeans. Trans. Ky. Acad. Sci. **38**:83-87.

Readio, P. A. 1924. Notes on the life history of a beneficial Reduviid, *Sinea diadema* (Fabr.), Heteroptera. J. Econ. Entomol. **17**:80-86.

Ridgway, R. L., and S. L. Jones. 1968. Plant feeding by *Geocoris pallens* and *Nabis americoferus*. Ann. Entomol. Soc. Amer. **61**:232-233.

Rudd, W. G., and R. L. Jensen. 1977. Sweep net and ground cloth sampling for insects in soybeans. J. Econ. Entomol. **70**:301-304.

Salas-Aquilar, J., and L. E. Ehler. 1977. Feeding habits of *Orius tristicolor*. Ann. Entomol. Soc. Amer. **70**:60-62.

Shepard, M., G. R. Carner, and S. G. Turnipseed. 1974a. A comparison of three sampling methods for arthropods in soybeans. Environ. Entomol. 3:227-232.

Shepard, M., G. R. Carner, and S. G. Turnipseed. 1974b. Seasonal abundance of predaceous arthropods in soybeans. Environ. Entomol. 3:985-988.

Shepard, M., V. Waddill, and S. G. Turnipseed. 1974c. Dispersal of *Geocoris* spp. in soybeans. J. Ga. Entomol. Soc. 9:120-126.

Shiyomi, M. 1974. Changes of distribution patterns of prey population caused by predation: I. Homogeneity of predator's ability to attack prey. Jap. J. Appl. Entomol. Zool. **18**:159-165.

Singh, Z., C. E. White, and W. H. Luckmann. 1973. Notes on *Amyotea malabarica*, a predator of *Nezara viridula* in India. J. Econ. Entomol. **66**:551-552.

Slater, J. A. 1964. A catalogue of the Lygaeidae of the world. vol. I, 778 p. Univ. Conn., Storrs.

Slater, J. A., and R. M. Baranowski. 1978. How to know the true bugs (Hemiptera: Heteroptera). Wm. C. Brown Co., Dubuque, Iowa. 256 p.

Sterling, W. L. 1977. Sequential sampling plans for management of cotton arthropods in south-east Queensland. Aust. J. Ecol. **1**:265-274.

Stoner, A. 1970. Plant feeding by a predaceous insect, *Geocoris punctipes*. J. Econ. Entomol. **63**:1911-1915.

Stoner, A., A. M. Metcalfe, and R. E. Weeks. 1975. Plant feeding by Reduviidae, a predaceous family (Hemiptera). J. Kans. Entomol. Soc. **48**:185-188.

Swadener, S. O., and T. R. Yonke. 1973. Immature stages and biology of *Zelus socius* (Hemiptera: Reduviidae). Can. Entomol. **105**:231-238.

Sweet, M. H. 1960. The seed bugs: a contribution to the feeding habits of the Lygaeidae (Hemiptera: Heteroptera). Ann. Entomol. Soc. Amer. **53**:317-321.

Tamaki, G., and R. E. Weeks. 1972. Biology and ecology of two predators *Geocoris pallens* Stål and *G. bullatus* (Say). USDA Tech. Bull. **1446**:46 p.

Tugwell, P., E. P. Rouse, and R. G. Thompson. 1973. Insects in soybeans and a weed host (*Desmodium* sp.). Ark. Agr. Exp. Sta. Rep. Ser. **214**:18 p.

Turnipseed, S. G. 1972. Management of insect pests of soybeans. Proc. Tall Timbers Conf. Ecol. Anim. Contr. Habitat Manage. **4**:187-203.

Turnipseed, S. G. 1974. Sampling soybean insects by various D-Vac, sweep and ground cloth methods. Fla. Entomol. **57**:219-223.

Usinger, R. L. 1936. Genus *Geocoris* in the Hawaiian Islands. Proc. Hawaiian Entomol. Soc. **9**:212-215.

Waddill, V., and M. Shepard. 1974. Potential of *Geocoris punctipes* (Hemiptera: Lygaeidae) and *Nabis* spp. (Hemiptera: Nabidae) as predators of *Epilachna varivestis* (Coleoptera: Coccinellidae). Entomophaga. **19**:421-426.

Waddill, V., and M. Shepard. 1975a. Dispersal of *Podisus maculiventris* nymphs in soybean. Environ. Entomol. **4**:233-234.

Waddill, V., and M. Shepard. 1975b. A comparison of predation by the pentatomids, *Podisus maculiventris* (Say) and *Stiretrus anchorago* (F.), on the Mexican bean beetle, *Epilachna varivestis* Mulsant. Ann. Entomol. Soc. Amer. **68**:1023-1027.

Waddill, V., M. Shepard, S. G. Turnipseed, and G. R. Carner. 1974. Sequential sampling plans for *Nabis* spp. and *Geocoris* spp. on soybeans. Environ. Entomol. **3**:415-419.

Walker, J. T., S. G. Turnipseed, and M. Shepard. 1974. Nymphal development and fecundity of *Geocoris* spp. surviving insecticide treatments to soybeans. Environ. Entomol. **3**:1036-1037.

Warren, L. O., and G. Wallis. 1971. Biology of the spined soldier bug, *Podisus maculiventris* (Hemiptera: Pentatomidae). J. Ga. Entomol. Soc. **6**:109-116.

Whitcomb, W. H., and K. Bell. 1964. Predaceous insects, spiders, and mites of Arkansas cotton fields. Ark. Agr. Exp. Sta. Bull. **690**:84 p.

Chapter 26

Sampling Ground Predators in Soybean Fields[1]

James F. Price and Merle Shepard

I. Introduction

Ground dwelling arthropod predators constitute a major part of the predatory complex in the soybean ecosystem. However, their role in the suppression of pest populations has not been clearly elucidated. Several studies have shown that many species of "ground predators" feed on and may be capable of reducing populations of certain insect pests of soybean. "Ground predators" are defined as those arthropods that spend a portion of their life histories on the soil surface and have the potential to destroy significant numbers of other arthropods in the soil, on the soil surface, or on the soybean plant. This group is composed primarily of certain ground beetles, earwigs, spiders, (Fig. 26-1) and ants.

Most ground predator species are nocturnal. Perhaps this is part of the reason why information about this group as a whole is lacking. Some species have been observed [e.g., the carabid beetle, *Calosoma sayi* Dejean, and the striped earwig, *Labidura riparia* (Pallas)] within the soybean canopy only at night; leaf litter or some other ground habitat usually harbors them during daylight hours. However, a related species of carabid (*Calosoma* sp.) has been observed feeding on velvetbean caterpillars, *Anticarsia gemmatalis* Hubner, in the soybean canopy during the daytime in Brazil (G. G. Newman and S. G. Turnipseed, personal communication).

---

[1] Technical contribution No. 1465 published by permission of the Director, South Carolina Agricultural Experiment Station.

II. Sampling Ground Predators

Sampling ground predators with a specified level of precision has usually been an enigma to most entomologists who have attempted it. Information gained from sampling ground predators may provide essential inputs for a totally integrated pest management program for soybean in the near future, although data obtained from ground predator studies may not be necessary for implementation of a program for management of soybean insect pests at the current level of understanding. Ground predator sampling schemes will certainly establish the relative abundance of these species. Season-long monitoring of their populations along with pest species may reveal certain important predator-pest interactions. Also, response of these predators to insecticide treatments, either by direct action of the chemical or indirectly by reduction of predator food resources, may influence decisions about certain chemical treatments or other pest management tactics.

As with any sampling program, the objectives dictate the details of the procedure. If the objective is simply to establish the presence of a certain predator, a cursory survey of the field may be sufficient. This may be accomplished by walking through the area and examining suspected resting habitats (i.e., leaf litter, etc.) that might harbor certain ground predators. For more detailed information, more refined sampling techniques are required. Often more than one procedure may be necessary, particularly if detailed studies of predator-prey interaction are to be carried out.

A. Sampling Predaceous Ground Fauna with Pitfall Traps

One of the most widely used means of sampling predatory ground fauna is the pitfall trap. It is a modification of an open pit and captures walking and crawling arthropods that attempt to traverse an area where the trap is placed. Excellent reviews of devices used in this kind of trapping are available (Muma 1975, Southwood 1978). Ecological aspects of many ground predators have been studied in soybean and other crops by the use of pitfall traps. Among those arthropods studied are wolf spiders (Whitcomb 1967), earwigs (Walker and Newman 1976, Price and Shepard 1977), carabids (Frank 1971a,b; Kirk 1971; Esau and Peters 1975; Price and Shepard 1978), ants (Wojcik et al. 1972), and other arthropod species (Hensley et al. 1961, Whitcomb and Bell 1964). Extensive accounts of predatory ground fauna from Florida soybean fields were produced through the use of pitfall trapping techniques (Hassee 1971, Neal 1974).

1. Advantages to the Use of Pitfall Traps

Given that the activity and density, and other factors, of species influence catches by pitfall traps, many workers have used this technique and have found that the resulting data are useful for quantitative analysis. Studies of ground beetles of

A

B

C

D

croplands have shown that pitfall traps were well suited for catching carabids (Kirk 1971). Data produced from his work with pitfall traps provided a reliable index of carabid population densities during each species' activity period. The advantages of pitfall trapping have been demonstrated in detecting the presence of species, estimating population densities by mark and recapture techniques, and determining the distribution of a species in a single vegetation type (Greenslade 1964). This method also provides information about the frequency of occurrence of species and may be the only practical method of obtaining carabid population data when densities are low (Greenslade 1964).

Pitfall traps were used in a study that provided evidence that *L. riparia* colonized new soybean habitats more slowly than did their prey (Price and Shepard 1977). Additionally, data derived from pitfall traps indicated that exposure of earwig populations to early season insecticidal treatments resulted in an increase in earwig numbers over those in plots where no chemicals were applied. The same method revealed that adult *C. sayi* immigrated to soybean habitats as lepidopterous larvae became plentiful there. This technique was also used to determine that larval *C. sayi* populations reached a peak density about two weeks following the adult immigration and then developed to the adult stage after two additional weeks in the soybean habitat (Price and Shepard 1978).

Pitfall trapping may be useful in estimating absolute densities of ground arthropods. The rate of capture of predaceous arthropods decreases as individuals are trapped and removed at regular intervals from a population whose size is otherwise constant (Gist and Crossley 1973). A relationship exists between successive capture rates and the total population. In a successful demonstration of this technique, pitfall traps were placed within an impounded area that prevented immigration and emigration of ground arthropods. Captured individuals were removed each day and a regression equation was developed and fitted to the data (Gist and Crossley 1973): $\hat{Y} = \alpha - \beta X$

$$\text{where } \beta = \frac{\Sigma XY - (\Sigma X \, \Sigma Y)/n}{\Sigma X^2 - (\Sigma X)^2/n} , \qquad (26\text{-}1)$$

$\alpha = \bar{Y} + \beta \bar{X}$, X = total number of arthropods removed, Y = number of arthropods captured each day, n = number of traps. The X-intercept, calculated by α/β, estimated the total population within the confinement.

Analyses of variance were used to compare population estimates from pitfall trap data to estimates obtained by hand sorting ground arthropods in similar plots. No significant difference between the two methods was indicated for estimates of spiders, beetles, and other arthropods present in the test plots (Gist and Crossley 1973).

Figure 26-1. (opposite) Common ground-dwelling predators of soybean pests in the southeastern United States: (A) the striped earwig, *Labidura riparia*; (B) a wolf spider, *Pardosa milvina*; (C) a larval caterpillar hunter, *Calosoma sayi*; (D) an adult caterpillar hunter.

The reliability of pitfall traps used for population studies has been shown to be comparable to that from hand sorting or from a Tullgren apparatus (Duffy 1962). Pitfall trapping is also a valuable method in determining the diel periodicity of arthropods (Southwood 1978) and in detecting differences in numbers of arthropods trapped at different times of the year and in different habitats (Williams 1958).

Pitfall trapping is instrumental in revealing elusive aspects of an arthropod's life history. For example, direct observation had indicated the presence on the desert surface of all life stages of the vejovid scorpion, *Anuroctonus phaeodactylus* (Wood). However, pitfall traps collected only males; thus, it was ascertained that only the males wandered (Williams 1966).

Procedures to estimate the numbers of pitfall traps required to derive population data that were reliable within specified bounds have been reported (Muma 1975). It was evident from this work that data from pitfall trapping are most reliable when the data are taken from many pitfall traps, over a long continuous period, and for more than one season. Extensive studies employing pitfall traps have led to the conclusion that these devices produce repeatable ecological data (Muma 1973).

2. Disadvantages of Pitfall Trapping

Data obtained from pitfall trapping programs provide an index over time of arthropod population density, but these data may be confounded by arthropod activity and the attractiveness of the trap (Grum 1959, Mitchell 1963, Greenslade 1964). The influence of these phenomena on trap collection was recognized almost 40 years ago by Williams (1940). He proposed that collections of traps were the product of insect activity and population size. The former component is principally a function of weather conditions. As such, insect activity may vary from hour to hour or day to day according to fluctuations of temperature, rainfall, and other weather parameters. Because populations generally change only slightly between two successive days, the variations in collections between consecutive days are effects of activity changes. By the same logic, differences between collections from two similar periods in different months are more probably due to population changes. Similarly, differences between average collections for the same month in different years may reflect differences in population density (Williams 1940).

The activity of arthropods, and thus collections in pitfall traps, may be further affected by feeding, mating, and egg-laying activities (Briggs 1961), soil moisture (Mitchell 1963), and time of day (Walker and Newman 1976). Differences in activity associated with mating can bias estimates of sex ratios obtained from pitfall data (Muma 1973). Thus, the problem of confounding by locomotor activity can be so great that a population's density may be only a minor function in determining the numbers caught (Briggs 1961). However, the joint action of population density and activity, as measured by a nonattracting or nonrepelling trap, is a measure of encounters with an object on the soil; for instance, a suitable prey. If the number of prey taken by ground predators is a function of individual encounters with prey, as has been modeled (Holling 1959), then this "encounter index" may be able, in some cases, to produce estimates of the net predatory effect of a population regardless of the confounding phenomenon.

Prolonged use of pitfall traps equipped with killing solutions may remove significant numbers of arthropods from a population; thus, it may be that removal of these arthropods by traps may alter their population density. The significance of any alteration should be evaluated, and if found to be important, then measures should be undertaken to keep it to a minimum. Research of several workers has not indicated that the removal of arthropods by pitfall trapping significantly alters populations. No differences were found to exist between numbers of scorpions caught in areas where they were caught and released and in areas where they were killed (Williams 1968). Arachnid populations were not significantly affected by removal sampling (Muma 1975) nor were populations of carabid beetles (Briggs 1961).

Price (1977) obtained similar results in soybean fields when he placed pitfall traps in centers of twelve 0.4 ha plots for 12 weeks, then placed other traps in areas about 40 m from each existing trap for 1 week. No significant differences were noted between numbers of *C. sayi* adults and larvae and *L. riparia* adults and nymphs caught from the heavily sampled areas and the unsampled areas.

Other possible limitations of pitfall trap data for deriving quantitative assessments of carabid fauna have been reported. The level of the soil profile at which traps were placed was found to affect the catches of the traps (Greenslade 1964). If trap openings were level with the surface of leaf litter, then certain species were excluded that would be caught when trap openings were level with the solum. Also, catches of a single species may vary between traps placed in different types of ground cover (Greenslade 1964). A trap placed in unobstructed soil would catch more of one species than a trap in heavily vegetated soil that restricted horizontal movement. Cultivated soybean, however, generally does not have an accumulation of organic debris above the solum except for leaf material that accumulates during the latter part of the growing season. Also, the surface of the soil beneath the plant canopy is usually free of vegetation that could restrict arthropod movement. As a consequence, variation in the composition of the catch from inconsistent positioning of the traps relative to the soil profile or from differences in ground cover is at a minimum in soybean fields.

The attractancy or repellency of a trap may be influenced by baits or behavior-modifying killing and preserving agents (Southwood 1978) or by the presence of signal-emitting species. A test was conducted in a soybean field that had substantial numbers of predatory earwigs, *L. riparia*, and carabid beetles, *C. sayi*, to determine whether traps containing ethylene glycol antifreeze captured greater or fewer numbers of these insects than traps containing a 0.5% detergent water solution (Price 1977). No significant differences in numbers of insects trapped were measured between the traps containing the two solutions. Consequently, the behavior of the two abundant species tested was probably unaffected by the presence of the antifreeze preservative.

3. Implementation of a Ground Predator Sampling Program

Reliable quantitative sampling with pitfall traps requires that many traps be employed. Accordingly, they should be economically and simply constructed. Traps placed in regularly cultivated soybean fields must be removed and replaced each time that cultivation occurs and, therefore, must be designed for quick installation. Pitfall traps that fully meet these requirements and are adapted to the special conditions of soybean fields have not been described previously. It is important also that procedures for recording trapped arthropods and handling resulting data be streamlined.

A sampling program based on the pitfall technique has been developed and implemented at Clemson University, South Carolina, to monitor ground predators in soybean. The program minimizes effort and resources and maximizes the amount of data useful in understanding the role of ground predators in soybean.

Traps are constructed from plastic "cottage cheese" containers (0.47 liter, 11.5 cm diameter rim, and 9 cm diameter base x 7.6 cm depth). This container and similar ones of other depths are available in 0.24 liter and 0.95 liter sizes from delicatessen supply firms. Cups may be placed into an 11 cm diameter hole formed with a lever action golf hole cutter (Fig. 26-2A) equipped with a 17.8 cm deep scalloped blade. A tight fit is formed when the cup is pushed until its rim is even with the soil surface. Since it is necessary to remove the cup to record the catch and since in loose or sandy soil the hole from which a cup has been removed may lose its integrity, it may be necessary to modify the above procedure when trapping in such soils. In that case, a 0.95 liter cup is placed into the hole, then a 0.47 liter cup, whose rim has been removed by cutting with a hot blade, is nested inside (Fig. 26-2B). The smaller cup collects the arthropods and is easily removed from the larger without collapsing the surrounding soil.

One hundred milliliters of ethylene glycol-base antifreeze (Muma 1975) are poured into the trap (Fig. 26-2C). Translucent antifreeze (such as Prestone II) enables clear observation of the contents of the trap. The ethylene glycol quickly kills the arthropods and prevents their decomposition for at least 2 weeks. A single application may be useful for a few weeks because the ethylene glycol does not readily evaporate. Fresh preservative may be required after dilution by heavy rains, after excessive contamination from foreign material, or during peak periods of arthropod activity.

To prevent dilution by rain, the 15.6 cm diameter lid of a no. 10 food can is supported about 3 cm above the trap by two 14 cm nails (Fig. 26-2D). Traps are not easily flooded if placed in the seed row that is usually higher than the row middle.

Studies of time of trap installation have shown that under usual conditions less than 20 sec are required to install a complete trap by this method. Of course soil type and moisture content determine to a large extent the time required. However, one person can install 500 traps in soybean fields, remove and replace traps during cultivation as required, and remove collections from traps each week.

Sampling Ground Predators in Soybean Fields

Figure 26-2. Materials required for pitfall trapping ground predators in soybean: (A) golf hole cutter to form pit, (B) outer cup for use in loose soil and inner cup to collect arthropods, (C) about 100 ml of a translucent ethylene glycol antifreeze to kill and preserve arthropods, (D) a no. 10 can lid placed above the trap to exclude rain.

Records of trap contents may be made at each trap, or arthropods may be removed from traps and counted in a laboratory. The former method may be more practical when few individuals are trapped. In that case arthropods may be removed individually from traps.

To facilitate data collection in the field, appropriate data sheets may be used; IBM coding forms are convenient. The contents of each trap are counted and recorded across the form on a single row. Specific vertical columns are reserved for like entries (e.g., all female *C. sayi*). Column headings need not be written onto each form but may be written once, then cut from the rest of the sheet and taped under a 5 cm x 32 cm plexiglass strip, thus forming a moveable set of column headings usable for each data sheet. This method works well with a legal size clipboard when the clip is used as a guide to align the columns with the column heading strip. Computer data cards can be key punched directly from the coding forms, and if cards are produced after each sampling date, data are then ready for immediate analysis on the last day of the experiment.

Often it may be advantageous to transfer samples from the fields to the laboratory before counting. Examination of pitfall samples in the laboratory results in a more detailed examination and fewer errors. Adverse weather conditions, problems associated with identification, or large numbers of arthropods may make data collection in the field impractical.

Gathering contents of traps in the field may be expedited by the following procedure: individual traps containing the samples are removed from the soil and replaced by another. The contents of each pitfall cup are poured through a straining cup that has a screen mesh small enough to prevent the smallest arthropod of interest from passing through. The ethylene glycol from each sample can be reused in the newly installed trap provided that it has not been diluted nor contaminated. Arthropods screened from each trap can be placed into 0.27 ℓ, styrofoam cups that are easily nested in one another for transport to the laboratory. The samples may be stored in a freezer for examination at a later date.

At the time of analysis, arthropods contained in styrofoam cups are returned to a screen bottom cup and are washed. To aid sorting, contents are then placed into a white enamel pan or similar container filled to about 3 cm with water.

An example of *L. riparia* nymphal collections recorded during a soybean growing season at the Edisto Research Station in Blackville, South Carolina, appears in Fig. 26-3. The data points presented are daily averages from 64 pitfall traps. The 95% confidence interval about each mean indicates the relatively low variation in the number of arthropods collected by traps.

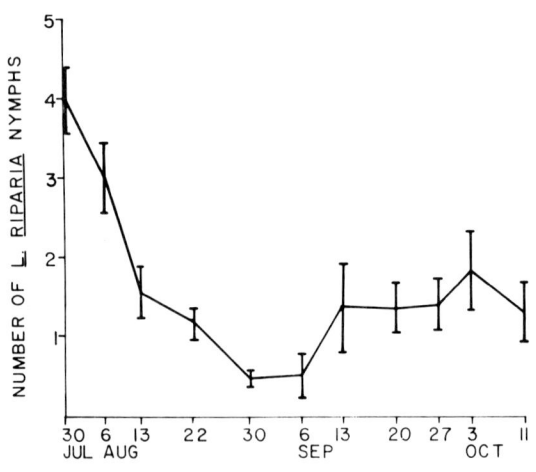

Figure 26-3. Numbers of *Labidura riparia* nymphs from pitfall traps located in a 1 ha soybean field. Data points represent average trap-day collections from 64 traps ± the 95% limits of confidence.

B. Other Methods of Sampling Predaceous Ground Fauna

Absolute populations of many ground predators could be estimated by the quadrat technique as practiced by Briggs (1961), Greenslade (1964), and Mitchell (1963) to determine populations of carabids. Random placement of a rectangular frame of a specified size over a portion of the sampled populations' habitat is employed in this method. The area encompassed by the frame is searched to find all the inhabitants of interest. This process is repeated until a specified total sample area has been searched. From this the total population can be estimated. Discussions of statistical procedures necessary for estimating population size with specified limits of confidence and the number of sampling units required for a sample are described in Chapter 3. The use of a frame for sampling ground predators has not been adapted for soybean fields and would probably be of limited value, largely due to the amount of effort involved in its implementation.

The complex of predators that habitually remain within the soil are not suited to pitfall trap techniques. Sampling of this group involves removing relatively small portions of their habitat and counting arthropods present. Habitat portions are usually removed from the soil by core cutting devices (Waldbauer and Kogan 1973). Hand sorting through core samples is usually sufficient to provide desired information, particularly if the arthropod subjects are large. Washing the entire core sample and resident arthropods through a series of increasing mesh screens is an effective and relatively simple means to remove many small arthropods that could escape detection by hand sorting. More laborious methods are employed to extract the smallest arthropods from soil and litter samples (Macfayden 1962, Edwards and Fletcher 1971). The techniques usually involve forcing arthropods to leave the soil sample by heat (such as used with a Berlese funnel) or by separating passive arthropods (eggs, pupae, dead insects, etc.) from soil and contaminants by flotation techniques.

III. Concluding Remarks

Current sampling techniques used to develop information concerning arthropods inhabiting the soybean canopy have led to significant advances in understanding the ecological role of arthropods inhabiting that stratum. Important among these advances are the establishment of economic injury levels and allied sequential sampling plans that enable sound decisions to be made concerning timing of pesticide applications. The development of efficient and reliable techniques to estimate population parameters for predators dwelling on the soil is an important first step toward bringing this group of arthropods within the focus of more comprehensive pest management programs.

References

Briggs, J. B. 1961. A comparison of pitfall trapping and soil sampling in assessing populations of two species of ground beetles (Col: Carabidae). E. Malling Res. Sta. Annu. Rep. **1960**:108-112.

Duffy, E. 1962. A population of spiders in limestone grassland. J. Anim. Ecol. **31**:571-599.

Edwards, C. A., and K. E. Fletcher. 1971. A comparison of extraction methods for terrestrial arthropods. pp. 150-185 in J. Phillipson, ed. Methods of study in quantitative soil ecology: population, production and energy flow. Blackwell Sci., Publ., Oxford.

Esau, K. L., and D. C. Peters. 1975. Carabidae collected in pitfall traps in Iowa cornfields, fencerows, and prairies. Environ. Entomol. **4**:509-513.

Frank, J. H. 1971a. Carabidae (Coleoptera) as predators of the red-backed cutworm (Lepidoptera: Noctuidae) in central Alberta. Can. Entomol. **103**:1039-1044.

Frank, J. H. 1971b. Carabidae (Coleoptera) of an arable field in central Alberta. Quaest. Entomol. **7**:237-252.

Gist, C. S., and D. A. Crossley. 1973. A method for quantifying pitfall trapping. Environ. Entomol. **2**:951-952.

Greenslade, P. J. M. 1964. Pitfall trapping as a method for studying populations of Carabidae (Coleoptera). J. Anim. Ecol. **33**:301-310.

Grum, L. 1959. Seasonal changes of activity of the Carabidae. Ekol. Polska-Ser. A. **7**:255-268.

Hasse, W. L. 1971. Predaceous arthropods of Florida soybean fields. M. S. thesis, University of Florida. 66 p.

Hensley, S. D., W. H. Long, L. R. Roddy, W. J. McCormick, and E. J. Concienne. 1961. Effects of insecticides on the predaceous arthropod fauna of Louisiana sugarcane fields. J. Econ. Entomol. **54**:146-149.

Holling, C. S. 1959. Some characteristics of simple types of predation and parasitism. Can. Entomol. **91**:385-398.

Kirk, V. M. 1971. Ground beetles in cropland in South Dakota. Ann. Entomol. Soc. Amer. **64**:238-241.

Macfayden, A. 1962. Soil arthropod sampling. Adv. Ecol. Res. **1**:1-34.

Mitchell, B. 1963. Ecology of two carabid beetles, *Bembidion lampros* (Herbst) and *Trechus quadristriatus* (Schrank). II. Studies of populations of adults in the field, with special reference to the technique of pitfall trapping. J. Anim. Ecol. **32**:377-392.

Muma, M. H. 1973. Comparison of ground-surface species in four central Florida ecosystems. Fla. Entomol. **56**:173-196.

Muma, M. H. 1975. Long term can trapping for population analysis of ground-surface, arid-land arachnids. Fla. Entomol. **58**:257-270.

Neal, T. M. 1974. Predaceous arthropods in the Florida soybean agroecosystem. M. S. thesis, University of Florida. 196 p.

Price, J. F. 1977. Arthropod predators of the soil surface in soybeans: Seasonal history, feeding behavior, and response to insecticides and tillage practices. Ph.D. diss., Clemson University. 90 p.

Price, J. F., and M. Shepard. 1977. Striped earwig, *Labidura riparia*, colonization of soybean fields and response to insecticides. Environ. Entomol. **6**:679-683.

Price, J. F., and M. Shepard. 1978. *Calosoma sayi*: seasonal history and response to insecticides in soybeans. Environ. Entomol. **7**:359-363.

Southwood, T. R. E. 1978. Ecological methods with particular reference to the study of insect populations. Halsted Press, N. Y. 524 p.

Waldbauer, G. P., and M. Kogan. 1973. Sampling for bean leaf beetle eggs: Extraction from the soil and location in relation to soybean plants. Environ. Entomol. **2**:441-446.

Walker, J. T., and G. G. Newman. 1976. Seasonal abundance, diel periodicity, and habitat preference of the striped earwig, *Labidura riparia* (Pallas) in the Coastal Plain of South Carolina. Ann. Entomol. Soc. Amer. **69**:571-573.

Whitcomb, W. H. 1967. Wolf and Lynx spider life histories. Terminal Rept. to Nat. Sci. Found. 142 p.

Whitcomb, W. H., and K. Bell. 1964. Predaceous insects, spiders, and mites of Arkansas cotton fields. Ark. Agr. Exp. Sta. Bull. **690**:84 p.

Williams, C. B. 1940. An analysis of four year's captures of insects in a light trap. Part II. The effect of weather conditions on insect activity; and the estimation and forecasting of changes in the insect population. Trans. Royal Entomol. Soc. London **90**:227-306.

Williams, G. 1958. Mechanical time-sorting of pitfall captures. J. Anim. Ecol. **27**:26-35.

Williams, S. C. 1966. Burrowing activities of the scorpion *Anuroctonus phaeodactylus* (Wood) (Scorpionida: Vejovidae). Proc. Calif. Acad. Sci. Fourth Ser. **34**:419-428.

Williams, S. C. 1968. Methods of sampling scorpion populations. Proc. Calif. Acad. Sci. Fourth Ser. **36**:221-230.

Wojcik, D. P., W. A. Banks, D. M. Hicks, and J. K. Plumley. 1972. A simple inexpensive pitfall trap for collecting arthropods. Fla. Entomol. **55**:115-116.

Chapter 27

Sampling Spiders in Soybean Fields

Willard H. Whitcomb

I. Introduction

Spiders, which belong to the arachnid order Araneae, are one of the more abundant groups of arthropods in agroecosystems. Always present in fields, spiders often outnumber predaceous insects in crops. However, the importance of spiders is seldom appreciated by economic entomologists. Araneae are often omitted from surveys of arthropod predators, and when included, usually all species sampled are lumped together under the group "spiders."

All spiders are obligate carnivores, feeding first by liquifying the soft tissues of their prey with enzymes secreted from the oral opening and then sucking the predigested broth into the gut. The mere fact that spiders are predators does not mean that they are entirely beneficial; spiders play four principal roles in the cultivated field (Whitcomb 1974). First, they prey on primary consumers, often destroying large numbers of immatures and adults of many pest species. Second, spiders are natural enemies of predatory insects, many feeding regularly on other entomophages; lady beetles, green lacewings, and various parasitoids are commonly found in theridiid and araneid webs. Third, spiders serve as food for other predators, a fact often overlooked. Araneid eggs and immatures are important prey for other beneficials. Only a few individuals, often less than 20, of hundreds of spiderlings from a single female attain maturity; the remainder serve as an important energy source for a wide variety of predators. The author has observed lacewing larvae feeding on immature green lynx spiders in Florida soybean on many occasions. In the laboratory, lacewing larvae mature quickly on a diet of first instar green lynx spiders (W. Whitcomb et al. unpublished). Fourth, spiders compete with predaceous insects; when prey becomes scarce, this role becomes increasingly important, since predator populations are usually food-limited. As a result of the four different roles of spiders, the same spider species may be highly beneficial in one field and pestiferous in another.

Spiders in an agroecosystem should be sampled in two steps—qualitative and quantitative sampling. Qualitative sampling must be done first, to determine which species are present, since only then can the workers ascertain which quantitative techniques will yield reliable results. The habits of spiders differ so greatly that no one method can sample adequately for all species. Nocturnal sampling with a headlamp (Wallace 1937) can yield good estimates of how many wolf spiders and other cursorial spiders are present, but this method produces poor results in sampling for jumping spiders. Kagan (1943) sampled spiders in Texas cotton fields and reported not a single wolf spider species, even though lycosids are known to be abundant in such fields. Kagan depended on the diurnal examination of plants, whereas wolf spiders are to be found only on the ground in daylight during a Texas summer and are most active at night. The hour of day in which spider populations are sampled is also critical (Gertsch and Riechert 1976) and influences the numbers of each species collected by any given technique. The time of year is important (Berry 1971) because many species are mature for only a few weeks, and immature spiders can only seldom be identified to species. To sample spider populations accurately, one must know what species are present and something about their habits.

Some of the regularly used insect sampling techniques (see Chapter 2) also give a fair idea of what spider species are present. Such approaches must be coupled with more specialized methods such as headlighting, close examination of webs for web symbiotes, and searching every nook and crevice for jumping and other hunting spiders.

II. Qualitative Surveys

Observation of the plant itself is the first step toward learning what spiders are present on the foliage (Whitcomb et al. 1963). A search between rows may disclose orb webs (Fig. 27-1) or tangle webs made by dictynid or theridiid spiders. At times, a spider can be found in the center of its web and, if a collecting vial is placed underneath and the web disturbed, the spider will drop into the vial. More often the spider will be hidden within a folded or dead leaf or other trash in the center or at the web periphery. Practice and patience in searching will lead to the spider. An experienced collector easily spots where to look; silken threads will often lead from the web to the spider's hiding place. Alternatively, the collector can strike a 380-512 Hertz tuning fork and place the vibrating end lightly against the web. If the owner is present, she will soon appear and attempt to wrap sticky silk around the vibrating "prey." This method is particularly useful in collecting black widow spiders (Fig. 27-2), which are numerous in some crop fields in abandoned mouse holes (Gowan, personal communication). Webs are most easily found during the night or early morning when covered by dew

Figure 27-1. Orb web of an araneid spider. (Photo by D. Richman).

(Fig. 27-3). By midmorning, dew will have disappeared and wind will have broken many webs.

The next step is to begin searching at the top of the plant, being careful to disturb it as little as possible. A crab or green lynx spider may be waiting for prey in the plant terminals. Carefully turn the leaves over or look at the undersides. The edge of a folded leaf or a tiny pile of trash beneath may conceal a theridiid female with her egg sac. If the heavy silken sac of a *Phidippus* (Fig. 27-4) or *Metaphidippus* jumping spider is located, place the beating net underneath the spider before disturbing it, for the spider will jump to the ground almost instantly. The same precaution is necessary when pulling apart leaves tied together with silk. The *Thiodina,* anyphaenid or clubionid within will leave quickly. If the fun-

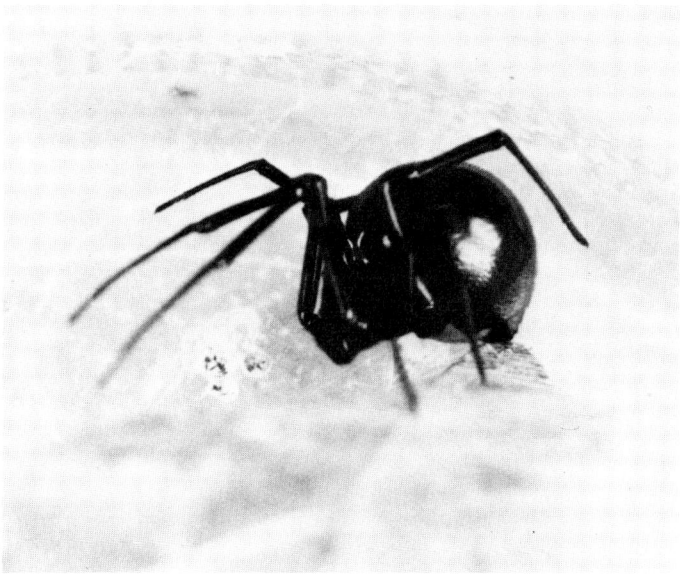

Figure 27-2. Black widow spider [*Latrodectus mactans* (F.)], a ground level web weaver. (Photo by D. Richman).

nel web of an agelenid is found, carefully pinch the escape tube at the bottom of the web closed before attempting to capture the specimen. Look for small harp-shaped tangle webs joining together upper and lower leaves; these are usually made by *Theridula opulenta,* another common theridiid spider in southern soybean fields. Search the main stem carefully, especially where branches join; tiny webs inhabited by theridiids or erigonids may easily be overlooked. Salticids and oxyopids move up and down the main stems and may easily be intercepted there. The base of the main stem, where it leaves the ground, usually has an extensive spider fauna associated with it. If this area is examined early in the morning, dew droplets will reveal the tiny webs of linyphiids, erigonids, and hahniids. Remove carefully any clods of dirt or any other debris found against the stem, which may hide the wolf spider, *Sosippus,* or other spiders.

The arachnologist, when searching for spiders on foliage, often lays out a light-colored cloth sheet and beats the foliage above with a stick, picking up the spiders as fast as they fall. Many cryptic forms not otherwise observed can be collected in this way. In soybean, the standard ground cloth serves the same purpose. One must move quickly, because once a spider has left the cloth, it can seldom be found. Spiders often "play dead," so one must not discard the trash on the cloth for at least five minutes. Ground cloth sampling should also be done at night. Anyphaenids, clubionids, and other cursorials that spend the day in a silken sac (Peck and Whitcomb 1970) and cannot be shaken loose during the day, move about on the foliage at night. The ground cloth technique is excellent

Figure 27-3. Sheet web of a linyphiid spider covered by dew. (Photo by D. Richman).

Figure 27-4. Jumping spider *Phidippus* sp. (Photo by D. Richman).

and, possibly, used more than any other spider sampling technique in soybean, especially where the foliage is heavy. Avoid over-reliance on the ground cloth, however, since the various behavior patterns of spiders require the use of a variety of sampling techniques.

For qualitative sampling of spiders on foliage, a sweep or beating net is essential. Sweeping often discloses species not taken on the ground cloth. Of course, much depends on the individual operator. The sweeper must give the net a special twist so that the spiders land in the net instead of dropping ahead to the ground. The handle should be thick and the cloth should be made of heavy canvas. Early in the season, when foliage is not so heavy, a lighter net may be used to avoid crushing specimens.

A D-Vac or other type of suction apparatus is useful for both qualitative or quantitative sampling of spiders. Heavy soybean foliage is difficult to sample with suction, but it can be as effective in sampling spiders as in collecting insects. Where the foliage is heavy, this suction apparatus is sometimes better than the ground cloth.

Direct examination of the soil surface in search of spiders is inefficient and seldom used except near the base of plants. By moving trash aside, a few spiders may be found, but they can usually be found by more efficient means. Searching under debris on a warm day in winter after the soybeans have been harvested can be productive, however, and can yield valuable information. If one enters the fields at dawn during the growing season, tiny webs across cracks in the soil surface and over lumps of dirt are revealed where the sun strikes the surface. Hahniid spiders are particularly abundant in these small webs in Florida, but other families, especially Linyphiidae, Erigonidae, and Theridiidae, are also represented.

Headlighting is the most effective way of capturing wolf spiders (Lycosidae) on the soil surface (Wallace 1927). For this method, a miner's lamp or similar apparatus is mounted over the eyes, or a flashlight is held at the tip of the nose under the eyes. As the light strikes the eyes of the wolf spider, it produces a gleam like the reflection of blue diamonds. The blue color distinguishes wolf spider eyes from water droplets and broken glass, both of which reflect a white light. Mammal eyes gleam red. Eyes of many lepidoptera are orange. The eyes of one family of neotropical spiders, the heteropodid "banana spiders," gleam red from the side, but are blue if viewed from directly behind or ahead. Once the collector spots the blue glimmer, which may be seen from as far away as 15 m, it is a simple matter to walk down the light beam to the spider and slip a snap-cap vial over it. The bare soil of a soybean field is ideal for this kind of collecting, since there is normally little trash for the spider to hide under. It is difficult to use this method when the soybean canopy is heavy or closed. Below about 16°C, nocturnal activity of most wolf spiders is limited and few can be taken. As the temperature rises, most lycosids become more active. In addition, the author has found the headlighting method particularly useful in searching for fishing spiders (Pisauridae) and giant crab spiders (Heteropodidae). Nocturnal web weavers, particularly large orb weavers, are also readily found by headlighting (although they do not have reflective eyes) because of the heavy dew load found on webs in evening and morning collecting.

Pitfall traps (see Chapters 16 and 26) are important for determining relative numbers of spiders on the soil surface (Muma 1973), although the method does not capture all of the species present there. Pitfalls trap only moving specimens—the faster and more active, the more likely they are to be caught. For this reason lycosids, ground-dwelling clubionids, and gnaphosids are especially abundant in pitfall catches. Web-spinners and other sessile spiders must be captured by other means.

There are almost as many different kinds of pitfall traps as there are researchers using them. For qualitative sampling of spiders in soybean fields, there are certain requirements that must be met. While it is desirable to keep out the rain, the creation of a distinct microhabitat above the trap should be avoided when possible. This means that no roof of metal, foil, or plexiglass or other such material should be used. A square of hardware cloth placed 5 cm above the trap is acceptable. In place of keeping out the rain altogether, some means of draining off rainwater before it reaches the collecting fluid is needed. If this cannot be accomplished easily, an overflow drain that does not permit the loss of specimens is acceptable. However, in this case the fluid must be changed after heavy precipitation. A slight hilling of the soil toward the trap does not affect its efficiency in collecting spiders and lessens the chance of flooding.

The soil next to the trap must not be disturbed when the trap is periodically emptied. For this reason, there should be a metal sleeve or similar device between the soil and collecting jar, so that the used jar can be lifted out and a fresh container substituted with minimal disturbance to the soil. Furthermore, the trap must be small enough to fit into a drill row to avoid damage by cultivation but wide enough so that the larger wolf spiders do not jump over it.

The only type of trap that meets all the above needs and does not require the spider to enter a solid cover is that of Fichter (1941) (Fig. 27-5), in which fine screening in the floor of the trap near the outside wall drains off rainwater without disturbing the collecting fluid. The only cover is over the mouth of the funnel, which the spider does not reach until after it is trapped. Unfortunately, it is difficult and expensive to make this type of pitfall small enough for a drill row. The pitfall designed by Uetz and Unzicker (1976) does not prevent the dilution of the collecting fluid, but is otherwise satisfactory.

The Tullgren apparatus (or Berlese funnel) and the heat table are important tools for the qualitative sampling of spiders. A modern version of the Tullgren apparatus has been described by Uetz and Unzicker (1976). Heat tables consist of a sheet of metal to which heat is applied underneath. Litter is emptied onto one end of the table and gradually worked in small samples over the hot center to the other end of the table. Spiders are collected as they move in response to the warmth. The sheet of metal must not be so hot as to injure the spiders, but hot enough to make them run.

In summation of qualitative sampling, four points must be stressed. First, the soybean field must be sampled at all hours of the day and night. Although this need not be done every week or every month on a regular schedule, it should be

Sampling Spiders in Soybean Fields 551

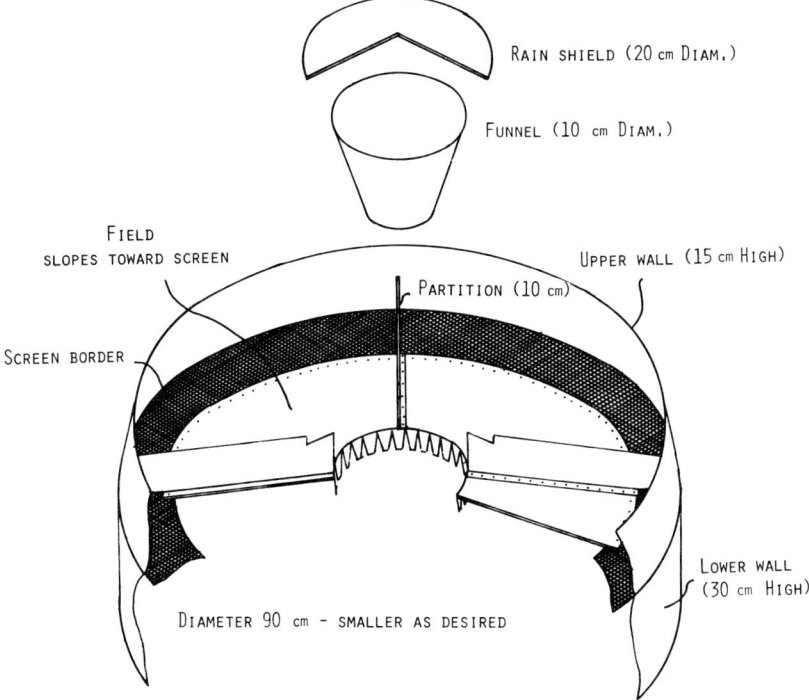

Figure 27-5. Rainproof pitfall trap consisting of a 90 x 45 cm can in two sections, the upper inclined inwards, an attached 5 cm screen border inside the can, a circular field 5 cm higher in the center and divided into four parts by four 10 cm high partitions, with a 10 cm hole in the center modified to hold an insert 50 to 5 cm taper funnel, and a 20 cm conical rain roof (the latter two removable for access to specimen jar below. (Modified from Fichter, 1941).

done often and consistently enough that species that are moving about at odd hours will not be missed completely. Second, the soybean field must be sampled for spiders at least biweekly from the time soybean is planted until harvest, since the species complex changes most during this growing season (Fig. 27-6). Third, the soybean field must be sampled after the growing season, and sampling should be continued at intervals until the field is plowed in the spring, because many spiders do not mature until late fall or early winter; some spiders mainly affect overwintering pest populations. Fourth, many and varied types of fields should be examined. Which species are present is influenced directly or indirectly by many ecological factors. It is difficult to predict which species can be found in a particular field unless a number of soybean fields in the area representing a variety of ecological conditions have been sampled.

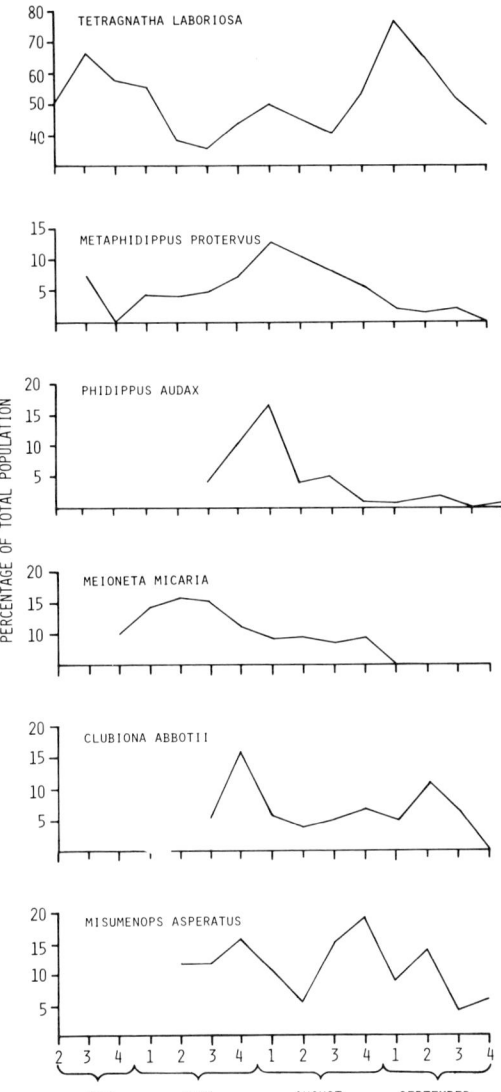

Figure 27-6. Times of colonization and population percentages of the six most common spider species found in Illinois soybean fields during 1975 and 1976. [From LeSar and Unzicker (1978) by courtesy of the authors and the Illinois Natural History Survey].

III. Quantitative Surveys

In quantitative sampling of spiders in soybean fields, the most difficult theoretical and practical problem faced is that of the clumped or clustered pattern of many spider species. In a diverse habitat where the flora and other ecological conditions are discontinuous and vary sharply in different parts of the area, a nonrandom spatial pattern would be expected, and such a condition is described superbly by Gertsch and Riechert (1976). Clumping is not confined to field edges or to other limited areas. In some fields a given species such as the cobweb weaver, *Theridion,* may be evenly dispersed, specimens of other species may be randomly dispersed and will be consistently taken in any large series of samples, while *clumps* of eight or ten specimens of other species will be taken in each of two or three widely separated samples. The causes of clustering are diverse, but the presence of prey especially suitable for a given spider species appears to be important in determining spatial patterns.

Because most species of spiders are not randomly dispersed, a precise estimate of the number of individuals per hectare would require a far greater number of samples than is generally feasible. For these nonrandomly dispersed species, quantitative results must be given as the range in numbers of individuals per hectare.

To determine the number (n) of samples needed to be taken for a fixed level of precision (to achieve a standard error, $S_{\bar{x}}$ that is D% of $\bar{x}$, the mean distribution in a preliminary sample), the author uses a modification of Southwood's (1978) formula $n = (t\, S_{\bar{x}}/D\, \bar{x})^2$ by Elliott (1971):

For species that tend to clump or aggregate,

$$n = \frac{1/\bar{x} + 1/k}{D^2} \qquad (27\text{-}1)$$

where k is the sample parameter of the negative binomial distribution (see Chapter 3).

For species that tend to have a random spatial pattern,

$$n = \left(\frac{S_{\bar{x}}^2}{D\,\bar{x}}\right) \qquad (27\text{-}2)$$

is used.

The tools most frequently employed in quantitative sampling of spiders in soybean are the ground cloth and the D-Vac. Both of these are excellent for sampling foliage for spiders. The author prefers the ground cloth in heavier foliage. These methods are effective for sampling a majority of species from most of the families occurring in soybean.

In a study conducted in central Illinois, LeSar and Unzicker (1978) compared the efficiency of the ground cloth, the sweep net, and the D-Vac in sampling spider populations in soybean fields. The sweep net yielded 34% fewer spiders and the ground cloth 19% fewer spiders than did D-Vac sampling. However, there was a significant correlation (r = 0.864) between ground cloth and D-Vac catches when numbers of spiders from sample units taken from 5.25 row m were compared (Fig. 27-7). It is, therefore, possible to establish a sampling program based on ground cloth methods that is comparable to a program based on D-Vac. In either case adequate calibration of the relative sampling methods by comparison to an absolute method is necessary.

Pitfall traps do not measure the absolute number of arthropods per hectare.

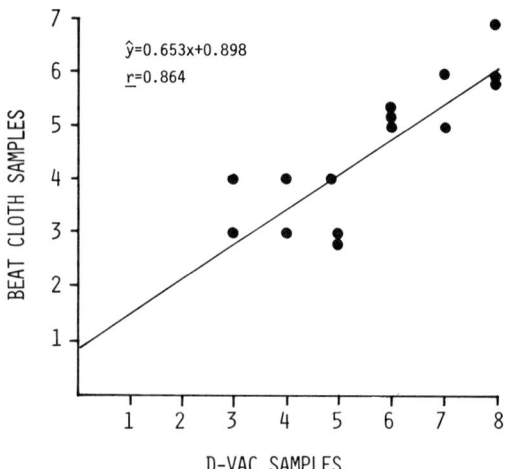

Figure 27-7. Correlation between the number of spiders collected by sweep net and D-Vac in soybean fields in Illinois. [From LeSar and Unzicker (1978) by courtesy of the authors and the Illinois Natural History Survey].

Only the quadrat method can do this with accuracy for soil-dwelling arthropods. Even comparisons of population densities from pitfall data can be of questionable validity, since the number of individuals trapped depends on activity as well as density (Greenslade 1964). However, Mitchell (1963), Harris and Whitcomb (1973), and Kay (1977) found that such comparisons could be useful for determining spatial patterns of highly mobile arthropods, especially in dense or similar vegetation types. Pitfall traps are quite effective in sampling lycosids, gnaphosids, pisaurids, and some clubionids.

To obtain estimates of the number of lycosids per hectare in soybean, the author delimits 1/10 hectare quadrats with fluorescent tape. Wolf spiders are then collected at night by headlighting. This method is only effective for those lycosids that do not tunnel or bury themselves in the ground.

The Tullgren apparatus and heat table are useful as long as the area from which the litter sample is taken is carefully measured. These methods are especially useful in quantitative sampling of oecobiids, some dictynids, some pholcids, some theridiids, some linyphiids, micryphantids, gnaphosids, and some clubionids and salticids.

For the sampling of some families in special situations, new methods must be developed or old ones specially adapted.

IV. Vertical Stratification

The following account on the vertical stratification of soybean spiders is based on a study conducted by LeSar and Unzicker (1978) in Illinois. It demonstrates the variation in species composition and density with time of collection and the stratum within the soybean plant canopy. The study was conducted with a D-Vac provided with a 22 cm diameter cone head. A 15.75 m length of row constituted a sample unit and the D-Vac head was kept at four levels of the canopy of ca. 1 m high soybean plants. Samples were taken from different rows at each canopy level to avoid the drifting of spiders from upper to lower levels due to the disturbance caused by the D-Vac. Samples were taken during four times of day: 0 800-1000, 1100-1 400, 1600-1800, and 2 000-2 100 hr.

Differences in number of spiders collected by time period and canopy stratum varied with species. Some species showed no quantitative differences in time, while others were more often collected at a certain time. Also, some species were randomly dispersed within all strata of the canopy, others were concentrated in certain zones. The location of 12 of the most abundant spider species on soybean plants is shown in Fig. 27-8. Table 27-1 shows that nearly 70% of all spiders collected at any time of the day were removed from the lower half of the canopy. This concentration suggests that prey abundance is probably greater on the lower portions of the plant during most of the day. This study revealed also that certain spiders seemed to prefer zone D (the upper fourth of the canopy) and moved down during the warmer times of the day.

These findings show how critical it is to develop sampling procedures that reliably account for these vertical microspatial patterns of spiders.

V. Concluding Remarks

Identification of spiders to species is time consuming. The few taxonomists in this country who are willing to identify spiders are already overburdened. There is no excuse for any competent entomologists not having the ability to identify the more common spiders to family. Kaston's *How to Know the Spiders* (1978) and Levi's *Spiders and Their Kin* (1968) are especially helpful. There is a workable key to spider families in both Kaston's book and the textbook by Borror et al. (1976). Once the spiders are keyed as nearly as possible to family, one can then ask an arachnologist to which specialists the different families should be sent. Keep on hand a synoptic collection of those already identified for reference, to avoid sending the same species in repeatedly. Between many groups, the differences are so fine that a careful examination of the female epigynum or the male palps is necessary to distinguish the species.

When the study is completed, material representatives of each of the spider species studied should be deposited in a permanent collection (such as in a museum) where it can be cared for. The taxonomy of spiders is still in some flux, and in 10 or 20 years specialists should be able to determine with what species you were working if the specimens are still available, even though the nomenclature may have changed.

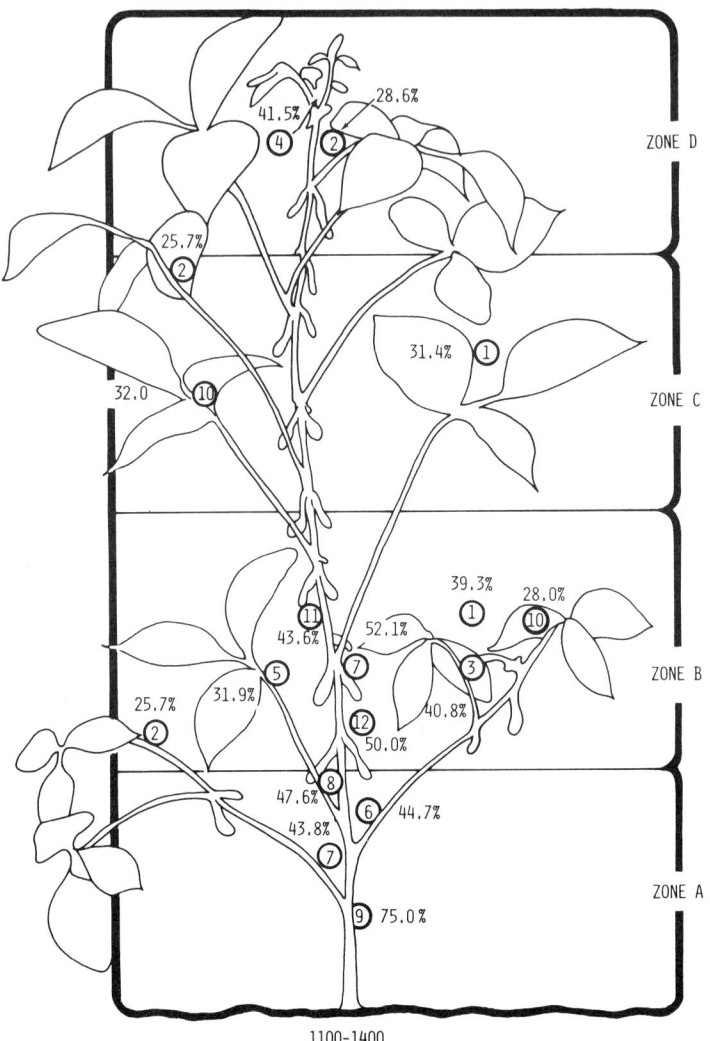

Figure 27-8. Locations of the 12 most abundant spider species on soybean plants during the 1100-1400 time period and the percentage found in each zone of the total population of each species. The species are: 1. *Tetragnatha laboriosa*, 2. *Misumenops asperatus*, 3. *Clubiona abbotii*, 4. *Philodromus aureolus*, 5. *Phidippus audax*, 6. *Metaphidippus protervus*, 7. *Microlinyphia pusilla*, 8. *Theridion neshamini*, 9. *Pardosa milvina*, 10. *Tibellus oblongus*, 11. *Oxyopes salticus*, 12. *Mimetus epeiroides*. [From LeSar and Unzicker (1978) by courtesy of the authors and the Illinois Natural History Survey].

Table 27-1. Vertical dispersion on soybean plants of all spiders collected from fields in Illinois. Zone A is at the bottom of the plant, and Zone D is at the top (see Fig. 27-8). (From LeSar and Unzicker 1978)

Time	Percentages of spiders collected in			
	Zone A	Zone B	Zone C	Zone D
0 800-1 000	35.7	32.7	21.9	9.7
1 100-1 400	38.4	32.8	21.9	6.9
1 600-1 800	35.5	33.6	20.6	10.3
2 000-2 100	39.3	26.0	20.3	14.4

References

Berry, J. W. 1971. Seasonal distribution of common spiders in the North Carolina piedmont. Amer. Midland Natur. **85**:526-531.

Borror, D. J., D. M. DeLong, and C. A. Triplehorn. 1976. An introduction to the study of insects. 4th ed. Holt, Rinehart and Winston, N. Y. 852 p.

Elliott, J. M. 1971. Some methods for the statistical analysis of samples of benthic invertebrates. Freshwater Biol. Assoc. Sci. Paper 25.

Fichter, E. 1941. Apparatus for the comparison of soil surface arthropod populations. Ecology **22**:338-339.

Gertsch, W. J., and S. E. Riechert. 1976. The spatial and temporal partitioning of a desert spider community, with descriptions of new species. Amer. Mus. Novitates **2604**:1-25.

Greenslade, P. J. M. 1964. Pitfall trapping as a method for studying populations of Carabidae (Coleoptera). J. Anim. Ecol. **33**:301-310.

Harris, D. L., and W. H. Whitcomb. 1973. Effects of fire on populations of certain species of ground beetles (Coleoptera: Carabidae). Fla. Entomol. **57**: 97-103.

Kagan, J. 1943. The Araneida found on cotton in central Texas. Ann. Entomol. Soc. Amer. **36**:257-258.

Kaston, B. J. 1978. How to know the spiders. 3rd ed. Wm. C. Brown Co., Dubuque, Iowa. 272 p.

Kay, C. A. R., J. N. Veazey, and W. H. Whitcomb. 1977. Effects of date of soil disturbance on numbers of adult field crickets (Orthoptera: Gryllidae), in Florida. Can. Entomol. **109**:721-726.

LeSar, C. D., and J. D. Unzicker. 1978. Soybean spiders: Species composition, population densities, and vertical distribution. Ill. Natur. Hist. Surv. Biol. Notes **107**:14 p.

Levi, H. W., L. R. Levi, and H. S. Zim. 1968. Spiders and their kin. Golden Press, N. Y. 160 p.

Mitchell, B. 1963. Ecology of two carabid beetles, *Bembidion lampros* (Herbst) and *Trechus quadristriatus* (Schrawk). II. Studies on populations of adults in the field, with special reference to the technique of pitfall trappings. J. Anim. Ecol. **32**:377-392.

Muma, M. H. 1973. Comparison of ground surface spiders in four central Florida ecosystems. Fla. Entomol. **56**:173-196.

Peck, W. B., and W. H. Whitcomb. 1970. Studies on the biology of a spider, *Chiracanthium inclusum* (Hentz). Ark. Agr. Exp. Sta. Bull. **753**:1-76.

Southwood, T. R. E. 1978. Ecological methods with particular reference to the study of insect populations. Halsted Press, N. Y. 524 p.

Uetz, G. W., and J. D. Unzicker. 1976. Pitfall trapping in ecological studies of wandering spiders. J. Arachnol. **3**:101-111.

Wallace, H. K. 1937. The use of the headlight in collecting nocturnal spiders. Entomol. News **48**:107-111.

Whitcomb, W. H. 1974. Natural populations of entomophagous arthropods and their effect on the agroecosystem. pp. 150-169 in F. G. Maxwell and F. A. Harris, eds. Proceedings of the Summer Institute on Biological Control of Plants Insects and Diseases. University Press of Mississippi, Jackson. 647 p.

Whitcomb, W. H., H. Exline, and R. C. Hunter. 1963. Spiders of the Arkansas cotton field. Ann. Entomol. Soc. Amer. **56**:653-660.

Chapter 28

Sampling Pathogens of Soybean Insect Pests

Gerald R. Carner

I. Introduction

Pathogens are generally considered to be one of the most important groups of natural regulating agents of arthropod pests of soybean, especially for the major species of lepidopterous pests. Dramatic epizootics of fungal pathogens on soybean pests were reported as early as 1931 when *Spicaria prasina* [=*Nomuraea rileyi* (Farlow)] was observed controlling late season infestations of the "soybean caterpillar" (*Anticarsia gemmatalis* Hübner) in Louisiana (Hinds and Osterberger 1931). Because of their ability to rapidly reduce insect pest populations, and because the visible evidence of their effects on insects is so obvious in fields where epizootics are in progress (dead larvae adhering to plants), pathogens have long been recognized for their potential as insect control agents in soybean.

Despite the recognized importance of these organisms in the soybean ecosystem, very little emphasis has been placed on development of standardized methods for sampling entomopathogens. Most field studies with insect pathogens have been directed toward determining their impact on pest populations. Live larvae have been collected from the field and held in the laboratory to determine disease incidence. Other studies have involved the sampling of foliage, soil, or air to recover the form of the pathogen that occurs outside the host.

The purpose of this chapter is to present a brief summary of background information on the major insect pathogens encountered in soybean, followed by a discussion of the various sampling methods that have been employed to determine the incidence of these pathogens in the soybean ecosystem. Techniques for the diagnosis of these insect pathogens are beyond the scope of this chapter but are available elsewhere (Weiser and Briggs 1971, Weiser 1977, Poinar and Thomas 1978).

A. Fungi

1. *Nomuraea rileyi* (Farlow) Samson

Fungi are the most obvious and widespread pathogens found in soybean fields, and *N. rileyi* is the predominant species of this group. This fungus has been reported from all the major soybean-producing areas of the United States and appears to be worldwide in its distribution. Allen et al. (1971) determined that *Spicaria rileyi* (=*N. rileyi*) was the major pathogen of *A. gemmatalis* in Florida, and Sprenkel and Brooks (1975) and Carner et al. (1975) found it to be the most abundant insect pathogen in soybean fields in North Carolina and South Carolina, respectively. Other studies have shown it to be a major component of the pathogen complex in Louisiana (Burleigh 1972), Missouri (Ignoffo et al. 1975), Alabama (Harper and Carner 1973), Brazil (Kogan et al. 1977), midwestern United States (Pedigo et al. 1973), and Australia (M. Shepard unpublished).

A possible explanation for the relative abundance of *N. rileyi* in soybean is that it affects a wide range of hosts including all the major lepidopterous pests. Puttler et al. (1976) list nine susceptible hosts. *N. rileyi* will usually build up in the field on early season pests such as the green cloverworm, *Plathypena scabra* (F.), and will maintain a high level of inoculum for infection of pest populations such as *Heliothis* spp., *Pseudoplusia includens* Walker, and *A. gemmatalis* as they appear later in the season. Larval cadavers resulting from *N. rileyi* infection will either be white or green depending on the stage of development of the fungus. As conidiophores develop on the outside of the larvae, they form a white mat covering the entire larva except for the head capsule (Fig. 28-1). If weather conditions are suitable this mat will change to a light green as spores develop on the cadaver.

More studies have been conducted with *N. rileyi* than with any other insect pathogen of soybean; the most extensive of these was by Kish and Allen (1978) in Florida. They conducted a detailed study of the environmental factors affecting the development of this fungal pathogen and constructed a model for predicting incidence of *N. rileyi* in *A. gemmatalis* populations.

Several recent studies have demonstrated that *N. rileyi* may have potential for use as a microbial insecticide. Sprenkel and Brooks (1975) in North Carolina and Ignoffo et al. (1976a) in Missouri were able to induce epizootics of this pathogen by field applications of the fungus. Also, Ignoffo et al. (1976b) found that certain strains of the fungus (e.g., a strain from Brazil) were much more pathogenic to some lepidopterous hosts than other strains, suggesting that strains can be selected for effectiveness in various geographical locations and

Figure 28-1. Cadaver of *Anticarsia gemmatalis* covered with conidiophores of *Nomuraea rileyi*.

with different pest complexes. *N. rileyi* is an effective pathogen of all of the major lepidopterous pests of soybean and plays an important role as a natural control agent.

2. Entomophthora gammae Weiser

This fungus is more limited in its host range than *N. rileyi*, affecting only larvae of *P. includens* and *Trichoplusia ni* (Hübner). Harper and Carner (1973) found this to be the predominant pathogen of soybean loopers in Alabama with infection levels of 70-100% at high population levels. In Louisiana, Burleigh (1972) observed that this fungus was important only when larval numbers reached high levels. Epizootics of this pathogen have also been reported from South Carolina (Newman and Carner 1975b) and Brazil (S. Turnipseed unpublished).

Cadavers resulting from *E. gammae* infection will take on two distinctly different forms depending on which spore form is present. Infection by the conidial form of the pathogen results in the looper becoming a discolored, shriveled cadaver on which external conidia are produced and forcibly ejected (Fig. 28-2A). The resting spore form results in the infected looper become a black (Fig. 28-2B), leathery cadaver filled with thick-walled resting spores (Newman and Carner 1975). Both forms are present during an epizootic with the resting spores becoming more predominant toward the end of the epizootic.

Figure 28-2. Cadavers of *Pseudoplusia includens* killed by *Entomophthora gammae:* (A) infection by the conidial form of the pathogen, (B) infection by the resting spores.

3. Other Species of *Entomophthora*

Numerous other species of *Entomophthora* have been observed affecting other soybean pests, but little is known of their abundance, frequency of occurrence, or impact on pest populations. Carner et al. (1979) reported *Entomophthora* spp. from *P. scabra* (Fig. 28-3), *Heliothis* spp. (Fig. 28-4), and *A. gemmatalis* (Fig. 28-5) in South Carolina. Other hosts affected by various species of *Entomophthora* were the Mexican bean beetle, southern green stink bug, green stink bug, and saltmarsh caterpillar (D. Montross unpublished). In Illinois numerous grasshoppers, *Melanoplus* spp., were killed by an *Entomophthora* sp. during an outbreak on soybean in 1978 (M. Kogan unpublished) (Fig. 28-6).

4. *Beauveria* spp.

Although this fungus is not a major pathogen of soybean insects, it has been observed at several locations affecting soybean pests. Pedigo et al. (1973) reported that *Beauveria bassiana* (Balsamo) was an important factor in the collapse of

Sampling Pathogens of Soybean Insect Pests 563

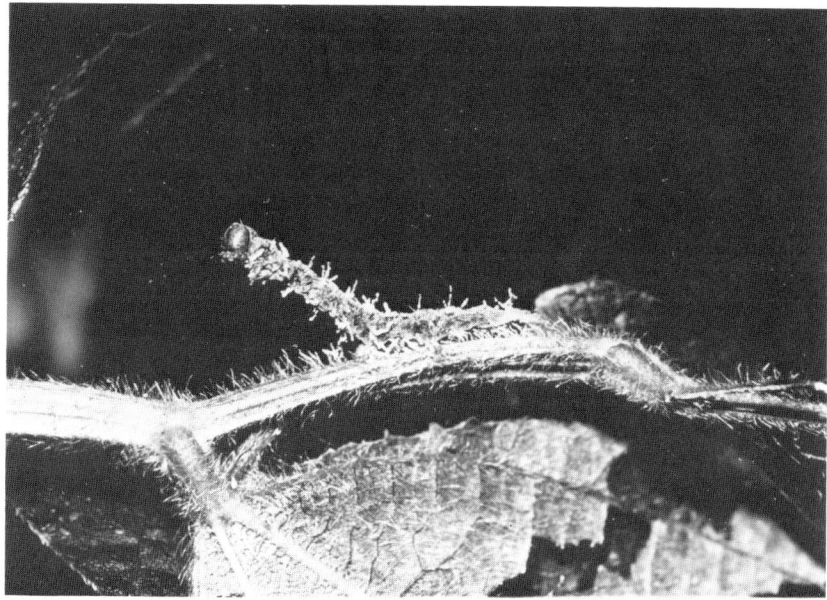

Figure 28-3. *Entomophthora* sp. infected green cloverworm, *Plathypena scabra*, larva.

outbreak populations of *P. scabra* in Iowa. This fungus has also been observed affecting southern green stink bugs and larvae of *A. gemmatalis* in Brazil and *A. gemmatalis* larvae in Louisiana.

Figure 28-4. *Entomophthora* sp. infected *Heliothis* sp. larva.

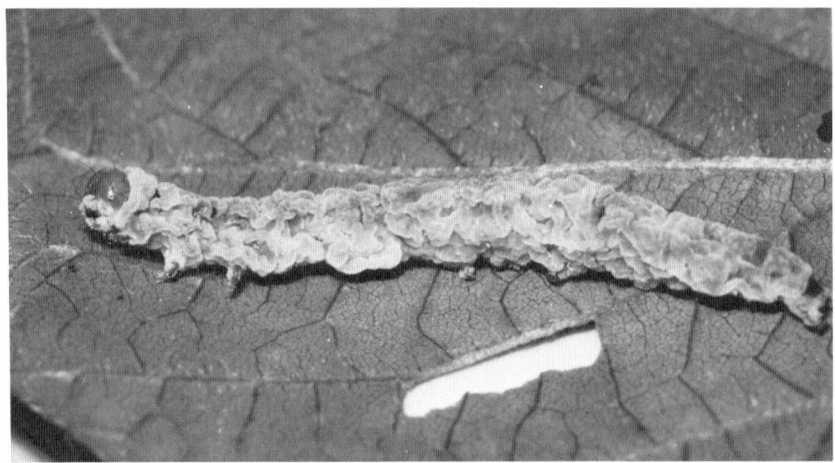

Figure 28-5. *Enthomophthora* sp. infected velvetbean caterpillar larva.

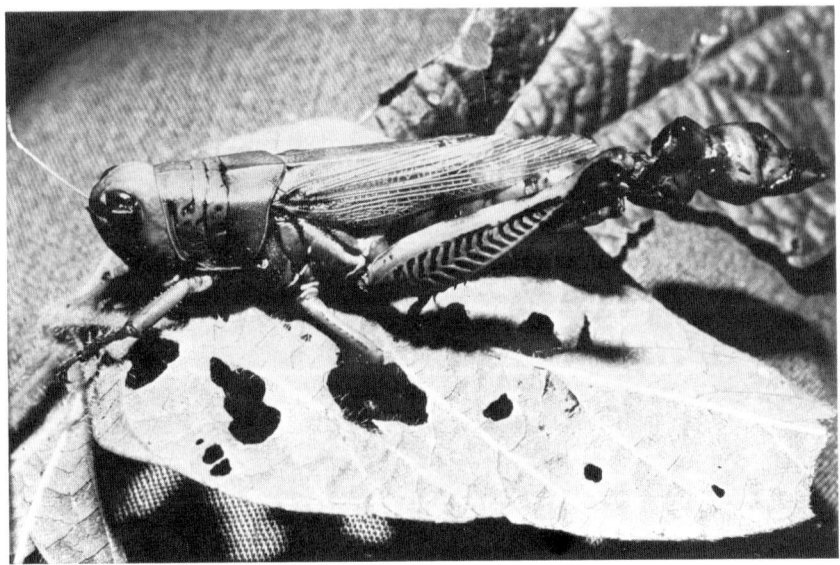

Figure 28-6. *Enthomophthora* sp. infected Melanoplus sp. (Courtesy of M. Kogan).

B. Viruses

1. *Heliothis* Nuclear Polyhedrosis Virus (NPV)

This virus has been studied extensively on other crops, and has been tested successfully as a microbial insecticide (Elcar) on soybean (Fig. 28-7). However, there have been few reports of naturally occurring epizootics of this virus in soybean. During 1977 and 1978 in South Carolina, *Heliothis* populations built up to high levels early in the season when *N. rileyi* was not present in the fields. During both years, these *Heliothis* populations were decimated by NPV epizootics, so the virus has become a major factor in the natural regulation of *Heliothis* populations.

2. *P. includens* NPV

This virus was first reported by Livingston and Yearian (1972) from soybean looper larvae collected in Central America. The virus has been effective in field tests against loopers in Arkansas and Louisiana (Livingston et al. 1979) and has become established in this region as a permanent natural control agent (W. Yearian unpublished).

Figure 28-7. Nuclear polyhedrosis virus of *Heliothis* sp.

3. *A. gemmatalis* NPV

No virus has been reported from velvetbean caterpillar populations in the United States, but an NPV does occur naturally in populations in Brazil (Gatti et al. 1977). The virus was described by Allen and Knell (1977) and Carner and Turnipseed (1977) and has been tested successfully as a microbial insecticide in Florida and South Carolina. The virus is an extremely virulent pathogen with an LD_{50} for crude preparations of 5 polyhedral inclusion bodies (PIB) per larva (Carner et al. 1979). Studies are currently being conducted to determine if the virus can become established in soybean ecosystems of the southeastern United States.

4. *P. scabra* Granulosis Virus

This virus has been reported from green cloverworm populations in Iowa (Pedigo et al. 1973) and South Carolina (Carner and Barnett 1975). Although it is not a major mortality factor in *P. scabra* populations, infection levels reached as high as 28% in some fields in South Carolina.

C. Other Pathogens

Bacterial pathogens have been studied very little in soybean and there are no reports of naturally-occurring bacterial pathogens in soybean insect pests. *Bacillus thuringiensis* Berliner is an effective pathogen of most lepidopterous pests of soybean and is recommended for control of *P. scabra, A. gemmatalis,* and loopers. Some studies with *B. thuringiensis* have involved sampling to determine the persistence of this pathogen in the field.

Protozoan pathogens have likewise received very little attention. *Nosema heliothidis* Lutz and Splendore and *Vairimorpha necatrix* (Kramer) are two microsporidian pathogens that infect *Heliothis* spp. larvae. *N. heliothidis* occurs naturally in larval populations and *V. necatrix* has shown promise as a short-term microbial control agent due to its virulence and relatively wide host range (Maddox 1966). Another microsporidian pathogen has been observed in larvae and adults of the Mexican bean beetle in North and South Carolina (W. Brooks unpublished). Field populations of *Acrosternum hilare* (Say) and *Cerotoma trifurcata* (Forster) in Illinois are frequently infected with microsporidia (J. Maddox personal communication).

II. Sampling for Disease Incidence in the Hosts

In most studies to determine disease incidence in field populations, the standard procedure has been to collect live insects from field populations and hold them in the laboratory on leaves or on an artificial diet until they die or complete their development. This type of sample will give an indication of the level of infection in the field population on the day the sample was taken. If these samples are taken periodically during the growing season, general patterns of enzootic and epizootic development of a pathogen in a field population can be followed.

Timing of these samples can be extremely important in accurately tracing the progress of an epizootic. Generally, when pest populations are low to moderate, weekly samples are sufficient to determine disease spread in a population. However, when pest populations begin a rapid increase and the pathogen is known to be present in a population, development of an epizootic can occur very rapidly. At this stage, sampling every few days may be necessary to keep track of developments in the epizootic. Figure 28-8 represents a typical population curve for an insect population affected by a disease epizootic. The population generally goes through a period of gradual increase during which the pathogen may be present at low levels. As the population begins its period of exponential increase, incidence of the pathogen will generally increase. However, there is usually a lag time between the increase in insect numbers and development of the epizootic. Because of this lag period, pest populations will often exceed the economic threshold before the effect of the pathogen can be realized.

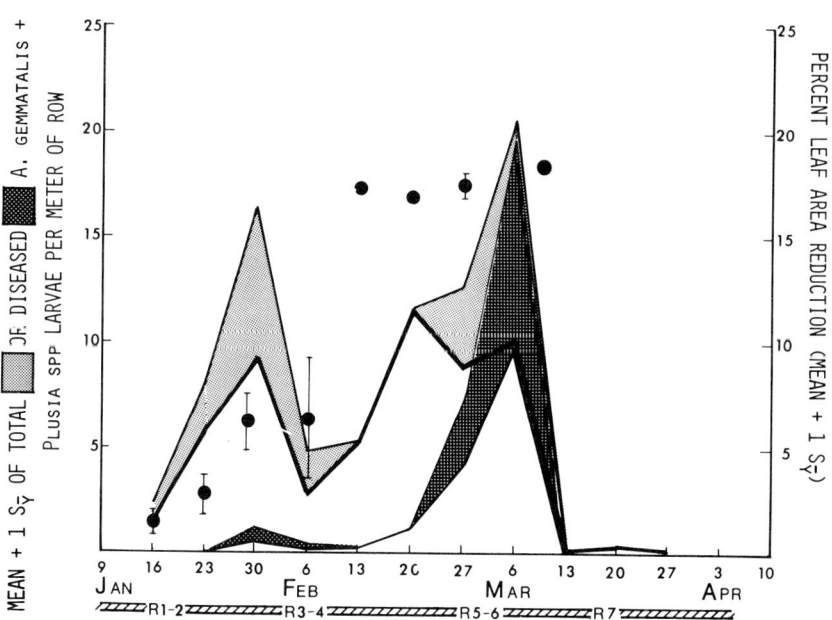

Figure 28-8. Fluctuation of populations of *A. gemmatalis* + *P. includens* on soybean in Brazil. The dark-shaded curve of infected caterpillars shows the delay in the epizootic that allowed populations to increase and produce the levels of defolation shown by the black dots. (Redrawn from Kogan et al. 1977).

When the pathogen reaches epizootic proportions, population levels of the pest will decrease dramatically. With certain pathogens such as *Entomophthora gammae* and some of the nuclear polyhedrosis viruses, this drop can occur within a period of 2-3 days. Thus, timing of samples can be crucial at this stage of the epizootic.

Sampling of live insects to determine disease incidence has several disadvantages. Many of the sources of bias already mentioned in Chapter 24 also apply to sampling for pathogens. These include: (1) bias due to interruption of exposure to pathogens, (2) bias due to temporal variation of infection, (3) bias due to spatial variation, (4) bias due to variation of infection with host age, (5) bias due to change in host longevity, and (6) bias due to competition with other diseases and parasites.

One other source of bias that applies specifically to pathogens is the problem of cross-contamination from pathogens during handling of samples. When live insects are collected in the field they are subjected to handling procedures that may increase their chances of becoming infected with certain pathogens. Collection techniques such as the sweep net and ground cloth bring larvae in close contact with each other and with debris from the plants. Insects in late stages of infection or insect cadavers in the plant debris may be sources of infection for healthy larvae collected in the samples. Also, the shaking of the plant during sampling may disseminate spores of fungi and increase chances of infection. If the live insects are transported together in a container to the laboratory, chances of cross-contamination are increased further, and if field-collected leaves are used as a food source for the larvae in the laboratory, the leaves present another possible source of contamination from pathogens.

Several steps can be taken to decrease the chances of contamination in field samples of larvae. A common practice is to utilize mortality data for only a specific period of time after the sample is taken. This period would depend on the development time of the pathogen at a specific temperature. Kish and Allen (1978) and Johnson et al. (1976) used this procedure for sampling *N. rileyi* in populations of the velvetbean caterpillar. Only fungal mortality that occurred during the first five days after collection was considered in determining disease incidence in the field. Any mortality that occurred after five days was considered to be a consequence of collection methods or laboratory contamination.

When samples contain several types of pathogens, each pathogen must be considered separately because development times will differ. For example, viruses that infect soybean Lepidoptera, will usually kill their host in a shorter period of time than will *Nomuraea*. Another disadvantage of using this procedure is that the development times for some pathogens (e.g., some species of *Entomophthora*) are not known.

Another procedure that will decrease contamination of field samples is separation of larvae as they are collected in the field instead of transporting the larvae together in a container to the laboratory. This procedure was evaluated in South Carolina by collecting loopers from a field in which an epizootic of *E. gammae* was occurring and collecting *Heliothis* larvae from a population heavily infected with virus. From each field one-half of the larvae were placed together in 3.5 ℓ cardboard cartons and were separated into individual containers after being taken to the laboratory. The rest of the larvae collected from each field were placed individually into petri dishes as they were collected. Infection by *E. gammae* was not increased when larvae were individually collected or held together. However, average infection from virus in *Heliothis* larvae was somewhat higher in larvae that were placed together in cartons.

Ignoffo et al. (1976a) used a different method to check for contamination during collection of live larvae from the field. In their study that involved monitoring of natural and induced epizootics of *N. rileyi*, whole soybean plants were cut and placed in plastic bags for transportation to the laboratory. Recovered larvae (*T. ni* and *P. scabra*) were then reared individually on a semi-synthetic diet. To assess contamination in the plastic bags, laboratory-reared first instars of *Heliothis zea* Boddie were added to each plastic bag at the time the plants were collected. These larvae were handled in the same way as the field-recovered larvae. None of these *H. zea* larvae became infected with *N. rileyi* and the authors concluded that contamination during collection procedures was not a problem with *N. rileyi*.

If field-collected leaves are used as a food source for larvae held in the laboratory, they should be washed in a 0.05% solution of sodium hypochlorite (Ignoffo and Dutky 1963) and rinsed thoroughly in distilled water. Ignoffo et al. (1975) and Kish and Allen (1978) have shown that soybean leaves are one of the primary sources of inoculum for *Nomuraea* and numerous studies have shown that contaminated leaves are the main source of infection for insect viruses.

In a test conducted in Missouri, field-collected soybean leaves were fed to first instar laboratory-reared *T. ni*. An average of 53% of these larvae became infected with *N. rileyi*. When leaves from the same location were treated with sodium hypochlorite, fungal infection was prevented completely (Ignoffo et al. 1975).

Another source of bias unique to sampling for pathogens is that some pathogens will alter the behavior of the host in such a way that infected larvae are collected at a different rate than uninfected larvae. Fungi in the genus *Entomophthora* are generally known to cause their host, during late stages of infection, to climb to the tops of plants (MacLeod 1963). Nuclear polyhedrosis viruses will cause the same behavior in infected hosts. This is generally considered to be of adaptive value to the pathogen since this behavioral change aids in the dissemination of the pathogen after the host has died.

This movement of the larvae to the tops of the plants may result in a greater proportion of infected larvae being collected in the samples than would normally be present in the population. This would especially be true if sweep net sampling were used. Newman and Carner (1975b) found that a greater proportion of soybean loopers collected by the sweep net method were infected with *Entomophthora gammae* than were loopers collected by the ground cloth method.

III. Sampling for Disease Agents in the Environment

A. Sampling from Foliage

Contaminated soybean foliage has been shown to be a major source of inoculum for most of the major pathogens of soybean insects. Field-collected leaves will induce fungal infection when fed to laboratory-reared larvae (Ignoffo et al. 1975, Kish and Allen 1978). Foliage also serves as a source of inoculum for viral pathogens (Ignoffo et al. 1974, Young and Yearian 1974). Because of the importance of soybean leaves in the infection cycle, a number of field studies have utilized leaf-sampling techniques to trace the build-up of a pathogen in soybean fields. Other studies have used leaf-sampling methods to follow the persistence of a pathogen on foliage after spray applications. In all these studies, pathogen levels were determined by bioassay of leaf samples with susceptible laboratory-reared hosts.

Fuxa and Brooks (1978) and Young and Yearian (1974) used a leaf-disc bioassay to trace the persistence of *V. necatrix* spores and *Heliothis* NPV on soybean foliage. Ignoffo et al. (1974, 1975) used whole leaves against first instars to sample for *B. thuringiensis, Heliothis* NPV, and *N. rileyi*. In addition, aliquots of the washings of soybean leaves treated with *B. thuringiensis* were assayed for viable spores by using standard pour-plate techniques. Gardner et al. (1977) used whole leaves against third instar hosts to bioassay for persistence of *B. bassiana, N. rileyi,* and *V. necatrix*.

B. Sampling from Soil

It is generally recognized that most insect pathogens can persist and overwinter in soil. Nuclear polyhedrosis viruses will remain viable in soil for long periods of time (Jaques 1964, 1967), and *N. rileyi* spores move from the soil to soybean seedlings (Ignoffo et al. 1977a). Soil sampling for insect pathogens can, therefore, play an important role in studies of the yearly development cycles of pathogens in soybean fields.

Ignoffo et al. (1977b) used a soil bioassay and counting technique to show that *N. rileyi* conidia remain in the upper 2 cm of loam soils by adhering to organic particles in the soil. Conidia were detected in the soil by counting with a hemacytometer and by feeding soil-treated leaves to *T. ni* larvae.

Jaques (1964) measured virus activity in the soil by diluting soil samples in water and dispensing these samples on leaf discs. These leaf discs were then fed to laboratory-reared *T. ni* larvae. Young and Yearian (1979) have also used this method to demonstrate overwintering of the *P. includens* NPV in the soil.

C. Sampling from Air

Most fungal pathogens will spread through a field by dispersal of conidia through the air. Some researchers have used spore sampling techniques developed by plant pathologists to monitor the buildup and movement of fungal spores in the air. Kish and Allen (1978) used a conidial trapping apparatus developed by Hirst (1952) to determine the relative density of *N. rileyi* conidia in the air above soybean fields in Florida. Conidial density was sampled continuously (hourly) throughout the growing season by trapping spores on microscope slides covered along the entire length of one side with a strip of clear dual adhesive tape. Each slide contained conidia sampled over 24 hr. recorded on 24 bands on the sticky surface. The slides were examined under a compound microscope and hourly conidial counts were made and plotted on a graph.

Garcia and Ignoffo (1977) trapped conidia of *N. rileyi* on glass slides covered with Sorbol to monitor dislodgment of conidia from larval cadavers. Newman and Carner (1975a) recorded the periodicity of sporulation of *E. gammae* by trapping conidia on plain glass slides in the field. No adhesive material was necessary on the slides because spores of *Entomophthora* have a sticky coating that causes them to adhere to any material with which they come in contact.

IV. Concluding Remarks

In recent years, increased emphasis has been placed on construction of predictive models for pest populations. To be accurate, especially for lepidopterous pests, these models must take into account the mortality caused by pathogens. Studies are currently underway to construct models of disease epizootics such as those developed for plant pathogens. With proper inputs, such as weather conditions and pest population levels, these models should be able to predict epizootic patterns and the decline of pest populations. For constructing these epizootic models, reliable sampling procedures will have to be developed.

Acknowledgments

Thanks are due to Dr. J. V. Maddox, Illinois Natural History Survey, Urbana, Illinois, for critically reviewing the manuscript and for making many valuable suggestions.

References

Allen, G. E., and J. D. Knell. 1977. A nuclear polyhedrosis virus of *Anticarsia gemmatalis*. I. Ultrastructure, replication, and pathogenicity. Fla. Entomol. **60**:233-240.

Allen, G. E., G. L. Greene, and W. H. Whitcomb. 1971. An epizootic of *Spicaria rileyi* on the velvetbean caterpillar, *Anticarsia gemmatalis* in Florida. Fla. Entomol. **54**:189-191.

Burleigh, J. G. 1972. Population dynamics and biotic controls of the soybean looper in Louisiana. Environ. Entomol. **1**:290-294.

Carner, G. R., and O. W. Barnett. 1975. A granulosis virus of the green cloverworm. J. Invertebr. Pathol. **25**:269-271.

Carner, G. R., and S. G. Turnipseed. 1977. Potential of a nuclear polyhedrosis virus for control of the velvetbean caterpillar in soybean. J. Econ. Entomol. **70**:608-610.

Carner, G. R., J. S. Hudson, and O. W. Barnett. 1979. The infectivity of a nuclear polyhedrosis virus of the velvetbean caterpillar for eight noctuid hosts. J. Invertebr. Pathol. **33**:211-216.

Carner, G. R., M. Shepard, and S. G. Turnipseed. 1975. Disease incidence in lepidopterous pests of soybeans. J. Ga. Entomol. Soc. **10**:99-105.

Fuxa, J. R., and W. M. Brooks. 1978. Persistence of spores of *Varimorpha necatrix* on tobacco, cotton, and soybean foliage. J. Econ. Entomol. **71**:169-171.

Garcia, C., and C. M. Ignoffo. 1977. Dislodgement of conidia of *N. rileyi* from cadavers of cabbage looper, *Trichoplusia ni*. J. Invertebr. Pathol. **30**:114-116.

Gardner, W. A., R. M. Sutton, and R. Noblet. 1977. Persistence of *Beauveria bassiana, Nomuraea rileyi*, and *Nosema necatrix* on soybean foliage. Environ. Entomol. **6**:616-618.

Gatti, I. M., D. M. Silva, and I. C. Corso. 1977. Polyhedrosis occurrence in caterpillars of *Anticarsia gemmatalis* (Hubner 1818) in the south of Brazil (IRCS Med. Sci.: Cell Membrane Biol.; Environ. Biol. Med.; Exp. Anim.; Microbiology; Parasitology) Infec. Dis. **5**:136.

Harper, J. D., and G. R. Carner. 1973. Incidence of *Entomophthora* sp. and other natural control agents in populations of *Pseudoplusia includens* and *Trichoplusia ni*. J. Invertebr. Pathol. **22**:80-85.

Hinds, W. E., and S. A. Osterberger. 1931. The soybean caterpillar in Louisiana. J. Econ. Entomol. **24**:1168-1173.

Hirst, J. M. 1952. An automatic volumetric spore trap. Ann. Appl. Biol. **39**:257-265.

Ignoffo, C. M., and S. R. Dutky. 1963. The effect of sodium hypochlorite on the viability and infectivity of *Bacillus* and *Beauveria* spores and cabbage looper nuclear polyhedrosis virus. J. Invertebr. Pathol. **5**:422-426.

Ignoffo, C. M., D. L. Hostetter, and R. E. Pinnell. 1974. Stability of *Bacillus thuringiensis* and *Baculovirus heliothis* on soybean foliage. Environ. Entomol. **3**:117-119.

Ignoffo, C. M., B. Puttler, N. L. Marston, D. L. Hostetter, and W. A. Dickerson. 1975. Seasonal incidence of the entomopathogenic fungus, *Spicaria rileyi* associated with noctuid pests of soybeans. J. Invertebr. Pathol. **25**:135-137.

Ignoffo, C. M., N. L. Marston, D. L. Hostetter, B. Puttler, and J. V. Bell. 1976a. Natural and induced epizootics of *Nomuraea rileyi* in soybean caterpillars. J. Invertebr. Pathol. **27**:191-198.

Ignoffo, C. M., B. Puttler, D. L. Hostetter, and W. A. Dickerson. 1976b. Susceptibility of the cabbage looper, *Trichoplusia ni* and the velvetbean caterpillar, *Anticarsia gemmatalis* to several isolates of the entomopathogenic fungus, *Nomuraea rileyi*. J. Invertebr. Pathol. **28**:259-262.

Ignoffo, C. M., C. Garcia, D. L. Hostetter, and R. E. Pinnell. 1977a. Laboratory studies of the pathogenic fungus, *Nomuraea rileyi* (Farlow) Samson: soil borne contamination of soybean seedlings and dispersal of diseased larvae of *Trichoplusia ni* (Hubner). J. Invertebr. Pathol. **29**:147-152.

Ignoffo, C. M., C. Garcia, D. L. Hostetter, and R. E. Pinnell. 1977b. Vertical movement of conidia of *Nomuraea rileyi* through sand and loam soils. J. Invertebr. Pathol. **70**:163-164.

Jaques, R. P. 1964. The persistence of a nuclear polyhedrosis virus in soil. J. Invertebr. Pathol. **6**:251-254.

Jacques, R. P. 1967. The persistence of a nuclear polyhedrosis virus in the habitat of the host insect. II. Polyhedra in soil. Can. Entomol. **99**:820-829.

Johnson, D. W., L. P. Kish, and G. E. Allen. 1976. Field evaluation of selected pesticides on the natural development of the entomopathogen, *Nomuraea rileyi*, on the velvetbean caterpillar in soybean. Environ. Entomol. **5**:964-969.

Kish, L. P., and G. E. Allen. 1978. The biology and ecology of *Nomuraea rileyi* and a program for predicting its incidence on *Anticarsia gemmatalis* in soybean. Univ. Fla. Agr. Exp. Sta. Tech. Bull. **795**:48 p.

Kogan, M., S. G. Turnipseed, M. Shepard, E. B. de Oliveria, and A. Borgo. 1977. A pilot pest management program for soybean in Brazil. J. Econ. Entomol. **70**:659-663.

Livingston, J. M., and W. C. Yearian. 1972. A nuclear polyhedrosis virus of *Pseudoplusia includens* (Lepidoptera: Noctuidae). J. Invertebr. Pathol. **19**:107-112.

Livingston, J. M., P. J. McLeod, W. C. Yearian, and S. Y. Young. 1979. Laboratory and field evaluation of a nuclear polyhedrosis virus of the soybean looper, *Pseudoplusia includens*. J. Ga. Entomol. Soc. (in press).

MacLeod, D. M. 1963. Entomophthorales infectious. pp. 189-231 in E. A. Steinhaus, ed. Insect pathology, an advanced treatise. vol. 2 Academic Press, N. Y. 689 p.

Maddox, J. V. 1966. Studies on a microsporidosis of the armyworm, *Pseudoletia unipuncta* (Haworth) Ph.D. diss. Univ. of Illinois, Urbana. 184 p.

Newman, G. G., and G. R. Carner. 1975a. Factors affecting the spore form of *Entomophthora gammae*. J. Invertebr. Pathol. **26**:29-34.

Newman, G. G., and G. R. Carner. 1975b. Disease incidence in soybean loopers collected by two sampling methods. Environ. Entomol. **4**:231-232.

Pedigo, L. P., J. D. Stone, and G. L. Lentz. 1973. Biological synopsis of the green cloverworm in Central Iowa. J. Econ. Entomol. **66**:665-673.

Poinar, G. O., Jr., and G. M. Thomas. 1978. Diagnostic manual for the identification of insect pathogens. Plenum Press, N. Y. 218 p.

Puttler, B., C. M. Ignoffo, and D. L. Hostetter. 1976. Relative susceptibility of nine caterpillar species to the fungus, *Nomuraea rileyi*. J. Invertebr. Pathol. **27**:269-270.

Sprenkel, R. K., and W. M. Brooks. 1975. Artificial dissemination and epizootic initiation of *Nomuraea rileyi*, an entomogenous fungus of lepidopterous pests of soybeans. J. Econ. Entomol. **68**:847-851.

Weiser, J. 1977. An atlas of insect disease. Academia, Prague. 321 p.
Weiser, J., and J. D. Briggs. 1971. Identification of pathogens. pp. 13-63 in H. D. Burges and N. W. Hussey, eds. Microbial control of insects and mites. Academic Press, London. 861 p.
Young, S. Y., and W. C. Yearian. 1974. Persistence of *Heliothis* NPV on foliage of cotton, soybeans, and tomato. Environ. Entomol. 3:253-255.
Young, S. Y., and W. C. Yearian. 1979. Soil application of *Pseudoplusia* NPV: Persistence and incidence of infection in soybean looper caged on soybean. Environ. Entomol. 8:860-864.

Index

Above-ground arthropod
    sampling 30–58
  absolute methods of 50–52
  conversion of relative to absolute
    population estimates in 53–55
  direct observations in 31–34
  fumigation cage in 50–51
  ground cloth sampling in 34–37
  principal methods of 30
  suction net sampling in 46–50
  sweep net sampling in 37–46
  whole plant harvest in 51–52
Absolute methods of sampling 61–62
  for above-ground arthropods 50–52
  for aphids 248–253
  conversion of relative measures and, *see*
    Conversion of relative measures to
    absolute measures
  for green cloverworm larvae 176–177
  for leafhopper 269–275
    adults 269–271
    nymphs 274–275
  for Pentatomidae 449
  for soybean looper 147–151
    adults 147–148
    eggs 150
    larvae 151
  for velvetbean caterpillar 120–122
    eggs 120
    larvae 121–122
Accuracy, definition of 63
Aerial nets 247–248
Aerial sampling
  for aphids 245–253; *see also* Aphids,
    in air
  for thrips 292–293, 299–302
Aggregation, measures of 65–68
Air, pathogens in 571
Air-conditioned tunnel extractor 342
Anthocoridae 511–513
Anticarsia gemmatalis, see Velvetbean
  caterpillar
Apanteles marginiventris 486–487
Aphids 239–256
  in air 245–253
    absolute measures of 248–253
    relative measures of 247–248
    skewed measure traps and 246
  alighting on canopy 245–253
    absolute measures of 250–253
  flight behavior of 241–242
  life cycle of 240–241
  live trapping of 254–255
  within plant canopy 242–245

Aphids [*cont.*]
　　direct observation of　244–245
　　plant clipping and extraction
　　　of　243–244
　　predicting abundance of　255–256
　　as vectors of viruses　254–255
Arithmetic mean, definition of　62
Arthropod pests
　　management of, *see* Pests, management
　　　of
　　parasitism of, *see* Parasitism of
　　　arthropod pests
　　pathogens of, *see* Pathogens of
　　　arthropods
　　predaceous, *see* Predators
　　in soil, *see* Soil arthropods
Asymmetrical intervals　64
Average sample number (ASN)
　　curve　85–87

Bag samples　177
Bagging techniques　274
Bait traps
　　molasses　427
　　for pod borer adults　427
　　for soil arthropods　345
Bean flies, *see* Stem flies
Bean leaf beetle　201–233, 337
　　adults　215–220
　　　sampling program for　218–220
　　　sequential sampling for　221–222
　　comparison of various sampling methods
　　　for　231–233
　　damage produced by　202–208
　　eggs　222–231
　　　extraction from soil of　224,
　　　　226–228
　　　sampling program for　231
　　　sequential sampling for　231–232
　　geographical distribution of　201–202
　　host plants of　202–208
　　　table of　204–206
　　larvae　222–231
　　　extraction from soil of　230–231
　　　sampling program for　231
　　life cycle and phenology of　209–215
　　pupae　222–231
　　　extraction from soil of　230–231
　　relative methods of sampling of　217
　　spatial patterns of　218–220, 231

Bean podworms　424, 426–427
Beat cloth sampling, *see* Ground cloth
　　sampling
Beating and sticky board technique　273
Beating foliage, mites and　320
Beauveria spp.　562–563
Berlese's funnel extraction technique　341
Binomial distribution
　　negative, *see* Negative binomial
　　　distribution
　　sequential sampling plan and　88–89
Blacklight traps
　　green cloverworm adults and　173
　　Heliothis adults and　414
　　Pentatomidae and　449
　　soybean looper adults and　149
　　velvetbean caterpillar and　117–119
Brachonid wasps　486–487
Brushing
　　for aphids　243
　　machine for　320
　　for mites　320

Cage-aerosol techniques, parasitoids
　　and　497–498
Cerotoma ruficornis　201, 210–211
　　marking patterns of　210–211
Cerotoma trifurcata　201–233, 337; *see
　　also* Bean leaf beetle
　　marking patterns of　209–210
Cicadellidae, *see* Leafhoppers
clam trap
　　leafhopper adults and　269–271
　　leafhopper nymphs and　274
Coefficient of variation of mean
　　definition of　63
　　sequential sampling plan with
　　　fixed　89–90
Colaspis brunnea　335
Coleoptera, parasitism of　494–495
Confidence interval, definition of　63
Conversion of relative measures to absolute
　　measures　53–55, 69–70
　　for bean leaf beetles　217
　　for ground cloth sampling　37
　　for Mexican bean beetle　194–197
　　for Pentatomidae　451–452
　　for soybean looper larvae　154–155
　　for suction net sampling　49–50
　　for sweep net sampling　46

Index 577

for velvetbean caterpillar 124–125
for visual counts 33–34
Cost of sampling 73–75
Crop loss, see Soybean, yield of
Cutworms 329–330, 345
Cylindrical impact trap, thrips and 292–293
Cylindrical sticky traps, aphids and 247–248

D-Vac suction net, see also Suction net sampling 46–47
adult parasitoids of soybean pests and 497–499
bean leaf beetle and 217
leafhopper adults and 268–269
Mexican bean beetle and 194–197
soybean looper larvae and 151–154
spiders and 549, 553
velvetbean caterpillar larvae and 122–124
Damage, see also Defoliation; Injury
assessment of 11–18
parameters of 11–12
bean leaf beetle produced 202–208
Epinotia aporema produced 374–375
foliage injury and 12–16
green cloverworm produced 169–170
Heliothis spp. produced 409
injury and 18–24
Mexican bean beetle produced 189–190
to pods 16–18, 22, 24
to seeds 16–18
sequential sampling plan and threshold of 83, 89
stem borer produced 358–361
stem fly produced 395–397
stink bug produced 467–468
threecornered alfalfa hopper produced 383–387
threshold of 83, 89
thrips produced 285–286
velvetbean caterpillar produced 107–109
Data transformation 68–69
Dectes, see Stem borers
Defoliation 12–16
depodding and 24

determinate and indeterminate soybeans and 20
leaf area and 12
leaf area index and yield reduction and 21–22
physiological effects of 21–22
seeds and 22
soybean yield and 19–22
equations for 21
by velvetbean caterpillar 134–135
visual estimation of 134–135, 160–161
Delayed counting methods 290–291
Depodding 22
defoliation and 24
Differential density method of seed injury measurement 17–18
Diptera, parasitism of 495–496
Direct count method 291–292
for mites 319–320
Direct observation, see also Visual observation
for aphids within plant canopy 244–245
for bean leaf beetle adults 215–216
characteristics of 32–33
counts over a measured length of row in 32
for green cloverworm larvae 176–177
for leafhopper
adults 268–269
nymphs 273, 274
for Mexican bean beetle eggs and pupae 198
single plant examination in 31
for spiders 549
for threecornered alfalfa hopper 390–391
Discard-count system for mites 320
Dry-count technique 290, 291
Dry funnel extraction procedure 341

Economic threshold 454–456
Efficient, definition of 64
Elasmopalpus lignosellus 333–334
larvae 345
Electric grid trap, soybean looper adults and 149
Electronic leaf area-meters 12–14

Emergence traps, leafhopper adults and 269–271
Empoasca fabae 260–276; *see also* Leafhoppers (Cicadellidae)
Entomophthora 562
Entomophthora gammae 561
Epilachna varvivestis, *see* Mexican bean beetle
Epinotia aporema 374–380
 adults 377
 damage produced by 374–375
 eggs 377–378
 larvae 378–379
 life cycle and phenology of 375–377
 pupae 379
Ermine lime trap 250–253
Etiella zinckenella see Limabean pod borer

Field conditions, sampling program and 3
Field survey
 for *Epinotia aporema* larvae 379
 for mites 318–319
Flotation method 343–344
 seed injury measurement by 17
 with sieve, *see* Sieve-flotation procedure
Foliage
 injury to 12–16
 sampling for disease agents in 570
Fumigation cage 50–51
 green cloverworm larvae and 177
 leafhoppers and
  adults 269
  nymphs 274
 Pentatomidae and 449–450
 relative sampling methods versus 450
 velvetbean caterpillar larvae and 121–122
Fungi 387, 560–563
Funnel extraction technique 341–342

Geocoris spp. 509–511
 key to identification of species of 525
Geometric leaf area measurement 15–16
Germination 18
Girdled stems, stem borers and 369–372
Granulosis virus of *Plathypena scabra* 566
Grape colaspis 335

Gravimetric leaf area measurement 14
Green cloverworm (*Plathypena scabra*) 169–185
 adults 173–176
 damage produced by 169–170
 eggs 176
 granulosis virus of 566
 host plants of 169–170
 larvae 176–180
  absolute sampling methods for 176–177
  behavioral characteristics of 176
  direct observation of 176–177
  relative sampling methods for 176–177
 life cycle and phenology of 171–173
 parasitism of 484
 sampling program for 180–183
 sequential sampling plan for 83, 182–183
Ground cloth sampling
 for above-ground arthropods 34–37
 for bean leaf beetle 210, 231–233
  adults 217
 characteristics of 36–37
 construction of ground cloth for 34–36
 conversion to absolute population estimates of 37
 for green cloverworm larvae 179
 for *Heliothis* larvae 415
 for Mexican bean beetle 194–197
 for mites 320
 for parasitoids of Lepidoptera 488–489
 for Pentatomidae 450–451, 462–463
 procedure for 36
 sample-unit size for 180–181
 sequential sampling and 88, 126–129, 131–132, 157
  plan for 127–129
 for soybean looper 151–154, 157
 for spiders 547, 549, 553
 for stem borers and 367
 sweep net sampling versus 462–463
 for threecornered alfalfa hoppers 391
 for velvetbean caterpillar 122–124
Ground dwelling predators, sampling of 533–541
 implementation of program

Index 579

of 538–540
pitfall traps in 533–540
  advantages of 533–536
  disadvantages of 536–537
quadrat technique in 541

Headlighting 549–550, 554
Heat table 550, 554
Heliothis spp. 407–419
  adults 414
  damage produced by 409
  eggs 414
  events simultaneous with outbreaks of 412
  geographical distribution of 407–408
  host plants of 409
  larvae 415–416
  life cycle and phenology of 409–414
  nuclear polyhedrosis virus of 565
  pupae 416
  sampling considerations for 417–419
  sequential sampling of 416
  "sugarline" sampling of 188–190
Hemiptera 505–524
  Anthocoridae 511–513
  key to common groups of 524–525
  Lygaeidae 509–511
  Nabidae 508–509
  parasitism of 496
  Pentatomidae 515–517
  pests and 523–524
  Reduviidae 513–515
  sampling methods for 520–521
  sequential sampling plans for 522
Homoptera, parasitism of 496–497
Horizontal ermine lime (sticky tile) traps 250–253, 292, 299–302
Host plants
  of bean leaf beetle 202–208
    table of 204–206
  of green cloverworm 169–170
  of *Heliothis* spp. 409
  of leafhoppers 263–265
    table of 262
  of Mexican bean beetle 189–190
  of mites 313–314
  of pod borers 422–424
  of southern green stink bug 440–443
  of soybean looper 141–142
    table of 144–146

  of stem borers 358–361
  of stem flies 395–399
    table of 398–399
  of threecornered alfalfa hopper 383–387
  of thrips 285–286
  of velvetbean caterpillar 107–111
    table of 110–111
  of whiteflies 306–307

Illinois Egg Separator 224, 226–228
Imprinting of mites 319
Index of infestation 309–310
Index of larval abundance, injury as 430
Infestation index 309–310
Injury 11, *see also* Damage; Defoliation; Depodding
  categories of 467–468
  damage and 18–24
  as index of larval abundance 430
  mite produced 313–314
  pod borer produced seed 434
  soybean looper produced 141–142
  velvetbean caterpillar produced 107–109
  whitefly produced 306–307

Johnson-Taylor suction trap 248–249

LAI (leaf area index) 21–22
Leaf abscission scar *Dectes* larvae and 369
Leaf area
  defoliation and 12
  measurement of 12–16
Leaf area index (LAI) 21–22
Leaf unit sampling, leafhopper nymphs and 273
Leaf washing, leafhopper nymphs and 274
Leafhoppers (Cicadellidae) 260–276
  adults 267–271
    absolute sampling methods for 269–271
    relative sampling methods for 268–269
    sampling program for 271
    spatial pattern of 271
  eggs 271
  geographical distribution of 261–263

Leafhoppers (Cicadellidae) [*cont.*]
  host plants of 262-265
    table of 262
  key to genera of 276-280
  life cycle and phenology of 265-266
  nymphs 273-275
    absolute sampling methods
      for 274-275
    relative sampling methods for 273
    sampling program for 275
  species of, most abundant 260
Leguminivora glycinivorella, see Pod
    borers
Lepidoptera, parasitoids of 488-494
Lesser cornstalk borer (*Elasmopalpus
    lignosellus*) 333-334
  larvae 345
Light traps
  Epinotia aporema and 377
  green cloverworm adults and 173
  pod borer adults and 427
Limabean pod borer (*Etiella zinckenella*)
  description of 424, 432
  eggs 428
  larvae 430
  life cycle 425-426
  pupae 430
Lodging resistance 11
Lygaeidae 509-511

Mathematical distribution, *see* Statistical
    distribution
Matsumuraeses spp. 424
  eggs 428
  larvae 430
Matsumuraeses falcana 424
Matsumuraeses phaseoli 424
Mean, arithmetic, definition of 62
Mean crowding, definition of 90
Melanagromyza sojae 395-397
Mexican bean beetle (*Epilachna
    varvivestis*) 189-199
  adults 193-197
  damage produced by 189-190
  eggs 198
  geographical distribution of 189
  host plants of 189-190
  larvae 193-197
  life cycle and phenology of 190-193
  pupae 198

  relative sampling methods
    for 194-197
  scouting programs for 199
  spatial pattern of 198
Mites 312-322
  adults 318-321
    field surveys for 318-319
    estimation of relative density
      of 319-320
  host plants of 313-314
    table of 313
  injury produced by 313-314
  juveniles 318-321
    field survey for 318-319
  life cycle and phenology of 314-318
  sampling program for 320-321
Molasses bait traps 427
Moth-flushing technique, green
    cloverworm adults and 175

Nabidae 508-509
Nabis spp. 508-509
Negative binomial distribution 66-68
  sequential estimation and 97, 99
  sequential sampling decision equation
    and 87
Nezara viridula, see Pentatomidae
Nodules, soil arthropods
    attacking 337-339, 345-348
Nomuraea rileyi 560-561
Normal distribution 68-69
Nuclear Polyhedrosis virus
  of *Anticarsia gemmatalis* 566
  of *Heliothis* spp. 565
  of *Pseudoplusia includens* 565

Oberea brevis 357, 361
Oil in seeds 18
Ophyomyia phaseoli 397
Orius spp. 511-513
  description for separation of species
    of 525

Pan-shake method, leafhopper nymphs
    and 273
Parasitoids of pests 481-499
  bias in estimating 481-488
    compensating for 484-488
  sources of 481-484
  brachonid wasps 486-487

Index 581

of Coleoptera 494-495
debris and separation of 498-499
of Diptera 495-496
estimating adult of green cloverworm 497-500
green cloverworm 484
of Hemiptera 496
of Homoptera 496-497
of Lepidoptera 488-494
  eggs 493-494
  larvae 488-493
  pupae 493
measurement of 488-497
  in laboratory 489-493
  Lepidoptera 488-494
tachinid fly 484-486
traps for adult 499
Voria ruralis 487
Pathogens of arthropod pests 559-571
in air 571
in environment 570-571
fungi 560-563
sampling for disease incidence and 566-569
in soil 570
viruses 565-566
Pentatomidae (stink bugs) 438-468, 515-517
adults and nymphs 449-463
  absolute sampling methods for 449
  relative sampling methods for 450-451
calibration of relative measures to absolute measures of 451-452
damage by 467-468
eggs 463, 467
geographical distribution of 439
host plants of 440-443
indirect sampling of 467-468
key to 525-527
  adult 469-474
life cycle and phenology of 443-446
sequential sampling of 454-463
  economic threshold of infestation and 454-455
  factors which influence 454-455
  level of precision and 458, 461-463
  management decision and 454-456
  population parameters and 456-463

simulation model and 456-458
spatial pattern of 453-454
sweep net versus ground cloth sampling for 462-463
Pests
Hemiptera effect on 523-524
management of
  Pentatomidae 454-456
  sampling method and 79
  scouting programs for 199
  sequential sampling and 454-456
  thrips 286
  velvetbean caterpillar larvae 125-132
parasitoids of 481-499; *see also* Parasitoids of pests
pathogens of 559-571; *see also* Pathogens of pests
sampling for disease incidence in 566-569
Pheromone trap
Heliothis adults and 414, sex, *see* Sex pheromone
Photoelectric leaf area-meters 12-14
Pitfall traps
green cloverworm larvae and 179-180
ground predators and 533-540
  advantages of 533-536
  disadvantages of 536-537
  spiders and 550, 553-554
Planimetric leaf area measurement 15
Plant clipping, aphids and 243-244
Plant density 11
Plant height 11
Plant observation, whole, *see* Whole plant examination
Plant shake sampling, *see* Ground cloth sampling
Planting dates 9
Plathypena scabra, *see* Green cloverworm
Plusiinae
larvae, key for identification of 164
moths, key for identification of 162-164
Pod
damage to 16-18
development of, assessment of 7
seed quality and injury to 22
yield and injury to 22

Pod borers (*Leguminivora glycinivorella*) 422–432
  adults 427
    identification of 435
  cocoons, description of 432
  eggs 428
    description of 432
  geographical distribution of 422–424
  host plants of 422–424
  key to injury by 434
  larvae 429–431
    description of 432
    key to 432–424
  life cycle and phenology of 424–427
  mortality factors of 429–430
  pupae 429–431
Poisson distribution 65–67
  sequential sampling plan and 82–88, 96–97
  table of values of 98–99
Population
  definition of 6?
  indices of 62
    for stem flies 401–402
  sampling for
    regional 75–77
    sampling theory for 61–77
  sequential sampling for 94–103, 456–463
    level of precision and 458, 461–463
    simulation models and 456–458
    for soybean looper 160
    for thrips 296–297
    for velvetbean caterpillar larvae 133–134
Power law 67
Precision
  definition of 63
  sequential sampling and level of 458, 461–463
Predators 505–524, 532–541
  Geocoris spp. 509–511
  ground dwelling, 532–541; *see also* Ground dwelling predators, sampling of
  Nabis spp. 508–509
  Orius spp. 511–513
  pests and 523–524
  sampling of 518–522
    comparison of 520–521
    factors which influence 520
    reasons for 518–519
    sequential sampling plans for 522
  Sinea spp. 514
  table of 506
  Zelus spp. 513
Protein in seeds 18
Protomicroplitis facetosa 487
Pseudoplusia includens, see Soybean looper
Pseudoplusia includens nuclear polyhedrosis virus 565

Quadrat technique
  ground predators and 541
  spiders and 554

Radar techniques, *Heliothis* adults and 414
Random stratification 120–121
Reduviidae 513–515
  key for genera of 525
Regional population estimation 75–77
Relative net precision, definition of 64
Relative sampling methods 62
  aphids in air and 247–248
  bean leaf beetles and 217
  comparison of 151–154, 194–197, 217
  conversion to absolute measures, *see* Conversion to relative measures to absolute measures
  fumigation cage versus 450
  green cloverworm larvae and 176–177
  leafhoppers and
    adults 268–269
    nymphs 273
  Mexican bean beetle and 194–197
  Pentatomidae and 450–451
  for pod borer adults 427
  for soybean looper
    adults 148–150
    eggs 150
    larvae 151–154
  for velvetbean caterpillar
    eggs 120–121
    larvae 122–124
Relative variation, definition of 63

Index 583

Reliability, definition of 63
Risk, sequential sampling plan and 89
Rivellia quadrifasciata 337–339
Rogas nolophanae 486–487
Roots
 development of, assessment of 7, 9
 soil arthropods attacking 335–337
 soil arthropods in 345–348
 soil types and 10
Rotary screen extractor 343
Rotary traps, aphids in air and 249–250
Row widths 9, 11

Sample, definition of 62
Sample size 70–73
 definition of 62
 standard error and 70–72
 statistical distribution and, 72–73
Sample unit, definition of 62
Sample-unit size, shakecloth method and 180–181
Sampling methods, *see also* specific methods
 comparison of 55–58
  average sample number curves and 159
  graphic 55–56
  for green cloverworm larvae 176–180
  statistical methods in 56–58
 pest management and 79
 selection of 73–75
 soybean crop characteristics affecting 4–11
 standardization of 126–132
 "sugarline" 119–120
Sampling program design 180–183
 field conditions and 3
 number of samples in 181
 pattern of sampling in 181–182
 pest management and 79
 population estimation and 61–77
 sample-unit size in 180–181
 sequential sampling, *see* Sequential sampling
 timing of sampling in 182
Sampling theory 61–77
 aggregation in, measures of 65–68
 conversion of relative to absolute estimates in 69–70

data transformation in 68–69
definitions of terms used in 62–64
normal distribution in 68–69
pest management and 79
population estimation and 61–77
reasons for sampling in 61
regional population estimation in 75–77
sample size determination in 70–73
sampling statistics in, characteristics of 62–65
selection of method of sampling in 73–75
spatial pattern in 65–68
statistical distribution in 65–68
Scouting programs 199
Seed(s)
 damage to 16–18
 development of, assessment of 7
 depodding and 22
 defoliation and 22
 injury of
  as index of larval abundance 430
  key to 434
 soil arthropods attacking 328–334
 soil arthropods in 344–345
Seedcorn maggot 328–329, 345
Sequential sampling 79–92
 advantages of 80
 binomial distribution and 88–89
 confidence intervals and 95–96
 design of plan for 80–89, 100–101
 damage threshold and 83, 89
 with fixed coefficient of variation of mean 89–90
 information necessary for 81
 mathematical distribution and 81–88, 97, 99, 100–101
 risk and 89
 for bean leaf beetle 221–222, 231
 for green cloverworm 182–183
 ground cloth method and 88, 127–129, 131–132, 157
 for *Heliothis* spp. 416
 for Hemiptera 522
 mathematical distribution and 81–82, 100–101
 negative binomial distribution and 87, 97, 99
 Poisson distribution and 96–99

584 Index

Sequential sampling [cont.]
　operating characteristic curves
　　and   85–87
　for *Plathypena scabra*   83
　for Pentatomidae   454–463
　　economic threshold of infestation
　　　and   454–456
　　factors which influence   454–455
　　management decision and   454–456
　for population estimation   94–103,
　　456–463
　　level of precision and   458,
　　　461–463
　　simulation models and   456–458
　for soybean looper   160
　for velvetbean caterpillar
　　larvae   133–134
　for soybean looper larvae
　　management   156–160
　　population estimation   160
　stopping rules for   101
　sweep net method and   88, 127, 129
　uses of   79–81, 100–101
　for velvetbean caterpillar   83–84,
　　126–132
　　larvae   126–132
　　population estimation   130–132
Sex pheromones in trapping
　Pentatomidae and   449–450
　soybean looper adults and   149
　velvetbean caterpillar and   118–119
Shake cloth sampling, *see* Ground cloth
　sampling
Sieve-flotation procedures   343, 345, 348
Sinea spp.   514
Single plant examination   31
Skewed measure traps in aerial aphid
　sampling   246
Soil
　pathogens in   570
　roots and types of   10
Soil arthropods   327–349
　extraction from soil of, *see* Soil samples,
　　extraction of arthropods from
　nodule attacking   337–339
　root attacking   335–337
　sampling techniques for   339–344
　seed attacking   328–334
　table of references on   346–347
Soil samples, extraction of arthropods

　from   224–230, 341–344, 541
　active methods of   341–342
　apparatus with rotatory screen
　　in   228–230
　comparison of methods of   344
　for eggs   224, 226–228
　Illinois Egg Separator in   224,
　　226–228
　for larvae   230–231
　passive methods of   343–344
　for pupae   230–231
　stationary nested screens in   224,
　　226–228
　for velvetbean caterpillar pupae   134
Southern green stink bug (*Nezara
　viridula*), *see* Pentatomidae
Soybean
　characteristics of crop of   4–11
　　agro-ecological dependent   9–11
　　growth and genetic dependent   4–9
　damage to, *see* Damage
　determinate and indeterminate   5
　defoliation and   20
　growth of   4–9
　　table of   8
　photoperiod and   4
　physiognomic description of   9–11
　pod development of   7
　root development of   7, 9
　soil type and   10
　seed development of   7
　yield of
　　defoliation and   19–22
　　loss of   11, 370, 372
　　pod injury and   22
　　stem borers and   370, 372
Soybean looper (*Pseudoplusia
　includens*)   141–161
　adults   147–150
　　absolute sampling methods
　　　for   147–148
　　relative sampling methods
　　　for   148–150
　eggs   150
　host plants of   141–142
　　table of   144–146
　indirect sampling procedures
　　for   160–161
　injury produced by   141–142
　larvae   150–160

Index 585

absolute sampling methods for   151
calibration of sampling procedures
    for   154–155
identification of species
    for   150–151
relative sampling methods
    for   151–154
sequential sampling plans for
    management of   156–160
spatial patterns of   155–156
life cycle and phenology of   143–147
population estimation of   160
pupae   160
sequential sampling for
    management of   156–160
population estimation of   160
Soybean nodule fly (*Rivellia
    quadrifasciata*)   337–339
Spatial pattern   65–68
of bean leaf beetle   231
    adults   218–220
of leafhopper
    adults   271
    nymphs   275
of Mexican bean beetle   198
of Pentatomidae   453–454
sequential sampling plan development
    and   81–88
of soybean looper larvae   155–156
of spiders   552–553
of thrips   296–297
of velvetbean caterpillar larvae   125
Spiders   544–555
qualitative sampling for   545–551
    search for spiders in   545–547
    summation of   550–551
    techniques for   547, 549–552
quantitative sampling for   552–554
spatial pattern of   552–553
vertical stratification of   554–555
Spissistilus festinus, see Threecornered
    alfalfa hopper
Standard deviation, definition of   62–63
Standard error
    definition of   63
    sample size and   70–72
Stationary nested screens   224, 226–228
Statistical distribution   65–68; *see also*
    Spatial pattern
    negative binomial, *see* Negative
        binomial distribution
    normal distribution   68–69
    Poisson, *see* Poisson distribution
    sample size and   72–73
    sequential sampling plan and   81–88,
        96–103
    for various species of insects   103
Stem borers (*Dectes*)   357–373
    adults   367–368
    damage produced by   358–361,
        370–372
    eggs   368–369
    geographical distribution of   358
    host plants of   358–361
    larvae   369–372
    life cycle and phenology of   361–364
    yield loss caused by   370–372
Stem flies   394–402
    adults   400
    damage produced by   395–397
    eggs   400–401
    geographical distribution of   394–395
    host plants of   395–397
        table of   398–399
    larvae   401
    life cycle and phenology of   397–400
    population indices for   401–402
    pupae   401
Stem girdling
    fungal infection and   387
    threecornered alfalfa hopper
        and   384–387, 391
Sticky coated screen, aphids in air
    and   247–248
Sticky traps
    green cloverworm adults and   174–175
    stem borers and   367–368
Stink bugs, *see* Pentatomidae
Suction net sampling
    for above-ground arthropods   46–50
    bean leaf beetle adults and   217
    characteristics of   49
    conversion to absolute population
        estimates of   49–50
    D-Vac, *see* D-Vac suction net sampling
    Johnson-Taylor   248–249
    of measured length of row   48–49
    procedures for   47–49
    spiders and   549
    spot sampling and   48

"Sugarline" method 119–120
  soybean looper adults and 150
Surveys
  for mites 318–319
  for spiders 545–551
Sweep net sampling
  above-ground arthropods and 37–46
  bean leaf beetle and 210, 231–233
  characteristics of 45–46
  comparison of sweeping methods
    for 389–390
  construction of sweep net for 38
  conversion to absolute population
    estimates of 46
  Epinotia aporema and 377
  green cloverworm larvae
    and 178–179
  ground cloth sampling
    versus 462–463
  "lazy-8" parallel sweep in 42
  leafhoppers and
    adults 268–269
    nymphs 273
  Mexican bean beetle and 194–197
  parasitoids of Lepidoptera and 488
  pendulum sweeps in 42
  Pentatomidae and 450–451, 462–463
  pod borer adult and 427
  procedures for 38, 40–44
  sequential sampling plan and 88, 129, 157
  soybean looper and 157
    larvae 151–154
  spiders and 553
  stem borers and 367
  sweeps across one row in 40–41
  sweeps across two rows in 42
  threecornered alfalfa hoppers
    and 389–390
  variability of results using 45–46
  velvetbean caterpillar larvae
    and 122–124
Systematic stratification 120–121

Tachinid fly (*Winthemia sinuata*) 484–486
Threecornered alfalfa hopper (*Spissistilus festinus*) 382–392
  adults 389–390
  damage produced by 383–387
  eggs 391
  host plants of 382–387
  indirect sampling of 391
  life cycle and phenology of 387–389
  nymphs 390–391
Thrips 283–302
  abundance of 284
  adults 290–293
    in air 292–293
    delayed count method for 290–291
    direct count method for 291–292
    on plants 290–292
  colonization patterns of 293–295
  damage produced by 285–286
  host plants of 285–286
  larvae 290–293
    delayed count method for 290–291
    direct count method for 291–292
    on plants 290–292
  life cycle and phenology of 286–288
  other pests and 286
  population density of 296–297
  predators and 286
  sampling program for 298–299
  sampling technique and species
    of 299–302
  species composition of 284
  as vectors of viruses 285
Tullgren funnel apparatus 341, 550, 554

Underground arthropods, *see* Soil
  arthropods

"V" caliper 9
Vacuum net, green cloverworm larvae
  and 179
Variance, definition of 63
Vectors of viruses
  aphids as 254–255
  thrips as 285
Velvetbean caterpillar (*Anticarsia gemmatalis*) 107–136
  adults 117–120
  blacklight traps and 117–119
  conversion of relative measures to
    absolute measures of 124–125
  distinguishing species of 118
  eggs 120–121
  host plants of 107–109
    table of 110–111

indirect sampling procedures
    for   134–135
injury caused by   107–109
larvae   121–134
    absolute methods of sampling
        for   121–122
    identification of species of   121
    population estimation of   133–134
    relative sampling methods
        for   122–124
    sequential sampling plan
        for   133–134
life cycle and phenology of
    109–116
nuclear polyhedrosis virus of   566
pest management decision
    for   126–132
pupae   134
sequential sampling plan for   83–84,
    126–132
    ground cloth method in   127–129,
        131–132
    pest management and   126–132
    sweep net method in   129
sex pheromone in sampling
    for   118–119
spatial pattern of   125
visual estimation of defoliation
    and   134–135
Vertical impact traps, aphids in air
    and   247–248
Vertical stratification of
    spiders   554–555
Viruses   565–566
    vectors of
        aphids as   254–255
        thrips as   285
Visual examination, see also Direct
    observation
    for above-ground arthropods   31–34
    conversion to absolute population
        estimates of   33–34

of defoliation   16
    by soybean looper   160–161
    by velvetbean caterpillar   134–135
    of seed injury   17
    for soybean looper eggs   150
    for threecornered alfalfa
        hoppers   391
Volumetric leaf area measurement   15
Voria ruralis   487

Washing, aphids and   243–244
Washing-flotation method   343–344
Waterpan traps, green cloverworm adults
    and   174
Weed density   11
Wet-count methods   291
Whirligig traps, aphids in air
    and   249–250
White grubs   336–337
Whiteflies   305–310
    absolute sampling methods for   309
    adults   309
    eggs   309–310
    geogaphical distribution of   305
    host plants of   306–307
    infestation index and   309–310
    injury produced by   306–307
    life cycle and phenology of   307–308
    nymphs   309–310
    pupae   309–310
Whole plant examination
    for leafhopper nymphs   273
    for parasitoids of Lepidoptera   488
    for pod borer eggs   428
Whole plant harvest   51–52
Winthemia asinuata (tachinid
    fly)   484–486
Wireworms   331

Yield, *see* Soybean, yield of

Zelus spp.   513